THE VERTICAL
TRANSPORTATION HANDBOOK

THE VERTICAL TRANSPORTATION HANDBOOK

THIRD EDITION

George R. Strakosch, Editor

JOHN WILEY & SONS, INC.

New York · Chichester · Weinheim · Brisbane · Singapore · Toronto

Copyright © 1998 by John Wiley & Sons, Inc. All rights reserved.

Published simultaneously in Canada.

This publication is designed to provide accurate and authoritative information in regard to the subject matter covered. It is sold with the understanding that the publisher is not engaged in rendering professional services. If professional advice or other expert assistance is required, the services of a competent professional person should be sought.

Library of Congress Cataloging-in-Publication Data:

The vertical transportation handbook / George R. Strakosch, editor. —
  3rd ed.
      p.   cm.
  Rev. ed. of: Vertical transportation / George R. Strakosch. 2nd
ed. ©1983.
  Includes bibliographical references and indexes.
  ISBN 0-471-16291-4 (cloth : alk. paper)
  1. Elevators.   2. Escalators.   I. Strakosch, George R. Vertical
transportation.
TJ1370.V476 1998
621.8'77—dc21   98-27340

Printed in the United States of America.

10 9 8 7 6 5 4 3

CONTENTS

PREFACE

Extensive review of the second edition of *Vertical Transportation,* published in 1983, indicated that a number of revisions and updates were needed. The vertical transportation industry, elevators and escalators, has undergone some significant changes both in equipment and in the nature of the specifiers and users of the equipment.

Comparison of this edition and the previous one will indicate that the basics have remained essentially the same and that the greater changes have taken place in the way elevators and escalators are installed, operated, regulated, specified, and maintained. The past decade or so has emphasized extensive application of the microprocessor to all aspects of the business. A generation of consultants has come forth and is becoming the main source of information and influence in application. Modernization of existing elevators has become the major activity of most elevator companies, which has fostered a generation of component suppliers rather than original equipment manufacturers. This later aspect has also opened up major markets for elevator contractors as opposed to equipment manufacturers who designed, built, and installed the equipment they originally built.

Those are but a few of the changes that have taken place and have led me to consider and institute an entirely new approach to this edition. Rather than try to do it all myself I have asked people who have special knowledge in the various aspects of the industry to be contributors. You will see who they are by reviewing various chapters, and their insight and commentary becomes a valuable part of this edition. I am grateful for their help in making this edition a reality.

Greater emphasis is placed on the need for considering accessibility in all aspects of application. Since building transportation systems are the most essential feature of any multiple-story building, this consideration must be foremost both in any new or existing building. The rules have been written and the revised examples in this book provide some of them.

It is interesting to note that changes have been so dramatic that the elevator or escalator ordered today may be obsolete by the time it is installed and running. Such has been the rapid adaptation of the latest microprocessor and software changes that take place on almost a daily basis. In addition, even as this is written, new developments are being introduced such as self propelled elevators, elevators that move both horizontally and vertically, equipment totally within the hoistway, and the revival of the roped hydraulic, plus a growing emphasis on automated horizontal transportation. Such will be the subjects for a fourth edition of *Vertical Transportation,* which I will have to leave for others to produce.

Writing and editing *Vertical Transportation* has been a labor of love for me. Many ask how it came about, and the story is relatively simple. Back in the mid-sixties, while employed by Otis, I was asked to revise a sales brochure entitled "Hints for Better Elevatoring" that they had published. The result was an unwieldy expanded version that became the basis of the first edition. Added to this was a summation of the answers to the many questions on application that I had answered as part of my position as traffic engineer for

Otis at that time. Through the blessing of Otis it was published and available to the industry at large. For the first time the obscure mystery of "elevatoring" became public and, in its small way, contributed to the growth of the industry. It helped people to help themselves and helped provide the confidence that many needed.

The second edition of *Vertical Transportation* was a reflection of my own growth in the industry in which I thoroughly enjoyed working. As a consultant, I was exposed to all aspects of elevatoring: the initial design; the layout; the specification; the bidding and contractor choice; approvals; expediting and field observation; as well as final acceptance and, if possible, subsequent reconciliation of problems. Many of the lessons I learned were included in the edition. My association with *Elevator World Magazine* as their technical consultant and director of education gave me the opportunity to share my experience with the industry at large.

Consolidation of much of my experience came forth as I conducted seminars throughout the country under the auspices of *Elevator World Magazine*. Readers of the magazine and of *Elevator World*'s other publications are aware of the hundreds of articles I contributed to the monthly issues and a number of their major publications. I am grateful to Bill Sturgeon, the publisher of the magazine, for having me as his associate and, most of all, for his friendship.

Looking back over the past fifty years or so, it has been fun and I intend to keep going as long as I can. Needless to say, I haven't done what I did alone and I will be eternally grateful to my associates, friends, contributors, and the many acquaintances I have made throughout the years. Most of all, I give my appreciation to my wife for her support and patience during the many hours that I've devoted to these efforts.

CONTRIBUTORS

ROBERT CAPORALE, Editor
Elevator World Magazine
354 Morgan Avenue, P.O. Box 6507
Mobile, AL 36606

EDWARD A. DONOGHUE, Edward A. Donoguhe Associates Inc.
1677 County Route 64, P.O. Box 201
Salem, NY 12865-0201

GEORGE GIBSON, G. Gibson & Associates
555 Deer Pass Drive
Sedona, AZ 86351

WAYNE A. GILCHRIST, TELE-Engineering
39 Oak Ridge Road
Newfoundland, NJ 07435

JON HALPERN, Millar Elevator Company
620 12th Avenue
New York, NY 10036

LEN LEVEE, Len Levee & Associates
P.O. Box 647
Bowie, MD 20718

JOSEPH MONTESANO, DTM Elevator Consultants
120-02 14th Road
College Point, NY 11356

LAWRENCE PIKE, J. Martin Associates
263 Molnar Drive
Elmwood Park, NJ 07407

AL SAXER
38 Schoville Road
Avon, CT 06001

WILLIAM C. STURGEON, Founder and Editor Emeritus
Elevator World Magazine
354 Morgan Avenue, P.O. Box 6507
Mobile, AL 36606

DAVIS L. TURNER
27615 Belmonte
Mission Viejo, CA 92692

1 The Essentials of Elevatoring

EARLY BEGINNINGS

Since the time man has occupied more than one floor of a building, he has given consideration to some form of vertical movement. The earliest forms were, of course, ladders, stairways, animal-powered hoists, and manually driven windlasses. Ancient Roman ruins show signs of shaftways where some guided movable platform type of hoist was installed. Guides or vertical rails are a characteristic of every modern elevator. In Tibet, people are transported up mountains in baskets drawn by pulley and rope and driven by a windlass and manpower. An ingenious form of elevator, vintage about the eighteenth century, is shown in Figure 1.1. (note the guides for the one "manpower"). In the early part of the nineteenth century, steam-driven hoists made their appearance, primarily for the vertical transportation of material but occasionally for people. Results often were disastrous, because the rope was of fiber and there was no means to stop the conveyance if the rope broke.

In the modern sense, an elevator[1] is defined as a conveyance designed to lift people and/or material vertically. The conveyance should include a device to prevent it from falling in the event the lifting means or linkage fails. Elevators with such safety devices did not exist until 1853, when Elisha Graves Otis invented the elevator safety device. This device was designed to prevent the free fall of the lifting platform if the hoisting rope parted. Guided hoisting platforms were common at that time, and Otis equipped one with a safety device that operated by causing a pair of spring-loaded dogs to engage the cog design of the guide rails when the tension of the hoisting rope was released (see Figure 1.2).

ELEVATOR SAFETY DEVICES

Although Otis's invention of the safety device improved the safety of elevators, it was not until 1857 that public acceptance of the elevator began. In that year the first passenger elevator was installed in the store of E. V. Haughwout & Company in New York. This elevator traveled five floors at the then breathtaking speed of 40 fpm (0.20 mps).[2] Public and architectural approval followed this introduction of the passenger elevator. Aiding the technical development of the elevator was the availability of improved wire rope and the rapid advances in steam motive power for hoisting. Spurring architectural development was an unprecedented demand for "downtown" space. The elevator, however, remained a

---

1. In England and other parts of the world the word "lift" is used. The legally recognized definition of an elevator can be found in ANSI/ASME A17.1, Safety Code for Elevators and Escalators.
2. Elevator speed is traditionally stated in fpm (feet per minute) or mps (meters per second).

The Vertical Transportation Handbook, Third Edition, Edited by George R. Strakosch
ISBN 0-471-16291-4   © 1998 John Wiley & Sons, Inc.

Figure 1.1. A very early type of vertical transportation.

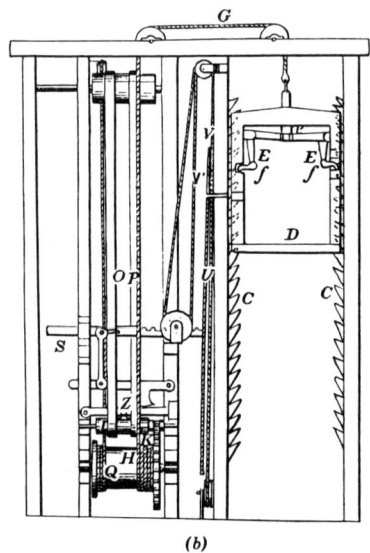

Figure 1.2. (*a*) Otis's demonstration, Crystal Palace, New York, 1853. (*b*) Otis's patent sketch for a safety device (Courtesy Otis Elevator).

slow vertical "cog" railway for quite a few years. The hydraulic elevator became the spur that made the upper floors of buildings more valuable through ease of access and egress. Taller buildings permitted the concentration of people of various disciplines in a single location and caused the cities to grow in their present form during the 1870s and 1880s.

HYDRAULIC ELEVATORS

The hydraulic elevator provided a technological plateau for quite a few years; it was capable of higher rises and higher speeds than the steam-driven hoist-type elevator, limited by its winding drums (Figure 1.3). The hydraulic elevator also evolved from the direct

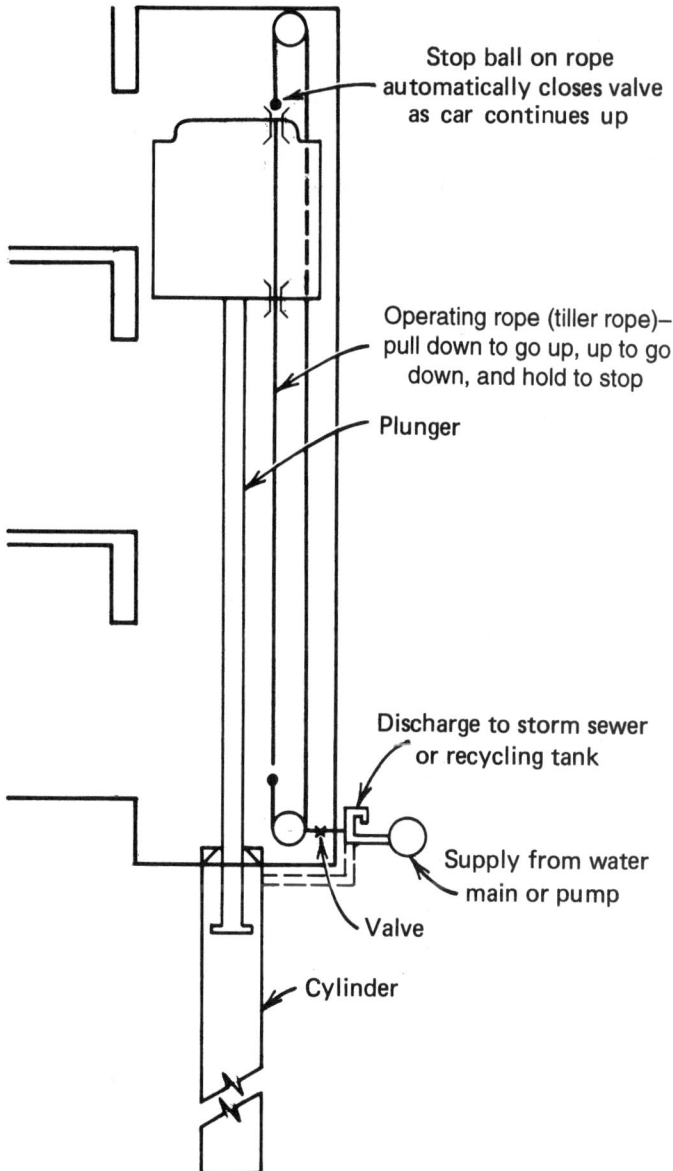

Stop ball on rope automatically closes valve as car continues up

Operating rope (tiller rope)– pull down to go up, up to go down, and hold to stop

Plunger

Discharge to storm sewer or recycling tank

Supply from water main or pump

Valve

Cylinder

Figure 1.3. Hydraulic elevator with handrope operation.

ram-driven elevator to the so-called geared or roped hydraulic (Figure 1.4) capable of speeds of up to 700 fpm (3.5 mps) and rises of 30 or more stories. The cylinder and sheave arrangement was developed to use multiple sheaves and was mounted vertically for the higher rises. The 30-story building did not appear until after 1900, well after steel-frame construction was introduced, but the hydraulic elevator served practically all of the 10- to 12-story buildings of the 1880 to 1900 era.

It was in this era that many of the aspects of elevators as we know them today were introduced. Hoistways became completely enclosed, and doors were installed at landings. Before that time many hoistways were simply holes cut in the floor—occasionally protected by railings or grillage. Simple signaling was introduced, using bells and buzzers with annunciators to register a call, which was manually canceled. Groups of elevators were installed, the first recorded group of four elevators being in the Boreel Building in New York City, and the "majordomo" of "elevator buildings"—the starter—entered the scene and was assigned to direct the elevator operators to serve the riding public.

The first electric elevator quietly made its appearance in 1889 at the Demarest Building in New York. This elevator was a modification of a steam-driven drum elevator, the electric motor simply replacing the steam engine. It continued in service until 1920 when the building was torn down. Electric power was here to stay, and the Otis Elevator Company installed the first automatic electric or push-button elevator in 1894.

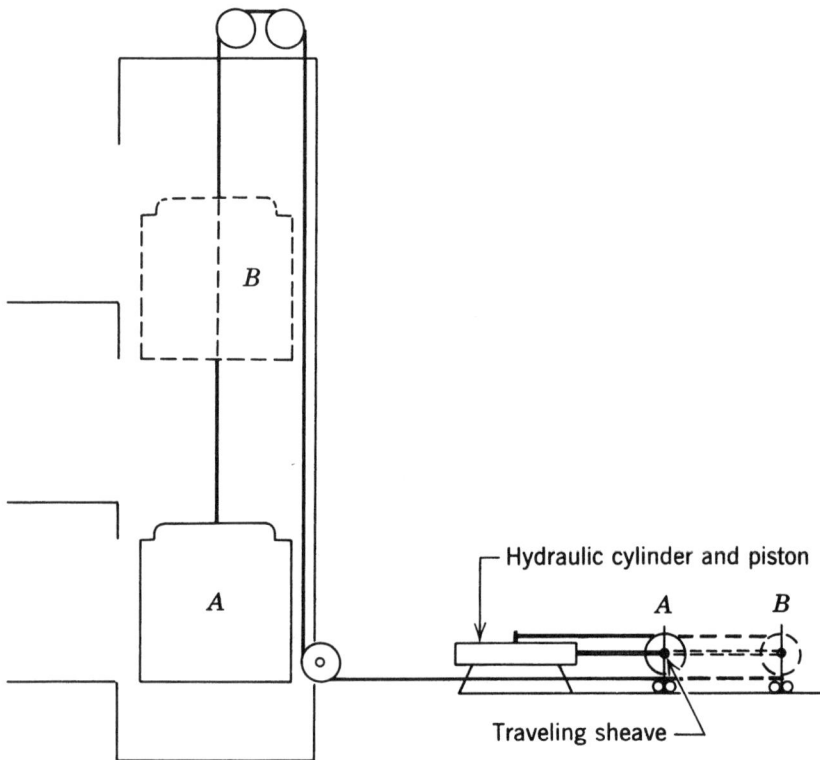

Figure 1.4. Roped hydraulic elevator.

With the tremendous building activity of the early 1900s and the increased size and height of buildings at that time, the questions of quantity, size, speed, and location of elevators began to arise. With these questions began the applied technology of elevatoring. A typical but wrong logic pattern of the time was, "Joe Doe has two elevators in his building and seems to be getting by all right. Since my building is twice as big, give me two twice the size." It rapidly became evident that people in the latter building had to wait twice as long for service as those in Joe Doe's building, and complaints and building vacancies reflected their dissatisfaction. The example is typical, and soon elevatoring emerged as a special design discipline.

ELEVATORING

Elevatoring is the technique of applying the available elevator technology to satisfy the traffic demands in multiple- and single-purpose multifloor buildings. It involves careful judgment in making assumptions as to the total population expected to occupy the upper floors and their traffic patterns, the appropriate calculation of the passenger elevator system performance, and a value judgment of the results so as to recommend the most cost-effective solution or solutions.

A major part of elevatoring is the understanding of pedestrian flow, pedestrian queuing, and the associated human engineering factors that will provide a nonirritating "lobby to lobby" experience. The traffic demands of passengers, service functions, and materials must be evaluated and simultaneously satisfied for an optimal solution.

Elevatoring, in the modern sense, is the process of applying elevators and the building interfaces necessary for the vertical transportation of personnel and material within buildings. Service should be provided in the minimum practical time, and equipment should occupy a minimum of the building's space. The need for refinement in this process became apparent in the early 1900s as the height and cost of buildings increased.

Elevators changed radically in the early 1900s. As electricity became common, and with the introduction of the traction elevator, the water hydraulic was rapidly superseded. Helping its demise was the rapid rise of building heights—the Singer Building, 612 ft (185 m); the Metropolitan Life Tower, 700 ft (212 m); the Woolworth Building, 780 ft (236 m); all in New York City and built by 1912. The roped hydraulic could not be stretched to compete with such rises, and the direct-plunger-driven elevator required a hole as high as the rise. Telescoping rams were tried and proved unsatisfactory. These buildings were made possible by the introduction of the traction elevator into commercial use in 1903.

TRACTION ELEVATORS

Description

Up until about 1903, either drum-type elevator machines wherein the rope was wound on a cylindrical drum, or the hydraulic-type elevator, either the direct plunger or the roped hydraulic machine, was the principal means of hoisting force. Both had severe rise limitations: the drum type, in the size of the drum; and the hydraulic type, in the length of the cylinder. The drum-type elevator had the further disadvantage of requiring mechanical stopping devices to shut off power to prevent the car from being drawn into the overhead if the machine failed to stop by normal electrical means. On a hydraulic machine this is prevented by a stop ring on the plunger.

The traction machine had none of the rise disadvantage of either the hydraulic or drum machines. The traction principle is a means of transmitting lifting force to the hoist ropes of an elevator by friction between the grooves in the machine drive sheave and the hoist ropes (Figure 1.5a and b). The ropes are simply connected from the car to the counter-

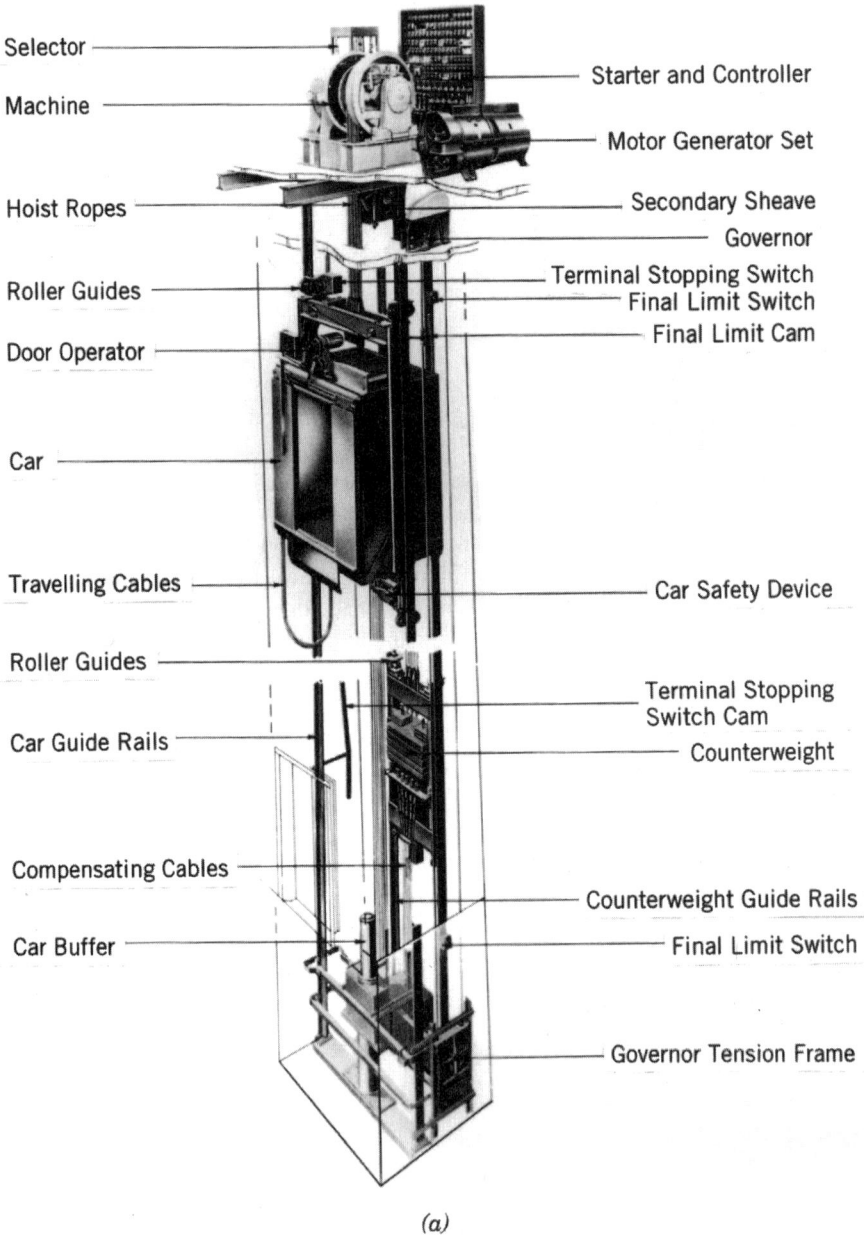

Selector — Starter and Controller
Machine — Motor Generator Set
Hoist Ropes — Secondary Sheave
— Governor
Roller Guides — Terminal Stopping Switch
— Final Limit Switch
Door Operator — Final Limit Cam
Car —
Travelling Cables — Car Safety Device
Roller Guides — Terminal Stopping Switch Cam
Car Guide Rails — Counterweight
Compensating Cables — Counterweight Guide Rails
Car Buffer — Final Limit Switch
— Governor Tension Frame

(a)

Figure 1.5. (a) Gearless elevator installation (Courtesy Otis Elevator). (b) Geared elevator installation (Courtesy Otis Elevator).

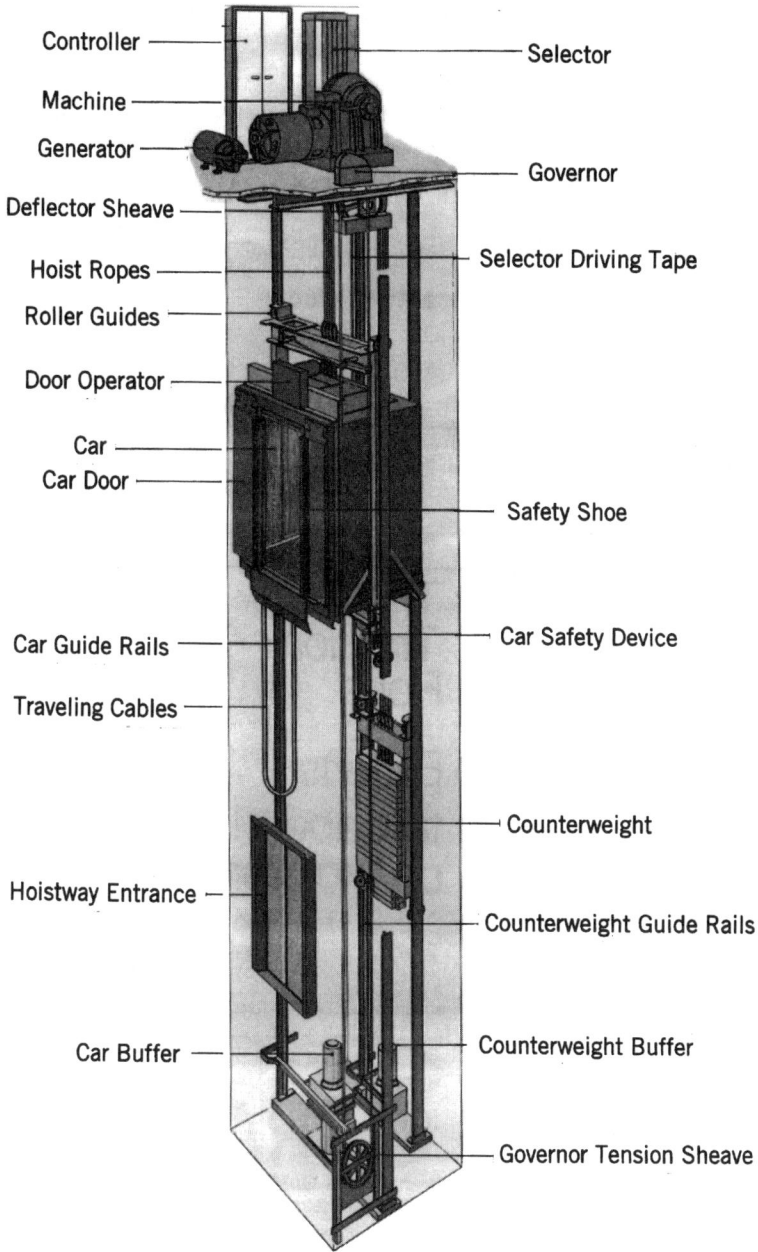

Controller — Selector

Machine —

Generator —

Deflector Sheave — Governor

Hoist Ropes — Selector Driving Tape

Roller Guides —

Door Operator —

Car —

Car Door — Safety Shoe

Car Guide Rails — Car Safety Device

Traveling Cables —

Counterweight

Hoistway Entrance —

Counterweight Guide Rails

Car Buffer — Counterweight Buffer

Governor Tension Sheave

(b)

Figure 1.5. *(Continued)*

weight and wrapped over the machine drive sheave in grooves. The weight of both the car and the counterweight ensure the seating of the ropes in the groove or, for higher-speed elevators, the ropes are double-wrapped; that is, they pass over the sheave twice.

The safety advantages of the traction-type elevator are manyfold: multiple ropes are used, each capable of supporting the weight of the elevator, which increases the suspension safety factor as well as improving traction. The drive sheave is intended to lost traction if the car or counterweight bottoms on the buffers in the pit. However, this is not universal and depends on the proper condition of ropes, sheave, loading, and so on. The possibility of the car or counterweight being drawn into the overhead in the event of electrical stopping switch failure is reduced.

Traction elevators are capable of exceedingly high rises, the highest (or lowest) being in a mine application in South Africa for a depth of 2000 ft (600 m). The critical factors become the weight of the ropes themselves and the load imposed on the sheave shaft and its bearings. It was the traction elevator, in addition to other advances in building technology, that made today's tall buildings of 100 or more stories practical.

The traction principle has been available for centuries. The capstan on a ship is an example. The first known elevator application was the "Teagle" hoist, which was present in England about 1845, as shown in Figure 1.6. This old print shows the traction drive and the counterweight. Motive force was provided by means of belts to the line shafting in the building where the lift was installed. The operation was by handrope, as described for the hydraulic elevator shown in Figure 1.3. The handrope acted to engaged the belt to the drive pulley, usually to the right or left of an idler pulley, to move the lift up or down.

Performance

With the application of electrical drives to elevators, the versatility of electrical versus mechanical controls allowed certain standards of elevator operation and control so that time-related factors in an elevator trip could be established. Speed no longer depended on varying water or steam pressure. The Ward–Leonard system of electric motor speed con-

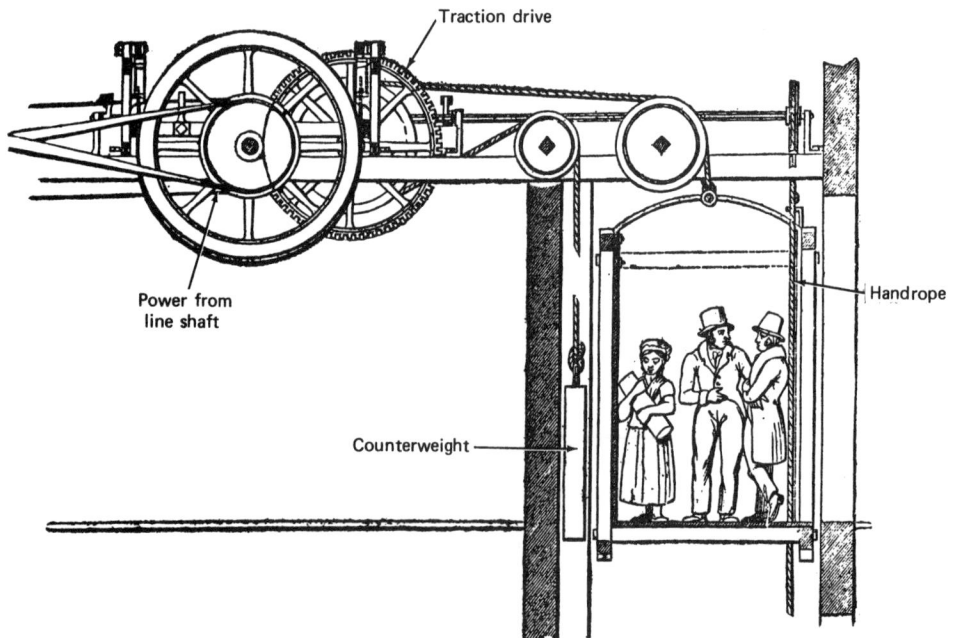

Figure 1.6. Teagle elevator (circa 1845) (Courtesy *Elevator World Magazine*).

trol was introduced early in the 1920s and allowed the smoothness of acceleration and deceleration common in elevators of today.

The Ward–Leonard system employs a motor generator driven by either an ac or dc motor, the output of the generator being directly connected to the armature of the dc hoisting motor. Varying the voltage on the field of the generator varies the dc voltage applied to the hoisting motor armature and, consequently, the speed and torque.

The Ward–Leonard motor-generator hoisting machine combination, generically known as "generator field control," was the quality standard for many decades from the 1920s through the 1980s. Thousands of elevators still employ it, and Figure 1.5a and b shows this equipment. The major change is in the machine room. The motor generator is gone, as is the selector shown in the background. The controller is no longer full of relays but replaced by a compact microprocessor and the SCR (silicon controlled rectifier) drive.

Replacement of the motor generator by solid-state control was introduced in the 1970s, which has superseded, in most instances, motor generators in both new installations and modernizations. One approach to a solid-state control system is to employ SCRs to convert the line ac into varying dc for the operation of a dc hoisting machine. Most of the higher-speed elevators, 500 fpm and above, use this approach, and it is favored for the modernization of existing elevators as well. A second approach is to employ SCRs to develop a varying voltage, varying frequency (VVVF) ac power to an ac driving machine. This is the favored approach for lower-speed (up to 500 fpm) geared machines.

Most new traction elevators are expected to have the VVVF control as further development of higher-speed gearless-type elevators proceeds. It is almost universally used with the new geared installations and is being applied as an upgrade to the thousands of single-speed, low-speed (100–150 fpm) ac machines that were widely used in the many six-story apartment buildings built in the late 1940s and through the 1950s.

The microprocessor has been a major innovation in the past decade, and practically all of the new and modernized installations employ one or more in both the control and operating systems. Details are to be found in Chapter 7 "Elevator Operation and Control." That chapter also includes many of the ramifications and disciplines needed to apply new technology. Greater emphasis on machine room environmental conditions such as air-conditioning and electromagnetic interference must be considered, as well as the quality of the incoming power supply both under normal conditions and when an emergency generator is used.

In the course of this book the operating characteristics of electric elevators are described and a basis for time study calculations of elevator trips are established. These time factors will become the basic tools in establishing the number of elevators necessary for any type of building and will be related to the speed at which people can be moved from place to place vertically. As a preliminary, familiarity with modern elevator types is necessary.

GEARLESS TRACTION ELEVATORS

Description

The preceding brief discussion of early elevator history introduced the traction-type elevator. The first high-rise application of this type of elevator was in the Beaver Building in New York City in 1903, followed by such notable installations as the Singer Building (demolished in 1972) and the Woolworth Building. These elevators were of the gearless traction type that are at present the accepted standard for the high-rise, high-speed [over 400 fpm (2.0 mps)], and high-quality elevator installation.

The gearless traction elevator consists of a large, slow-speed (50 to 200 rpm) dc motor of four to eight poles directly connected to a drive sheave of about 30 to 48 in. (750 to 1200 mm) in diameter. An electrically released, spring-applied brake is arranged to apply stopping to the drive sheave. Slow-speed dc motors and ac motors (being introduced), though expensive and massive, are necessary to maintain the necessary torque to directly drive large-diameter sheaves. The larger-diameter sheaves also conform to the bending radius of elevator steel ropes. A limitation is imposed by safety codes as good practice for long rope life and is generally established at a minimum of 40 times the diameter of the wire rope used. For example, a ½-in. (13-mm) wire rope would require a minimum sheave size of 20 in. (500 mm).

The slow speed of the direct drive gearless traction machine is necessitated by the speed of the elevator it serves. For example, for a 500 fpm (2.5 mps) elevator and sheave diameter of 30 in. (750 mm), a top speed of 86 rpm is required. To level this elevator to a landing at a maximum speed of 25 fpm (0.125 mps), 4.3 rpm is necessary. Gearing with higher-speed motors has been introduced by at least one major manufacturer to gain these higher speeds. The continuous operation of elevators [up to 25,000 mi (40,000 km) per year] and the relative ease of maintenance of the gearless machines, as well as their dependability, makes them the preferred type for higher speeds.

On higher-speed gearless traction machines of 800 fpm (4.0 mps) or more, the double-wrap principle is generally applied to obtain traction and to minimize rope wear. The ropes from the car are wrapped around the drive sheave, around a secondary or idler sheave, around the drive sheave, and down to the counterweight (Figure 1.7a–c). The groove seats are round, providing support on the full half of the rope, thus eliminating pinching action and minimizing wear. Traction is obtained by the pressure of the ropes on the sheave. As may be noted, increasing the weight on the car or counterweight increases the force so that friction between the ropes and the sheave increases traction.

Elevator machines are also roped with a single-wrap arrangement, which is applied to both gearless and geared machines. The single-wrap arrangement provides traction by the use of grooves that will pinch the ropes with varying degrees of pressure depending on the shape of the groove and its undercutting (see Figure 1.7 and later discussion). The most effective single-wrap arrangement provides 180 degrees of rope contact with the sheave without a deflecting sheave, as shown in Figure 1.8 [for single-wrap traction (SWT), 2:1 roping].

Conventional elevators are roped either 1:1 or 2:1 (Figure 1.8) for both car and counterweight. In some unusual installations and special applications, 1:1 car and 2:1 counterweight roping has been used. In that event the counterweight must be at least twice as heavy as the weight of the car. The 1:1 arrangement is the most popular for higher speeds and has been used for a load and speed of 10,000 lb (4500 kg) at 1600 fpm (8 mps). The 2:1 arrangement allows the use of a higher-speed, and therefore a smaller but faster, elevator motor. The mechanical advantage of 2:1 roping requires that only half the weight be lifted, so 2:1 is generally used whenever loads in excess of 4000 lb (1600 kg) must be lifted. The economy of the faster motor, which can be built smaller and lighter than lower-speed dc motors, also makes 2:1 roping attractive for a full range of speed requirements from 100 to 700 fpm (0.5 to 3.5 mps) or more and for any lifting capacity.

Any of the aforementioned 1:1 and 2:1 roping arrangements can be provided with the elevator machine in the basement or at a lower level. The appropriate sheaves are installed in the overhead space to direct the ropes from the machine to the car and counterweight. The preferred arrangement is the single wrap traction type. A foundation must be provided for the machine that will overcome the uplift and solidly anchor the machine under all conditions of operation and safety application.

The long life, smoothness, and high horsepower of gearless traction elevators provide a durable elevator service that can outlive the building itself. The original gearless machines in the Woolworth Building were reused when that building's elevators were modernized in 1950, again in 1970, and for a third time in 1990. The gearless machine not only provides speed, if necessary, but is also capable of performance essential to any well-elevatored building.

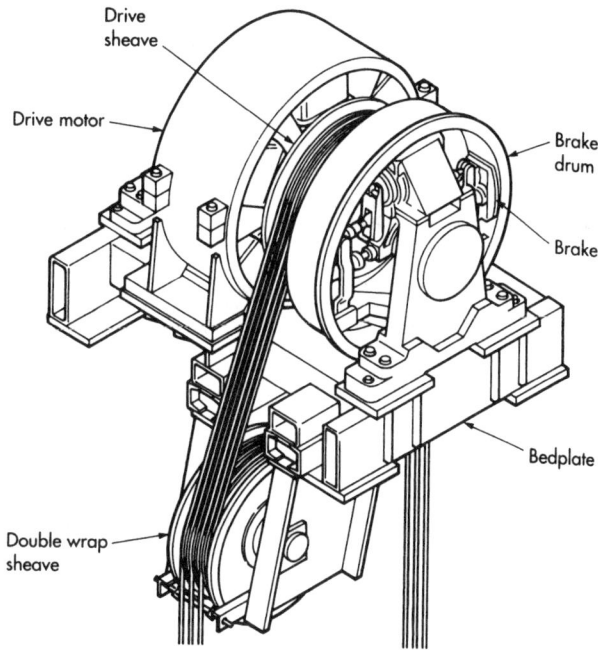

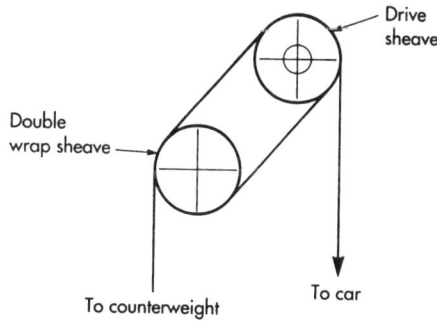

Figure 1.7. (*a*) Double-wrap gearless machine—Otis type 219HT with internal brake (Courtesy Otis Elevator).

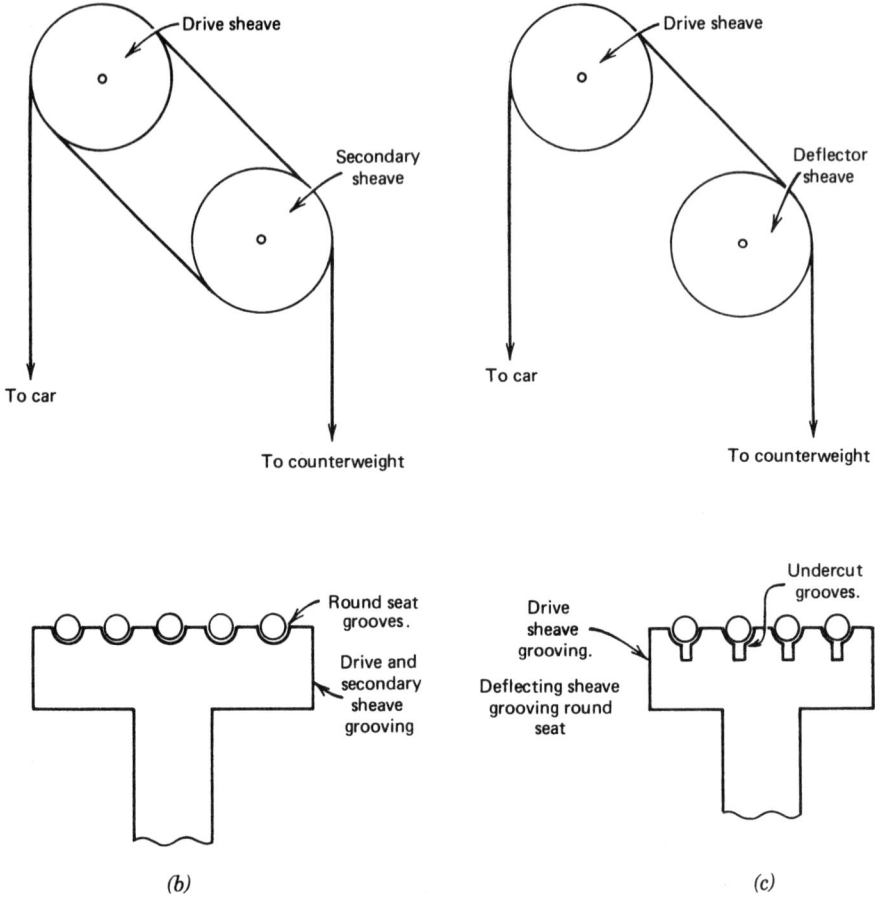

Figure 1.7. *(Continued)* *(b)* Double-wrap traction arrangement. *(c)* Single-wrap traction arrangement.

Gearless Machines—Performances

Essential to elevatoring considerations is the requirement that a gearless traction machine, no matter what its lifting capacity or speed, must be capable of optimum floor-to-floor operating time commensurate with passenger comfort. Stated another way, the machine must be capable of starting a filled elevator car, accelerating to a maximum speed for the distance traveled, and slowing to a stop in a minimum time of about 4.5 to 5.0 sec. This must be performed under all conditions of loading, either up or down. The elevator system must be so arranged that such acceleration and deceleration take place without discomfort to the passenger from a too rapid change in the rate of acceleration or deceleration (with optimum jerk). Furthermore, the elevator must be capable of releveling, while passenger load is changing at a floor (correcting for rope stretch), with almost imperceptible movement. The aspects of performance are discussed further in a later chapter.

GEARED TRACTION MACHINES

As the name implies, the geared traction elevator machine utilizes a reduction gear with a high-speed motor to drive the traction sheave. A high-speed ac or dc motor drives a worm

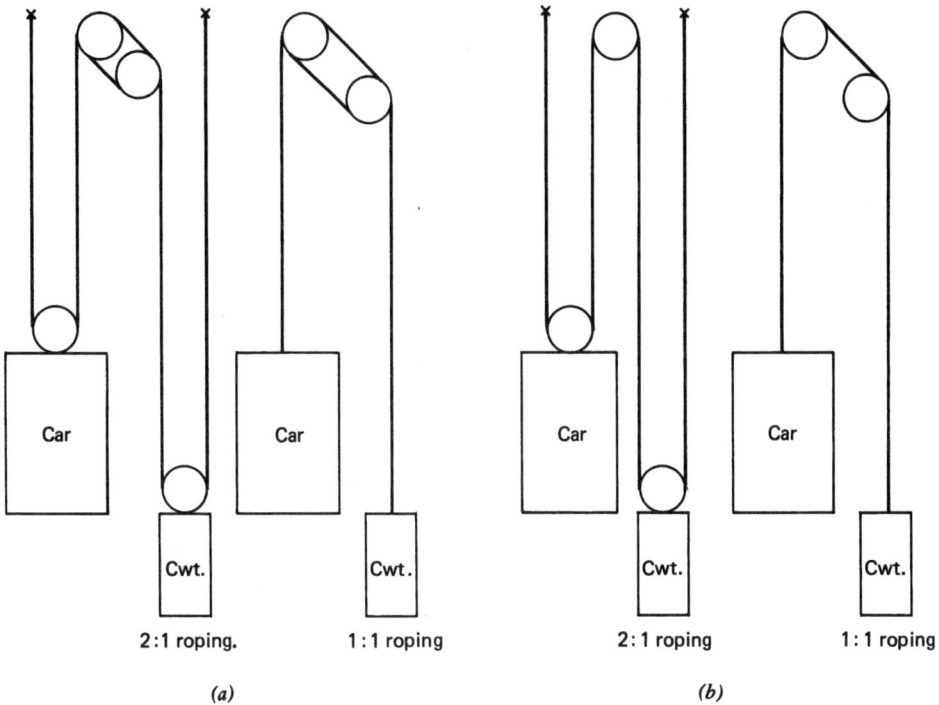

Figure 1.8. (*a*) Double-wrap roping. (*b*) Single-wrap traction roping.

and gear reduction unit which in turn drives the hoisting sheave, the net result being the slow sheave speed and high torque necessary for elevator work. A brake is applied by spring to stop the elevator and/or hold the car at a floor level. Recent (1990s) introductions have been planetary gearing and helical gearing to replace the traditional worm gear approach.

The geared traction machine is used for elevators and dumbwaiters of all capacities from 25 to 30,000 lb (10 to 14,000 kg) or more, and speeds from 25 to 450 fpm (0.125 to 2.3 mps). The complete flexibility of worm gear ratios and motor speeds and horsepowers, as well as drive sheave diameters and roping arrangements (1:1, 2:1, and, sometimes, 3:1), makes this vast range of application practical. In some materials handling applications, geared machines are used for speeds of 600 fpm or more (3.0 mps) with excellent results.

The geared traction elevator is an outgrowth of the earlier drum-type elevators. The steam engine gave way to the electric motor and gear (Figure 1.9), and the drum gave way to the drive sheave (Figure 1.10). The grooved drive sheave was an outgrowth of the traction principle applied to gearless elevators; instead of ropes being wrapped around the sheave, grooves were cut into the sheave and the necessary friction was created by the pinching action of the grooves on the rope (Figure 1.11). Various types of grooving are used for different loads and traction requirements. Generally, the sharper the undercut angle, the greater the traction (and, usually, the greater rope and sheave wear).

Polyurethane groove liners capable of providing greater traction and less rope wear are being used by at least one manufacturer.

Geared machines have been driven by either one-speed or two-speed ac motors, by dc motors utilizing the Ward–Leonard means of control, or by ac or dc motors with SCR,

Figure 1.9. (*a*) Early steam-driven hoisting machine. (*b*) Early electric-driven hoisting machine.

VVVF, or other solid-state control. Ac motor machines are often used for speeds from 25 to 150 fpm (0.125 to 0.75 mps) with single- or two-speed motors or with solid-state drives to 450 fpm (2.25 mps). Stopping with single-speed motors is accomplished by disconnecting the power from the motor and stopping the car by a combination of slide and brake action. Two-speed ac operation employs a double-wound motor, a fast-speed wind-

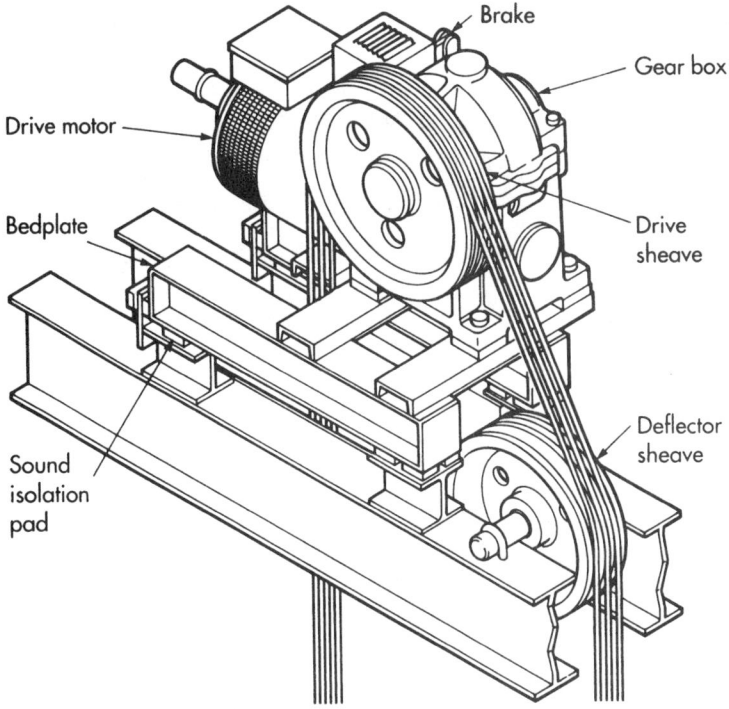

(a)

Figure 1.10. (a) and (b) Typical geared machines (Courtesy Titan Machine).

(c)

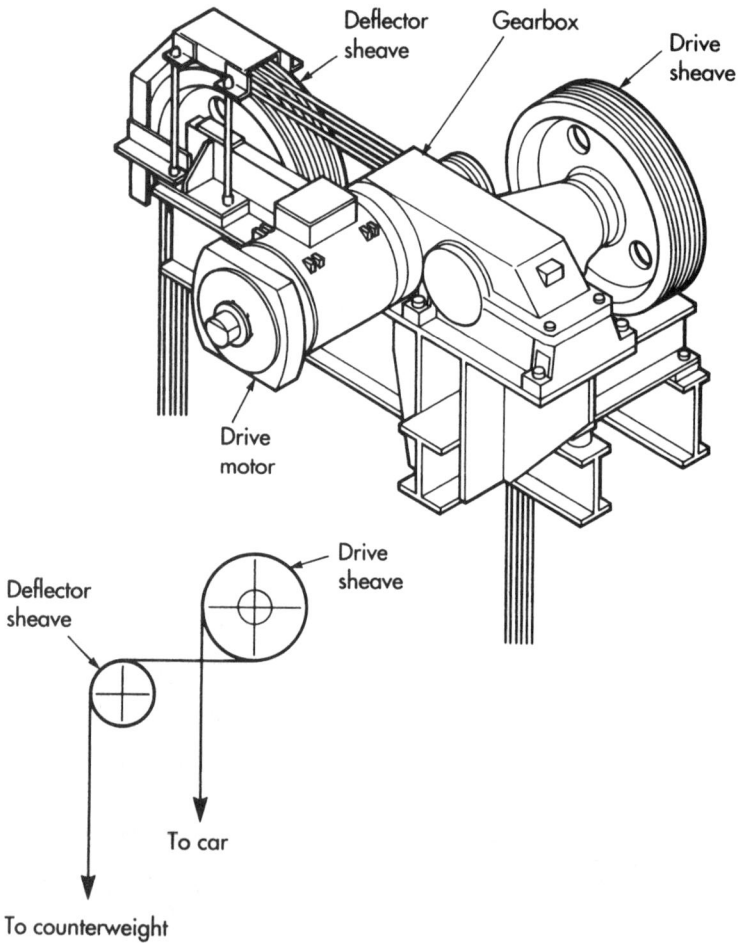

(d)

Figure 1.10. *(Continued)* (*c*) Worm gear machine (Courtesy Hollister Whitney). (*d*) "Longwrap" traction machine with helical gear.

Figure 1.10. *(Continued)* (*e*) Planetary geared machine (Courtesy Hollister Whitney).

ing for full-speed running, and a slow-speed winding (which can be any ratio as high as 6:1, i.e., the slow speed being ⅙ full speed) for stopping, leveling, and, if required, releveling. Operation is generally to start at full speed, run, switch to low speed at a measured distance from the stop, and accomplish the final stop by combination of brake and slide. The floor-level accuracy of plus or minus ½ to 1 in. (13 to 24 mm) can be obtained under all conditions of load, as contrasted with one-speed accuracy of 1 to 3 in. (24 to 75 mm), which will vary with load. Much greater accuracy can be obtained when solid-state ac motor drives are employed, and various upgrades can be retrofitted to existing single-speed elevators. In contrast, the dc Ward–Leonard drive or a solid-state motor drive

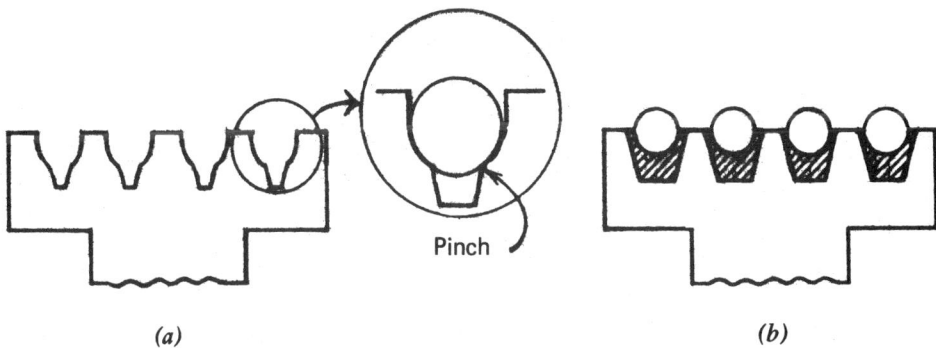

Pinch

(a)   *(b)*

Figure 1.11. (*a*) Undercut sheave groove. (*b*) Sheave groove with polyurethane liner.

allows the car to be stopped electrically before the brake is applied, resulting in leveling accuracy from ¼ to ½ in. (6 to 13 mm) under all conditions of load, and much softer stops than produced by the ac machine.

With either ac or dc geared elevators, the floor-to-floor performance can be established, which is essential in calculations in estimating the numbers of elevators for a particular building.

HYDRAULIC ELEVATORS

A third major type of elevator in use today is a modern version of the hydraulic elevator. Most hydraulics are direct-plunger-driven from below (the cylinder extending into the ground as high as the elevator rises), and the operating fluid is oil moved by high-speed pumps rather than water under pressure (Figure 1.12a). Rapidly gaining favor are roped or indirect hydraulic elevators as well as many "holeless" types, some utilizing telescoping pistons. Hydraulic elevators today are used for both passenger and freight service in buildings from two to six stories high and for speeds from 25 to 200 fpm (0.125 to 1.0 mps). Single-ram capacities will range from 2000 to 20,000 lbs (1000 to 10,000 kg) or more. Multiple rams are used for high capacities of 20,000 to 100,000 lb (10,000 to 50,000 kg). Varied speeds and high capacities are obtained through multiple pumps. Elevatoring performance time considerations of hydraulic elevators are slightly slower than that for geared elevators.

Drilling a hole for a direct plunger hydraulic elevator has always been of concern since underground conditions are often unknown. This concern has led to a number of equipment variations, namely, the "holeless" hydraulic and the indirect drive (roped) hydraulic. The former, shown in Figure 1.12b, is favored for the lower rise; the plunger (also called a jack or a ram) is mounted on the side of the car and connected to the top of the car structure. A variation is the use of a telescoping plunger (Figure 1.12c), which allows extended travel. Heavier cars can be accommodated by two plungers, one on each side of the car.

The indirect hydraulic (currently termed as "roped hydraulic") consists of an underslung roping arrangement, with the ropes driven by a vertically traveling sheave mounted on a jack located at the side of the car. (See Figure 1.12d; a sheave, A, is mounted on top of the plunger, B, and ropes, C, are slung over the sheave and hitched to the car.) This is a 1-to-2 roping arrangement, since 1 ft of plunger travel allows 2 ft of car travel. As with any suspended elevators, an elevator code requirement is to provide a safety device to prevent falling if the hoisting rope fails.

A growing concern regarding direct plunger hydraulic elevators is the condition of the cylinder after being buried in the ground for a number of years. Earlier units had a minimum of external protection, whereas those installed recently are usually encased in PVC or other anticorrosion material. Any loss of oil in the pump unit must be thoroughly investigated, and an underground leak correction often involves total replacement.

ESCALATORS AND MOVING WALKS

A very important factor in vertical transportation is the escalator (moving stairway) and moving inclined walks. Before the 1950s, escalators were mainly found in stores and

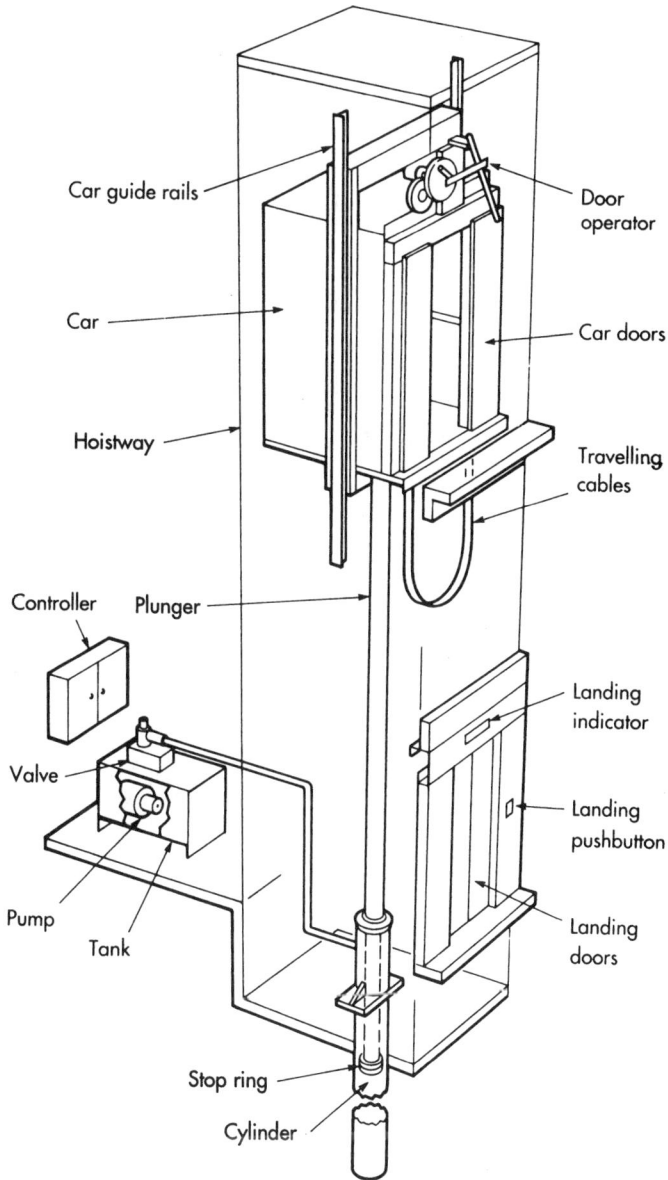

Figure 1.12. (*a*) Hydraulic passenger elevator—direct plunger (Courtesy Otis Elevator).

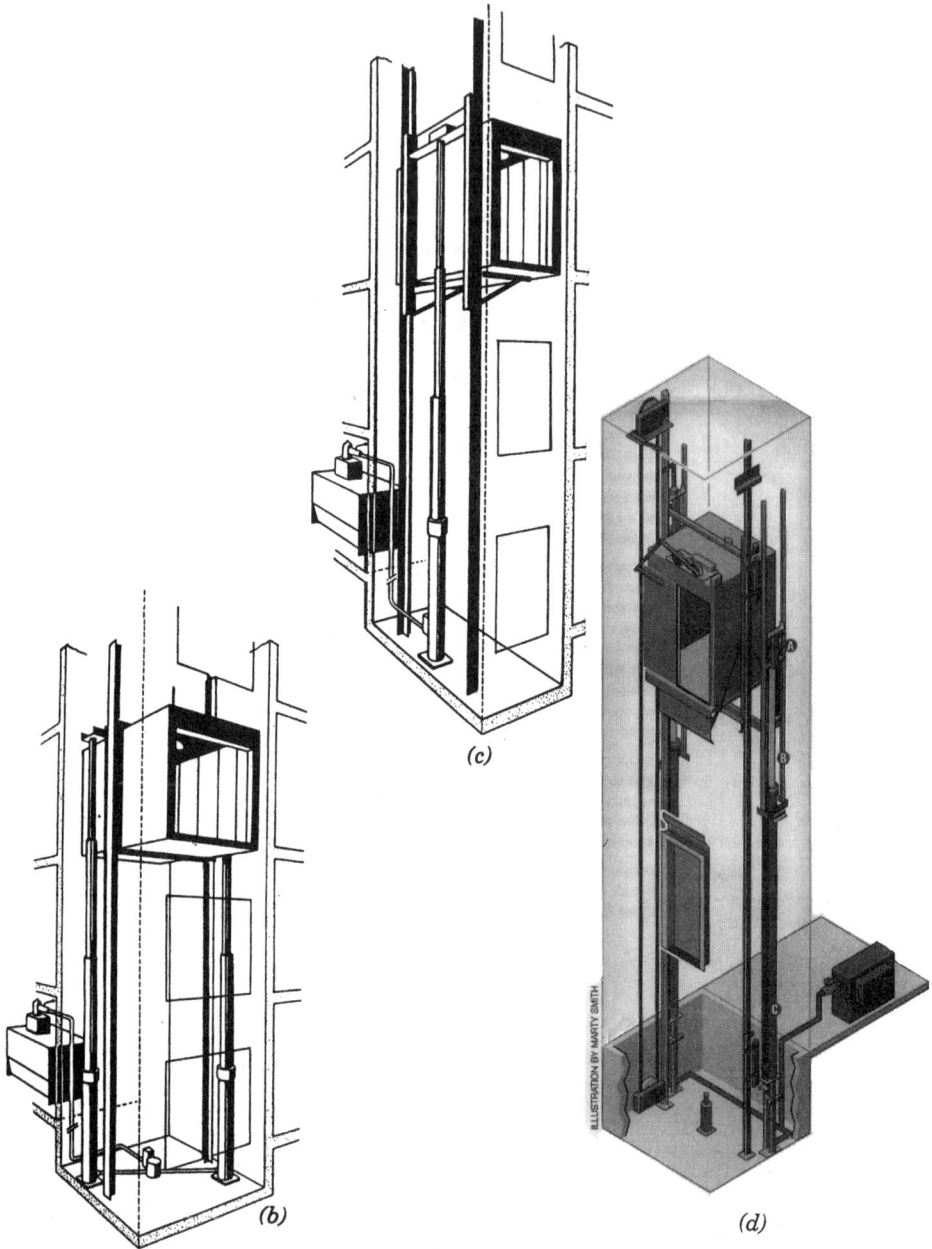

Figure 1.12. *(Continued)* (*b*) Holeless hydraulic elevator (dual plungers) (Courtesy *Elevator World Magazine*. Reproduced from *The Guide to Elevatoring*). (*c*) Telescoping "holeless" hydraulic elevator. (Courtesy of *Elevator World Magazine*. Reproduced from *The Guide to Elevatoring*). (*d*) Roped indirect hydraulic elevator (Courtesy Otis Elevator).

transportation terminals. Today their use has expanded to office buildings, schools, hospitals, banks, and other places where large flows of people are expected or it is desired to direct people vertically in a certain path. Escalators and elevators are often used in combination either to provide necessary traffic-handling capacity or to improve elevator operation by directing people to one elevator loading level. They are essential in the lobby of a building with double-deck elevators. Many office buildings and some factories have found escalators ideal for rapid shift changes or rapid floor-to-floor communication.

Inclined, flat, or contoured moving walkways are closely related to the escalator form of vertical transportation (Figure 1.13). The passenger-handling ability of such

Figure 1.13. Moving walkways: (*a*) inclined; (*b*) flat (Courtesy Montgomery KONE).

conveyances is based on speed and density of passenger loading per step. Nominal ratings are in passengers per 5 mins. Qualifications of capacities and application of the moving stairway or walk are discussed in later chapters.

As a historical note, the flat step escalator was first introduced by the Otis Elevator Company at the Paris Exposition in 1900. This was a "one-way" escalator arranged for either up or down travel. A person could walk on directly but upon exiting encountered a deflecting barrier, forcing the person to step off to the right or to the left (Figure 1.14). This design, known as a "Seeberger" was preceded by the "Reno" type, which was an endless series of inclined "indentations" that were boarded at the same angle of rise. Major development began in 1920, the initial features being flat steps with cleats, followed by flat boarding and debarking areas, narrower step cleats and combs, extended newels, and glass balustrading (Figure 1.15).

Because a number of people, especially those with impaired mobility, either refuse or are incapable of using escalators, it has become essential that alternate vertical transportation in the form of a two-stop elevator be provided within sight of the entrance to an escalator. Another reason is that any wheeled vehicle such as a stroller, baggage cart, hand truck, or the like, presents a hazard if its movement is attempted on an escalator.

Figure 1.14. "Seeberger" flat step escalator. Note barrier deflector (Reproduced from *Elevator World Magazine,* January 1992).

Figure 1.15. Glass balustrade escalator (Courtesy Montgomery KONE).

The value of escalators in vertical transportation is in providing a continuous flow of people as contrasted with the batch approach of elevators. This continuous flow principle proves valuable where the movement of large numbers of people is required, such as in an airport when an airplane unloads or at a sports event both prior to its start and at the end. A detailed discussion of the application of escalators is presented in Chapter 9.

DUMBWAITERS AND MATERIALS-HANDLING SYSTEMS

Other forms of vertical transportation, discussed in Chapter 15, of this book are dumbwaiters and materials-handling systems. Modern buildings use these devices for a variety of purposes: delivery of books in libraries, distribution of mail in office buildings, delivery of food and supplies in hospitals, and so on. The dumbwaiter (Figure 1.16) is actually a small elevator, which can have all the performance characteristics of an elevator.

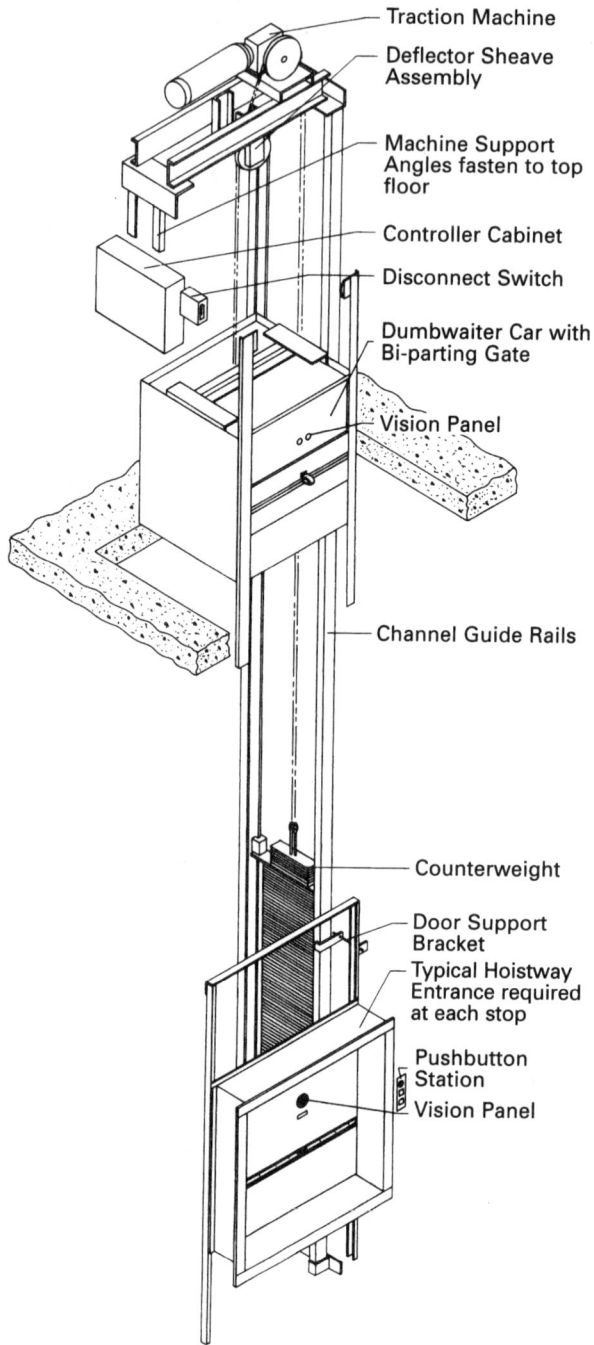

Figure 1.16. Ambassador TR dumbwaiter (Courtesy D.A. Matot, Inc.).

Loading and unloading can be either at counter or floor level, and either manual or automatic. Size can vary from letter size to car sizes consisting of any arrangement of 9 ft^2 (0.9 m^2) or less of platform area, and a car with an effective height of no more than 4 ft (1200 mm). This is a limitation imposed by elevator safety codes (covered in detail in Chapter 16), and anything over that size must be classified as an elevator. Dumbwaiters need not have safeties and are strictly for material handling. They are always operated from the landing, not from within the cab as in an elevator.

Other forms of vertical materials-handling systems include tote box conveyors, automatic loading and unloading cart lift systems, and self-propelled vehicle systems either with a dedicated rail system or following a guide path on the floor. These are discussed in detail in a later chapter.

HANDICAPPED LIFTS

A growing segment of the vertical transportation industry is the introduction and development of a special means to accommodate persons in wheelchairs and others with limited mobility. Elevators are the universal means, but retrofitting them to an existing building is either a disproportionate solution or impossible. Ramps are a traditional means; however, they require extended horizontal space to limit the slope. An alternative has been to develop a platform-type lift that can be installed adjacent to the common stairs or, in some applications, within the stairs themselves.

The approaches have a range of application, from a short vertical rise of a few feet to extended rises of a floor height or so to provide access to a mezzanine or balcony (Figure 1.17). Equipment is available in a variety of types: vertical platform lifts with open enclosures, those serving two or three levels, and enclosed, inclined platform lifts either fixed adjacent to a stairs or capable of folding and using the same path as a stairs. A simple variation is a lift that is set in an existing stairway and has a fold-down seat for the user (Figure 1.18).

Figure 1.17. A platform-type lift installed adjacent to common stairs (Courtesy The National Wheel-O-Vator Co., Inc.).

Figure 1.18. An inclined lift set in an existing stairway with a fold-down seat for the user (Courtesy The National Wheel-O-Vator Co., Inc.).

The impetus for accelerated development has been the passage of the Americans with Disabilities legislation, which mandated, in some form or another, access to the various floors in any building used by the public, as well as to work spaces. Although elevators can be used and are essentially mandated for new construction, retrofitting to an existing building is often impossible or extremely costly. This is especially true in infrequently used buildings such as churches or where the only barrier is a set of stairs leading to an entrance. The handicapped lift is a solution. Any new building being constructed must recognize the need for access and is an unlikely candidate for this lift. The thousands of existing buildings, the aging and increasingly mobility-limited population, as well as economics, have created the demand. Although these lifts are not a major factor in the application of elevators, their use and limitations should be recognized as a means for needed vertical transportation.

Elevator codes have limited size, capacity, and speed and include rules for personnel protection and emergency considerations. In general, rise is limited to 12 ft (3600 mm), speed no greater than 30 fpm (0.15 mps), and capacity no greater than 750 lb (340 kg). Additional code rules prescribe application limits, as well as runway design and hoistway protection. Code references are found in Chapter 16.

Recent (1996) ASME A17.1 Code rules have been promulgated to recognize a new class of elevators, limited in use and application to accommodate the mobility-impaired, but not restricted to that use alone as are the aforementioned "handicapped lifts." These are referred to as "limited use/limited application" (LU/LA) elevators. They are restricted in size, rise, and speed but are universal in application and usable for any vertical transportation requirements. They have all the attributes of a regular elevator, such as powered landing and car doors, fully enclosed hoistways, and safety features common to all elevators. Space requirements have been minimized, making them attractive and more economical for limited application.

Residential-type elevators have been installed for use in single-family residences from about the late 1880s. In fact, the first of the completely automatic elevators was designed for such application, but was initially too costly for the average residence. In recent years the design of this type of elevator has been refined to a point where it is popular for residential use and its cost is seldom more than that of an expensive automobile. Recognizing its limited use, a part of the ASME A17.1 code is devoted to residential elevators and allows more liberal requirements than those established for conventional elevators. This elevator, like the LU/LA, is limited in size, speed, and rise but is ample for expected use by family members, including accommodation for wheelchairs. Figure 1.19 gives the layout details of such an elevator, and a number of manufacturers offer a variety of designs to fit most applications. That shown in the illustration is a "holeless, roped hydraulic"; other models include traction drives.

The foregoing represent the principal forms of vertical transportation, correctly applying these forms is the major thrust of elevatoring. Our earlier simplified definition of elevatoring is restated in the following section.

STUDY OF ELEVATORING

Elevatoring is the analysis of the requirements of vertical transportation of people and materials in a building, under all operating conditions. Such transportation requirements may be studied from a compatible aspect, as in an office building; from a function aspect, as in a hospital, or from a merchandising point of view in a department store.

The first essential step in elevatoring is pedestrian planning; that is, determining how many people will require transportation, what will be the peak traffic, and how will it occur—all up, up with partially down traffic, or equally up and down simultaneously. Some well-established guidelines may be available, as will be demonstrated in the various chapters on commercial, institutional, and residential buildings. For some projects, extensive study of the expected pedestrian movements must be made before the process of determining the requirements for elevators or escalators can start.

Once the critical pedestrian traffic is established or estimated, the next steps can be taken.

Elevatoring requires consideration of all the time factors and movements that take place during the operations providing transportation for people and/or materials. These time factors must be related to a total time required for the service based on the actual or estimated demands. Efficient elevatoring requires minimizing the time factors to maximize service.

The time components of an elevator round trip that will be studied and evaluated are as follows:

Loading Time. The time required for a number of people to board an elevator car, moving stairway or walk, or the time required to load material or a vehicle onto an elevator or lift. Loading time must be considered under many conditions of operation, consisting of narrow or wide elevator cars, wide doors, narrow doors, arrangement of elevators, and partially filled or empty elevators.

Transfer Time. The time to unload (or reload) an elevator at a local stop above the main landing. Transfer time is based on all the considerations of loading time plus, essentially, the density of the passenger or other load remaining on the elevator, and the direction of the transfer either entering or leaving.

ELEVATOR CAR AND FLOOR PLAN
NO SCALE

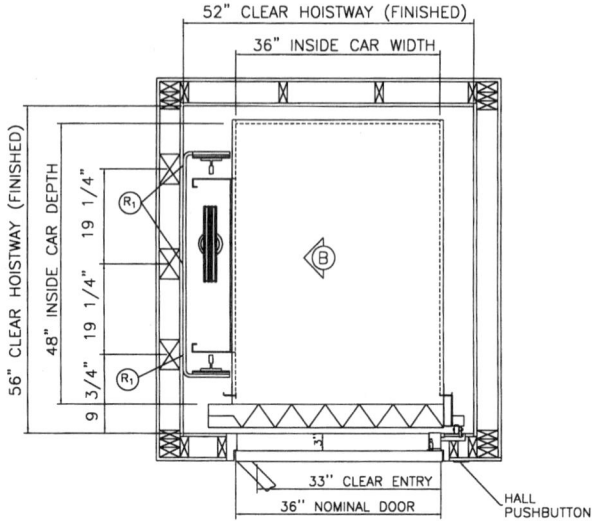

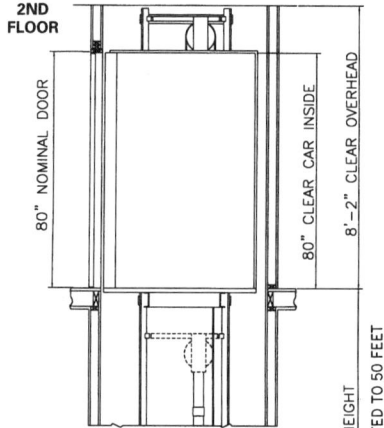

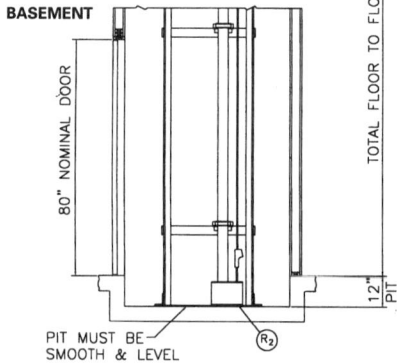

SHAFT SECTION
NO SCALE

Figure 1.19. Residential elevator—holeless, roped hydraulic (Courtesy D. A. Matot, Inc.).

These two elements, loading and transfer time, are the most difficult to quantify because, in general, these times are based on the interaction of people. Estimates have been made based on hundreds of field traffic studies of human behavior, and the conclusions are reluctantly (because of the doubt that such a person exists) based on "the average person."

Transfer time is mitigated by both legislation and environmental considerations. Most elevators are held at a floor for a minimum period of time based on the time it takes to exit and a separate time allowed for entry. The Americans with Disabilities legislation has mandated a minimum of 3 seconds for a person to exit and an extended time to enter, based on the location of the landing call button in proximity to the entrance to an elevator. These factors must be considered in calculating total transfer time. Extended discussion is given in specific application examples. Typical opening and closing times for various types and widths of entrances are found in Chapter 4, Table 4.3.

The other factors in an elevator or escalator trip are the mechanical times, which can be established accurately and assured by specification that can be developed before installing an elevator or escalator. These time factors are as follows:

Powered Door-Closing Time. A function of door weight (mass). Width of opening and type of opening for horizontally sliding doors—center opening (Figure 1.20c & d),

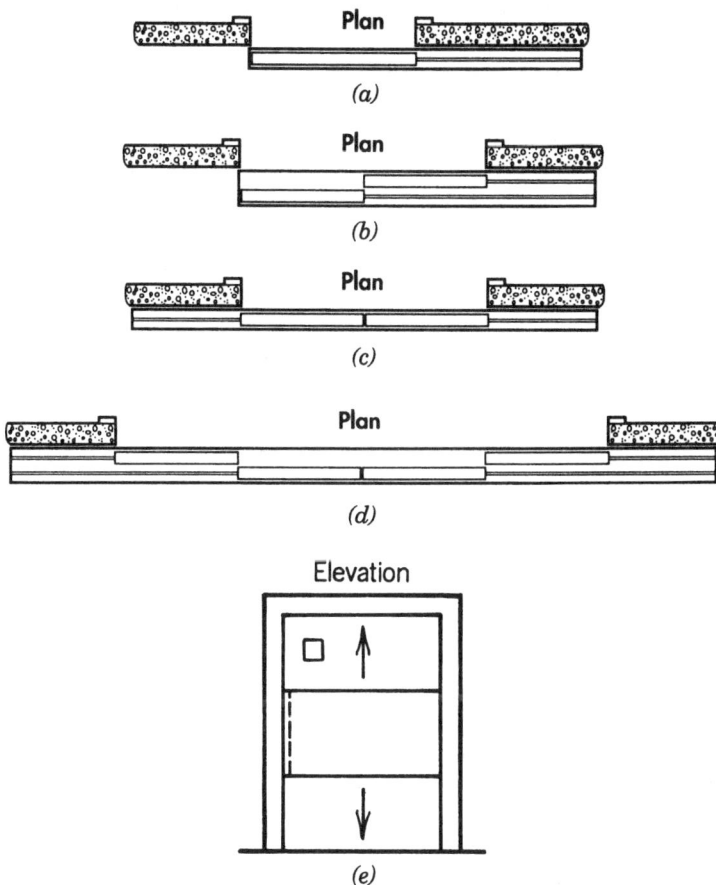

Figure 1.20. Horizontal power-operated sliding doors; (*a*) single-slide, (*b*) two-speed side-opening, (*c*) single-speed center-opening, (*d*) two-speed center-opening, (*e*) vertical biparting.

single-slide (1.20*a*), two-speed (1.20*b*), or the height of the opening for vertical biparting doors (Figure 1.20*e*) for freight application—involve different masses that affect closing speed. The kinetic energy of closing doors is limited by elevator safety codes and is usually established at no more than 7 ft poundal (0.29 joules). In practical terms, this means that the familiar 48-in. (1200-mm) center-opening sliding door will require about 3 sec to close. Closing and opening time is a vital consideration in elevatoring, for the door operation on a typical elevator occurs hundreds of times a day. An elevator cannot leave a floor until the doors are closed and locked, and passengers do not transfer until the doors are essentially fully opened.

Powered Door-Opening Time. This can be minimized by proper arrangement. The door-opening time can and should be much faster than the door-closing time. Doors can start to open while an elevator is leveling under certain conditions. This preopening must be limited so that the opening is not wide enough to allow passenger transfer before the elevator is sufficiently level with the landing to avoid a tripping hazard. The time necessary to open the doors will vary with the width and the type of doors. For example, center-opening doors take less time than single-slide or other types of the same width, and wide openings require more time than narrower openings.

Operating Time. A function of the speed control arrangement of the elevator and the number of stops the car will make in a round trip. The considerations necessary are the time required for a one-floor run, a two-floor run, and full-speed operation. The operating speed of escalators or moving ramps is constant, and maximum movement capacity is fixed.

Because the floor-to-floor operations of elevators are repeated over and over again, an estimate of the probable number of stops an elevator will make in the course of a single trip is required. Knowing or estimating the number of stops provides a means to calculate the total time for functions and leads to the cycle time or round-trip time of a single elevator. The number of stops can be established by applying a statistical formula, by inspecting the various attractions of each floor at which an elevator is required to stop, or a combination of statistical and logical determinations. The various approaches are discussed in Chapter 19, on evaluating elevator performance.

On a single elevator or escalator, once the time required to serve a given number of people is determined, the number and size of elevators or escalators that will serve the critical pedestrian traffic can be established. Although this is the essence of our studies, elevatoring cannot end here.

The grouping and operation, as well as location, of those elevators (or escalators) must be also established so that the installation will provide the expected service. The following chapters establish principles of arrangement and location of elevators and escalators. In addition, a discussion of elevator grouping, stops served, lobby arrangements, skip-stopping, operation, and all of the aspects of elevatoring are presented.

MODERNIZATION

Many elevators installed 50 or more years ago are still in active service. It is estimated that there are at least 300,000 such units with an age of 20 to 25 years in the United States and, perhaps, a million or more in other parts of the world. A well-maintained elevator can easily have a life expectancy of more than 50 years; however, changing social and economic conditions usually demand that such equipment be replaced or upgraded.

The basic structure of the elevator, such as the rails, pit equipment, car frame and counterweight, safety system and hoisting machine, can often be rehabilitated and reused. Electrical systems, wiring, operating fixtures, door equipment, control equipment and, perhaps, the motor itself are the prime candidates for replacement. The tenants of a building will not be appreciative of any upgrade unless a new cab and improved door operation are included. Suppliers of equipment and elevator contractors have responded, and there are today a variety of approaches readily available. Later chapters will discuss these in detail.

Modernization in an existing building is a much more labor-intensive project than construction of a new building. A major objective is to maintain a reasonable level of service while a car is out of action for the necessary work. Of serious concern is the single-elevator building where the only alternative is to walk the stairs. The information in this book can be used to evaluate the impact of the change in service while the elevator installation is operating with "N − 1" units and can only leave it to the reader's imagination how "−1" impacts ("−1" is one less than the number of units normally in service).

2 The Basis of Elevatoring a Building

EARLY POPULATION

Before any thought is given to the elevators in a building, a thorough and detailed study must be made of how people will arrive at the building, occupy that building, and move about the building. Occupancy is an obvious prerequisite to the design and size of the building itself. It must be expected to perform its function to provide a living or working environment and to have an economical, physical, and functional life span.

Basic factors in elevatoring a building include the number of occupants and visitors, their distribution by floors, and the times and rates of arrival, departure, and movements. We can determine the population in an existing building by census and the average population in a new building based on so many square feet of space per person. We can also determine that people, in arriving at or departing from an existing building, relate to existing vertical transportation systems and will require an estimated quantity of vertical transportation in a new building. Surveys of existing buildings of similar nature can help confirm those estimates.

The population factors and the intensity of pedestrian traffic for all types of buildings are suggested in later chapters of this book. You will note the term "suggested"; the actual use of the building is beyond the control of the architect and the elevator engineer. The basic population and usage estimates should be a consensus of the assumptions of the entire building team in order to properly elevator a building. When the level of uncertainty is high, the basis of elevatoring must be conservative, since the expense of adding additional equipment is often prohibitive.

ELEVATOR TRAFFIC

In every type of building there is a critical elevator traffic period. The type, direction, and intensity of elevator traffic during this period determines the quantity of elevator service for the building. If the elevators serve traffic well during the critical time, they should be capable of satisfying traffic at all other times. The quality of elevator service during this critical period will be set by the class of occupancy.

Critical traffic periods vary with with building types and in various areas of the United States, Canada, and the rest of the world. For example, in office buildings in downtown areas served by mass transit, the critical traffic period is often the morning in-rush, complicated by persons who have arrived early and are traveling down. If elevators are sufficient to serve the peak of that in-rush period, the rest of the day usually does not present a problem. In other cities the critical traffic in an office building may be the noontime

The Vertical Transportation Handbook, Third Edition, Edited by George R. Strakosch
ISBN 0-471-16291-4 © 1998 John Wiley & Sons, Inc.

period, when lunch hours may be standardized and the entire building may go to and return from lunch at designated times.

In recent years there has been an introduction of staggered work hours and "flextime" wherein employees set their own arrival and departure times. This practice has reduced the intensity of the morning peak arrival rate, but has added an opposing down traffic to the predominant up traffic.

The critical traffic period in some hospitals may be in the forenoon when doctors are visiting patients, transfers for treatment are being made, operations are performed, and essential hospital traffic reaches a peak. In other hospitals, critical traffic may occur during visiting hours or the afternoon shift change period.

In apartment houses the critical traffic is usually found in the late afternoon or early evening period when tenants are returning from work, children are coming home from school, shoppers are returning from stores, and other are leaving for evening entertainment. A downtown apartment with a predominant business-person tenancy may find that the critical traffic period is in the morning when practically everyone is leaving for work.

Once the critical traffic period for a building is determined and evaluated in terms of required elevator handling capacity, the choice of the proper number, speed, size, and location of elevators may proceed.

Critical traffic periods for various types of buildings have been determined by observations, traffic studies and tests, discussions with building managers and owners, and by research in occupancy and population use requirements. The estimations of these traffic factors are reviewed in the chapters of this book dealing with various building types.

PEDESTRIAN PLANNING CONSIDERATIONS IN ELEVATORING

The pedestrian planning considerations in elevatoring cover the process of locating elevators in a building, providing proper access and queuing space to passenger elevators, designing and shaping them to best accommodate people, and determining door sizes and lobby arrangements, to make sure the optimum use and benefit is gained from the total elevator plant in a building.

The direct capital cost of elevators and the indirect cost of the space dedicated to their function is a major cost component of any building. Elevators must be placed, arranged, and designed to provide the most cost-effective performance. Unsatisfactory elevator service can damage a building's reputation and cause loss in the productivity of its occupants.

Elevator Platform Shape

The elevator platform—the area on which passengers ride—must be large enough to accommodate a passenger (or freight) load without undue crowding and allow each passenger ready access to the elevator doors.

An average person will require about 3 ft^2 (0.28 m^2) of floor area to feel comfortable (touch zone) (Figure 2.1). Passengers can be crowded, however, to a minimum of about 2 ft^2 (0.19 m^2) for the average person. If no one is pushing or forcing people to crowd, each person will take close to the 3 ft^2 (0.28 m^2). There are exceptions: at office building quitting time, or if passengers know each other, densities of 1.5 ft^2 (0.14 m^2) per person have been observed.

12in. (300mm) radius—touch zone

Body Ellipse

18 in. Body depth
(450mm)

24 in.
(600mm)
Shoulder
breadth

Pedestrian area 3 ft² (0.28 m²)

(a)   *(b)*

18 in. (450mm) no touch zone

Lateral
passage
restricted

Pedestrian area 7 ft² (0.65 m²)

(c)

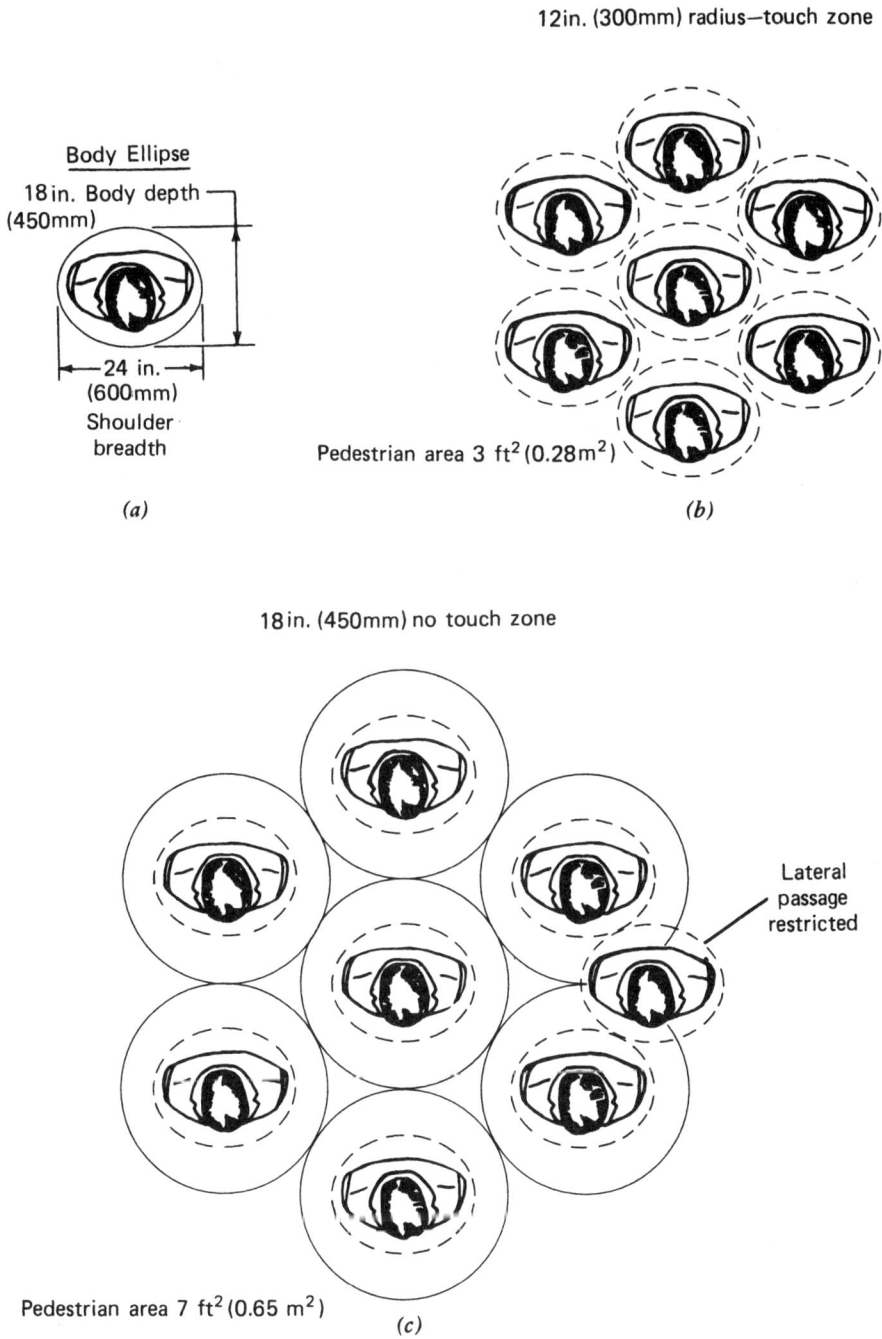

Figure 2.1. Pedestrian queuing: (*a*) body ellipse; (*b*) crowded; (*c*) nominal (Courtesy *Pedestrians*, by J. Fruin, currently published by *Elevator World Magazine*).

The average space per passenger in elevator cars means that the elevator capacity—expressed in pounds (kilograms) and translated to square feet (square meters) so that the car will not exceed its rated load if packed full—must be arranged in the best dimensions to accommodate the shape of people. The arrangement of ranks and files has been found to be best, and the inside car dimensions shown in Table 2.1 for common-size passenger elevators have been adopted partially as industry standards.

For example, the average loading of a 3500-lb (1600-kg) elevator is 16 passengers. You will note from Figure 2.2 how these 16 people may arrange themselves inside the elevator car.

Of course, the neat rank and file is idealistic. In actual practice, the initial passengers generally arrange themselves with their backs to the walls and subsequent passengers fill up the center.

In Figure 2.3 note how the same area, with a different width and depth, may lead to awkward unloading situations. More ranks of passengers now make access to the door difficult and generally require someone to step out of the car to let others out. These complications add a time delay to each elevator stop, which accumulates during the total trip

TABLE 2.1. Area Versus Capacity—Suggested Standards

1. Inch/Pound Units

Capacity (lb)	Car Inside (in.) Wide	Deep	Area (ft²)	A17.1 Area Allowed (ft²)	Observed Loading (people)
2000	68	51	24.0	25.4	8
2500	82	51	29.0	30.5	10
3000	82	57	32.5	35.4	12
3500	82	66	37.5	38.0	16
3500 (alt.)	92	57	36.4	38.0	16
4000	82	73	41.6	42.2	19
4000 (alt.)	92	66	42.2	42.2	19
4500	92	73	46.6	46.2	21
5000	92	77	48.2	50.0	23
6000	92	90	57.5	57.7	27

2. Metric Units

Capacity, kg (lb)	Car Inside (mm) Wide	Deep	Area (m²)	Code Area Allowed (m²)	Observed Loading (people)
1200 (2640)	2100	1300	2.7	2.8	10
1400 (3080)	2100	1450	3.05	3.24	12
1600 (3520)	2100	1650	3.5	3.56	16
1600 (alt.)	2350	1450	3.4	3.56	16
1800 (3960)	2100	1800	3.8	3.88	18/19
1800 (alt.)	2350	1650	3.9	3.88	18/19
2000 (4400)	2350	1800	4.2	4.2	20
2250 (4950)	2350	1950	4.6	4.6	22
2700 (5940)	2350	2150	5.1	5.32	25

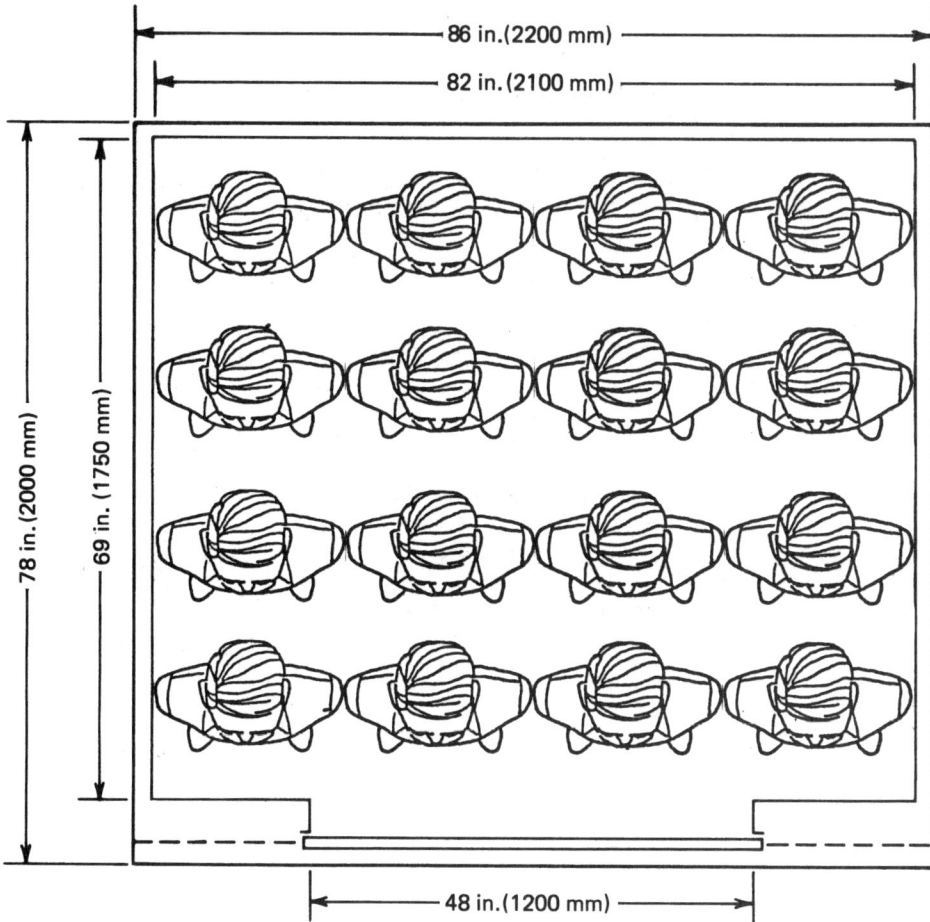

Figure 2.2. Nominal loading, 3500-lb (1600-kg) "passenger-shaped" elevator.

and seriously reduces efficiency. The deep, narrow arrangement also leads to loss in passenger capacity—15 passengers versus 16 passengers shown in Figure 2.2. Part of this loss is due to the extra space required for the car doors, but most is caused by platform shape.

Study of the two illustrations suggests the conclusion that the most efficient elevator car is only one person deep! This is true but not practical, because efficient door arrangement must also be considered.

Door Arrangement

The most efficient door is one that opens and closes in minimum time and allows two person to enter or leave an elevator simultaneously. The doors must also be reasonably economical and adaptable to efficient platform sizes. The 48-in. (1200-mm) center-opening door meets most of these requirements and is recommended for high-quality elevators when optimum peformance is required (Figure 2.4). It can fit the average 86-in. (2200-mm)-wide platform and can be opened in slightly less than 2 sec. Closing speed, as

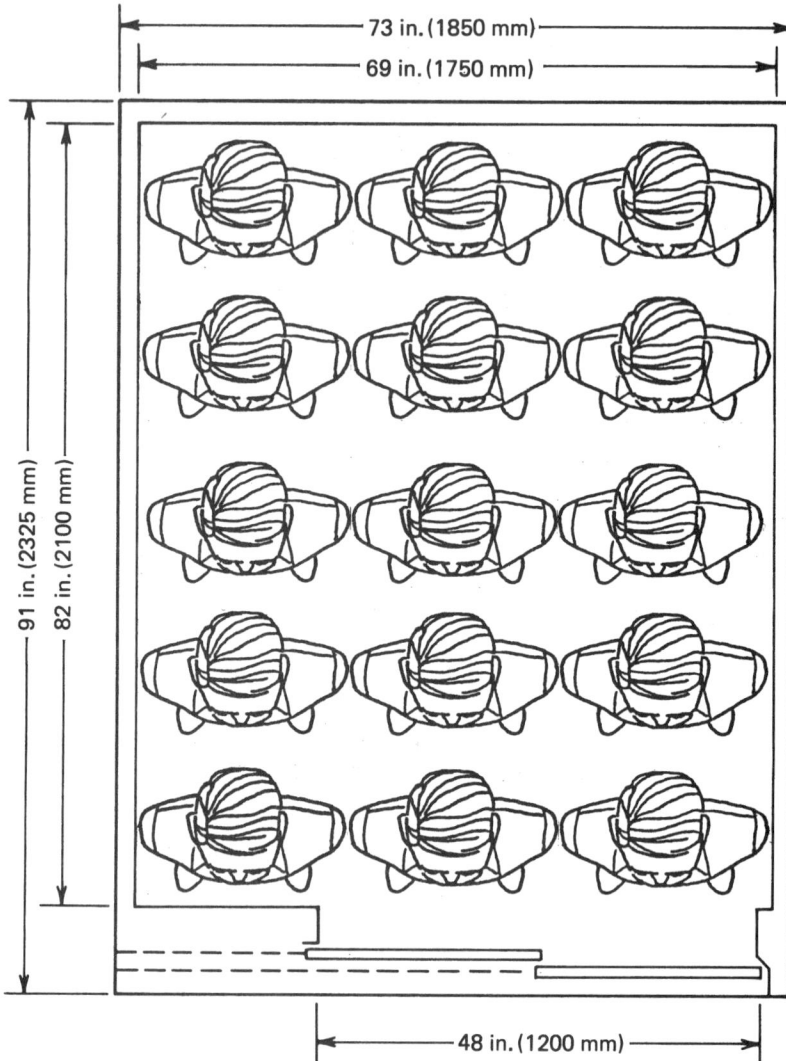

Figure 2.3. Nominal loading, 3500-lb (1600-kg) "stretcher-shaped" elevator.

mentioned previously, must be within the 7-ft poundal (0.29 joules) kinetic energy limitation. Because each panel of the door is half the weight of the entire door [no more than 100 lb (45 kg) or so per panel] and the distance traveled is only half the opening width, the 48-in. (1200-mm) center-opening door can be closed in 2.9 sec within this kinetic energy limitation.

Doors 42 in. (1100 mm) and less can be considered on-person doors. Note how awkward it becomes for two people to pass each other (Figure 2.4); the natural tendency is to allow one person to leave while the other holds up elevator service until he or she can enter.

Wider doors are often necessary for special purposes, such as the 60-in. (1550-mm) door on a hospital elevator that must accommodate a hospital bed with an attendant or on a service elevator that must accommodate wide containers or carts. In these cases the efficiency of the door is secondary to the function the elevator must perform.

(a)

(b)

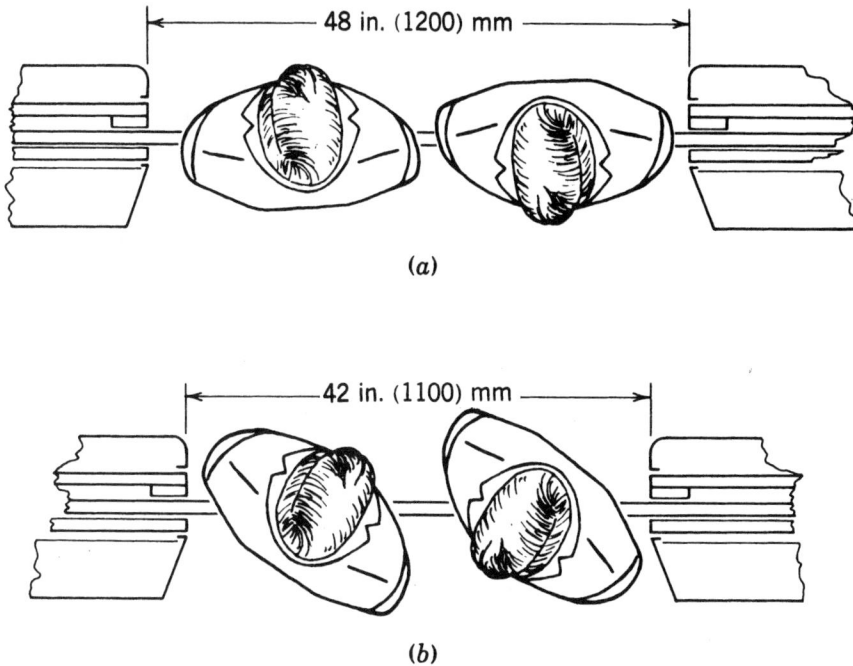

Figure 2.4. Passenger transfer.

In apartment houses the economy of the single-slide door prevails. Because passengers are expected to move at a somewhat leisurely pace, some efficiency may be justifiably sacrificed for economy. To effectively allow a wheelchair or ambulance stretcher to enter or leave an elevator, door width should be a minimum of 42 in. (1100 mm).

Lighting and Signals

The interior of an elevator car should be well lighted and the lights arranged so that they cannot be turned off by unauthorized persons. To this end there should be a key-operated light switch. The car threshold should have a minimum of 5 ft-candles (50 lux) of light without the benefit of outside lighting, as prescribed by elevator safety codes, since inaccuracy in floor-level stops can create a possible tripping hazard.

A person should be able to look at an elevator and tell whether it will travel up or down when it leaves the floor. With a single-car installation in a normally quiet building this is not serious, for if the person is the only one on the floor, the car will generally travel in the direction chosen. Considerations for disabled persons require a lantern to sound a gong or chime when illuminated, one stroke for an up traveling car and two strokes for down.

With more than one car serving a floor, some form of directional indication becomes essential, for a car going either way may stop. Lighted directional arrows in both door jambs of the arriving elevator offer a simple, relatively inexpensive solution to this dilemma, but may not be totally desirable when efficient elevator performance is wanted. The car direction is not visible until the doors are open, which may be too late for the eager passenger, resulting in unnecessary passenger movement (Figure 2.5).

Figure 2.5. "In-car" directional lantern (Courtesy G.A.L. Manufacturing Corp.).

The most effective indication to the waiting passenger is provided by lanterns installed over or next to each hoistway entrance, which will inform the prospective passenger of the next car to arrive at the floor and the direction it will travel when it leaves the floor (Figure 2.6). Such lanterns should be arranged to light up sufficiently in advance (about 4 sec) of a car's arrival at each floor to give passengers time to walk to the entrance of the arriving car and be ready to board it. This is not always accomplished, but, as will be noted later, each second thus saved in passenger boarding time is worth extra elevator capacity. A lantern must also be prominent enough that it may be seen from any point where passengers are likely to stand. Note, in Figure 2.6, arrangement *C*, that a lantern on a transom or in a depression cannot be clearly seen. Arrangement *B* is the most acceptable.

Landing call buttons that light when touched or pressed inform the prospective passenger that the call for service has been acknowledged, and are arranged to extinguish when the car arrives. For efficient service, the button should not hold the elevator, lest a stuck button cause an elevator to stay at a floor, or people to keep an elevator at a floor until their friends arrive, thus delaying service.

Once passengers are in the elevator car, they should be readily able to indicate where they are headed and be promptly informed when they arrive. The car call buttons should be conveniently located so passengers can register their calls as they enter an elevator, the side locations proving to be the most acceptable. The car call button should also light to acknowledge the call, which will help improve service. Entering passengers can see that their floor stop will be made if someone has registered it. Numerals or symbols at least $\frac{5}{8}$ in. (16 mm) high and raised 0.030 in. (0.8 mm) adjacent to each car button are being required as an aid to visually disabled persons.

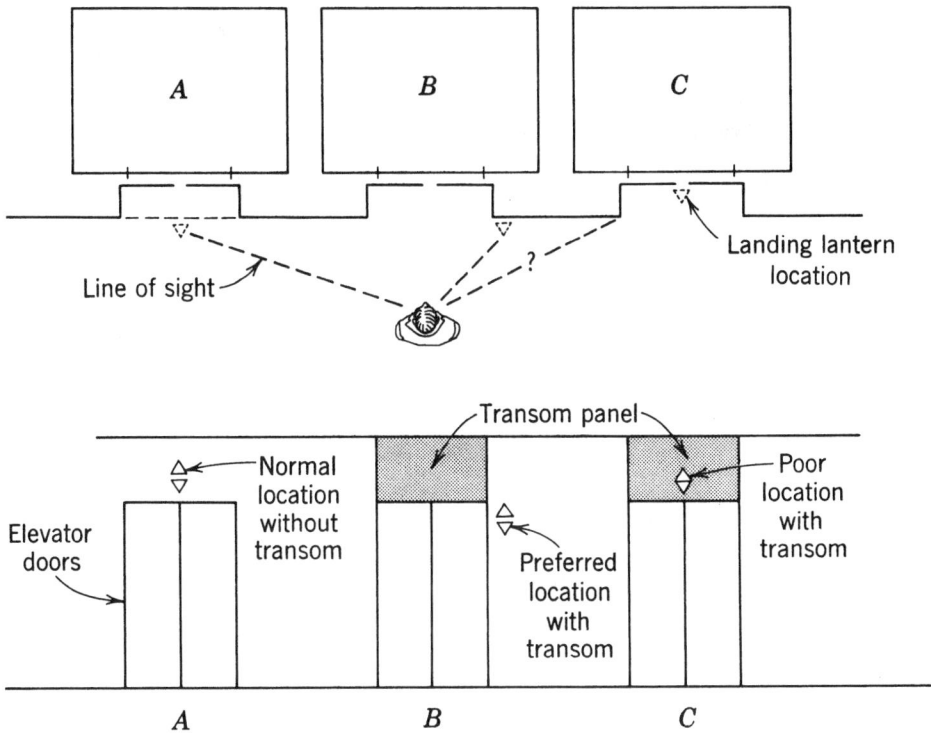

Figure 2.6. Landing lanterns should be located to be seen clearly.

Symbols are required to indicate the floor served, the main entry floor, stop switch, alarm button, and buttons that will hold the doors open or expedite their closing. Legislation in behalf of disabled persons has mandated the height of the car buttons to be a minimum of 35 in. (890 mm) above the floor and no higher than 54 in. (1370 mm) to accommodate a person in a wheelchair. Nearsighted persons and those with back problems are not considered in this legislation. Figures 2.7 and 2.8 show many of the informational and safety devices incorporated in modern elevators. Figure 2.7 illustrates the devices that are directly associated with the use of the elevator. The prospective passenger operates the landing button and, when the car arrives, the door operator opens the car door and the landing door only at the floor where the car stops. This is ensured by a clutching device on the car door, which engages only that landing door. On some installations a lantern in the hall will indicate the imminent arrival of the car and the direction in which it is expected to leave the floor.

The doors are timed to stay open for a predetermined period and, if the passenger is still in the entrance, the door protective device will either reverse the closing doors or maintain them in the open position until the entrance is clear.

Some form of position indicator in the car showing the location of the elevator in relation to the floors served is usually provided. This may be a series of lights over the interior entrance of the car or, as is more common with modern elevators, a digital device either over the entrance or integral with the car operating panel. The car operating panel is the main means of passenger communication with the elevator system. The destination floor must be indicated by operating the button provided for that floor, and if it is necessary to

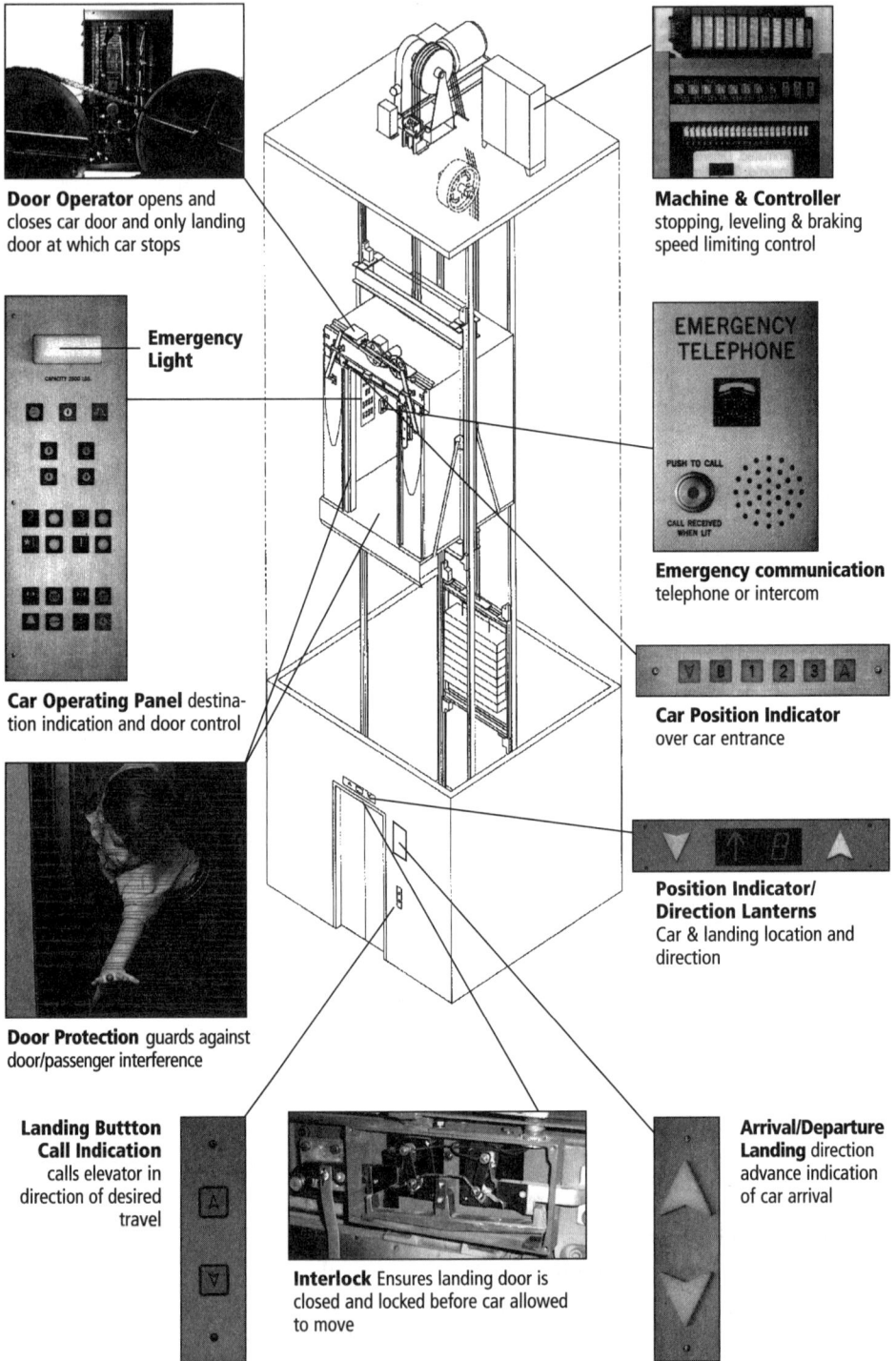

Door Operator opens and closes car door and only landing door at which car stops

Machine & Controller stopping, leveling & braking speed limiting control

Emergency Light

EMERGENCY TELEPHONE

PUSH TO CALL

CALL RECEIVED WHEN LIT

Emergency communication telephone or intercom

Car Operating Panel destination indication and door control

Car Position Indicator over car entrance

Position Indicator/ Direction Lanterns Car & landing location and direction

Door Protection guards against door/passenger interference

Landing Buttton Call Indication calls elevator in direction of desired travel

Interlock Ensures landing door is closed and locked before car allowed to move

Arrival/Departure Landing direction advance indication of car arrival

Figure 2.7. The human interface with elevators: passenger operating and safety features (Courtesy G.A.L. Manufacturing Corp. and Hollister-Whitney Corp. (photos), and Leibowitz Communications).

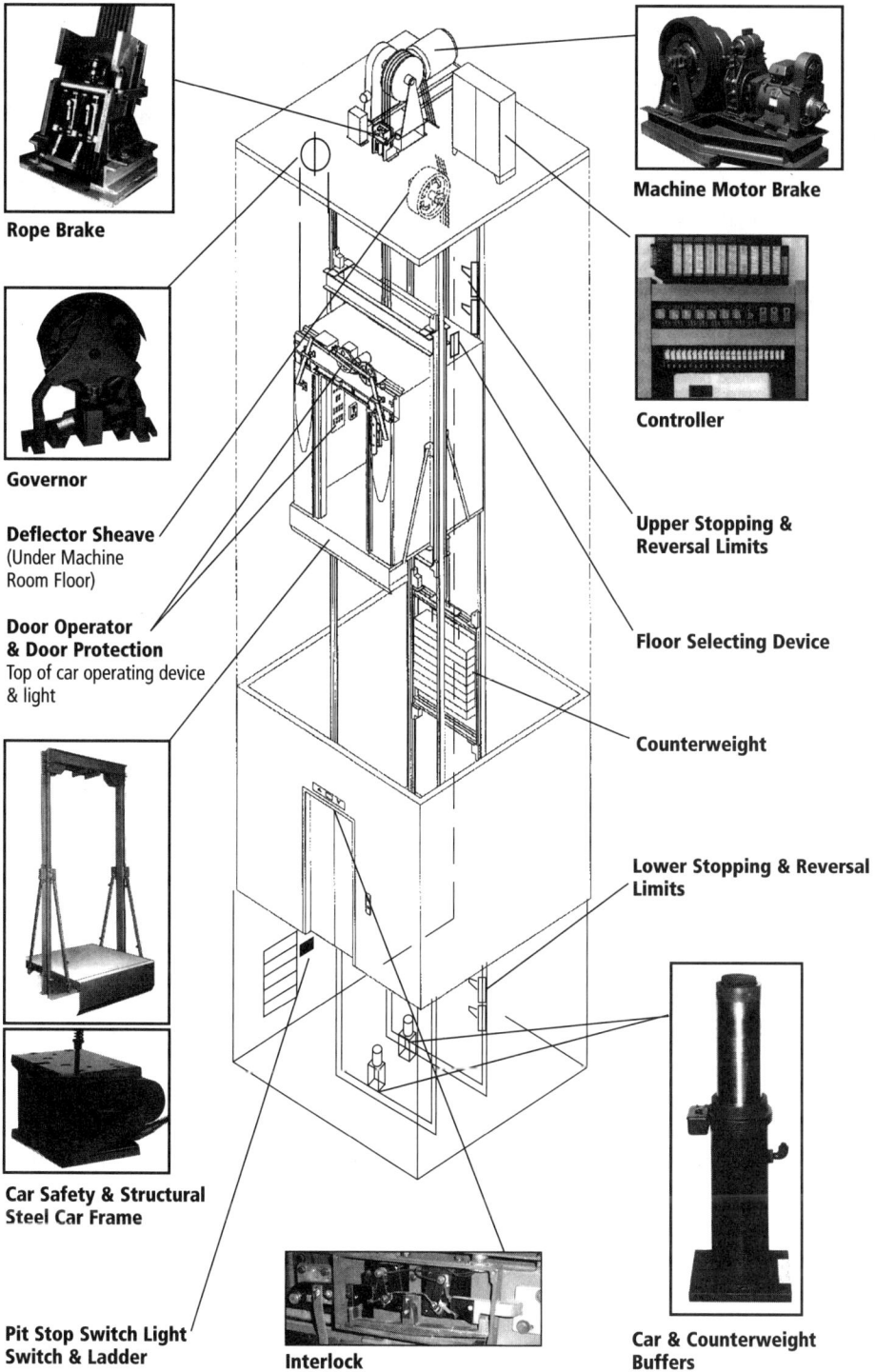

Rope Brake

Machine Motor Brake

Governor

Controller

Deflector Sheave
(Under Machine
Room Floor)

**Upper Stopping &
Reversal Limits**

**Door Operator
& Door Protection**
Top of car operating device
& light

Floor Selecting Device

Counterweight

**Lower Stopping & Reversal
Limits**

**Car Safety & Structural
Steel Car Frame**

**Pit Stop Switch Light
Switch & Ladder**

Interlock

**Car & Counterweight
Buffers**

Figure 2.8. Safety features of elevators: mechanical and electrical (Courtesy G.A.L. Manufacturing Corp. and Hollister-Whitney Corp. (photos), and Leibowitz Communications).

hold the door open, a "door-open button" is provided. A door-close button is also provided which, in some installations, will shorten the time the door remains open at a floor. In other applications the door-close button is needed when the car is equipped with either an attendant or an independent service operation.

In both new buildings and buildings where the elevators have been modernized, which either are classified as "high rise" or are required by local regulations, various switches for fire fighters' operation are included in the car operating panel, plus indicator lights to show that an emergency exists. They are inoperative during normal passenger use and are activated if the car is placed on fire fighters' service, as described in Chapter 7.

Also included is an emergency light, which will illuminate the space if the ordinary car lighting fails. It is mandatory that there be some form of communication with an attended station in the event the car stops without door operation or between floors. This feature is in addition to the alarm button located in the car operating panel, which sounds an alarm audible in the building and in the hoistway.

The elevator will travel to the desired floor once the car and landing doors are closed and locked. A lantern may be provided in the car doorjamb to indicate its stop, and reference to the car position indicator is needed to determine whether this is the desired floor. If not, the action is repeated for any subsequent stops.

Figure 2.8 shows the various protective and safety devices the passenger does not see, but which are required to qualify an elevator as code compliant. These consist of a variety of mechanical and electrical devices located in the machine room, in the hoistway, on the car, and in the pit. Overspeed in the down direction is monitored by the governor, which cuts off power if a certain speed is exceeded and causes the mechanical safety device located on the car frame to actuate and lock the car to the guide rails if the speed continues to increase. Overspeed in both the up and down directions is sensed by the rope brake device located in the machine room, which clamps the hoist ropes and stops the elevator. This is a recent innovation and is being applied in numerous installations.

There is concern not only with overspeed, but with failure to stop at the limits of travel either up or down. The hoistway limit switches operate to cut off power and apply the brake to the machine. Continued travel of the car into the pit is arrested by the buffer, as is continued travel of the car into the overhead wherein the counterweight buffer is employed.

As can be noted from the illustration, the car and counterweight are connected by the hoist ropes. Each rope (a minimum of three is required) is capable of supporting the weight of the elevator with a safety factor of about 5. Multiple ropes are needed to create the traction between the drive sheave on the machine and the ropes, which is prescribed by strict rules. The counterweight balances the weight of the car plus about 40% of its capacity load. An elevator actually generates electrical power through its machine and motor if the weight in the car is more than about 40% of the indicated pound (kilogram) capacity. The code requires the area of the platform to be of such a size that it must be more than "packed" with people before that weight is exceeded (which is explained in detail in Chapter 6).

The controller in the machine room processes all the electrical signals given to the system by passengers operating the elevator and ensures, through electrical circuits, that the doors are closed and locked before the car can run, that power is sufficient, that no overload exists, and that all the safety switches and circuits are intact. The floor-selecting device, located on the car, provides electrical signals to the controller as to the location of the elevator and the floor at which it may be stopped.

The foregoing is a simplification of a complex mechanism. Succeeding chapters provide the details, as well as insight into the many regulations and practices that ensure safe and efficient vertical transportation.

In many buildings, it is desired to have doors 8 ft (2400 mm) or higher. With high doors, car position indicators should be of the digital type and placed over each car operating panel so they are visible from any location within the car. The readout should be distinct, with sufficient illumination, so that the car lighting does not overwhelm the indications.

In many buildings, especially hotels and hospitals, in which the usual passenger may not be accustomed to riding in elevators, the floor designation may be helpfully placed on the jamb of the hoistway door so it may be seen when the door opens. This is required for disabled persons, and the jamb-mounted markings using 2-in. (50-mm) high numerals raised 0.030 in. (0.8 mm) are a necessity.

POSITIONING OF A BUILDING'S VERTICAL TRANSPORTATION SYSTEM

Elevators and escalators should be accessible and centrally located. All entrances should lead to the vertical transportation nodes, which should be near the main entrance of the building. If a parking lot or subway entrance is near the building, it is reasonable to expect that a significant portion of the passenger traffic will come from that direction. If the flow is expected to be heavy, escalators or moving walkways should be employed to deliver passengers to the elevator lobby.

Locating elevators in the geometric center of the population on each floor in a building allows all part of each floor to be equally accessible to the elevator core. The location of elevators at one end of a building detracts from the desirability of the other end and adds considerable extra time for passengers to cover the horizontal distance (Figure 2.9).

It is often proposed to provide two elevator cores, serving the same floors at separate locations, to avoid a long horizontal distance to a central elevator core. This is practical; however, additional elevator capacity must be provided in each core since the imbalance of traffic can be totally unpredictable. In addition, the creation of an attraction, such as a coffee shop or shopping center, near one core can totally change the nature of traffic and may cause excessive loading on one core.

Experience has shown that the walking distance from the elevators to the farthest office or suite should not exceed 200 ft (60 m), with a preferred maximum distance of about 150 ft (45 m). For example, in a building 300 ft (90 m) long, the elevators should be located at the center point (Figure 2.10). Buildings with X, Y, or T floor layouts have a natural center point.

In some buildings, where the busiest entrance is near one end rather than the middle, a central location for elevators may still be desirable. If people use the main entrance only when entering or leaving the building but use elevators repeatedly during the day, locating them near the center of the floor population may achieve greater net savings of time and energy for all users. This saving of time accrues from the fact that persons usually walk 2 to 4 fps (0.6 to 1.2 mps).

Grouping of Elevators

If a building requires more than one passenger elevator, all the passenger elevators should be grouped. Single elevators in various parts of a building have serious disadvantages and are generally unsatisfactory.

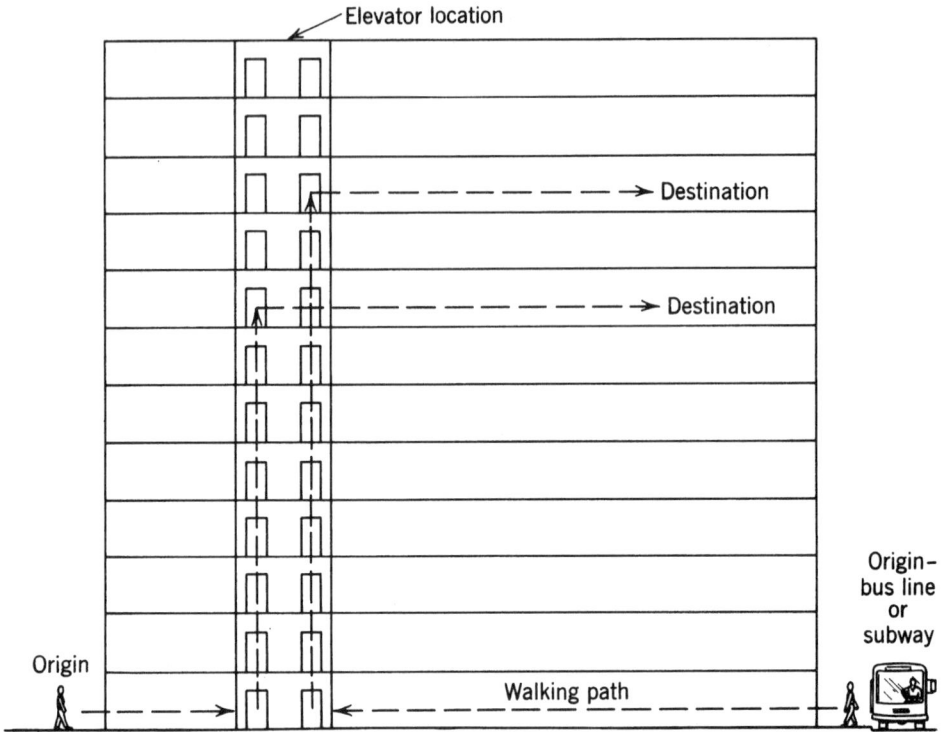

Figure 2.9. Noncentralized elevators mean additional walking time.

With a single-car installation, a passenger who just misses an elevator must wait for that car to return, for a period based on the round-trip time. With two or more elevators in a group, the reduction in waiting time is proportional to a factor related to the average round trip time of each elevator divided by the number of elevators in that group. As an example, if four elevators serving the same floors are together in a group and it takes, say, 120 seconds for each elevator to make a trip, a person calling for service may wait between 0 and 30 seconds (120 seconds divided by 4). One person may get service in 0 seconds whereas another may wait the full 30 seconds or, simply, an average of 15 seconds.

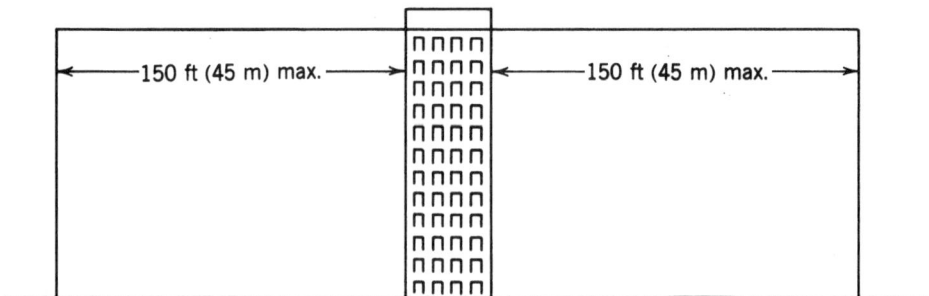

Figure 2.10. A distance of 150 ft (45 m) should be the maximum distance people have to walk to an elevator on any floor.

This is an over simplification of the concept of interval versus waiting time. This concept is fully explained in Chapter 5, Figure 5.6 and is illustrated simply in Figure 2.11*a* and *b*.

Any elevator requires periodic servicing and replacement of wearing parts. During that time a person who depends on a single car is totally without service and must walk. Making repairs at night is extremely costly and not favored. A group of elevators minimizes this problem and permits periodic repairs without total passenger inconvenience. A well-designed elevator plant will always have excess capacity so that vital traffic is maintained even though one car is out of service.

If a tenant is moving into or out of a building, or if building renovations are scheduled, one car of a group of elevators can be assigned to these tasks. With a single car such moving or renovation is a weekend or nighttime proposition or, again, the passenger must walk (Figure 2.11*c* and *d*).

Notable exceptions where elevators are not grouped (if there is more than one elevator in a building) are service elevators designed for specialized use in apartment houses, motels, nursing homes, or office buildings. These service elevators are mainly for building maintenance purposes, moving operations, or food service and, as such, are generally subject to marring, scuffing, and other damages. They should not be considered for general passenger use, and the car interiors should be of durable material. If passengers must be carried, removable pads can be used during service operations and building services made to wait until off-peak passenger travel times. Food service elevators for upper-floor dining facilities are in the category of possible acceptable single-car installations. A garage shuttle elevator serving the basement garage in an office building is also a possible exception—provided no more than a few stops are served, handling capacity is sufficient, and there is a convenient stairway. The safe-deposit elevator in a bank is a logical single-car installation. Providing access for disabled persons from grade level to a raised or lower main elevator lobby or where escalators are the main access to a building lobby is an essential building service, and more than one elevator should be considered.

As a general rule, if elevator service is essential to the building operation, two elevators should be considered as the minimum equipment in a vertical transportation node.[1]

Serving Floors. All elevators in a group should serve the same floors. This is a common-sense rule that is often violated for false economy. If, for example, only one car out of a group of three serves the basement, people wishing to go to the basement from an upper floor have only one chance out of three that the next elevator that comes along will take them to the basement. Conversely, people in the basement must wait three times as long for elevator service than upper floor passengers. Ideally, all cars should serve the basement, but if not, a special shuttle elevator must be considered, to run only between the main floor and the basement. This latter scheme is favored if multiple basements exist and a large group of main passenger elevators is required. The difference in the cost of providing entrances on all the main elevators at the basement levels and the cost of the shuttle elevator or elevators may be negligible when general construction costs are considered in addition to elevator costs.

The expedient of providing a separate call button at an upper floor to call the single car that serves the basement has often been tried and never proved satisfactory. The average person will operate both the normal call button and the basement call button, take the first

1. William S. Lewis: "Two elevators are a system, one is a toy." Williams S. Lewis (deceased) was formerly a partner in the engineering firm of Jaros, Baum & Bolles. He was in charge of their vertical transportation activity until his retirement, and I was an associate to him during the time I spent with that firm.

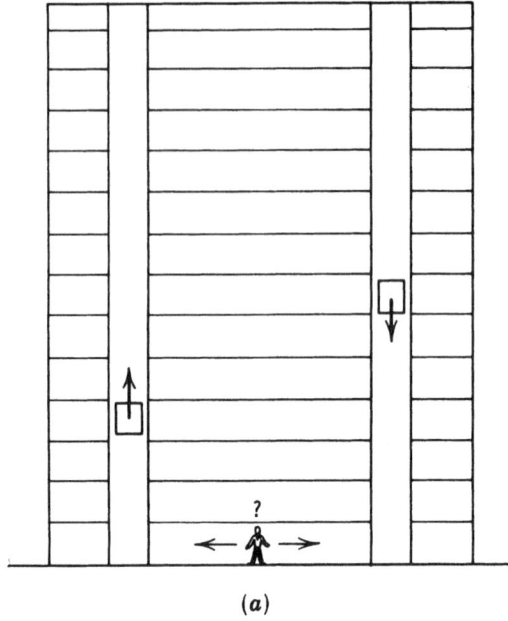

(a)

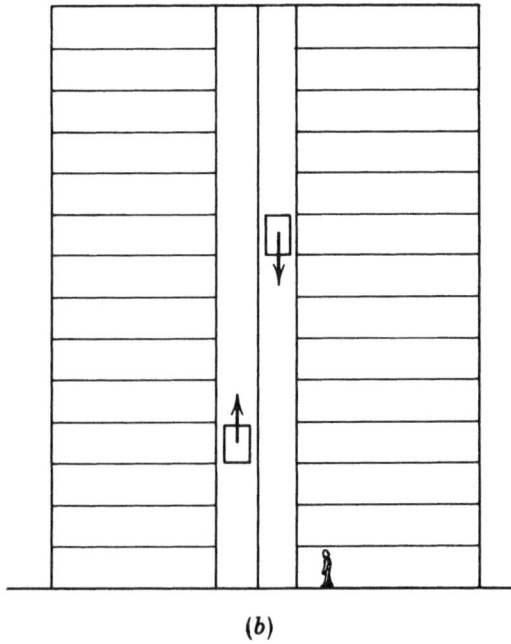

(b)

Figure 2.11. (a) Separated elevators reduce transportation efficiency. (b) Grouped elevators avoid passenger indecision and improve service. (c) Grouped elevators afford continuity of service. (d) Separated elevators can deny service.

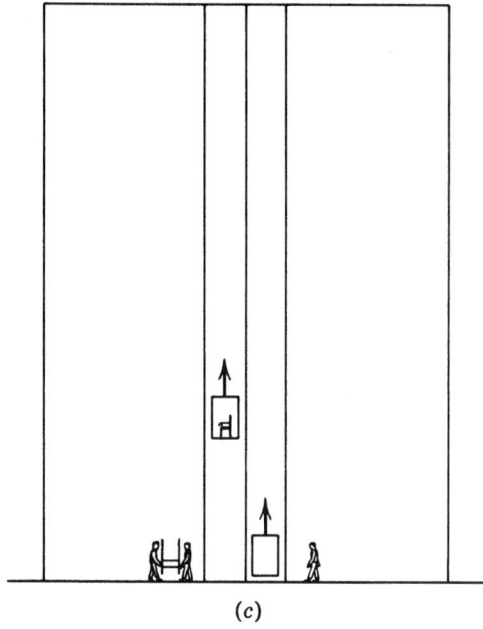

(c)

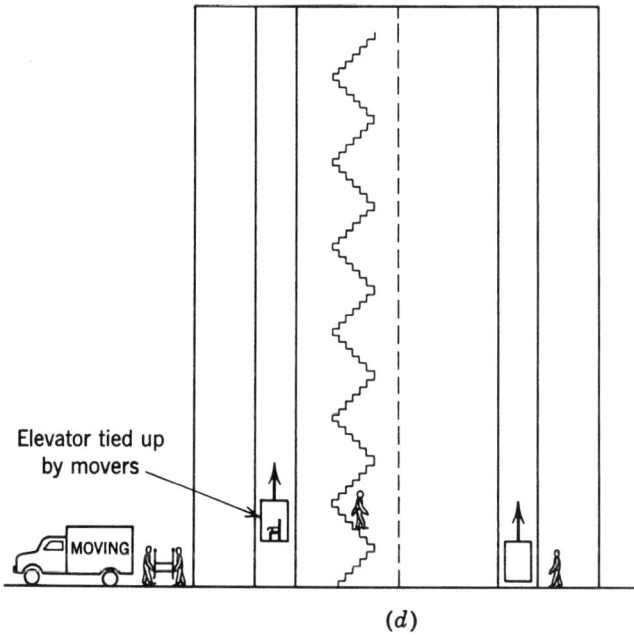

Elevator tied up
by movers

MOVING

(d)

Figure 2.11. *(Continued)*

car that comes along, and cause the basement car to make a false stop. Such false stops will add up in lost elevator efficiency over the years to more than pay for the cost of the extra entrances on all elevators.

Skip-Stops. Many schemes have been tried to reduce the number of stops an elevator will make, thereby attempting to improve operating efficiency and to reduce equipment costs. Each has disadvantages with respect to floor access and is not legal where legislation for disabled persons has been enacted. The most notable are schemes involving skipped stops on either a single elevator or groups of elevators.

Single-car skip-stop schemes are elementary; the elevators serves alternate floors or every third floor (Figure 2.12*a* and *b*). The disadvantage is obvious: with the single-skip scheme persons on the floor skipped must walk either up one floor or down two. In apartment buildings elderly or disabled persons cannot be asked to walk up stairs, baby carriages become difficult for people on the nonelevator floors to manipulate, and moving operations are extremely difficult. The passenger load remains but is concentrated on fewer floors, and the intended increase in elevator capacity owing to the limited stops becomes dubious because more persons move in and out at each stop, thus delaying the elevator. The main saving is in the lower cost of fewer total elevator entrances, which may be from 5 to 10% of the total installation cost.

In two-car installations, one car may serve odd floors and the other, even floors, thereby saving one entrance per floor (Figure 2.12*a*). The total installation is equivalent to two single elevators side by side, with most of the disadvantages of a single car and the sole advantage of having to walk only one floor if an elevator is shut down. Service is not improved as it would be in a conventional, all-stop, two-car installation.

Another skip-stop variation is a two-car installation with both cars serving every other floor (Figure 2.12*b*), whereby waiting time is reduced but half the passengers must walk. Still another variation is the intermediate elevator landing scheme with the cars serving every other floor (Figure 2.12*c*); passengers must walk up or down only half a floor.

Skip-stop elevators may be used advantageously in school dormitory buildings if furniture moving is minimal, or may be used to create duplex apartments in apartment buildings where the feeling of luxury offsets the inconvenience of the alternate-floor elevator stopping.

Arrangements

As essential as grouping elevators and the proper location of the group in a building is the arrangement of the elevators in the group. As a rule, elevators should be arranged to minimize walking distance between cars.

With any group of elevators, one spot in the lobby is usually favored by waiting passengers, often near the location of the elevator call button. As an elevator arrives at that floor, waiting passengers must react and walk to the elevator while it waits for them. Because an average passenger walks at a rate between 2 and 4 fps (0.6 to 1.2 mps), the time spent by the elevator at each stop must be adjusted to this rate of walking.

Advance landing lantern operation will help minimize that waiting time. Excessive distances between the cars in a group results in longer time delays at each stop, which cannot be overcome by lantern operation. Excessive delay caused by this situation, coupled with other inefficiencies resulting from poor group arrangement, may require the installation of additional elevators.

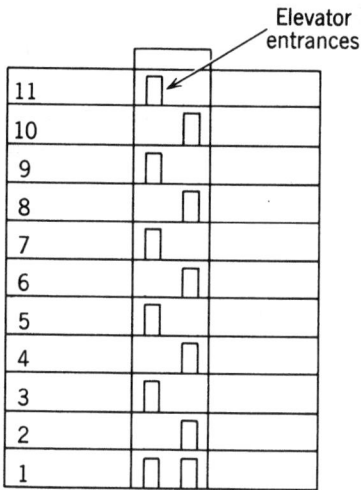

(a)

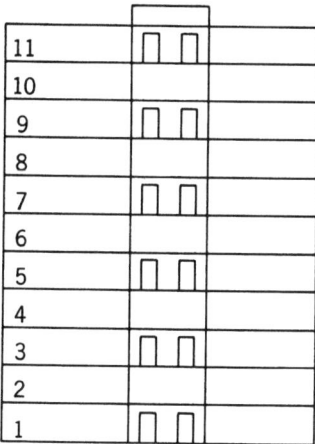

(b)

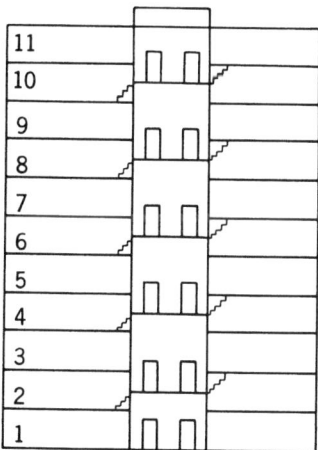

(c)

Figure 2.12. (a) Odd-even skip-stopping—single elevator performance on each floor. (b) Alternate floor skip-stopping—dual elevator performance on half the floors, no elevator service on the other half. (c) Intermediate landings—dual elevator performance to intermediate landings; everybody must use the stairs.

This is especially true for physically and visually impaired persons. Added time at each elevator stop must be considered to provide ample time for a disabled person to move to an elevator after the landing lantern operates. Handicapped codes and requirements are defining the minimum time between a landing lantern actuation of an elevator and the closing of the elevator doors to allow sufficient time for a disabled person standing near the landing call button to travel to the stopped elevator.

The specific requirements of the Americans with Disabilities Act Accessibility Guidelines (ADAAG) are as follows:

4.10.7 Door and Signal Timing for Hall Calls. The minimum acceptable time from notification that a car is answering a call until the doors of that car start to close shall be calculated from the following equation:

$$T = D/(1.5 \text{ ft/s}) \text{ or } T = D/(445 \text{ mm/s})$$

Where T total time in seconds and D distance (in feet or millimeters) from a point in the lobby or corridor 60 in. (1525 mm) directly in front of the farthest call button controlling that car to the centerline of its hoistway door (see Figure 2.13). For cars with in-car lanterns, T begins when the lantern is visible from the vicinity of hall call buttons and an audible signal is sounded. The minimum acceptable notification time shall be 5 seconds.

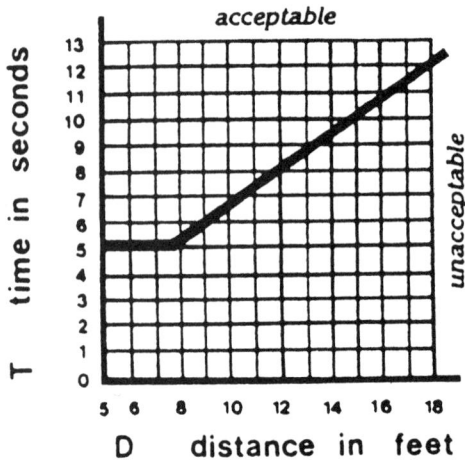

Figure 2.13. Graph of timing equation (Reproduced from ADAAG).

Note that providing a landing lantern with advance signaling of a car's arrival and subsequent direction allows a greatly reduced time to hold the car at that floor. This is an important consideration in calculating elevatoring requirements, as will be discussed in the next chapter.

Experience has demonstrated the desirability of the elevator arrangements illustrated and discussed in the following pages.

Two-car Groupings. For a two-car group, side-by-side arrangement is best. Passengers face both cars and react immediately to a direction lantern or arriving car. Two cars facing each other constitute an acceptable alternative, as the passenger need only turn around to be facing an elevator.

Separation of the elevators should be avoided. The greater the separation, the longer each elevator must be held until a passenger can arrive at that car. Excessive separation tends to destroy the advantages of group operation; passengers will wait at the call button, and, rather than run for the second car if it is too far away, will let it go and reregister a call. Adding a second call button will not relieve this situation but will only result in the effect of two individual elevators serving the same floors, each providing only half the service of which the group is capable.

The lobby in front of the elevators on upper floors should, as a minimum, be as wide as the elevators are deep if the elevators are side by side, usually from 4 to 6 ft (1.2 to 1.6 m), and from one and one-half to two times as deep, 8 to 10 ft (2.4 to 3 m) if the cars are opposite each other (Figure 2.14). The side-by-side arrangement requires more space in the main floor lobby because more people are expected to wait in this area. With the cars opposite each other, an assembly area should be provided and the elevator lobby maybe dead-ended (alcove arrangement).

In hospitals or other buildings where elevators carry vehicles, the lobby must be wide enough to accommodate the vehicles as they are turned; hence, the diagonal of the vehicle must be considered as the major vehicle dimension. Vehicles must be pushed straight in or out of the elevator to avoid hitting the protective edge on the car door. About two additional feet (0.6 m) should be allowed to accommodate an attendant.

Three-car Groupings. The arrangement of three cars in a row is preferable, and two cars opposite one is acceptable, the main problem being the location of the elevator call button. The type of elevator door may influence the choice. With center-opening doors, three cars in a row may be preferable, as this will give the elevators a balanced appearance

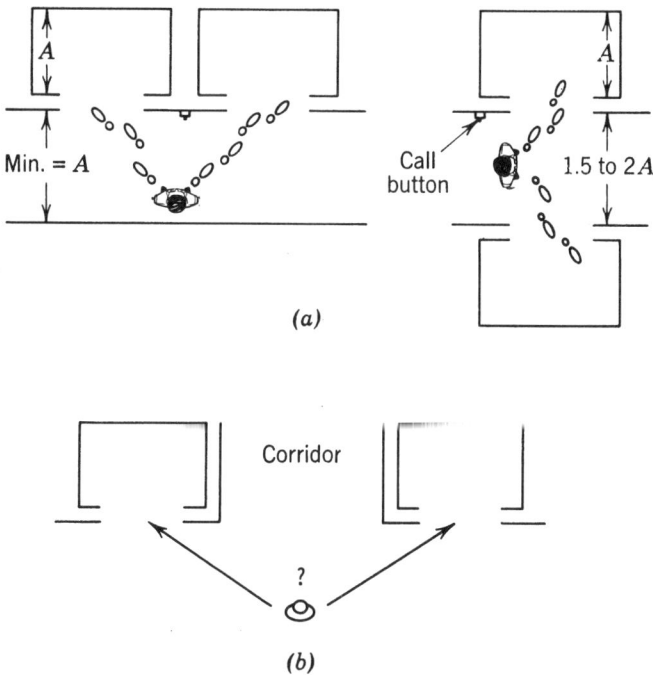

Figure 2.14. Two-car arrangements: (*a*) preferred; (*b*) wrong.

(Figure 2.15). With two-speed or single-slide doors, unequal space between elevator entrances (Figure 2.15) may make the two-opposite-one arrangement desirable.

There is very little difference in walking time with either of the three-car arrangements, as the turnaround reaction time offsets the shorter distance. Lobby widths must be slightly greater than for two-car arrangements. With the row arrangement, a person should be able to stand back far enough to see the entrances of all three cars; thus the lobby width should be about one and one-half times the car depth or 6 ft (1.6 m) minimum. With the one-opposite-two arrangement, lobby width should be from one and one-half to two times the car depth, or a minimum of about 8 ft (2.4 m).

Four-car Groupings. Four elevators in a group are common in larger, busier buildings. Experience has shown that a two-opposite-two arrangement is the most efficient.

The alternative arrangement, four cars in a row, has some disadvantage because of the increased distance between the landing call button and the last car in the row. With average-sized elevators, this walking distance is about 12 ft (3.7 m).

The preferred two-opposite-two arrangement (Figure 2.16) should have a lobby from one and one-half to two times the depth of an individual elevator but no less than 10 ft (3 m) for those times when all cars arrive at the main lobby filled and passengers entering or leaving must pass each other. In addition, 10 ft (3 m) is about the minimum space required for elevator machinery at the machine room level. The closed-end alcove arrangement is acceptable, for even if people wait at the end of the alcove they are only a car length away from the next car to arrive or depart.

If architectural factors necessitate four cars in a row, the lobby should be at least one and one-half times the depth of an individual elevator but no less than 8 ft (2.4 m). This width will allow a person to stand back far enough to see the directional lantern of any elevator. At the main floor the lobby should be wider and longer to accommodate the assembly of people waiting to board the elevators. This is especially necessary during the incoming rush, and to provide quick exit for outgoing passengers.

Six-car Groupings. Groups of six elevators are often found in large office buildings, public buildings, and large hospitals. Six elevators frequently provide the combination of quantity and quality of elevator service required in these busy buildings. The arrangement of six cars, three opposite three, is the preferred architectural core scheme.

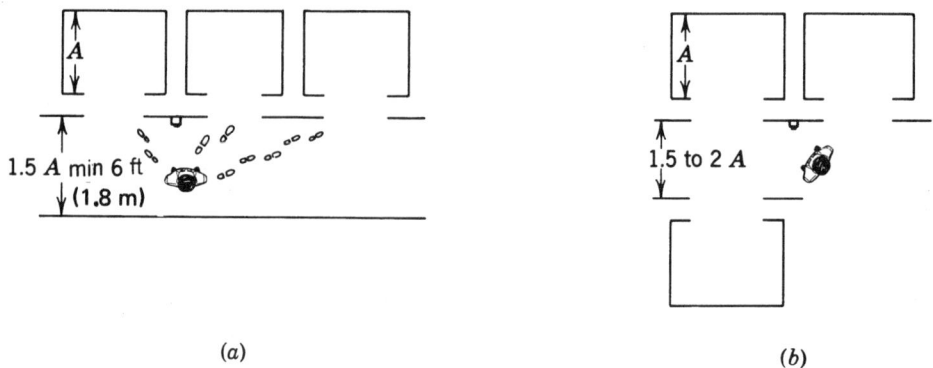

(a)   (b)

Figure 2.15. Three-car arrangements: (*a*) preferred; (*b*) acceptable.

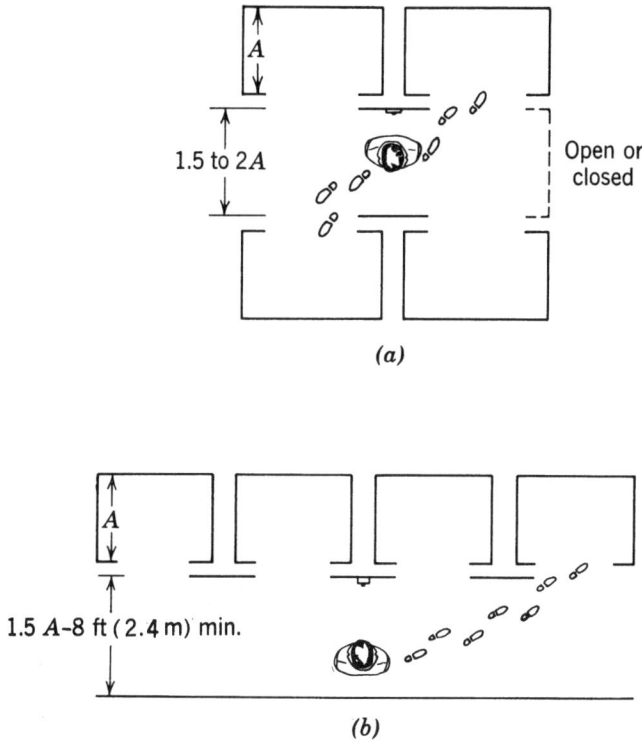

1.5 to 2A

Open or closed

(a)

1.5 A–8 ft (2.4 m) min.

(b)

Figure 2.16. Four-car arrangements: (*a*) preferred; (*b*) acceptable.

The waiting passenger can see all six elevators simply by turning around. The distance to the next arriving elevator is a minimum, and the car need be held at a stop for a minimum time. The main floor lobby should be open at both ends but may be alcoved if necessary.

The lobby width for a six-car group of elevators should be from one and three-fourths to two times the depth of an individual elevator but no less than 10 ft (3 m). If the lobby is to be used as a passage for other than elevator passengers (never recommended at the main floor), its width should be no less than 12 ft (3.6 m).

An acceptable arrangement other than three cars opposite three, is two cars opposite four (Figure 2.17). However, passenger response time now becomes appreciable, and, finally, with six cars in a row, becomes totally unacceptable. It was formerly believed necessary in a department store to have all the elevators visible to passengers. This may have been true so that attendants of manually operated elevators could see the approaching passengers, but in-line arrangement is hard to justify in view of the universality of automatic operation today. Moreover, because the time per stop is allowed for the slowest passenger, the random rotation of elevator arrivals at a floor may result in the two end cars arriving simultaneously. If one fills quickly, the passengers who could not board must run to the other end and risk missing that car also (Figure 2.18).

Eight-car Groupings. The largest practical group of elevators in a building is eight cars—four opposite four. The main lobby is required to be open at both ends, each of

(a)

(b)

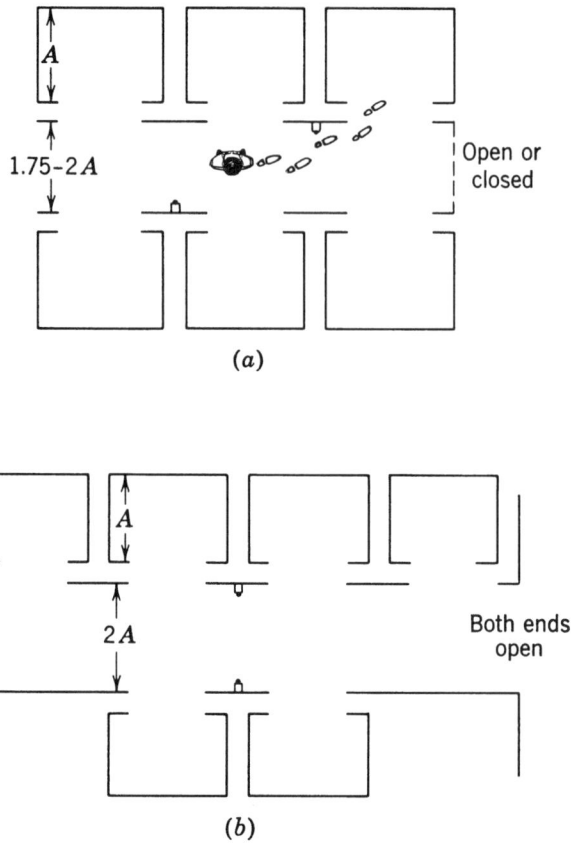

Figure 2.17. Six-car arrangements: (*a*) preferred; (*b*) acceptable.

which should be equally accessible to a main entrance to the building, for the handling capacity of eight elevators will require passengers to assemble both in the elevator lobby and in the space beyond both ends of the lobby (Figure 2.19). Equally important is that departing passengers must be able to leave the lobby without having to pass other elevators, which will soon be arriving with capacity passenger loads.

The lobby width of an eight-car group should be about two times the depth of an individual elevator and never less than 10 ft (3 m). The maximum width of the lobby should never exceed 14 ft (4.3 m), and even then the passenger response time to an elevator will be long enough to require extra waiting time at landing calls for some of the cars.

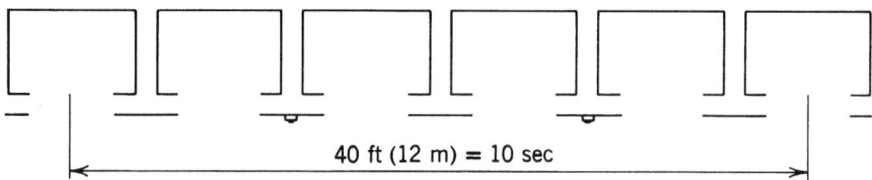

Figure 2.18. Unacceptable six-car arrangement.

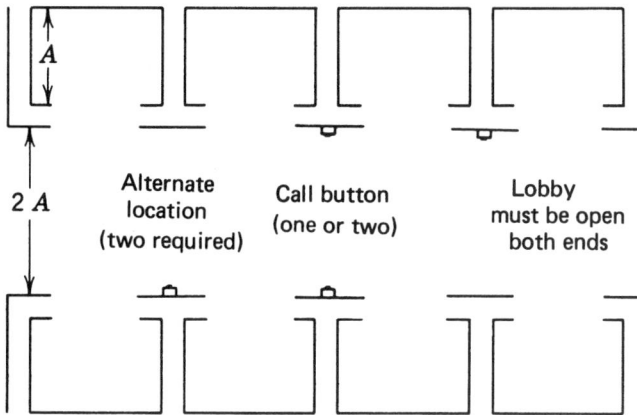

Figure 2.19. Eight-car arrangement.

Landing call buttons should be located at the center of the lobby, on opposite sides. The best place for persons to wait and, where they usually do wait, is dead center so they may see each elevator. Passengers can then respond as quickly as possible, with minimum walking distance.

Using the main floor elevator lobby as a passage is not recommended, as accommodating any kind of traffic other than elevator traffic would require too wide a lobby. At the upper floors such an arrangement is possible because traffic is expected to be minimal. A main floor lobby, which is open-ended, and upper-floor lobbies in an alcove arrangement is completely acceptable.

Additional considerations for the arrangement of elevators, and especially for lobby arrangements, are the concern of the structural engineer as well as the architect and elevator engineer. The building columns should be located at the back of the elevator shafts. If they are placed at the front, the columns can interfere with the elevator door size or, if placed in the elevator lobby space, can obscure the landing lanterns and inhibit free passage (Figure 2.20).

Unique Arrangements. Certain distinctive elevator arrangements have either unique merit or serious disadvantages. The architecture of particular buildings, rather than economic considerations, is usually responsible for their use.

The angular arrangement of elevators is feasible with a single group of cars serving a building, but has certain inherent disadvantages that must be compensated for by additional elevator service. As can be noted in Figure 2.21*a*, cars at the narrow end of the lobby must be a minimum distance apart, which tends to make the wide-end cars too far apart. This leads to extra time losses per upper-floor landing call stop.

The cornered arrangement in Figure 2.21*b* consolidates the elevator lobby but leads to interference between passengers entering or leaving the corner cars. This is serious if the full capacity of the elevators is required, and is not recommended for that reason.

The circular arrangement (Figure 2.21*c*) is a variation of the angular and cornered, with a premium in cost because of the shape and special mechanical features of the elevator car and doors. The closeness of the elevator entrances requires either special door arrangements or substantial extra hoistway space to accommodate conventional elevator

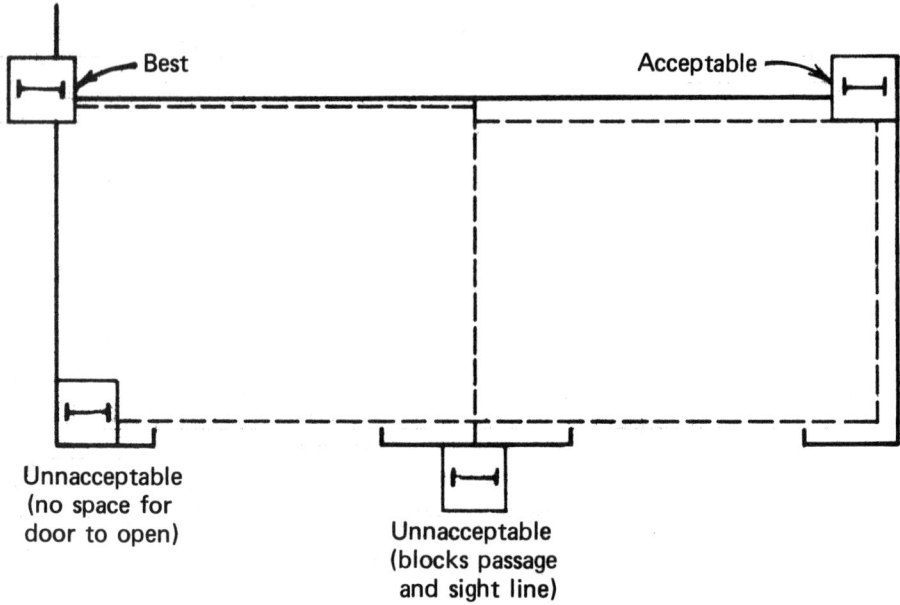

Figure 2.20. Building column placement.

doors. Traffic congestion possible with such an arrangement of elevators is an equally important discouraging factor.

Odd-shaped elevator cars, glass elevators, atrium building elevators, and other unique arrangements of elevators are discussed in the chapter on elevators for special applications.

Front and Rear Entrances. One of the most important of the special arrangements of elevators is the use of both front and rear entrances on one or more elevators in a group. This is one of the areas of elevatoring in which the operation of the elevators must be closely coordinated with their arrangement.

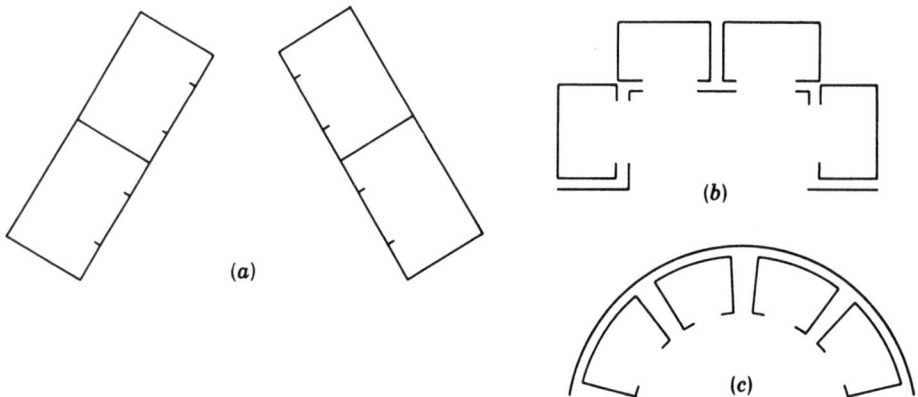

Figure 2.21. Unique elevator core arrangements: (*a*) angle; (*b*) alcoved; (*c*) circular.

As a general rule, when two sets of entrances are furnished on one elevator in a group but are not on all of the elevators, one set or another but never both should be in operation at one time. The reason is the same as if only one car in a group serves an odd floor: special controls must be furnished to intercept that car for the rear entrance, and a person's wait at a rear-entrance landing may become intolerable.

A suitable arrangement for rear entrances is the use of one of the cars in a group as a part-time service elevator. The elevator is then disconnected from group operation and made to serve a separate landing call button for rear service (Figure 2.22). The rear en-

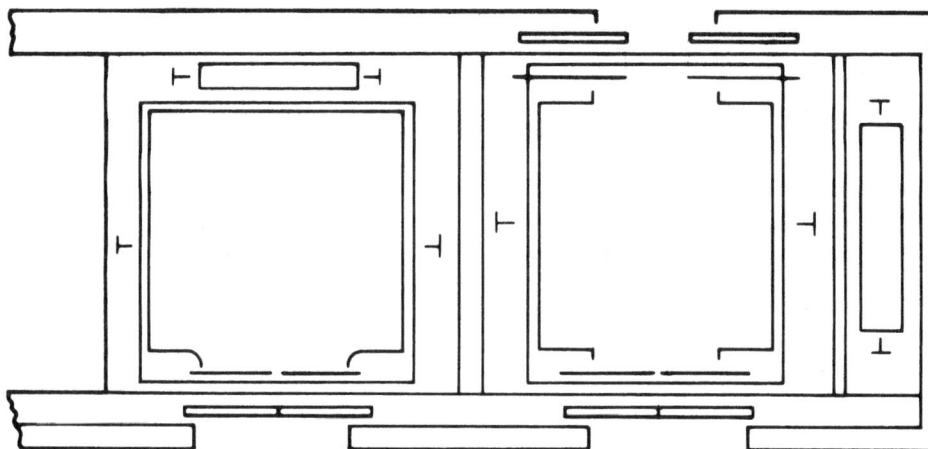

(a)

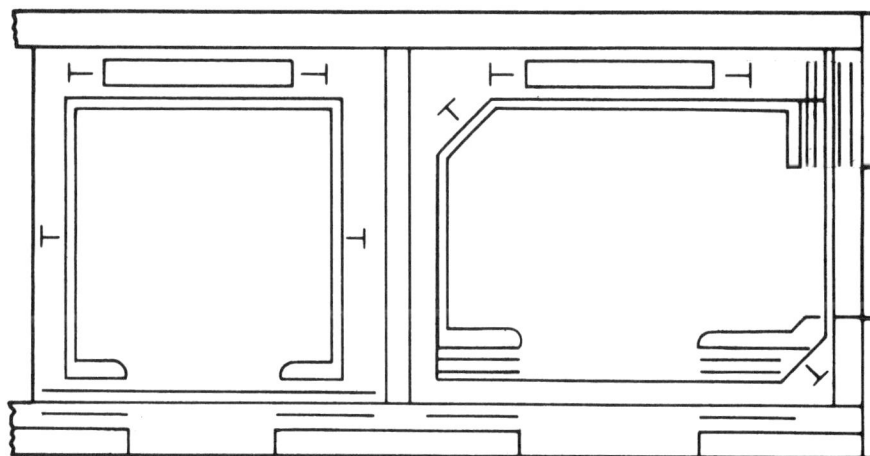

(b)

Figure 2.22. Two entrances on an elevator: (*a*) preferred arrangement—front and rear entrance; (*b*) awkward and costly arrangement—side and front entrances require corner-post-type elevator.

trance arrangement is the acceptable arrangement for new buildings. In the past, side entrances were offered and the elevator car frame and rails had to be arranged "corner-post." At present this arrangement has become prohibitively expensive and should not be considered except under extremely special circumstances.

Another acceptable application of front and rear entrances is the use of a group of elevators between old and new structures to bridge uneven floors. Here all elevators in a group should serve both the front and rear entrances at all times and car call buttons should be provided for each opening. For example, if the third floor rear is a different level from the third floor front, a person should be able to choose the level required. Separate buttons should be provided even if both front and rear are on the same level. Front and rear openings served in this way make for a building with elevator requirements almost equivalent to one with as many floors as the total number of openings. Because most of an elevator's trip time in a low building is spent in stopping, the increased number of possible stops tends to increase total trip time. Serving both front and rear entrances generally requires additional elevator capacity (Figure 2.23).

When a passenger elevator serves only two stops and heavy pedestrian traffic is expected, front and rear openings allow more prompt loading or unloading of the car. Passengers can walk directly into the elevator and face the entrance from which they will leave, and a loading group at the opposite entrance can follow them in as they are leaving (Figure 2.24).

Front and rear entrances on automobile and freight elevators may expedite loading and unloading. An auto that can be driven off without backing can often be unloaded in half the time.

Other special elevator arrangements are discussed in the sections dealing with specific building types and special applications.

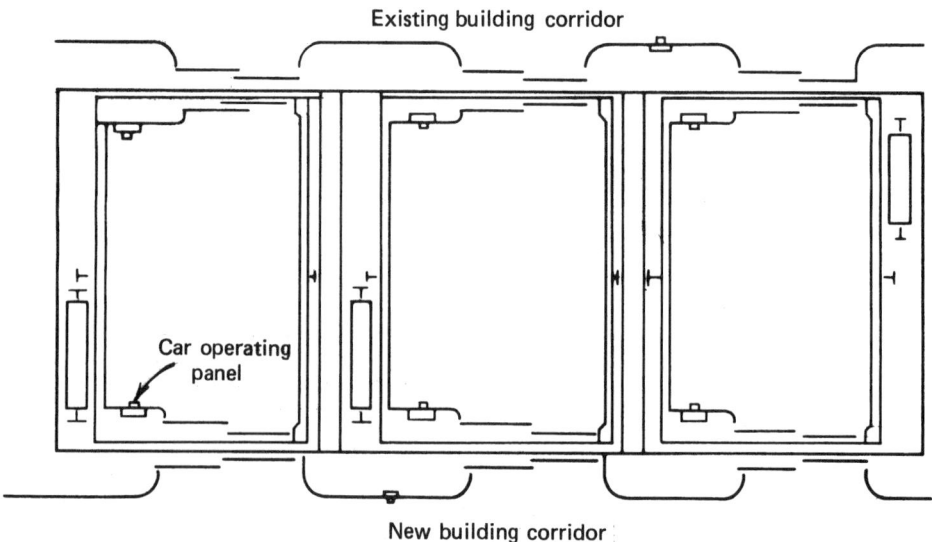

Figure 2.23. Three cars—connecting an existing building to a new building.

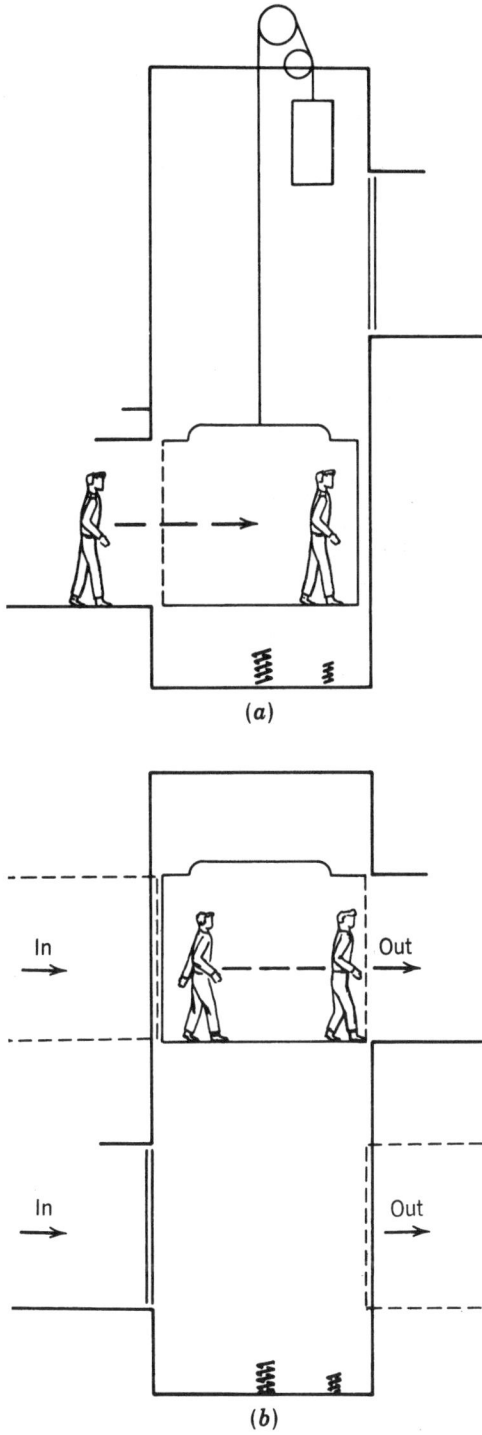

Figure 2.24. Pedestrian movement on elevators: (*a*) transfer on; (*b*) transfer off.

MODERNIZATION

The recommendations included in this chapter for a new installation must be evaluated if modernization of an existing building is planned. Such aspects as the location of the equipment in relation to the entrance, any later changes to the lobby or changes that can be made to improve the access to equipment, the size and adequacy of the existing total installation, the changing nature of tenancy, future plans for the building, the changing nature of the neighborhood, and competition between buildings, are all considerations in modernization. There are many others.

An important aspect is the size and shape of the elevator platforms plus the door arrangement. These are the prime factors in expected loading and efficiency of transfer, and they must be considered in any analysis of service potential. Most elevators are installed on the basis of pounds (kg) capacity, which indicates a certain maximum platform area. It is often the practice in car design to use more space for aesthetic features than necessary, which results in a usable interior area of the car well below the maximum allowed for weight rating. It is to the building's advantage if the capacity in pounds (kg) can be legally reduced, because the impact is far-reaching in both building structural load and the loads on the equipment and electrical power. This should always be an area of investigation.

An essential part of the modernization process is to make a traffic study of the existing plant and use that as a projection of how such modernization will improve the response of the elevators to traffic and whether the capacity will be at least equal to, if not more than, that presently experienced. Any plans for change in the nature of the building can be included in that study, as well as the possibility that the present elevator plant is more than adequate and the building can provide quality service with fewer units. That was often the case during the 1950s when the major conversion from attendant elevators to automatic took place.

Not to be ignored are the service needs of the building regarding movement of material in and out. Older buildings are often undergoing renovations (as are newer ones), and this becomes an important consideration. Other factors, such as the location of food service areas, the need to accommodate physically impaired people, the impact of multiple entrances, garage floors, and building security, as well as emergency services, are all part of the modernization process and just as essential as when new elevators are planned. Details are discussed in many of the later chapters in this book.

3 Passenger Traffic Requirements

If we know how many people will require elevator service within a given period of time, the task of providing vertical transportation is one of time and motion study. It embraces many variables, the most important being the human engineering factors and the human reactions to multiple elevators.

THE ELEVATOR TRIP

In modern buildings people are accustomed to operating a call button to summon an elevator and will move to an elevator that opens its doors to offer them service. Usually, but not universally, passengers will note a lighted lantern showing direction of car travel and will choose the car corresponding to the desired direction. In a building with light traffic and a tendency for only one elevator to stop at a floor at one time, passengers may ignore the lantern and get into the next elevator that arrives. In busier buildings there is a strong possibility of two cars stopping at a floor at one time. If passnegers have taken the wrong car once or twice they become conditioned to be cognizant of the directional signal.

Once passengers board the elevator car they are expected to operate car buttons for the destination floors. Failure to do so may take them where they do not want to go, which quickly teaches them to register car calls each time. The passengers must do one more thing before their trip is complete: to get off at their floor, they must note where the elevator is stopping as shown by the car position indicator mounted on the car front and leave the car at their stop.

REQUIREMENTS OF GOOD ELEVATOR SERVICE

The elevator engineer is interested in the passengers' impatience while waiting as well as during their trip. While they are waiting at some intermediate floor, their impatience is growing. In a commercial environment, they are less tolerant of waiting than in a residential or recreational environment. Frequent studies have indicated that passengers become impatient after waiting about 30 sec in a commercial building, and about 60 sec in a residential building.

From these observations come the first requirement for good elevator service—the elevator system must be designed to provide average waiting time of less than 30 sec in commercial buildings and less than 60 sec in residential buildings.

The second requirement is to provide sufficient quantity of elevator service for the maximum passenger arrival or departure rate expected within a peak traffic period. This

The Vertical Transportation Handbook, Third Edition, Edited by George R. Strakosch
ISBN 0-471-16291-4   © 1998 John Wiley & Sons, Inc.

can be accomplished by either a platform of sufficient area to accomodate all persons waiting to ride or, alternatively, a sufficient number of smaller platforms. The alternative of more platforms is usually preferred because it reduces the waiting time.

A good analogy is to compare vertical transportation to a batch and continuous-flow conveyor system. The contiuous-flow system transports material from a reservoir into a stream or hose in which the material is moved to its destination. A batch system moves measured quantities to the destination from the reservoir, where they accumulate until another batch is moved, usually in a bucket. An elevator can be compared to a batch conveyor. The arrival of people at a building is in a continual flow and the elevator system is the batch conveyor moving these people from the reservoir (lobby) to their destination. The ideal elevator arrangement is to have the multiplicity of elevators approximate the continuous-flow process so that the lobby (reservoir) is never filled in excess of the quantity of people a single elevator will transport.

A good example of continuous flow in vertical transportation is the escalator. Platforms (steps) are provided with minimum waiting time (usually 0 sec waiting because the steps are constantly moving) so that a person has immediate access to vertical transportation. Because the platforms are large enough to accommodate only one or two persons at a time, if more than one or two require service at the same instant, someone must wait. The wait is short since the prospective passengers can see the escalator is in service and the extent of their wait. On the other hand, a person waiting for an elevator at an upper floor may not be able to see whether the cars are in service and therefore becomes impatient while waiting.

Once people board an escalator they know they will be delivered to the next floor in a relatively short period of time, and other than on the extreme high-rise escalators in some subway stations, they can see the top landing. Elevator passengers often do not know how long they will be in the car. If it serves many floors in a busy building and the number of elevators is limited, a person may be on an elevator for a considerable period of time.

Studies have found that about 100 sec becomes the limit of tolerance for people in an elevator making several stops, each for one person. Tolerance will lengthen to about 150 sec if a few people are being served at each stop; the "average person" feels more tolerant if two people are being served at a time. Finally, if monotony is relieved by a changing scene, our passenger may tolerate a ride as long as 180 sec. These time factors are necessarily approximate since an individual's tolerance varies with the urgency of mission or other factors affecting feelings or atmosphere.

The third requirement of good elevator service, therefore, is to design the system so that a person will not be required to ride a car longer than a "reasonable" time. If the first two requirements are met, the third is usually satisfied as a natural consequence.

Three more considerations are necessary to develop a "quality" elevator installation as opposed to an ordinary or utilitarian installation. One, the platform areas as indicated by the capacity of the elevator should be large enough to allow a comfortable area of about 2.5 to 3 ft^2 (0.19 to 0.28 m^2) per person. Two, the door width should be wide enough, 48 in. (1200 mm) is recommended, to allow ease of transfer on and off the elevator. And three, a study should be made of the effect of one elevator out of service, and, if critical, an additional elevator or larger platform areas and capacities should be recommended.

CALCULATING THE TIME FACTORS

Two-stop Elevators

To calculate the total time for an elevator trip, a practical procedure is to break the trip down into its components. A simple example of a two-stop elevator will be followed by analysis of more complex and multiple-stop trips.

Suppose we have an elevator that makes two stops about 10 ft (3m) apart and wish to calculate how long it will take a person to ride to the higher or lower landing.

When the passengers arrive at the landing and operate the elevator call button the trip is, in effect, starting. When they leave the elevator at the other landing they have completed their trip. Once the call is registered the elevator is serving the passenger and the time factors will be as follows.

Referring to Figure 3.1, if a car is at the lower landing when the passenger arrives and operates the call button, the elevator doors need only open (*a*). A typical door requires about 2 to 3 sec to open, depending on the width and type of door. About 2 sec must be allowed for the passenger to enter the car and operate the car button (*b*). The doors must close again (about 3 sec), and the car must travel the 10 ft to the next landing (about 7.5 sec) (*c*). The doors must again open and will take 2 sec, with another 2 sec for the passenger to leave (*d*). The total time consumed by that passenger is 19.5 sec.

Before another person can get service, more time must necessarily elapse. The doors must close again (3 sec), and the car return to the opposite landing (7.5 sec). At this point the cycle can be repeated. The elevator's total cycle time or round-trip time has been 30 sec. Thus, 30 sec is the approximate time a person who just missed the elevator at the first floor will have to wait for it to return and give service. This is called the "interval" between elevator service at a floor. If we view this as a continuous process with a stream of passengers moving in one direction, the average passenger can expect to wait an average of one-half the round-trip time of the elevator. Some will arrive just before the elevator leaves and will have to wait 0 sec, whereas others will just miss the elevator and will have a 30-sec wait, so the average wait is 15 sec.

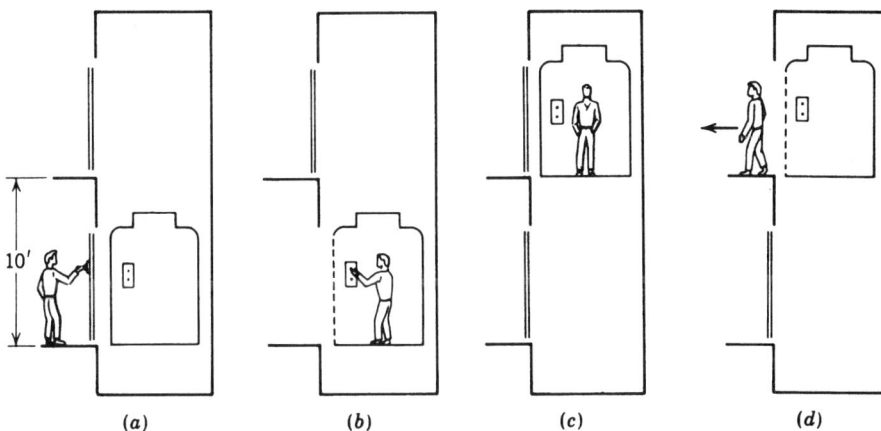

Figure 3.1. Pedestrian-elevator interaction.

If two elevators are side by side, each serving two stops and operating as described (each with a round trip time of 30 sec), the interval will be one-half the round trip time, or 15 sec, and the average wait will be about 7.5 sec. This is established by the expected operation of the elevators; when one is at the upper floor the other is at the lower floor and vice versa.

The foregoing example serves to show two important considerations in elevatoring: the total time required to serve a person and its relation to elevator handling capacity, and the wait for an elevator as indicated by interval and its direct relationship to round-trip time.

Handling capacity—the number of people served in a given period of time—is calculated from the round-trip time of an elevator. The basic time period is generally established as 5 min for the following reasons.

Five-minute Peaks. Peak requirements in an office building usually occur in the morning when people are trying to be at their desks by a certain starting time. Human nature being what it is, many employees arrive at the building within a few minutes of the deadline.

Figure 3.2 shows a typical arrival rate at an office building lobby where prospective elevator passengers peak in the 5 min preceding starting time. If there are not sufficient elevators to serve this peak, lobby congestion persists past the deadline. Some conscientious people arrive earlier to avoid being late, but, in general, office workers consider themselves on the job if they arrive at the building just prior to the normal starting time. It is the responsibility of their employer to transport them quickly up from the lobby to their office floors.

Because elevators for office buildings received the most attention when formal study of elevatoring began, the standard of a 5-min peak has persisted. It has been found that 5 min is a convenient time period to measure peak traffic on elevators in any type of building. For that reason our calculations are concerned with critical 5-min traffic peaks and 5-min elevator handling capacity.

Handling Capacity. Translating our first example into 5-min handling capacity by means of the formula, we find that the elevator has a 5-min handling capacity of 10 people:

$$5\text{-min handling capacity} = \frac{\text{number of passengers per trip} \times 300 \text{ sec}}{\text{round-trip time in seconds}} \qquad (3.1)$$

$$\frac{(1) \text{ passenger per trip} \times 300}{30} = 10 \text{ passengers}/5 \text{ min}$$

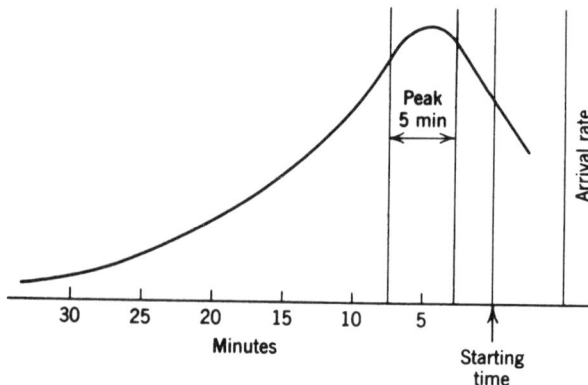

Figure 3.2. Typical arrival rate at an office building.

If only one passenger is served per trip in one direction, our 5-min handling capacity is only 10 people; with two passengers per trip (or one person up and one person down per trip) in the same time, handling capacity would be 20, and so on. If more passengers are served with each trip, however, the time factors, especially entering and exiting transfer time, would increase.

Interval. Interval or the time between elevators is determined directly from round-trip time and is inversely proportional to the number of elevators in a group. It is calculated by the formula:

$$\text{Interval} = \frac{\text{round-trip time of an elevator}}{\text{number of elevators in the group}} \tag{3.2}$$

In our example, with only one elevator, interval is equal to round-trip time. Obviously, the interval measures the theoretical longest time a person should have to wait for an elevator. In practice the actual interval varies from trip to trip, because of passenger delays or random traffic on an elevator, and should be stated as an average for a given period of time.

Waiting Time. Here we introduce the concept of waiting time related to interval. The theoretical average wait of all persons is one-half the interval, the interval being the theoretical longest wait of any person. The term "theoretical" is used for two reasons: (a) intervals must be considered as averages for a period of time and (b) the operating system must maintain uniform spacing of elevators. With the best available operating systems, average waiting times should be about 55 to 60% of the interval, depending on the refinement of the operating system.

For example, if four elevators are provided and the average round trip over a period of time is 120 sec, the average interval for that period is 30 sec. Some people will get immediate service, 0 sec wait, and others will wait the full 30 sec. The average waiting time is therefore 0 sec plus 30 sec divided by 2 for an average of 15 sec. The maximum waiting time should be 30 sec. However, if during the time period an elevator is filled and arranged to bypass waiting landing calls for that trip, only three elevators are available and the interval is 40 sec. In the same period, someone may delay an elevator and the average round-trip time for the four elevators can increase to 140 sec so the interval becomes 35 sec. If all the incremental trips are calculated and an average taken, the average interval may still be 30 sec (some trips will be faster than the 120-sec round-trip time) but the average waiting time will be longer than 15 sec, probably in the range of 18 to 20 sec.

Multistop Elevators

A two-stop elevator is the simplest and most efficient elevator system. Everyone who gets on at one landing is expected to get off at the other landing. Transfer time is minimized and no question of probable stops occurs since there is only one possible stop.

Even groups of two-stop elevators have very little complexity. Scheduling can be simple: if traffic is two-way, one car should be at the top and another car at the bottom; with one-way traffic, elevators should be concentrated top or bottom.

Elevatoring calculations become more complex with three or more stops. With three stops, for example, a number of elevator trips are possible: a person at the lower landing may wish to go to either the middle or top landing; a person at the second landing may wish to go to either the top or bottom landing; or a person at the top may wish to go to either the second or the lower landing.

The time required for a typical trip may be calculated on either of two bases. The more conservative basis assumes that the elevator will make every stop up and down. A second, more typical and more complex, method determines the probable stops the elevator will make related to the traffic it is expected to handle. We will investigate the latter approach in detail.

Probable Stops

The first step in calculating the time required for a typical trip is to determine the number of stops an elevator will make in an average round trip. This depends on the following factors:

1. The number of people entering the elevator at the lobby floor greatly influences the number of stops the elevator makes. If the elevator serves more floors than the number of passengers on the car during that particular trip, assuming passengers enter at the ground floor only, the elevator is not expected to make more stops than passengers carried. The probability is it will make fewer stops than the number of passengers carried. Viewed differently, the number of floors the elevator serves influences the number of stops an elevator is expected to make with a given passenger load. For example, an elevator that serves 20 floors and carries 10 passengers per trip is more likely to make 10 stops each trip, than an elevator that serves 10 floors and carries 10 passengers per trip, which will probably make fewer than 10 stops per trip.

2. The normal population of each floor an elevator serves also influences the number of stops the elevator makes. For example, if some floor served by the elevator is a storage or mechanical floor, the elevator's tendency to stop there is zero. If one floor has 100 persons and the other floors have only 10, the tendency to stop at the 100-person floor is greatly increased.

3. The expected direction of traffic imposed on an elevator makes predominantly up car stops and returns to the lobby floor with very few down landing stops.

   a. When everyone is coming into the building, the elevator makes predominantly up car stops and returns to the lobby floor with very few down landing stops.

   b. When persons are traveling between upper floors in a building, each person causes the elevator to make two stops: an up or down landing call for that person to board the elevator and a subsequent car call for that person to leave.

   c. When visitors are coming to and going from a building or when the occupants are leaving and returning during lunch or other times, up car stops are generated by people entering the elevator at the lobby and down landing stops by persons on upper floors. In extremely busy situations and with inadequate elevators, it is possible for an elevator to make every stop up and every stop down, resulting in intolerably long round-trip times in a building of any height. This is especially true if elevators have large capacities and carry many people, which can cause many stops.

   d. When many people wish to leave the building within a short period of time, as at quitting time in an office building, a different pattern of stops is likely. The elevators are expected to make predominantly down landing stops and to fill quickly at each stop. Because at least a carload of people is eagerly awaiting the elevator and each person is concentrating on leaving the building as quickly as possible, the number of down stops in a typical trip may be far fewer than an equivalent up trip for the same number of people.

As may be seen, determining probable stops requires a knowledge of critical traffic in the building under consideration. Timing and the nature of the critical traffic for each type of building is discussed in later chapters of this book. One form of traffic may be critical in many types of buildings, or many types of traffic may ocur in one building during various times of the day, so that probable stops will be related to the various types of elevator traffic.

In the following chapters we discuss the major traffic periods and the necessary time required for elevators to serve passengers during those periods. With this information and with a proposed or existing building, and an estimate of the type of tenants and their population, a reasonable estimate of the elevator system performance for a critical traffic period can be calculated. If elevators can handle the critical traffic, and if they are properly operated, they will also be adequate for the other traffic periods.

4 Incoming Traffic

INTRODUCTION

Incoming traffic provides one of the heavier traffic periods in office buildings and occurs to a greater or lesser degree in any building. The number of elevators required to serve a given number of people during this period is calculated and their operation discussed. In succeeding chapters two-way and outgoing traffic are similarly discussed.

Incoming traffic calculations are discussed in detail, as this traffic type is important in any building. In addition, the method of calculation that is established in our study of incoming traffic can be extended to include any type of traffic. Providing sufficient handling capacity is the major consideration during incoming traffic periods.

Incoming or up peak traffic exists when everyone arriving at a lobby floor is seeking transportation to upper floors (Figure 4.1). Minimum down traffic is expected during this period and usually amounts to about 10% of the up traffic during an observed 5-min time period. A single elevator trip during this period consists of the following elements:

1. Loading time at the lobby
2. Door-closing time and running time to the next stop
3. Door-opening time and time to transfer part of the passenger load at that floor
4. Door-closing time and running time to the next stop
5. Door-opening time, transfer time, etc., until the highest stop is reached
6. Door-closing time and running time to a down stop
7. One or two down stops and running time to the lobby
8. Door-opening time at lobby unload, and repeat loading time

As may be seen, the elevator trip is made up of various time elements related to such factors as the number of people entering and leaving, the speed of the elevator, and the time to open and close the doors. Of prime importance is the number of stops the elevator will make on its up trip. Two methods are suggested for estimating the number of stops. The method selected will depend on the information available about the distribution of the building's population, working hours, and people's activities within the building. The more that is known about these factors, the more practical and accurate the probable stop value will be. The less that is known, the more theoretical the probable stop value will be.

In a building in which a great deal is known about the per floor population it is possible to assign a stop value to each floor. Table 4.1 is based on an 11-story building. The population per floor is as indicated, and everyone is expected to report for work at the same time. We assume that each elevator in this building has a nominal capacity of 10

The Vertical Transportation Handbook, Third Edition, Edited by George R. Strakosch
ISBN 0-471-16291-4   © 1998 John Wiley & Sons, Inc.

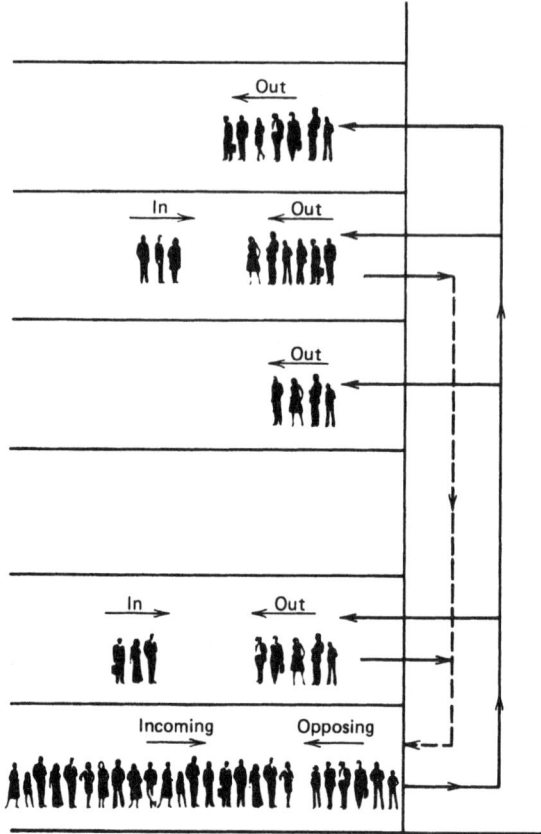

Figure 4.1. Elevator people handling—typical incoming traffic flow.

TABLE 4.1.

Floor Number	Population per Floor	Expected Stop
11	100	1
10	100	1
9	100	1
8	100	1
7	50	$\frac{1}{2}$
6	50	$\frac{1}{2}$
5	25	$\frac{1}{4}$
4	25	$\frac{1}{4}$
3	10	$\frac{1}{10}$
2	5	$\frac{1}{20}$
		Expected stops $\overline{5.65}$

people. For this table we have made the reasonable assumption that the expected number of stops (probable stops) is proportional to the number of people an elevator can carry. If we assume the top floor population as 5 instead of 100 as shown, an elevator would stop there only once out of every 20 trips instead of every trip. If people on the tenth floor, for example, started work, say 15 min later, the expected stopping at the tenth floor would be greatly reduced during the time the other floors required service.

Seldom is as much information available as is shown in Table 4.1. If it were, the task of elevatoring could be reduced, as is possible with a known building in which the elevators are manually operated and there is a desire to automate them. A study of the existing conditions can determine the average number of stops made each trip and, if sufficient elevators available, the degree of improvement possible with automatic operation.

In estimating probable stops for a building under design we must give due regard to the real-life situation. We will not know that all starting times are the same, we will know only that so many square feet of area are available for the per-floor population and we will not know the difference in attraction for each floor.

We can be reasonably certain, however, that in any elevator trip the elevator will stop at upper floors in proportion to the number of people in a car and the number of floors that elevator serves. This forms the basis of a second method of estimating probable stops.

If we assume that population on each floor is equal and that persons are entering a building in a random fashion so that all must be in place by a given time, there is a distinct possibility that one or more people will get off the elevator at the same floor within the same trip. A statistical calculation of the probable number of passengers leaving the elevator at a given floor at the same time provides the following formula:

$$\text{Probable stops} = S - S\left(\frac{S-1}{S}\right)^p \tag{4.1}$$

which becomes the formula for probable stopping of an elevator with a given passenger load wherein

S = the number of possible stops above the lobby

p = the number of passengers carried on each trip

Formula 4.1 is the one used in planning elevators for buildings when little is known about the distribution of future population and each floor is assumed equal in population.

Applying the formula to a 12-passenger, 12-upper-floor elevator trip, a probable stop value of approximately 7.8 stops is obtained. Chart 4.1 shows approximate probable stop values for various numbers of passengers per trip and various numbers of upper floor stops.

Transfer Time

Table 4.2 shows the time required to transfer people from the lobby to a waiting elevator based on various numbers of people loading. There is usually a fixed time the lobby doors are held open, often referred to as dispatch time. In some operating systems this time is established after the first car call is registered, and in other systems it is established after the car arrives and its doors are opened. The times represented are nominal and will be used in our calculations.

CHART 4.1. Probable Stop Table

Passengers per Trip

Upper Floors Served	2	4	6	8	10	12	14	16	18	20	22	24	26	28	30
30	2	4	5.7	7.6	9.5	10.5	11.7	12.8	13.8	14.8	16.0	17.2	18.0	19.0	19.5
28	2	3.9	5.5	7.2	9.0	10.1	11.6	12.5	13.5	14.6	15.6	16.6	17.6	18.1	18.4
26	2	3.8	5.5	7.0	8.5	9.8	11.2	12.2	13.1	14.1	15.1	16.0	16.8	17.4	17.7
24	2	3.8	5.4	6.9	8.3	9.6	10.8	11.9	12.8	13.8	14.6	15.4	16.1	16.7	17.3
22	2	3.7	5.4	6.8	8.2	9.4	10.5	11.6	12.5	13.3	14.1	14.8	15.4	16.0	17.0
20	2	3.7	5.3	6.7	8.0	9.2	10.3	11.2	12.1	12.8	13.5	14.2	14.7	15.3	16.0
18	2	3.7	5.2	6.6	7.8	8.9	9.9	10.8	11.6	12.3	12.9	13.4	13.9	14.4	15.0
16	2	3.6	5.1	6.5	7.6	8.6	9.5	10.3	11.0	11.6	12.1	12.6	13.0	13.4	13.9
14	2	3.6	5.0	6.3	7.3	8.3	9.0	9.7	10.3	10.8	11.3	11.6	12.0	12.2	12.5
12	2	3.5	4.9	6.0	7.0	7.8	8.5	9.0	9.5	9.9	10.2	10.5	10.8	11.0	11.3
10	2	3.4	4.7	5.8	6.5	7.2	7.7	8.2	8.5	8.8	9.0	9.2	9.4	9.5	9.5
8	2	3.3	4.4	5.3	5.9	6.4	6.8	7.0	7.3	7.5	7.6	7.7	7.8	7.8	8
6	2	3.1	4.0	4.6	5.0	5.3	5.5	5.7	5.8	5.8	5.9	5.9	6	6	6
4	2	2.7	3.3	3.6	3.8	3.9	3.9	4	4	4	4	4	4	4	4
2	1.5	2	2	2	2	2	2	2	2	2	2	2	2	2	2

TABLE 4.2. Transfer Times

Lobby: Minimum 8 sec plus 0.8 sec per passenger over 8 passengers in, in and out, or out only.

Number of Passengers	8	10	12	14	16	18	20
Lobby Time (sec)	8	10	11	13	14	16	18

Car Calls	Dwell-time	3 sec per stop
	Transfer time	Use 3 sec for first two passengers (dwell-time) + 1 sec per passenger over 2
		Example: 4 passengers, 1 stop time = 3 + 2 = 5 sec
Landing Calls	Dwell-time	4 sec per stop*
	Transfer time	Use dwell time + 1 sec per passenger over 1 passenger entering
		Example: 4 passengers, 1 stop time = 4 + 3 = 7 sec

*Based on providing a landing lantern for each elevator, arranged to signal a minimum of 1 sec prior to the car's arrival at a floor with a maximum of four elevators in a row and the landing call button located in the middle. Without a landing lantern or with an in-car lantern, 5 sec initial dwell-time must be used for two cars side by side and increased by 4 sec for each additional car in a row. (See Americans with Disabilities Act Accessibility Guidelines (ADAAG), paragraph 4.10.7, reproduced in Chapter 2, page 52).
Note: Based on nominal 48-in. (1200-mm)-wide center-opening entrances. For other widths and types of entrances, see inefficiency associated with the various entrance widths and types, Table 4.3.

At an upper floor, if the elevator stops only for a car call, a certain predetermined dwell-time is established between the time the doors are fully opened and when they start to close. This time can be extended or shortened by the interception of a light ray when a passenger enters or leaves the car or when the doors are reopened by operation of the door safety edge. The car dwell-time is, initially, about 3 sec and can be adjusted to be longer. A landing call dwell-time is usually about 4 sec and should be adjusted to reflect the time a person will require to walk from a likely waiting position near a landing button to the individual elevator. For our calculations we are assuming a minimum of 3 sec per car call and 4 sec per landing call. These times are based on providing a landing lantern as prescribed in Table 4.2

The times given are based on a 48-in. (1200-mm)-wide center-opening door. In Table 4.3, which lists door times, a list of increased (or decreased) transfer time and door time inefficiencies are given which should be applied to the total standing time (door time plus transfer time) an elevator spends at a particular stop.

Requirements for disabled persons have been adopted as law or are reflected in building codes in many areas. Many of these requirements establish the amount of time that must expire between the time an elevator arrives at a floor or, a landing lantern lights, and the time the elevator leaves. These times are designed to allow a slow-moving person to travel from a point near the landing call button to the waiting elevator and should be reflected in calculated transfer time if applicable.

We will use the 3-sec car call dwell-time and the 4-sec landing call dwell-time in our calculations since observations have indicated that most passengers will accept them. It is interesting to note the impatience of some passengers who, upon perceiving they are waiting too long, will push the car call button or the door close button in an attempt to shorten the remaining time. Observations also indicate that many will hold the car door or operate the door-open button if they see a slow-moving person wishing to board the car.

TABLE 4.3. Door Operating Time

Door Type	Width in. (mm)	Open (sec)	Close (sec)	Total[a] (sec)	Transfer Inefficiency[b] (%)
Single-slide	36 (900)	2.5	3.6	6.6	10
Two-speed	36 (900)	2.1	3.3	5.9	10
Center-opening	36 (900)	1.5	2.1	4.1[c]	8
Single-slide	42 (1100)	2.7	3.8	7.0	7
Two-speed	42 (1100)	2.4	3.7	6.6	7
Center-opening	42 (1100)	1.7	2.4	4.6[c]	5
Two-speed	48 (1200)	2.7	4.5	7.7	2
Center-opening	48 (1200)	1.9	2.9	5.3[c]	0
Two-speed	54 (1400)	3.3	5.0	8.8	2
Center-opening	54 (1400)	2.3	3.2	6.0[c]	0
Two-speed	60 (1600)	3.9	5.5	9.9	2
Center-opening	60 (1600)	2.5	3.5	6.5[c]	0
Two-speed, center-opening	60 (1600)	2.5	3.0	6.0[c]	0

[a]Includes 0.5-sec car start.
[b]Transfer inefficiency: Increase normal standing time inefficiency by this percentage to reflect delay in passengers passing through doors.
[c]When preopening can be used, these values can be reduced by 1 sec.

After the elevator loads at the lobby floor it takes time for the doors to close and the car to start moving to the next landing. This time varies with the size and type of entrances as shown in Table 4.3. Repeated for each stop, these times become an important factor in round-trip calculations.

Once the door is closed it takes a bit more time to ensure the door is locked and for the elevator motor to "build up" to run to the next floor. For this requirement, 0.5 sec is added to the door operating time in the table. More time should be allowed if it is known that door mechanisms are slower or if build-up time is longer than the value used.

Door-closing time is established by safety codes based on kinetic energy, and the values shown are within the 7-ft.-poundal (0.29 joule) kinetic energy code limitations for average-weight, hollow metal doors.

Ignore for the moment the time it takes an elevator to travel from floor to floor; the next element in an elevator trip is the door opening at the next stop. Many elevators have doors that start opening as the elevator levels to the floor (preopening)[1] and open wide enough for passengers to start moving in or out by the time the car stops.

This procedure saves a small amount of time per stop. If adjustment is not maintained, it can present a tripping hazard; the present trend is toward requiring the elevator to be stopped before the doors start to open. This is especially true with solid-state motor drives wherein various safety circuits ensure that the elevator is stopped before doors are allowed to open. Preopening is usually limited to center-opening doors and reduces door

1. Preopening is often erroneously called "premature."

time about 1 sec per stop. Properly adjusted preopening limits the start of door opening so that the car is fully stopped before the doors are open wide enough for passenger transfer to start.

Table 4.3 shows the total door opening and door closing time for various widths and types of elevator entrances. The total time includes 0.5 sec for car start. A table of various inefficiencies based on different types of entrances is given. This value is in addition to the expected inefficiency of an elevatoring study based on a particular type of building, which will be discussed in general in this chapter and specifically in later chapters.

In addition to the loading and unloading and door opening and closing, running time is the other element in the round-trip time of an elevator. Part of the running time is spent in accelerating the elevator up to speed and decelerating it to a stop.

Persons can feel changes of acceleration and deceleration but are not very conscious of a constant acceleration or deceleration; therefore elevator equipment must be able to overcome inertia and accelerate and decelerate the elevator smoothly. The machinery must also be capable of moving the car a distance of 11 to 12 ft (3m), a floor height, in minimum time. Heavy-duty equipment such as a gearless machine can move an elevator from floor to floor in about 4 to 5 sec. Lighter equipment, such as a geared or hydraulic machine, will require about 6 sec for the same distance, depending on the ultimate speed the machine can develop and its accelerating capability.

Floor-to-floor speed seldom exceeds about 350 to 400 fpm (1.8 to 2 mps) regardless of the ultimate speed of the elevator. The heavy-duty gearless equipment is necessary for minimum floor-to-floor time since average acceleration of 5 fps^2 (1.5 mps^2) can be obtained versus 3.5 to 3.4 fps^2 (1 to 1.2 mps^2) with geared equipment.

A sample set of time-distance curves for various speeds of elevators is shown in Chapter 7, Figure 7.4. Note that the curve includes both acceleration and deceleration time so that the total time to travel the distance shown is indicated by a point on a curve.

Since time-distance curves vary with all elevator speeds, a means to approximate running time will be given which can be used easily for any speed and elevatoring situation.

Running Time

Typical floor-to-floor traveling time for elevators of various speeds and with average acceleration rates of 3.5 fps^2 (1.07 mps^2) for 400 fpm (2.0 mps) and under, and of 4.5 fps^2 (1.37 mps^2) for 500 fpm (2.5 mps) and above, are given in Chart 4.2.

The aforementioned time is measured beginning when the car starts from a floor until the elevator stops level at the next adjacent floor. It includes acceleration, running at the maximum speed that can be attained in the given distance, and deceleration plus an allowance, as shown, for leveling. The formula for the development of these times is given in detail on page 142 (Chapter 7).

To calculate the time for a distance greater than those given previously, the following formula is used.

$$\frac{(\text{Total travel} - \text{distance } a \text{ from Chart 4.2}) \times 60}{\text{ultimate elevator speed in fpm}}$$

$$+ \text{ time from Chart 4.2 for distance } a = \text{total time}$$

CHART 4.2. Running Times, Car Start to Car Stop, Seconds

Floor Heights: Feet	9	10	11	12	13	14	15	20	30	Each Additional 10 ft	Notes
Floor Heights: Meters	2.7	3.0	3.35	3.65	4.0	4.3	4.6	6.1	9.1	Each Additional 3 m	
Elevator speed											
100 fpm (0.5 mps)	7.6	8.2	8.8	9.4	10.0	10.6	11.2	14.2	20.2	6.0	a
150 fpm (0.75 mps)	6.7	7.1	7.5	7.9	8.3	8.7	9.1	11.1	15.1	4.0	
200 fpm (1 mps)	5.8	6.1	6.4	6.7	7.0	7.3	7.6	9.1	12.1	3.0	b
300 fpm (1.5 mps)	5.2	5.4	5.6	5.8	6.0	6.2	6.4	7.4	9.4	2.0	
400 fpm (2 mps)	4.8	5.0	5.1	5.2	5.4	5.6	5.7	6.5	7.0	1.5	
500 fpm (2.5 mps)	—	—	4.3	4.4	4.5	4.6	4.7	5.2	6.4	1.2	
700 fpm (3.5 mps)	—	—	4.3	4.4	4.5	4.6	4.7	5.2	6.1	0.86	
1000 fpm (5 mps)	—	—	4.3	4.4	4.5	4.6	4.7	5.2	5.8	0.6	

[a] Speeds of 100 fpm and 150 fpm include 0.75 sec for leveling.
[b] Speeds of 200 fpm and above include 0.5 sec for leveling.

For example: What is the time required to travel 100 ft (30.5m) at 300 fpm (1.5 mps)?

1. From Chart 4.2, 30 ft at 300 fpm = 9.4 sec.

2. $\dfrac{(100 - 30) \times 60}{300} + 9.4 = 14 + 9.4 = 23.4$ sec.

3. Total time to travel 100 ft at 300 fpm is 23.4 sec.

If the equipment under consideration for a particular installation requires longer for acceleration for floor-to-floor travel, an extra time value must be considered. If equipment levels at slow speed, such as two-speed ac elevators, or if the leveling speed operation is too long, floor-to-floor operating time may be as long as 8, 10, or more seconds.

Incoming Traffic Calculations

Suppose we want to know how many people a 16-passenger elevator at 500 fpm (2.5 mps) with 48-in. (1200-mm) center-opening doors, in an 11-story building with 12-ft (3.65-m) floor heights, can serve during a 5-min incoming traffic peak period.

1. Chart 4.1 shows that 16 passengers will make approximately 8.6 stops on the 10 upper floors in this building.
2. Time to load 16 passengers (Table 4.2) = 14 sec
3. Time to open and close 48-in. (1200-mm) center-opening doors and start car (Table 4.3) = 5.3 sec
4. Time to run from floor to floor in the up direction:
   10 floors × 12 ft = 120 ft
   120 ft ÷ 8.6 probable stops = 14 ft
   From Chart 4.2, 14 ft at 500 fpm = 4.6 sec
5. Time to run down from top floor to the lobby

   $\dfrac{(120 - 20) \times 60}{500} + 5.2$ (time to run 20 ft) = 17.2 sec

6. Transfer time at upper floors
   8.6 probable stops × 3 sec per stop = 25.8 sec
   versus 16 passengers at 1 sec each, use 26 sec
7. Adding all the above time factors together:
   a. Standing time

Lobby transfer time (2)	14	sec
Upper floor transfer time (6)	26	sec
Door operation upper floors, 8.6 times		
plus lobby door operation, 1 time		
Therefore 9.6 = 5.3 sec (3)	51	sec
Total time spent at floors	91	sec

   b. Normal standing time inefficiency 10%

	9.1	
Adjusted standing time	100.1	sec

(Total from page 79, 100.1 sec)

c. Running time
   run from floor to floor up
     4.6×8.6 (4)   39.6 sec
   run down (5)   17.2 sec
Total round-trip time   156.9 sec

$$\text{Elevator 5-min capacity:} \frac{16 \text{ passengers per trip} \times 300 \text{ sec}}{\text{round-trip time } 160 \text{ sec}} = 30 \text{ people.}$$

In other words, the single elevator in our example can serve 16 passengers in 160 sec or a total of 30 passengers in 5 min.

Item 7 (b), a 10% factor for inefficiency, is added to compensate for the rounding off of probable stops, door time, transfer time, and starting and stopping times, as well as the unpredictability of people.

If a two-speed door, 48 in. (1200 mm)-wide was used, the door time would be 7.7 sec per stop and the inefficiency would be the normal 10% plus 2% for the two-speed door, or a total of 12% (see Table 4.3).

The elevator calculation format just developed will be used in Example 4.1 to show its application as follows:

Example 4.1. Total Incoming Traffic Calculation

Given: 15-story building, lobby 1 to 2, 20 ft; typical floor 12 ft
Required: elevatoring to accommodate 110 people during 5-min peak morning in-rush (assume 11% of 1000-person population)
Procedure:
Assume: four 500-fpm elevators, serving floors 1 to 15, 17 passengers per trip = 10 probable stops
Time to run up, per stop

$$\frac{20 \text{ ft (lobby)} + 13 \times 12 \text{ (to top floor)}}{10} = \text{rise per stop}$$

$$\frac{176}{10} = 17.6 \text{ ft} = 5 \text{ sec (from Chart 4.2)}$$

Time to run down,

$$\frac{(176 - 17.6) \times 60 \text{ sec}}{500 \text{ fpm}} + 5 = 24 \text{ sec}$$

Elevator performance calculations:
Standing time
  Lobby time   =   15   sec
  Transfer time, up stops 3 sec × 10 stops   =   30
  Door time, up stops (10 + 1) stops × 5.3   =   53
  Total standing time   98   sec
Inefficiency, 10%   9.8
                                     Total 107.8 sec

(Total from page 80, 107.8 sec)

Running time
　Run up (10 × 5)   =　50
　Run down   =　24
　Total round-trip time   181.8 sec

$$\text{Five-minute handling capacity (HC)} = \frac{17 \times 300}{182} = 28 \text{ people}$$

$$\frac{110}{28} = 4 \text{ elevators required for handling capacity}$$

Interval: 182/4 = 46 sec. Too long. Should be 30 to 35 sec.
Observe: four cars can carry load but time between cars (interval) too long, need fifth car to improve interval.
Recalculate:
Assume: five 500-fpm elevators, 11 passengers per trip = 7.8 probable stops
Time to run up, per stop

$$\frac{176}{7.8} = 22.6\text{-ft rise per stop}$$

$$22.6 \text{ ft} = 5.4 \text{ sec}$$

Time to run down   = 24 sec
Elevator performance calculations:
Standing time
　Lobby time   =　11　sec
　Transfer time, up stops 3 sec × 7.8 stops   =　23.4
　Door time, up stops (7.8 + 1) stops × 5.3   =　47
　Total standing time   81.4 sec
Inefficiency, 10%   8.1
                                                                  Total 89.5 sec

Running time
　Run up (7.8 × 5.4)   =　42
　Run down   =　24
                                                                        155.5 sec

$$\text{HC} = \frac{11 \times 300}{156} = 21 \text{ people}$$

$$\frac{110}{21} = 5 \text{ elevators required for handling capacity}$$

Interval: 156/5 = 31 sec

Full-trip Probability and High Call Reversal

In the discussion of probable stops earlier in this chapter it was indicated that stopping is a function of the attraction each floor has for the passengers in an elevator. The probable stop chart (Chart 4.1) is based on an equal attraction per floor. Attraction can also be considered proportional to the population on each floor and experience has shown that an elevator, especially one operating during an incoming period, does not travel to the top floor each trip. The average trip will be a percentage shorter than a full trip to the top and advantage may be taken of this in elevatoring calculations.

As an arbitrary rule, if the population of any office building floor is less than 10% of the total population on the floors served by a group of elevators, the average trip can be reduced proportionally. For example, if the group of elevators is expected to serve 1500 people on 15 upper floors, the per floor population is 100. A 10% reduction is 150 people, hence the elevator trip and the basis of probable stops can be reduced from 15 floors to 14 floors. The total population served by the group of elevators will remain constant at 1500 people and the required percentage handling capacity sought will also remain the same.

If the top floor to be served contains some special attraction such as a cafeteria or restaurant, the advantage of this high call reversal should not be taken.

In the residential buildings such as apartments and hotels, similar advantages of high call reversal can be taken. The percentage will be greater than in office buildings and will be between 15 and 25% of the population served. Additional discussion will take place in the chapter on residential buildings.

If Example 4.1 is recalculated on the basis of short trips, the following results would be obtained (see Examples 4.1A and 4.1B).

Example 4.1A. Incoming Traffic Calculations (Short Trips)

Probable stops, 13 floors, 17 passengers = 9.6
Time to run up, per stop

$$\frac{20 \text{ ft (lobby)} + 12 \times 12}{9.6} = \frac{164}{9.6} = 17\text{-ft rise per stop}$$

$$17 \text{ ft (from Chart 4.2)} \qquad = 5 \text{ sec}$$

Time to run down,

$$\frac{(164 - 17) \times 60}{500} + 5 = 22.6 \text{ sec}$$

Elevator performance calculations:
Standing time

Lobby time, 17 passengers up	=	15 sec
Transfer time, up stops 9.6 × 3	=	28.8
Door time, up stops (9.6 + 1) stops × 5.3	=	52.2
Total standing time		96 sec
Inefficiency, 10%		9.6
		Total 105.6 sec

Running time

Run up (9.6 × 5)	=	48.0
Run down	=	22.6
Total round-trip time		176.2 sec

Four 17-passenger @ 500 fpm elevators serving floors 1 to 15

	Calculated	Required
5-min capacity per elevator:	29 people	27.5 people
4 elevators; interval:	44 sec	30 to 35 sec

The four elevators do not provide an adequate interval even though handling capacity is sufficient.

Example 4.1B. Incoming Traffic Calculations (Short Trips)

Probable stops, 13 floors, 11 passengers $= 7.6$
Time to run up, per stop

$$\frac{20 \text{ ft (lobby)} + 12 \times 12}{7.6} = \frac{164}{7.6} = 21.6\text{-ft rise per stop}$$

21.6 ft from Chart 4.2 $= 5.2$ sec

Time to run down,

$$\frac{(164 - 21.6) \times 60}{500} = 5.2 = 22.3 \text{ sec}$$

Elevator performance calculations:
Standing time
　　Lobby time, 11 passengers up $= \quad 11 \quad$ sec
　　Transfer time, up stops 7.6×3 $= \quad 22.8$
　　Door time, up stops $(7.6 + 1) \times 5.3$ $= \quad \underline{45.6}$
　　Total standing time 79.4 sec
Inefficiency, 10% $= \quad \underline{\quad 7.9}$
Total 87.3 sec

Running time
　　Run up 7.6×5.2 $= \quad 39.5$
　　Run down $= \quad \underline{22.3}$
　　Total round-trip time $= 149.1$ sec

Five 11-passenger @ 500 fpm elevators serving floors 1 to 15

	Calculated	Required
5-min capacity per elevator:	22 people	22 people
5 elevators; interval:	30 sec	30 to 35 sec

When all other economic factors are considered, the foregoing may encourage the building of one more floor if five elevators are provided.

Choosing the Proper Elevator Capacity

The choice of the proper elevator capacity is a major decision in the design of the elevators in a particular building. There will always be conflict between the space the architect wants to allow and the requirements of a properly sized elevator platform and the necessary space for the hoistway to contain the mechanical equipment to lift and guide the elevator.

The essential feature of a proper elevator is the interior square foot (or square meter) area of the car. This is based on the number of people required to be carried per trip as established by elevating calculations and a minimum allowance of square feet (square meters) per person. The relation between area and load is established by elevator codes and is usually based on a maximum density of people in a given area. Actual loading is less than the weight allowed since people will not crowd that closely together. Table 4.4, which is an expansion of Table 2.1, gives the relationship between weight and area versus a people density of 2.3 ft^2 (0.22 m^2) per person for normal (observed) loading and 1.5 ft^2 (0.14m^2) for maximum loading. The relation given is linear for 4000 lb and above; that is, the same density is used for the smallest to the largest elevator. In practice, the actual loading of a smaller elevator, 3500 lb (1600 kg) and below, is somewhat lower and is indicated as observed.

When elevator capacities are chosen for a particular building, consideration must always be given to the possibility of one elevator being out of service. For example, if calculations indicate that four 10-passenger elevators are required, which, by the table, would be four 2500-lb elevators, consideration of one car out of service is made by a simple calculation as follows:

$$4 \times 2500 = 10,000 \text{ lb}$$

If one elevator is out of service, $3 \times 3500 = 10,500$ lb would be required; therefore, 3500-lb elevators should be used instead of 2500-lb elevators. This is especially true where elevator service is vital such as in a hospital. The requirement can be modified if alternate transportation to the floors, such as a service elevator, escalators, or overlapping elevators, is available, so judgment must be employed. We will call this consideration the (N-1) rule.

The normal capacity shown in Table 4.4 is based on expected office-building-type loading. For apartments, hospitals, or hotels, where two-way traffic is generally the rule, normal loading is about one-half the value given. This allows for people with baggage, accompanying carts, packages, or other items that may be carried in addition to the passengers. The reduced loading also provides additional space for people to satisfy the one-elevator-out-of-service consideration. This will be specifically discussed in various examples.

INCOMING TRAFFIC PERIOD INTERVAL

Calculations of elevator requirements for incoming traffic yield an average loading interval or average time between elevators leaving the lobby floor. This is the average time a person who just misses an elevator has to wait before the next one leaves. As can be seen from the calculations, the interval includes the loading time of the elevator.

For efficient loading, a carload of people should arrive and enter the elevator during the time allowed for loading. Visualize the dynamic situation: people are arriving at random and waiting for an elevator. When an elevator arrives, a car will fill and additional people arrive so that, on the average, a carload of people are waiting and their waiting time approximates the interval between car departures.

TABLE 4.4. Area Versus Capacity—Suggested Standards

1. Inch/Pound Units

Capacity (lb)	Platform (in.)		Car Inside (in.)		Area (ft²)	A17.1 Area Allowed (ft²)	Observed Loading (people)[a]	Maximum Loading (people)[b]
	Wide	Deep	Wide	Deep				
2000	72	60	68	51	24.0	24.2	8	16
2500	36	60	82	51	29.0	29.1	10	19
3000	36	66	82	57	32.5	33.7	12	22
3500	36	75	82	66	37.5	38.0	16	26
3500 (alt.)	96	56	92	57	36.4	38.0	16	26
4000	96	82	82	73	41.6	42.2	19	28
4000 (alt.)	96	75	92	66	42.2	42.2	19	28
4500	96	82	92	73	46.6	46.2	21	33
5000	96	86	92	77	48.2	50.0	23	35
6000	96	99	92	90	57.5	57.7	27	41

2. Metric Units

Capacity, kg (lb)	Platform (mm)		Car Inside (mm)		Area (m²)	Code Area Allowed (m²)	Observed Loading (people)[a]	Maximum Loading (people)[b]
	Wide	Deep	Wide	Deep				
1200 (2640)	2200	1550	2100	1300	2.7	2.8	10	19 (16)
1400 (3080)	2200	1700	2100	1450	3.05	3.24	12	22 (18)
1600 (3520)	2200	1900	2100	1650	3.5	3.56	16	26 (21)
1600 (alt.)	2450	1700	2350	1450	3.4	3.56	16	26 (21)
1800 (3960)	2200	2050	2100	1800	3.8	3.88	18 to 19	29 (24)
1800 (alt.)	2450	1900	2350	1650	3.9	3.88	18 to 19	28 (24)
2000 (4400)	2450	2050	2350	1800	4.2	4.2	20	32 (27)
2250 (4950)	2450	2200	2350	1950	4.6	4.6	22	34 (30)
2700 (5940)	2450	2500	2350	2150	5.1	5.32	25	39 (36)

[a] Based on 2.3 ft² (0.22 m²) per person for 4000 lb and above.
[b] Based on 1.5 ft² (0.14 m²) per person.
Number in parentheses in the metric maximum loading column are established in European Code EN81.

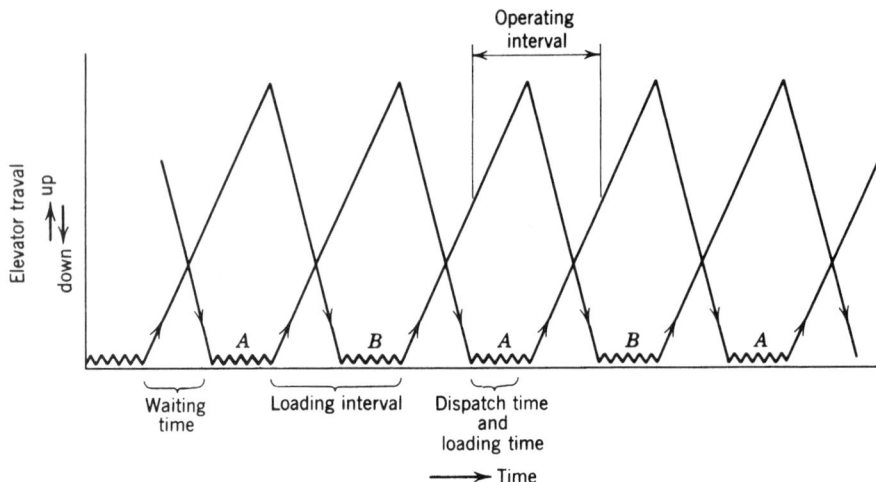

Figure 4.2. Loading and operating interval.

Incoming traffic interval differs somewhat from operating interval as described earlier in the discussion on a two-stop installation. With incoming traffic, loading interval represents the average wait for service at the lobby; operating interval is related to the wait for service at an upper floor. Loading interval includes lobby waiting time, and loading and operating interval are approximately equal. Operating interval, theoretically, is twice waiting time at an upper floor but in practice is about 60 to 70% of the interval. This distinction is shown graphically in Figure 4.2.

Figure 4.2. shows a two-car operation. With four cars, all other time factors remaining equal, loading time and loading interval are almost equal. The next car arrives at the lobby as the last car leaves and a person's wait for service is spent in the car. With eight cars and all other time factors remaining equal, two cars must be loaded simultaneously to avoid wasting carrying capacity.

INCOMING TRAFFIC OPERATION OF ELEVATORS

To provide a sufficient quantity and quality of elevator service during incoming traffic periods, elevator operation must conform to that traffic. For proper operation elevators must be controlled as follows:

1. Each car must depart from the lobby as soon as it is filled to an optimum percentage of capacity.
2. Loading should start as soon as the doors are opened and the first call is registered in the car, assuming the first person who enters operates a car button.
3. The elevator should travel no higher than required by calls registered in the car or at the highest landing.
4. The elevator should return to the lobby promptly to the extent that some landing calls may be bypassed if there is no elevator at the lobby.
5. Passengers should be encouraged to load as promptly as possible.

6. If the interval between elevators is less than the average loading time or if random distribution causes more than one car to be at the lobby at one time, more than one car should be loaded at one time.
7. Cars should not be held for loading beyond the time allowed for dispatching.

These requirements must be translated into automatic operations by the elevator operating system. Modern microprocessor operating systems do this with the necessary speed and accuracy. Automatic operations should include the features discussed in the following paragraphs.

In office buildings especially and in other buildings as well, the incoming traffic peak usually begins after a long period of quiet. If the operating system has shut down the elevators, that is, parked them with doors closed, they must be activated as traffic occurs in the building. Because traffic is now incoming, someone entering the building will operate a call button at the lobby floor to summon a car or activate a car parked there with the doors closed.

When the elevator responds, its doors open and one or more passengers operate call buttons in the car for their floors. If the car does not fill, these passengers should not be held at the lobby more than 10 to 15 sec. If the car fills quickly, as indicated by a load-weighing device on the elevator, the car should depart immediately and another elevator become activated since quick loading of a car signifies continuing incoming traffic.

An extremely important aspect of incoming traffic operation of elevators is the use of a switch or device that can measure the loading of an elevator car. Present practice is to measure the deflection of the elevator platform and trip a switch or sensor when a certain load in pounds (or kilograms) is attained. At this setting, the elevator is practically filled and should leave since additional people would disproportionally delay elevator service.

The load sensor should also initiate actions to return additional elevators to the lobby floor since a filled elevator is an indication that possibly more people are waiting. Continued load dispatch should also maintain a programmed up-peak operation of the elevators since such continuation is a good sign that this is major traffic, especially when compared with a minimum number of down landing calls.

Development work is required to create load weighing or sensing devices that can determine whether the elevator is filled on the basis of volume. Present designs are simply based on the assumption that the average person weighs about 150 to 175 lb (68 to 80 kg), and do not consider baggage carried or mail carts.

Once the car has started it should serve its car calls and return to the lobby immediately after serving the highest car or down landing call. A continuing flow of traffic into the lobby will be indicated by filled cars or a multiplicity of car calls in cars that are at the lobby, and by no appreciable landing call activity.

Cars should be controlled so that they depart from the lobby after a period of time equal to the expected loading time or depart as soon as they are filled to an adjustable percentage of load.

A lighted sign in the lobby should indicate to passengers the next car to depart. Even if it has not yet arrived at the lobby, its sign should be lighted so people gather at the entrance, ready to load promptly when the car does arrive. If more than one car is expected to load at one time, all cars that return to the lobby should have their doors open to encourage simultaneous loading, but only one car should have a lighted sign.

Under some circumstances, especially when car loading is relatively light, that is, with four to eight people per car, it may be desirable to close the doors of an elevator that is not selected for dispatching. This will concentrate the loading in one elevator and improve efficiency. If this is done, and a person enters the nonselected car before the doors close, the doors should remain open if a car call is operated or reopen if a car call is operated after the doors close. The car should then become the next to be selected and dispatch timing should start when selection takes place. If the car becomes filled as measured by the load sensor, whether selected or not, it should depart.

If there are no cars at the lobby and previous cars have left the lobby filled, a heavy incoming rush is indicated. During this period the down traveling elevators should automatically bypass down landing calls until at least one car has returned to the lobby for incoming passengers. Elevators can bypass landing calls wherever lobby traffic requires this priority, or bypassing can be restricted to certain periods of the day. At lunchtime, when both down traffic and incoming traffic become heavy, elevator service should be shared equally and the lobby should not have priority.

Restricting operations and giving priority to incoming traffic during certain periods of the day can be accomplished by a time clock. Such clocks can be set to recognize days of the week and restrict operations to certain times on weekdays only. The clock can have the elevators in position for a known recurring incoming rush and also sustain this operation during momentary lulls.

Once rapid loading and a multiplicity of car calls at the lobby ceases for a time or upper-floor landing calls become numerous, the incoming rush has subsided and the mode of elevator operation should change. Depending on the intensity of this subsequent two-way or interfloor traffic, elevator operation should be changed, as is discussed in the following chapter.

Handling Incoming Traffic Problems

If service demand in a building has changed or if it had an inadequate number of elevators from the start, certain measures may improve the handling capacity of its elevators during the incoming traffic period.

The first step in such an improvement is to determine whether any factors are interfering with the handling of incoming traffic. Such factors may include an upper-floor eating facility to which building occupants travel by elevator to have breakfast or coffee after checking in at their floors. Basements that are secondary entrances or contain eating facilities can also detract from proper handling of incoming traffic. If management solutions, for example, restricting cafeteria hours until after rush hour or eliminating secondary entrances, do not help, other means such as staggering of working hours may have to be employed.

Another measure is the restriction of landing call service during certain periods. Only one or two elevators of the group would be allowed to pick up landing calls during a clock-controlled period, thus concentrating most of the elevator service at the lobby entrance floor.

Up-peak Zoning

A major and drastic step, recommended only as a last resort, is up-peak zoning. This is accomplished by reducing the number of stops each elevator makes. If enough elevators are in a group, they may be zoned. Some of the elevators may be designated and operated to serve the lower floors, and the others the upper floors. Separate means of operation must be provided for each subgroup and stopping for landing calls should be restricted. Special

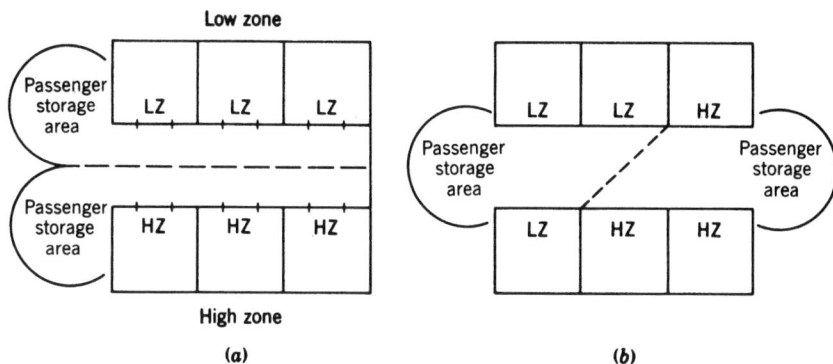

Figure 4.3. Lobby zoning arrangement for incoming traffic: (*a*) poor arrangement; (*b*) acceptable arrangement.

signs should be provided at the lobbies, and the car buttons of the floors not served must be deactivated. Zoning should be restricted to specified periods and for incoming passengers at the lobby only, with the elevators arranged for ready access from the passenger waiting space. Figure 4.3*b* shows a preferable elevator arrangement for zoning.

Zoning increases handling capacity at the expense of interval. For example, one six-car group would be replaced by two three-car groups. From the calculation in Example 4.2, the handling capacity of a typical group of elevators is increased by 41% at the expense of a greatly increased (46%) loading interval. Zoning should never be used as the basis of establishing the number of elevators in a building.

Example 4.2. Effect of Incoming Traffic Zoning

Original: six 3500-lb elevators @ 500 fpm, serving floors 1 to 12, rise 132 ft (12-ft floors)
Probable stops 8.6
Time to run up, per stop

$$\frac{132}{8.6} = 15.3\text{-ft rise per stop}$$

(from Chart 4.2) 15.3 ft = 4.7 sec

Time to run down,

$$\frac{(132 - 15.3) \times 60}{500} + 4.7 = 18.7 \text{ sec}$$

Elevator performance calculations:
Standing time

Lobby time	=	14 sec
Transfer time, up stops 8.6×3	=	25.8
Door time, up stops $(8.6 + 1) \times 5.3$	=	50.9
Total standing time		90.7 sec
Inefficiency, 10%		9.1
	Total	99.8 sec

(Total from page 89, 99.8 sec)

Running time
    Run up 8.6 × 4.7   =   40.4
    Run down   =   18.7
    Total round-trip time   158.9 sec
Interval: 159/6   =   26   sec
HC: 6 cars = 6 × (16 × 300)/159   = 181 people

With zoning:
Low group: Three 3500-lb elevators @ 500 fpm, serving floors 1 to 7
Probable stops 5.7
Time to run up, per stop

$$\frac{6 \text{ floors} \times 12}{5.7} = \frac{72}{5.7} = 12.6\text{-ft rise per stop}$$

(from Chart 4.2) 12.6 ft = 4.5 sec

Time to run down

$$\frac{(72 - 12.6) \times 60}{500} + 4.5 = 11.6 \text{ sec}$$

Elevator performance calculations:
Standing time
    Lobby time   =   14   sec
    Transfer time, up stops 5.7 × 3   =   17.1
    Door time, up stops (5.7 + 1) × 5.3   =   35.5
    Total standing time   66.6 sec
Inefficiency, 10%   6.7
                                         Total 73.3 sec

Running time
    Run up (5.7 × 4.5)   =   25.7
    Run down   =   11.6
    Total round-trip time   110.6 sec
Interval: 111/3   =   37   sec

HC: 3 cars = 3 × (16 × 300)/111   = 130 people

High group: Three 3500-lb elevators @ 500 fpm, serving floors 1, 8 to 12
Probable stops 4.8
Time to run up, per stop

$$\frac{4 \text{ floors} \times 12}{(4.8 - 1)^*} = \frac{48}{3.8} = 12.6\text{-ft rise per stop}$$

(from Chart 4.2) 12.6 ft = 4.5 sec

---

* Note: With the express run, i.e., traveling nonstop from the lobby to the eighth floor, the travel time to the first probable stop in the up direction is accounted for by the time to travel the express portion of the trip. To compensate, the number of probable up stops is reduced by one as shown.

Up express run

$$\frac{(84 - 12.6) \times 60}{500} + 4.5 = 13 \text{ sec}$$

Time to run down

$$\frac{(132 - 12.6) \times 60}{500} + 4.5 = 18.8 \text{ sec}$$

Elevator performance calculations:
Standing time

Lobby time	= 14 sec
Transfer time, up stops 4.8×3	= 14.4
Door time, up stops $(4.8 + 1)\times 5.3$	= 30.7
Total standing time	59.1 sec
Inefficiency, 10%	= (5.9)
	Total 65 sec

Running time

Express run up	= 13.0 sec
Run up (local 3.8×4.5)	= 17.1
Run down	= 18.8
Total round-trip time	113.9 sec
Interval: 114/3	= 38 sec

HC: 3 cars $= 3 \times (16 \times 300)/114$	= 126 people
Group capacity:	= 256 people
Gain by zoning: 75 people or 41%	
Interval increase: 46% longer	

Incoming traffic zoning has other limitations. Once a group of elevators is arranged for zoning, people must be conditioned as to what to expect. Once zoning is initiated it must be maintained for a period of time, since people do not pay close attention to signs and become confused if zoned and unzoned operations alternate spasmodically. One sub-group, even though it may momentarily have idle elevator capacity, cannot temporarily aid another subgroup. An alert lobby attendant can only temporarily control initiation or discontinuance of zoning where it is required.

An additional negative aspect of up-peak zoning is that interfloor traffic, that is, traffic between high- and low-zone floors, requires two elevator trips whereas only one would be required without zoning. A person wishing to go to a low-zone floor from a high-zone floor, or vice versa, must travel to the lobby. If the high-zone cars are allowed to stop at the low-zone floors, or vice versa, when car calls are registered, the purpose of zoning would be defeated and only restricted landing call service would result. A transfer floor, say at the top stop of the low zone and the lowest stop of the high zone, would have doubtful value since an extra stop would probably be created on both groups of elevators, therefore detracting from the expected gain.

Spotting

A step beyond zoning is an operation known as "spotting," wherein each elevator in a group is designated to serve a limited number of floors. Prospective passengers are made to line up in front of individual elevators, and the elevator signage indicates which floors a particular elevator serves. There are severe limitations to this operation; one is the need for lobby space to accommodate the separated queuing, and another is the need to limit any interfloor traffic during this operation. Persons on upper floors wishing to go down must travel to the lobby and join the queue for another destination.

This type of operation has been popular for schools and larger single-purpose office buildings and was limited to given time periods each day. The advantage was that it increased the handling capacity of the system, yet at the major expense of passenger waiting time. However, it was a means to handle the expected overwhelming traffic in, for example, a school where classes started at given hours. Although some installations still employ this approach, it has gone out of favor and remains only as an expedient if all else fails to solve the problem.

A modern version, designated as "channeling," or "predestination" has been introduced by a number of companies. In this approach, and with the aid of microprocessors, the popularity of the various floors over a period of time is determined, and elevators are designated to respond to this popularity by means of an illuminated sign over each entrance in the main lobby. One car may be designated for one trip to serve a group of floors, and passengers are conditioned to look for and observe those signs. Those who fail to do so are penalized in that they are unable to access their floors on that car and must return to the lobby or get off at an upper floor and wait for a car that will respond to their call and is free to make all stops.

Chapter 7 discusses this operation as well as other approaches to improve handling capacity. Doing so without a severe time penalty, that is, maintaining the quality of service, is the goal.

High- and Low-rise Elevators

It is a consequence of zoning that the concept of high- and low-rise elevators came into being, but with the criteria for both interval and handling capacity being met. Instead of having all the elevators in a building serve all the floors, the elevators were divided into permanent groups, one group serving the lower floors of the building, a second group the next higher floors, and so on. The grouping of elevators is a characteristic of almost every major building of any height built today.

Each group of elevators has its own operating system and is calculated in a manner similar to the up-peak calculations. The essential difference is that sufficient elevators must be provided in each group to maintain a good operating interval. To serve the higher floors in the building, the time required to transverse the express run of the hoistway is reduced by higher-speed elevators. In 40- and 50- story buildings the highest-rise groups of elevators (those serving the uppermost floors) may have speeds of 1000 fpm (5.0 mps) or 1200 fpm (6.0 mps).

Utmost efficiency in providing elevators may be gained by establishing various-sized groups of elevators in a given building. The performance of each should be equalized by adjusting the number of elevators and floors served so that the interval and handling capacity is essentially equal among all groups.

Multiple-Entrance Floors

A building with multiple-entrance of lobby floors requires extra elevator capacity. If ample capacity is not available, limiting incoming traffic to one entrance floor can improve elevator service.

Reviewing the calculations for incoming traffic handling capacity, it can be noted that only one time allowance was included for the lobby or loading floor. If passengers are expected to board elevators at more than one floor, this time allowance must be increased in proportion to the number of passengers expected to enter at each floor.

For example, if a building has two entrance floors and half the incoming passengers are expected to arrive at each, two lobby stops are necessary. The time to make a round trip will automatically be increased by at least 15 sec in order to operate to the lower level. To ensure full use of elevator capacity, each elevator must stop at both loading levels. The loading time spent at each loading level will be equal and will represent the time needed to load about one-half the capacity of the elevator. The trip back up will require additional time, so a minimum of 15 sec extra is required to serve the second terminal.

If elevators are filled at the lower level and bypass people on the upper level, the latter will soon discover that they must go to the lower level for reasonable service. Escalators or shuttle elevators should then be provided. Preferably, the escalators or shuttle elevators should carry all passengers to the upper level, thus eliminating an extra elevator stop, reducing the distance traveled, and increasing the efficiency of the entire system. If stairs or escalators are used between building entrance levels, a shuttle elevator must be used to provide the necessary service for disabled persons.

Garage Floors

Similar shuttle elevator arrangements are helpful for multilevel garages above or below the main entrance level. To provide good service for passengers to and from the garage levels, all elevators must stop at those levels. Each stop adds 15 to 20 sec to the normal incoming traffic round-trip time, but a shuttle elevator system would avoid this loss in elevator efficiency. Passengers could then enjoy the convenience of parking in the building and the benefit of improved elevator service to their floors. Building security is enhanced since nontenant parkers do not need to use the tenant elevators.

Parking floors above the lobby level but below the office floors or other usable space create more serious problems. If parking is used by tenants in the building, each parking level represents a potential loading level in calculating elevator round-trip time and handling capacity. A time factor proportional to the number of stops expected at parking levels must be added. If parking is open to both tenants and transients, still another time factor must be included proportional to the number of people expected to leave the building from the parking levels.

For example, if floors 2 through 6 are parking floors and about 50% of the spaces are to be used by transients, and remaining parking patrons represent 50% of the tenant population, a number of additional stops may be expected. If each elevator is expected to carry 12 people on a trip, 6 people who work in the building are expected to enter the elevator at the ground floor and 6 parkers who work in the building are expected to enter on floors 2 through 6. The latter 6 people will create about three additional stops during the up elevator trip. In addition, as the elevator returns to the lobby floor it must stop at least two or three more times to pick up the 6 transient parkers leaving the building. These people must be unloaded before the elevator can begin to serve incoming passengers again. In all, the five garage levels add about five additional stops (about 50 sec) to the round trip of an elevator, plus extra unloading time at the lobby.

The best solution is shuttle elevators between the lobby and parking levels, with main elevators running directly from the lobby to the office floors. This improves tenant security as well as elevator service. People cannot enter the building or leave without passing through the main lobby. The parking area can be used in the evening whereby access to the upper floors of the building is controlled and use of its main elevators is minimized.

If the parking areas must be served, the controls for all elevators should be designed to minimize the impact of the extra stops. The number of stops any elevator makes at the garage levels should be limited to only a proportional measure of service. Limiting the number of elevators that will answer down calls at garage levels is a means of limiting elevator service in that area. How extensive such limitations are depends on the particular building.

DOWN TRAFFIC DURING UP-PEAK

Because of increased horizontal transportation congestion and as a convenience to employees, many organizations are employing "flextime," or staggering starting and quitting times. Flextime is a concept whereby employees can establish their own starting and quitting times, within limits, provided that they put in the prescribed hours per day and all personnel are at work between fixed times. This may naturally happen in multitenanted buildings where employees soon learn to change their starting and quitting times to avoid the rush on the elevators.

This practice imposes a critical aspect on the operation of the elevators during the incoming traffic peak, since, while people are coming in, others are about their early chores and traveling either down or between floors on the elevators. No longer can the landing calls be temporarily ignored to solve a traffic problem, nor can use of up-peak zoning be considered to reduce lobby congestion. The building must accommodate the opposing traffic and make provisions to serve it.

When traffic calculations are made to determine the number of elevators for a particular building, a factor for down stops during up-peak traffic periods should be built in. Observations have shown that this amounts to about 10% of the peak up traffic. For example, if an elevator carries 16 people on each up trip, an average of 1.6 persons will probably ride on its return trip to the lobby. Such traffic will create two to three additional elevator stops and a subsequent increase in round-trip time. This situation often has an effect on the total number of elevators needed for the building.

Redoing Example 4.1 with the 10% down traffic consideration will change the results of the calculation as follows.

Inspecting the results of Example 4.1, it can be noted that the five elevators each with an 11-passenger capacity were equal to the required handling capacity. By adding down stops, the handling capacity will be reduced; hence, the example will be redone as Example 4.3, including elevators each with a 14-passenger capacity.

Example 4.3. Incoming Traffic with Down Traffic

Given: five elevators at 500 fpm, 14 passengers up per trip, 1.4 passengers down per trip, 9 probable stops up, assume 1.4 probable stops down

Time to run up, per stop

$$\frac{176}{9.} = 19.5\text{-ft rise per stop}$$

(from Chart 4.2) 19.5 ft = 5.2 sec

Probable stops down 1.4
Time to run down

$$\frac{176}{(1 + 1.4)^*} = \frac{176}{2.4} \qquad = 73.3 \text{ ft}$$

$$\frac{(73.3 - 19.5) \times 60}{500} + 5.2 = 11.7 \text{ sec}$$

Elevator performance calculations:
Standing time

Lobby time 13 + (1.4 × 0.8)	= 14.1 sec
Transfer time, up stops (9 × 3) = 27	= 27.0
Door time, up stops (9 + 1) × 5.3	= 53.0
Transfer time, down stops 1.4 stops × (4 + 1)	= 7.0
Door time, down stops 1.4 × 5.3	= 7.4
Total standing time	108.5 sec
Inefficiency, 10%	10.8
	Total 119.3 sec

Running time

Run up 9 × 5.2	= 46.8
Run down 2.4 × 11.7	= 28.1
Total round-trip time	194.2 sec

$$HC = \frac{(14 + 1.4) \times 300}{194} = 24 \text{ people per elevator}$$

Elevators required: (110 + 11)/24 = 5.0
Interval: 194/5 = 38.8 sec versus a 35-sec criterion

The five elevators will just meet the handling capacity requirements. The interval is deficient, but close enough to the 35-sec maximum criterion when considered together with the ample handling capacity. The final elevator design should include elevators ot at least 19-passenger capacity so that sufficient handling capacity would be available if one car is out of service. The final recommendation then becomes five 4000-lb (1800-kg) elevators at 500 fpm (2.5 mps).

---

* Note: The uppermost "up" stop is also the highest "down" stop. To compensate, one (1) is added to the probable down stops to obtain a correct down time-distance calculation.

GENERAL CONCLUSIONS

Incoming traffic problems are usually those of handling capacity and often result in increasing time delays. The foregoing sections gave examples of means to reduce time delay and increase handling of incoming passengers.

But maximizing incoming traffic handling by zoning or other means that reduce the number of elevators may create a critical traffic situation at some other time of the day. Calculations of incoming traffic must be compared with other types of traffic in reaching final elevatoring decisions. Down traffic during peak incoming traffic periods is especially prevalent and must be a major elevatoring consideration.

This chapter has introduced elevator time-study calculations. Comparing time factors in a calculation with those observed in actual installations aids in a personal evaluation of elevatoring problems. Means can be discovered to speed up a trip: better door-opening operation, faster floor-to-floor time, means to get doors closed and a car started sooner, or eliminating unnecessary stops.

Each opportunity should be utilized to investigate various phases of the elevator trip as well as the loading and unloading of people from the elevator car. The result will be improved pedestrian transportation.

Analysis was begun with incoming traffic because it is not complex and its time factors can be calculated with relative ease. The following chapters will discuss two-way, including interfloor, and outgoing traffic, rounding out the treatment of elevatoring calculation methods presented in this chapter.

5 Two-Way Traffic

IMPORTANCE OF TWO-WAY TRAFFIC CONSIDERATIONS

Once a building is occupied, its residential or working population will want to enter or leave the building at various times during the day. Some people may be entering at the same time others are leaving, as at lunchtime (two-way lobby traffic), or people will be going from one floor to another (two-way interfloor traffic), or there will be a combination of these traffic patterns.

In many buildings the peak elevator traffic period is two-way traffic. In a hotel guests are often checking out or going to activities while other groups are checking in or returning from meetings.

In buildings other than hotels the peak traffic may often be between floors. In some hospitals, for example, traffic can be heaviest when patients are being transferred for treatment, doctors are making visits, volunteers are tending to patients' needs, and other personnel are moving from floor to floor. In other hospitals, peak traffic may occur when one shift is leaving and another is taking its place, together with visitor traffic to and from the lobby.

Traffic situations like these require elevators to make stops in both the up and down directions during the same round trip.

Two-way lobby traffic may be relatively simple; people enter an elevator at the lobby floor and leave at various stops during the up trip. On its down trip the elevator picks up passengers at various floors and lets them out at the lobby floor. Compared with purely incoming traffic, the round trip will now be considerably longer but the elevator will serve many more people, the up passengers plus the down passengers on each round trip.

The two-way interfloor traffic trip is more complex. During an up trip, for example, an elevator makes one stop to pick up a passenger and another stop to let the passenger off. Each landing stop for a waiting passenger on a typical floor usually requires two stops of the elevator. The round-trip time is therefore long in relation to the number of passengers served.

During extremely heavy interfloor traffic, as in schools during class change periods and in department stores during rush buying seasons, it is conceivable that each elevator will make every stop up and every stop down!

Calculations for two-way lobby and two-way interfloor traffic follow the course outlined for incoming traffic. During these periods elevators should be operated to equalize waiting time for all elevator users, both on the upper floors and at the lobby.

Fortunately, during periods of two-way traffic people will generally tolerate longer waiting times than during the up-peak incoming traffic periods. This subject is discussed in detail in a subsequent section, "Interval and Waiting Time."

The Vertical Transportation Handbook, Third Edition, Edited by George R. Strakosch
ISBN 0-471-16291-4   © 1998 John Wiley & Sons, Inc.

TWO-WAY AND INTERFLOOR TRAFFIC REQUIREMENTS

To calculate how many elevators are required to serve a two-way or interfloor traffic situation, we begin by estimating passenger demand. Intensity of demand is a function of the building population, whereas its complexity depends on the distribution of that population among the floors of the building.

Determining population should be an easy matter. The building has been designed to accommodate so many office personnel, so many hospital beds and related staff, so many students, or so many sleeping rooms. These factors determine the population.

Next, we need to know how many of these people will require elevator service during the critical period.

As with incoming elevator traffic, we are concerned with a critical 5-min period. A 5-min period is convenient to use; it has been evaluated in all types of buildings and, if longer periods of traffic demand persist, 5 min can indicate average intensity over the longer period. In schools the entire peak may occur during a 5-min period since this is all the time the students may be allowed to change classes. Should 10 min be allowed, elevator traffic will peak in the middle 5 min of the longer period.

For each building type discussed later in this book, a qualifying percentage of population is given for the critical traffic period. The percentage, representing expected elevator passenger demand during a critical 5-min period, may vary from a low of about 6% of the resident population for apartments to a high of about 40% for classroom buildings.

Characteristics of two-way or interfloor traffic depend not only on the building types but also on the relative attraction of each floor in a building. In a hotel, for example, the location of meeting rooms or ballroom floors and their relation to the lobby or guest rooms have a direct influence on the number of stops an elevator makes. In a school, hospital, or office building the cafeteria location may greatly influence the stopping on each elevator trip. This chapter gives guidance of a general nature; more specific considerations of various building types appear in later chapters.

Interfloor traffic can be a problem in an office building. If a single tenant or organization occupies many floors, traffic between various divisions of such an organization may adversely affect elevator service.

Under certain conditions, a separate interdepartmental special service elevator may prevent overburdening of the main building elevators. For example, mail distribution in a large organization may require almost continuous use of an elevator or cart lift. Mail-handling conveyors may also be used to reduce elevator use. Chapter 15 gives details of various systems.

Changing habits are making two-way traffic more critical than the incoming peak in office buildings in many areas. When people drive to the office their arrival is influenced by the accessibility of parking rather than by the time they are due at work. Arrivals may be spread over a longer period, which reduces up-peak elevator traffic. During the lunch period half the building population may be leaving while the other half is returning. The percentage of building population the elevators must serve during a peak 5 min of the luncheon period may be far greater than during a 5-min incoming traffic period.

Factors like these must be considered in calculating elevator requirements for two-way and interfloor traffic. The consequence of insufficient elevators for this period is impaired service for passengers. They must wait too long for an elevator and, once aboard, may face an unduly long trip marked by excessive stops for entering and leaving passengers.

Interval and Waiting Time

Capacity is seldom the problem; waiting time and riding time are the more critical aspects, so elevators must necessarily pass the floors in a building with sufficient frequency in order to provide prompt service to all passengers.

An earlier European device, the paternoster, has a series of continuously moving up and down platforms in a dual hoistway (Figure 5.1). Passengers must leap onto a moving platform and off at their floor of destination. If they pass their floor they can stay on and make the full cycle up and over or down and under, or get off at the next floor and take the platform in the opposite direction. Although the paternoster is a relatively slow-speed

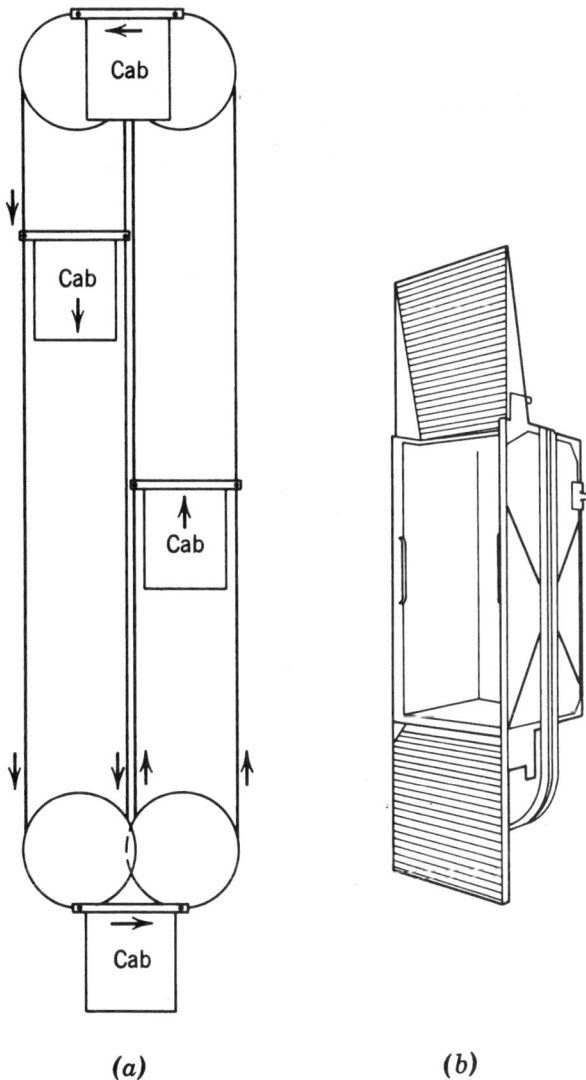

(a)   (b)

Figure 5.1. "Paternoster elevator" (Courtesy *Elevator World Magazine*). (*a*) Schematic; (*b*) paternoster car.

Figure 5.1. *(Continued)* (*c*) Actual installation.

device [about 60 fpm (0.3 mps)], its use demands agility and its installation is no longer allowed; however, some are still in operation in Europe.

With conventional elevators passengers can be transported only if the cars are intercepted, slowed, and stopped. A passenger's wait for service becomes a function of the elevator operating interval. To provide service with acceptable average waits, the operating interval must be established based on the maximum wait that passengers will tolerate.

In hurried commercial environments this waiting tolerance seldom exceeds about 40 to 50 sec. In a more relaxed residential environment, a wait of about 60 to 75 sec is tolerated before tempers rise and people complain. Interval is related to average and maximum waiting time in the following manner. If we establish a desired maximum waiting time of 40 sec, the average wait will be one-half or 20 sec. Waiting time is about 60% of interval, as discussed in Chapter 2; therefore, the interval should be 20 ÷ 0.60 = 33 sec. The operating or two-way interval in a diversified office building should therefore be about 35 to 40 sec and about 65 to 70 sec in a residential building. It also must be recognized that some waits will exceed the desired maximum; this aspect will be discussed later in this chapter.

Calculations for Two-Way Traffic

Calculating the round-trip time of elevators serving two-way lobby and two-way interfloor traffic can follow the format used for incoming traffic. The time for each stop is established and multiplied by the number of expected stops, to which is added lobby time and running time. The total is an approximate round-trip time.

To establish the number of expected stops, a passenger load per trip in each direction must be determined. The total of up passengers plus down passengers times 300 sec (for 5 min) divided by the round-trip time per trip gives an elevator's 5-min capacity:

$$\text{5-min handling capacity} = \frac{\text{(up plus down traveling passengers)} \times 300 \text{ sec}}{\text{round-trip time in seconds}}$$

Operating interval or average time between elevators passing a given upper floor in a building in either direction is found by dividing the round-trip time of a single elevator by the number of elevators in the group.

$$\text{Interval} = \frac{\text{average round-trip time per elevator}}{\text{number of elevators in group}}$$

Probable stops. The number of stops an elevator makes in any trip is a function of the number of passengers it carries, the number of floors the elevator serves, the relative attraction of each floor, and the relationship of each floor to the others in the building. Establishing probable stops for a building requires a great deal of evaluation and judgment.

Prevailing occupancies and observations of people in known buildings can help in this evaluation and give an indication of the number of stops elevators will make during two-way trips. For example, people usually go to lunch at noon or to a coffee break in mid-morning. In specialized buildings, such as hospitals or hotels, activities take place that can be measured, evaluated, and projected to the next hospital or hotel to be built.

In any building certain floors will be expected stops for each elevator trip. The lobby is one such floor; the cafeteria floor is another.

Entrances to the building from parking areas on several levels will cause elevator stopping relative to the attraction of each such floor. During two-way traffic, the probability of passengers going to or from the top floor of a building is proportional to the relative population attraction on that floor.

As a beginning, we have at least two certain stops in each two-way or interfloor elevator trip—the lobby and some upper floor in the building. Stopping at other floors is a function of the number of people using the car during each trip. For two-way lobby traffic, because we are concerned with people boarding the car at the entrance floor or floors, the number of up stops is approximately the same as for incoming traffic. Provisions for highest call reversal should be made based on the relative attraction of the upper floors.

On the down trip, during periods when there is approximately equal traffic up and down, the number of probable stops is usually less than the number made during the up trip. This is especially true in office buildings, where people usually go to lunch with companions or there is a general accumulation of people at various upper floors waiting to go down. The percentage of stops in the down direction as compared with up stops may be arbitrarily taken as 75%, based on actual values from 70 to 80%.

The probable stop values for two-way traffic then become:

Two-way probable up stops = probable up stops, from Chart 4.1

Two-way probable down stops = 0.75(probable up stops)

When many floors are to be served, the likelihood of each passenger's causing a stop is increased. If, for example, an elevator is expected to carry 10 passengers in a trip and can make, say, 30 stops, it is almost certain to make 10 stops. Similarly, if an elevator is ex-pected to carry 10 passengers in each direction and can make only 5 stops, it is almost certain to make every stop up and down. These later considerations will prove especially valuable in approximately two-way trips in apartment buildings or any situation in which minimum car loading is expected.

Elevator capacity must be ample for two-way lobby and two-way interfloor traffic. We can assume that the elevator car is filled equally in both directions, although this is sel-dom the case in actual situations. Essentially, the platform must be large enough to serve the required number of people in each direction in each trip. As a rule, the car should not be filled to more than about one-half to two-thirds of its observed capacity, as shown in Table 4.4 in two-way traffic calculations. With full loads the trip time usually becomes much too long, the transfer time at each stop is unduly lengthened, and no reserve capac-ity remains for momentary traffic surges or imbalance between up and down traffic.

Chart 4.1, developed for up-peak probable stops, can also be used for two-way proba-ble stops. Based on the given number of floors in a building and a certain number of pas-sengers per trip either up or down, the probable stop value is chosen. Interpolation is used for odd numbers of passengers or odd numbers of floors. It is assumed that each floor has an equal attraction and that the elevators do not invariably travel to the top floor on each trip.

Earlier versions of two-way probable stop tables were predicted for elevators making full trips from the bottom to the top terminal and based on the formula

$$\text{Probable stops} = S - (S - 1)\left[\frac{S-1}{S}\right]^p$$

where S is equal to the number of possible stops above the lobby and p is equal to the number of passengers traveling in a single direction. The formula in Chapter 4 changed the first $S - 1$ term to S, which had the effect of reducing the number of stops and elimi-nating the constant top-floor stop.

For interfloor traffic, determining probable stops under extreme conditions is quite simple. Because each passenger requires two elevator stops, one to board and one to dis-embark, the number of stops will be twice the number of passengers in each direction less one for the highest stop. For example, if two people are expected to travel up and two people down, an elevator will make about seven stops. We say "about" because one or more of those stops may be coincidental—that is, a person will get on where a passenger gets off. This can occur when both the car call and landing call for the same floor in the direction the car is traveling. Stated otherwise, some landing calls are expected to be co-incidental with car calls. A coincidence factor can be established, but because it must ap-ply to only a particular building we will be conservative and assume none.

As may be seen, interfloor traffic trips can be time-consuming and if a large number of people are traveling up and down, the elevators will make almost every stop in each direc-tion. This is particularly true in educational situations during class changes. In offices and hospitals the general average is about every other stop up and down with the elevator filled to about 50% of its nominal capacity.

Transfer Time. Approximately the same time is required for passengers transferring at the lobby or at the upper floors as was discussed for incoming traffic.

At the lobby, the transfer time will consist of both passengers unloading from the down trip and loading for the up trip. The average of 0.8 sec per passenger is valid. At an upper floor, transfer out of the elevator at a stop for which a car call is registered will include the same minimum door-hold open time (dwell-time) of 3 sec. If the stop is for a landing call in the up direction, as would possibly occur during periods of interfloor traffic, the dwell-time for a landing call of 4 sec must be used. Likewise, dwell-time in the down direction must be based on whether a car call or landing call is registered, as well as the number of people expected to transfer at that floor.

During interfloor traffic trips there will be occasions when people are transferring out of the elevator and others are entering. In that event, the transfer time will be the longer landing call transfer time plus the additional time required for the number of passengers in and out. Care must be taken to consider local code requirements for disabled people, which may call for transfer time to be considered from the time the elevator stops and the doors are fully opened plus a certain minimum time, depending on the location of the elevator in relation to the landing call button location. This minimum time is for a walking time of 2 fps, which must be added after the car has stopped and car doors are opened. In some code jurisdictions, this time is measured from the time the landing lantern illuminates and a gong sounds. This advance lantern time may be as long as 4 sec before the elevator doors are fully opened, depending on company design, and will serve to reduce the total time spent at the floor.

As discussed for the incoming traffic situation, time compensation adjustment should also be allowed for narrow or wide doors different from the 48-in. (1200-mm) standard center-opening entrances. Additional time should be allowed for other than standard-shaped platforms. If elevator lobbies are exceedingly wide, still additional transfer time will be required.

This is accomplished by increasing the percentage of inefficiency over the 10% allowed in our calculations for office buildings, as shown in various examples that follow. As an approximate guide, the door time table (Table 4.3) shows certain additional inefficiencies for different widths and types of doors. For narrow and deep platforms the inefficiency should be increased from 5 to 10%, depending on the increase in platform depth. In addition, if the elevator lobby is excessively wide (or narrow with entry restrictions) another 5% inefficiency should be added.

SAMPLE CALCULATIONS

Examples 5.1 and 5.2 show the complete calculations for two-way and interfloor traffic. Both moderate car loadings and full car loadings are shown.

For two-way traffic, handling capacity is simply defined, as all people have obvious destinations. Handling capacity is measured in terms of people traveling to and from the lobby.

In the interfloor traffic examples, handling capacity is the net number of people transferred to and from the lobby and does not account for additional people carried from floor to floor.

It is essential that the standard of elevator demand be established.

If the demand is defined as people carried from floor to floor, elevators effectively carry many more people during interfloor traffic than is shown by only a lobby count. This is best explained by a table:

	Up	Down
Fl. 4	20 ←	→ 20
	20 ←	← 20
Fl. 3	20 →	→ 20
	20 →	→ 20
Fl. 2	20 ←	← 20
	20 pass.	20 pass.
Fl. 1	in →	out →

In and out at first floor per trip,
40 people.
In and out at all floors per trip,
120 people.

Note that the total interfloor trip carried only 40 people to and from the lobby but 120 people from floor to floor. The essential point is to define and measure demand in terms of the problem.

For simplicity, demands for interfloor traffic will be expressed in terms of lobby traffic in and out of the building. If between-floor traffic must be analyzed, percentages of passengers traveling one, two, three, or more floors, as well as the percentage of passengers traveling to and from the lobby, must be ascertained. Generally, if a high interfloor traffic demand is expected, escalators are the best way to serve that demand.

Example 5.1. Two-Way Lobby Traffic

A. Given: two-way traffic peak of 120 people in 5 min, 12-story building, 10-ft floor heights

Assume: four 500-fpm elevators—8 passengers up, 8 passengers down

Probable stops up 5.9

Probable stops down $0.75 \times 5.9 = 4.4$

Time to run up, per stop $\dfrac{11 \times 10}{5.9} = 18.6$-ft rise per stop

Time to run up 18.6 ft. = 5 sec

Time to run down $\dfrac{11 \times 10}{(4.4 + 1)} = \dfrac{110}{5.4} = 20.4$ ft = 5.2 sec

Elevator performance calculations:		Up Transit
Standing time		Time
Lobby time 8 in + 8 out	= 14 sec	8 sec
Transfer time, up stops (5.9×3)	= 17.7	17.7
Door time, up stops ($5.9 + 1) \times 5.3$	= 33.4	33.4
Transfer time, down stops (4.4×4)	= 17.6	—
Door time, down stops (4.4×5.3)	= 23.3	—
Total standing time	106.0 sec	59.1
Inefficiency 10%	10.6	5.9
	Total 116.6 sec	65.0 sec

Totals from page 104 (116.6 sec)   *(65.0 sec)*

Running time
    Run up 5.9 × 5   = 29.5   29.5
    Run down 5.4 × 5.2   = 28.1   ——
    Total round-trip time   174.2 sec   94.5 sec

HC: 4 cars $= 4 \times \dfrac{(8 + 8) \times 300}{174}$

$= 110$ people in 5 min vs. 120 desired

Interval: $174/4 = 43.5$ sec vs. 35 to 40 sec desired

B.   Given: two-way traffic peak of 250 people in 5 min, 12-story building, 10-ft floor heights
    Assume: four 500-fpm elevators, 16 passengers up, 16 passengers down
    Probable stops up 8.6
    Probable stops down $0.75 \times 8.6 = 6.5$

Time to run up, per stop $\dfrac{11 \times 10}{8.6} = 12.8$-ft rise per stop

Time to run up 12.8 ft = 4.5 sec

Time to run down $\dfrac{11 \times 10}{(6.5 + 1)} = \dfrac{110}{7.5} = 14.7$ ft $= 4.7$ sec

Elevator performance calculations: Up Transit
Standing time   Time
    Lobby time 16 in + 16 out $(8 + .08) \times 24$ = 27.2   16.0
    Transfer time, up stops (8.6×3) = 25.8   25.8
    Door time, up stops $(8.6 + 1) \times 5.3$ = 50.9   50.9
    Transfer time, down stops $(4 \times 6.5) = 26 + 9.5*$ = 35.5   ——
    Door time, down stops (6.5×5.3) = 34.5   ——
    Total standing time   173.9 sec   92.7
Inefficiency, 10%   17.4   9.3
                                               Total 191.3 sec   102.0 sec

Running time
    Run up 8.6×4.5 = 38.7   38.7
    Run down 7.5×4.7 = 35.3   ——
    Total round-trip time   265.3 sec   140.7 sec

HC $= \dfrac{(16 + 16) \times 300}{265} = 36$ per elevator

Elevators required $= 250/36 = 6.9$, use 7

Interval: $265/7 = 37.8$ sec vs. 35 to 40 sec desired

In Example 5.1A and B we introduce the concept of up transit time by separating all the time factors involved for the theoretical last passenger to exit on an up trip of the elevator. Up transit time is the third factor in judging good elevator service, as described in Chapter 3. In Example 5.1A, this time is calculated to be 95 sec, which is quite acceptable. In Example 5.1B, the up transit time is 141 sec, which is over 2 min and approaching an arbitrary limit of 2.5 to 3 min (150 to 180 sec), which is about the longest a passenger will tolerate riding an elevator in a single trip without a degree of irritation, depending on the individual's patience.

*Rule: 4 sec for first passenger per stop plus 1 sec per passenger over 1. 16 passengers, 6.5 stops = $(4 \times 6.5) = 26$ for first 6.5 passengers plus 9.5 sec for remaining 9.5 passengers.

If the situation depicted in Example 5.1B were to be faced, judgment may suggest that rather than a single group of seven elevators, two groups of four elevators, low-rise and high-rise, be considered. Other factors, such as incoming traffic, population, building design, and the economics of eight elevators versus seven, as well as the architectural treatment of the core area, would be among the judgment factors affecting the final recommendations.

Example 5.2. Two-Way Interfloor Traffic

A.   Given: two-way traffic requirements 90 people in 5 min measured to and from the lobby, 8-story hospital, 12-ft floor heights, 48 in two-speed doors
Assume: four 400-fpm elevators, 6 passengers up, 6 down
Assume: elevators make every other stop
Probable stops up: assume every other floor, 7 upper floors, use 4 probable stops
Probable stops down: 4 every other floor = 3 probable stops

Time to run up, per stop $\dfrac{(7 \times 12)}{4} = \dfrac{84}{4} = 21$-ft rise per stop

(from Chart 4.2) 21 ft at 400 fpm = 6.7 sec

Time to run down, $\dfrac{84}{(3 + 1)} = \dfrac{84}{4} = 21$ ft = 6.7 sec

Elevator performance calculations:			Up Transit
Standing time			Time
Lobby time (6 in + 6 out)	=	11	6
Transfer time, up stops (4 × 3)	=	12	12
Door time, up stops (4 × 1) × 7.7	=	38.5	38.5 sec
Transfer time, down stops (3 × 4) = 12 + 3	=	15.0	—
(The additional 3 sec is for the top down stop)			
Door time, down stops (3 × 7.7)	=	23.1	—
Total standing time		99.6 sec	56.5 sec
Inefficiency, 10% normal + 2% doors + 5%			
Hospital = 17%	=	16.9	9.6
		Total 116.5 sec	66.1
Running time			
Run up 4 × 6.7	=	26.8	26.8
Run down 4 × 6.7	=	26.8	—
Total round-trip time		170.1 sec	92.9 sec

HC: 4 cars = $4 \times \dfrac{(6 + 6) \times 300}{170}$

= 85 people vs. 90 required

Interval: 170/4 = 42.5 sec vs. 40 sec desired

B.   Given: interfloor traffic requirement 250 persons in 5 min (lobby count), 6-story school, 12-ft floor heights, 54-in. center-opening doors
Assume: 300-fpm elevators, 23 passengers up, 23 down
Probable stops: each floor up and down, 5 up and 4 down

Time to run up, per stop $\dfrac{(5 \times 12)}{5} = \dfrac{60}{5} = 12$-ft rise per stop

(from Chart 4.2) 12 ft at 300 fpm = 5.8 sec

Time to run down, $\dfrac{(5 \times 12)}{(4 + 1)} = \dfrac{60}{5} = 12$ ft = 5.8 sec

Elevator performance calculations: Up Transit
Standing time   Time

Lobby time 23 in + 23 out (8 + 38) × 0.8	= 38.4 sec	19.2 sec
Transfer time, up stops (5× 3) = 15 + 13	= 28	28
Door time, up stops (5 + 1) × 6	= 36	36
Transfer time, down stops (4 × 4) = 16 + 19	= 35	—
Door time, down stops (4 × 6)	= 24	—
Total standing time	161.4 sec	83.2 sec
Inefficiency, 10% normal + 10% school = 20%	32.3	16.6
	Total 193.7 sec	99.8

Running time

Run up 5 × 4.8	= 29	29
Run down 5 × 5.8	= 29	—
Total round-trip time	251.7 sec	128.8 sec

$$HC = \frac{(23 + 23) \times 300}{252} = 55 \text{ people per elevator}$$

Elevators required = 250/55 = 4.5, use 5
Interval: 252/5 = 50.4 sec vs. 50 sec acceptable for school
Five elevators required for acceptable handling capacity and interval, 5000-lb capacity minimum.

ELEVATOR OPERATION DURING TWO-WAY AND INTERFLOOR TRAFFIC

When the number of elevators is calculated for two-way interfloor and lobby two-way traffic situations, a primary objective is to minimize waiting time by providing a short interval of service. Interval depends largely on the effectiveness of elevator operation and is mainly the result of the sophistication of the group supervisory system provided.

Minimizing the time an average passenger would spend on the elevator is of secondary concern. Riding time is a function of elevator layout and arrangement as well as the number of floors served and the size of each car.

With modern automatic elevators all decisions to start or "dispatch" an elevator are made by electronic devices. The simplest will start an elevator whenever a call is registered with little regard for closeness to the car ahead or behind or "elevator spacing." As a result, cars often operate in close proximity, and the group of elevators offers service not much better than one large elevator.

An on-call operation of this type can be effective only when traffic is light and elevators are sufficient so that each car can operate almost independently of the others. Our calculations have indicated that if only a few people are seeking service, even the most elementary operating system will suffice.

Where many people are seeking service at a given time, elevators require more intricate and educated automatic operation to maintain a proper spacing in time and the ability to respond to a myriad of traffic conditions to achieve an optimum mode of operation that is constantly changing. Each prospective passenger will then have an equal opportunity to board the elevators and the individual's average wait will not exceed that of any other passenger.

To begin our discussion of elevator operation, let us first classify elevator traffic into three categories as follows:

1. Light traffic, when the number of passengers seeking service at a given time is no more than two or three times the number of elevators available to give service
2. Moderate traffic, when the number of passengers seeking service at one time will fill the elevators to less than one-half their nominal capacity
3. Heavy traffic, when the number of people seeking service will fill elevators beyond 50% of their nominal capacity

By ensuring the most efficient operation during moderate and heavy traffic situations, the number of elevators required is minimized.

Light Traffic

During light traffic elevators may often park and wait for the next landing call. The cars should be distributed among the floors of the building so that they are within minimum response time of a next landing call. It is also essential that the cars move promptly to serve an existing call and to park at a strategic location in anticipation of subsequent calls.

An arrangement for light traffic operation is shown in Figure 5.2*a* and *b*. Waiting cars are parked in zones (1 and 2) that have been established on the basis of population of the various floors or the expected activity caused by a single tenant on a group of floors. Each car should move to the next call in its zone with minimum delay and be free to travel either up or down from that call as passengers desire.

One car is stationed at the lobby, where the probability of a call is high. If that car leaves, another car should promptly take its place and the former lobby car should park, after completing its car call, in the zone of the car it displaced or to occupy a vacant zone.

Because elevators can give individual service, each car should operate only to the highest or lowest call in the zone it covers, remaining in that zone to answer the next call. If a car is taken from one zone to another, and the car in the latter zone is also moved to another zone, the former car should assume responsibility for the new zone. If for any reason a zone is temporarily unoccupied, an elevator in an adjacent zone should be free to answer calls in the unoccupied zone.

Moderate Traffic

When traffic increases and elevators must be deployed to answer frequent landing calls at various floors, there is little opportunity to park a car in a zone (such as shown in Figure 5.2) and allow it to remain idle. Various strategies must be employed to reduce the overall waiting time between the registration of a landing call and the prompt response of an elevator to that call.

A historical note: Prior to the 1980s, there were many attempts to accommodate two-way and interfloor traffic requirements. Operation of the elevators was limited by the ability of the relay logic and its time delays to develop the necessary prompt response. A number of strategies, such as cruising the elevators throughout the building in anticipation of traffic, were tried. As sophistication increased, the initial application of "zoning," that is, placing cars parked throughout the floors served, was developed and adapted by many suppliers. Zoning is described in the following paragraphs. Although many such installa-

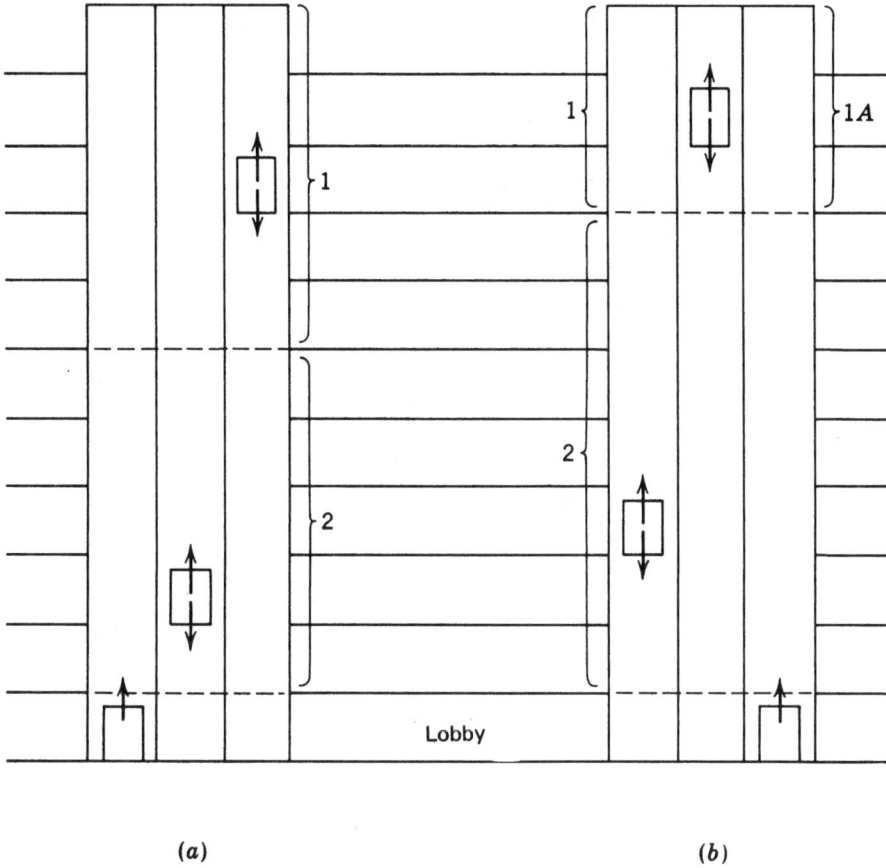

(a)   (b)

Figure 5.2. Distribution of population for effective zoning: (a) population zone 1 = population zone 2; (b) population zone 1 = population zone 2 or zone 1.A is a single tenant.

tions are still in use, they are rapidly being superseded by microprocessor-based systems, which are introduced later in this chapter and discussed in detail in Chapter 7.

The light traffic parking strategy tends to space elevators throughout a building in the best locations to answer the prejudged next landing call for service. When there are one or two calls per zone, an elevator can serve them and quickly be restored to the strategic parking position. When there are a number of calls, the elevators are moving, answering calls, and the tendency increases for calls to become "long wait." In anticipation of this condition, the elevator operating system has to take certain actions to prevent excessively long waits.

One such action is to activate any parked elevator to help out in a zone when either a certain number of calls are waiting or the number of landing calls and their accumulated waiting time reaches a predetermined maximum. Another action is to prevent any excessive travel to the main floor unless a car or landing call is registered for that floor.

There is a tendency for elevator operating systems to park at least one car at the main entrance floor. This practice should be temporarily abandoned until such time as the level of traffic is reduced to a minimum to allow such parking.

Figure 5.3 shows the dynamics of a moderate elevator traffic situation. In general, elevator traffic is not continuous but comes in spurts. During lulls the elevators should be parked in the various zones throughout the building to be in a better position to serve the next traffic requirement. A typical parking arrangement is shown in Figure 5.3*a*. This is in opposition to some elevator operating systems which allow last call parking, which is often at the lobby floor since most elevator trips terminate there. Figure 5.3*b* shows a wave of traffic in two parts. The symbols designated "1" are the initial calls, and those designated "2" the subsequent wave. As can be appreciated, the pattern of calls and elevator movements are constantly changing and we can view only a "snapshot."

The various actions that take place are as follows: Car A, being parked at the floor with a down landing call, immediately answers that call. Car B responds to the car call registered in it for the third floor. When car B enters the zone where car D is parked, car D is immediately released to answer a call in car A's zone or to replace car B at the main lobby. We show it answering the down landing call at the fifth floor. Car C moves to answer the up landing call behind car A and, in doing so, is in a position to answer the subsequent down landing call on floor 10.

Any variety of traffic situations can occur, and as long as the elevators can answer all the calls within moderate time of an arbitrary 30 sec or less, there is little need for sophisticated operating systems. It is when calls start intensifying and waiting times increase that the need for sophisticated operations is apparent. This can occur in the best elevatored building when, for example, one car is out of service for repair or is handling an alteration or moving job. In other buildings such as hospitals and schools, the elevators are frequently overwhelmed and special operating considerations need to be made for this heavy traffic.

Modern approaches to elevator group supervisory systems include microprocessors, which can gather all the available information such as: the location, direction, and number of landing calls; the number of car calls in each elevator; the loading of each elevator as indicated by the load sensors; the direction each elevator is traveling or whether it is stopped at a floor to discharge or receive passengers, or free to be sent to another floor. This information is processed and various strategic decisions are made, which should be designed to minimize the waiting time of landing calls and to cause elevators to travel to areas where they are needed to serve the current traffic. If the supervisory system fails to do this, the elevators can be hopelessly overwhelmed and heavy traffic will exist where only moderate traffic existed before. As can be appreciated, elevator traffic is dynamic—it is always changing—and if it is not served when it occurs passengers will back up at every floor, where they will register both up and down calls in frustration and make the situation worse. A few seconds can change the entire aspect of a system from good to bad; hence, everything possible must be done to serve such traffic and ensure the response of the elevators to it.

A summary of the operations required for moderate two-way and interfloor traffic should include the following items.

The supervisory system must have the ability to measure the traffic a group of elevators is expected to encounter and dispatch each elevator accordingly in the proper direction. Such traffic factors measured would include car calls in each car, the loading of each car, the number of landing calls that each car may encounter, the spacing between one car and the preceding car, the relation between up and down traffic, the location of all cars in the system, and the time to service the existing car calls to the destination floor or floors.

In addition, the group supervisory system must decide whether the elevators should be quickly changed from concentration on down or up traffic to traffic in the opposite direc-

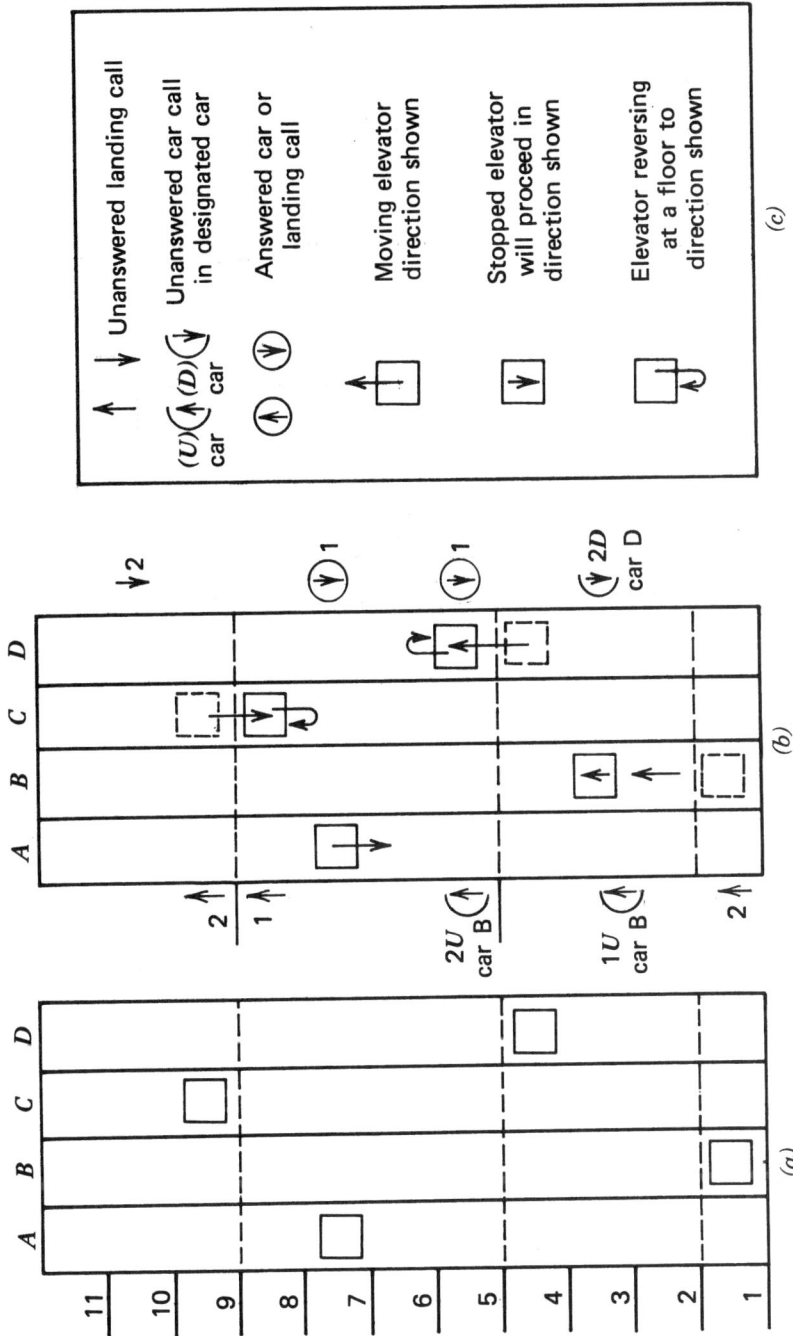

Figure 5.3. Elevator deployment during zoning: (*a*) elevators parked during a traffic lull; (*b*) elevator movements: 1, first calls; 2, second calls; (*c*) key to symbols used.

111

tion, whether to travel an elevator through to the main landing or to short-trip that car, where to send cars to be fed into prevalent traffic demand, and when to concentrate elevators in zones where long-wait calls may exist.

All of the foregoing constitutes a formidable task that microprocessors can accomplish by gathering, storing, and processing all the necessary information. Once the calls are counted, they must be weighted. Elevators are not traveling at constant speed but rather running, stopping, or starting; people's habits are not predictable, and one call can change the entire aspect of an elevator system in less than a second. We have dealt with averages in our computation. In real life discrete trips and individual people must be accommodated.

Heavy Traffic

Once traffic becomes exceedingly heavy in both directions, we find the cars filling on the up trip (either at various floors or at the lobby, similar to up-peak operation) and filling in the down direction. The cars will probably make every stop up, and they should go no higher than the highest up landing call or car call, and be immediately reversed. On the down trip the cars may make every stop down and perhaps bypass the lower of the down landing calls because of capacity load in the car and the load bypass arrangement. The time the elevator spends at the lobby should be the minimum required for passenger transfer.

Landing calls that are bypassed should be given a measure of preference. An upper limit of waiting time can be established, and any car in the proper direction, with capacity and without intervening car calls, can be bypassed to the call that has waited for an overlong period of time.

Bypassing must be used with discretion. Any bypassed call becomes a potential long-wait call—with indiscriminate bypassing, many calls could wait for an overlong period and a hopeless situation could develop. To avoid this it is necessary to do everything possible to deploy elevators at the first indication of anticipated traffic, rather than wait for traffic demand to increase to such proportions.

Overwhelming situations can occur in any building. It is most noticeable in schools during class change periods and in hotels during the time when convention meetings break up. Everyone in the building wants to go someplace at the same time, and if enough elevators were provided there would be little room for usable area on the building's floors. The potential problem should be recognized when the pedestrian planning is done for the building and means incorporated to alleviate the possible problem.

One of the most appropriate means is to use supplementary continuous-flow systems such as escalators or large stairways. Only one floor should be designated to load the heavy incoming crowd and all people directed to that floor.

Additional efficiency can be gained by providing means to call more than one elevator to a designated landing floor if it is other than the main dispatching floor of the group of elevators. On other than the main dispatching floor, if one elevator stops for a landing call, all other elevators are electronically prevented from stopping unless someone registers a car call for that floor. If it is known that at certain times a floor other than the main dispatching floor is going to have excessive crowds, electrical provisions can be made to have more than one elevator stop or to temporarily change the system so that a secondary dispatching floor is created at that busy floor.

Another solution to an overwhelming traffic problem is to restrict service to incidental floors such as basements or penthouses. These floors could be manually or automatically cut out, and lighted signs could inform passengers to use stairs. In a building that is initially underelevatored or where a change in building use has created handling capacity problems, the feasibility of adding more elevators should be considered. Permanently removing time-consuming and poorly productive elevator stops such as basements or garage floors should be accomplished by installing an additional single shuttle elevator to assume that function.

Waiting Time for Landing Calls

The effectiveness of an elevator group supervisory system's operation during periods of two-way lobby and two-way interfloor traffic is measured by the waiting time for landing calls. Before the introduction of microprocessors, recording devices producing records as shown in Figure 5.4 were employed. Countless hours were needed to analyze and chart the results, but the information was discrete and could be used to pinpoint troublesome floors if they existed. The recording meter employed pens actuated to mark a continually running calibrated paper chart to indicate when a call was actuated and when response took place. The paper chart moved at a given speed, usually about $\frac{1}{2}$ in. (13 mm) per minute, so that each $\frac{1}{10}$ in. (2.5 mm) represented 12 sec (see Figure 5.4). One pen was provided for each landing call in each direction on each floor.

With the introduction of microprocessor operating systems, newer or modernized elevator installations have such recording capability built in, or microprocessor-based recorders can be attached to a system to provide waiting time data. Chapter 19, on traffic studies, describes these systems in detail.

By tabulating the number and duration of landing calls a chart of the distribution of landing call waiting times can be developed (see Figure 5.5). With well-designed elevator systems most calls are expected to be answered within the design interval for the average two-way traffic. An excess percentage of calls over that time is cause for concern and further investigation. Such investigation may determine faults with the supervisory system or discover elevators being excessively held out of service.

The relation between operating interval and the normal expected excess waiting time for landing calls is given by the following example.

As discussed earlier in the chapter, average waiting time is about 60% of the operating interval. Assume that it is desired to know the percentage of landing calls that will wait an excessive amount of time when the elevator system is designed for a certain operating interval. The distribution of waiting times will generally follow a curve developed on a basis of the natural number e, which is derived from an infinite series and is equal to 2.71828 (to 5 decimal places).

For a given operating interval, the percentage of landing calls expected to wait over a given limit can be calculated by the formula

$$P = e^{-t/T}$$

where P is a number less than 1 which can be converted to a number representing the percentage of calls waiting over a given time t.

t = the arbitrary time limit
T = the average waiting time, assumed to be 60% × the operating interval

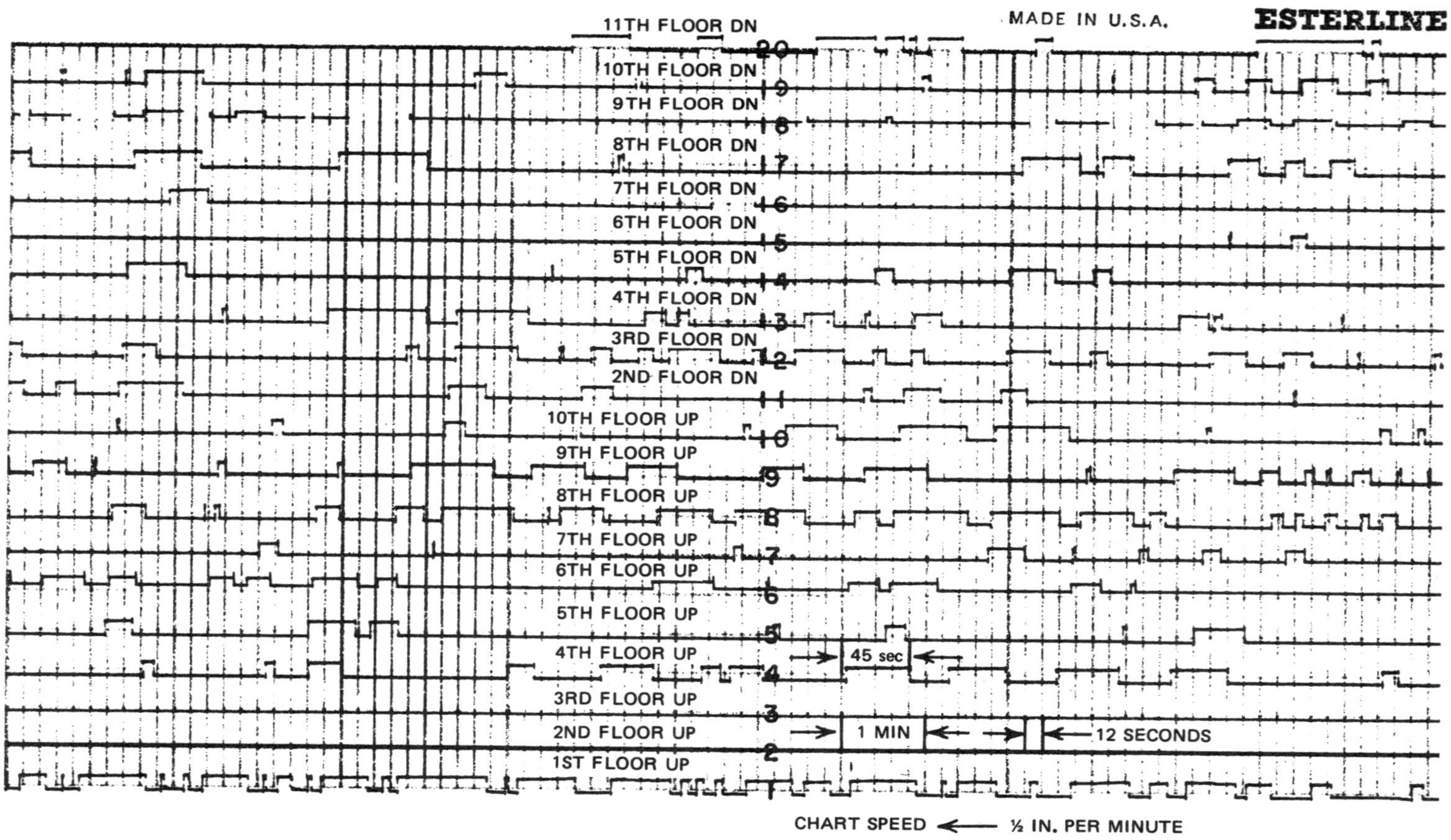

Figure 5.4. Sample of landing call waiting time recording.

114

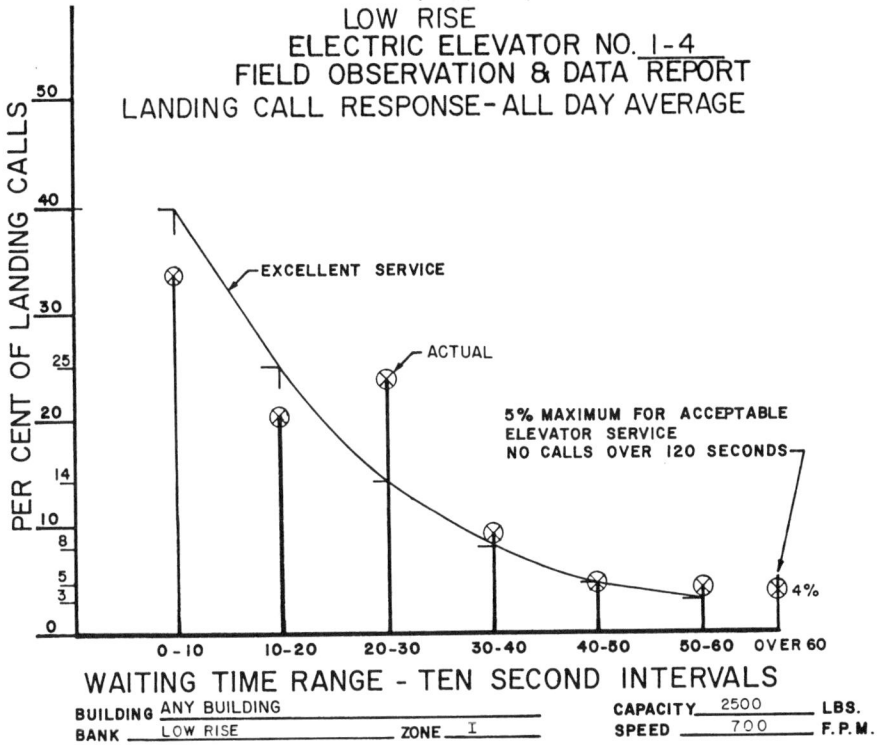

Figure 5.5. Actual observed waiting time distribution (Courtesy Jaros, Baum & Bolles).

For example, if the interval is 40 sec, and $T = 24$ sec, what percentage of landing calls will wait more than 60 sec (t)?

$$P = e^{-t/T}$$

$$P = e^{-60/24} = e^{-2.5}$$

$$P = 0.0821 \times 100 = 8.21\% \text{ of the landing calls wait longer than 60 sec}$$

Figure 5.6 shows various values of P calculations for intervals of 20, 30, and 40 sec.

The actual length of time a landing call has been in registration before an elevator responds is subject to a variety of interpretations, depending on the operating system of the elevators. Some systems consider a response to occur when a call is assigned to a particular elevator, whereas others will count response time as the time when the doors of the elevator open at the particular floor. A correction factor may have to be added, depending on the nature of the installation. What is most important is how long the passenger actually waits to get service after registering his or her call.

These aspects are discussed in Chapter 7. It may be found that with the most sophisticated of operating systems, the actual distribution of waiting time may approach a bell curve. This would be a distortion of the distribution shown in Figure 5.6. The operating system will favor calls with potential for a longer wait over those that are registered after an existing call—first in-first out type of response rather than a possible last-in first-out.

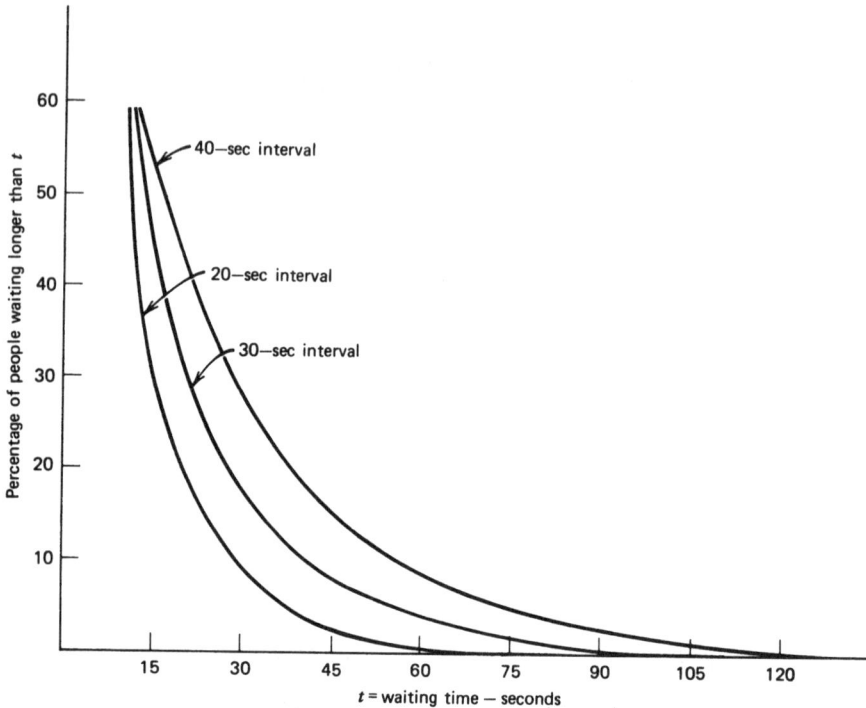

Figure 5.6. Distribution of waiting times—various intervals.

Extensive discussion of the various expected distribution of waiting times of landing calls can be found in a paper by Jon B. Halpern entitled "Statistical Analysis of Modern Elevator Dispatch Hall Call Response Time." This source can be found in the appendix.

GENERAL CONCLUSIONS

Two-way lobby and two-way interfloor traffic can be the most serious of the traffic situations encountered in any building. not only must sufficient elevators be provided to serve the expected traffic, but they must also be operated in such a way that their most efficient use is ensured. Failure to do so will create a temporary heavy traffic condition where, perhaps, only an extended moderate traffic should exist.

All the inefficiencies of poor elevator shape, size, layout, and so on, as described for incoming traffic, will be amplified during the heavier two-way traffic situations. Poor operating intervals will cause immeasurable employee time loss by riding and waiting passengers. When we consider the number of passengers riding and waiting at a given time, we see that 10 to 20 sec extra time spent waiting and on each trip will add up to a considerable number of man-hours lost by inefficient elevator service each year. This cost can be avoided only by proper initial design of the elevator system along with subsequent concern about the installed operating system.

As with all the examples and calculations given in this book, these are offered as a guide to the philosophy of elevatoring.

6 Outgoing Traffic

IMPORTANCE OF OUTGOING TRAFFIC

When the quitting bell sounds or, in some buildings, when lunch time arrives, a building's occupants are ready to leave and will demand elevator service. During these outgoing traffic peaks passengers, by sheer numbers alone, may measurably overwhelm an elevator system unless provisions are made for this contingency.

The most elementary of these provisions is sufficient elevator capacity to serve the outgoing traffic demand expected. In office buildings especially, the outgoing 5-min peak traffic may exceed any other traffic peak by 40 or 50%. Proper design and operation of an elevator system can provide this capacity without more elevators than are normally required to serve other traffic periods.

Proper design requires a reliable means to prevent an elevator filled to capacity from responding to landing calls. For this load weighing and bypassing operation, automatic elevators can be equipped with any of several sensing devices, the present most sophisticated of which will measure only the live load on the platform. Devices of this nature should be capable of adjustment; average capacity load in one building may never be attained in another, butt the problem of efficient outgoing traffic operation will remain. Future development should include capacity loading based on volume rather than weight.

Since demand is usually scattered among the various floors and an elevator is loaded promptly when it arrives at each floor, it is possible to take advantage of these factors in calculating outgoing traffic performance. Because people are on their own time, so to speak, and anxious to get out of a building, they willingly load elevators to near, and often beyond, normal capacity. Loads can exceed the average nominal loading experienced during incoming traffic periods, so much so that elevator safety codes require that the platform area in relation to elevator capacity in pounds or kilograms be designed to accommodate 125 lb (60 kg) per square foot (0.093 m^2). In addition, the elevator brake must be able to slow down and stop a down traveling car filled to 125% of rated elevator capacity. Figure 6.1 shows heavy loading of an elevator during a down-peak period. The elevator shown [3000 lb (1400 kg)] in the photo has a nominal capacity of 12 to 14 people, but 22 people are overcrowded into the car at approximately 1.5 ft^2 (0.14 m^2) per person.

OUTGOING TRAFFIC REQUIREMENTS

Normally an elevator system should be capable of evacuating the population of a building within 25 to 40 min. This period may be lengthened or shortened depending on particular tenant needs. Although the primary need is quantity of service (handling capacity), and qual-

The Vertical Transportation Handbook, Third Edition, Edited by George R. Strakosch
ISBN 0-471-16291-4 © 1998 John Wiley & Sons, Inc.

Figure 6.1. Heavy elevator car loading.

ity of service (interval) is secondary, service should be available at every floor within 60 sec. A sophisticated elevator operating system will have a means to provide equal service to all floors once a multiplicity of down landing calls and an absence of up landing calls or cars loading at the main lobby is apparent. The sequence of operations might be as follows:

If all floors are receiving outgoing traffic demands, the elevators should be equally spaced in time in the down direction, unloaded in minimum time at the lobby, and immediately returned up into the building to serve additional outgoing traffic. Elevators should be allowed to serve up landing calls as long as these calls do not interfere with service to the outgoing rush.

As the cars fill at upper floors they tend to bypass landing calls in the lower section of the building. When this occurs, all but one or two elevators in the group should then bypass up landing calls.

As the waiting time of calls in the lower section of the building increases beyond a predetermined value, some elevators should be assigned to serve only the lower portion of the building.

This can be accomplished by various means, one of which is to have the first up traveling car reverse in the next upper zone, and so on. Another means is to use a timed-out zone approach, whereby the waiting time is either measured in each zone as a total or totalized by the number of down landing calls times the number of seconds each call is waiting, and to assign a car to that timed-out zone. Up landing calls can be restricted to those cars that are assigned to the higher zones. These procedures will tend to equalize the elevator capacity available to all floors of the building.

If traffic continues to increase until landing calls tend to wait beyond a satisfactory period of time, a priority operation should be instituted. After elevators are unloaded at the lobby they are directed to any floor at which passengers (as indicated by a down landing call) have waited beyond a predetermined period of time. At that call the elevator reverses and, if it is not filled, it continues to pick up down landing calls until it fills or reaches the lobby. Under extreme conditions with this operation, elevators may become, in effect, two-stop elevators, completely loaded at one floor and completely unloaded at another. This is the most efficient operation that can be gained form any elevator.

Obviously, if extreme conditions for down traffic occur with any frequency in a building, service to any unusual floors, up landing calls, basements, and so on, should be temporarily suspended. Any traffic opposed to the outward rush will reduce the system's efficiency and should be considered in that light. Naturally, if up service is necessary and outgoing traffic demands all the available elevator-handling capacity, staggered quitting times or other approaches to the problem, which will be discussed later in this chapter, are necessary.

CALCULATIONS FOR OUTGOING TRAFFIC

Intensity of outgoing traffic is seldom the primary consideration in elevatoring a building. Under certain circumstances, however, it can be exceedingly important. For example, if elevators serve an upper-floor meeting room or auditorium, they require sufficient capacity for the orderly evacuation of that facility. In theaters, classroom buildings, stores, and so forth, evacuation ability may be of prime interest in elevatoring. For upper-floor "sky lobby" arrangements (discussed in a later chapter), shuttle elevators should be configured based on the down-peak capacity of the local elevators they serve.

Determining the capacity of an elevator system during outgoing traffic periods can proceed in a manner similar to that used for incoming traffic, with the following essential differences.

The mode of operation must be established. Will the peak be severe enough to warrant two-stop operation (enough elevators must be provided)—loading at one floor and unloading at a lower floor, or are many stops in the down direction expected?

Elevators may carry greater loads than during the incoming peak; an increase by one-eighth to one-third of the nominal car loading in Table 4.4 can be considered.

Loading can be efficient. An average of 0.6 sec per person can be used rather than the 1 sec established for incoming traffic period landing calls, provided there is sufficient door width and a reasonably shaped platform. We presume passengers will be attentive to hall lanterns and be at the entrance when an elevator arrives at a floor.

The time that passengers take to exit at the lobby will be correspondingly shorter and will be based on a value of 0.6 sec per person (or about 75% of the incoming traffic transfer time), as shown in Table 6.1.

Times in Table 6.1 are based on 48-in. (1200-mm) center-opening doors. Use the inefficiency in door time shown in Table 4.3 plus normal inefficiency.

Other time factors that are used in calculations, as well as calculations for outgoing handling capacity and interval, will be the same as used for incoming traffic.

$$\text{5-min handling capacity} = \frac{\text{number of passengers down} \times 300 \text{ sec}}{\text{round-trip time}}$$

TABLE 6.1.

Lobby:	Minimum 6 sec plus 0.6 sec per passenger over six passengers out only						
Number of Passengers	8	10	12	14	16	18	20
Lobby Time (sec)	7	8	9	10	11	13	14

Car Calls	(Same as incoming and two-way traffic)
	Dwell time 3 sec per stop.
	Transfer time. Use 3 sec for first two passengers (dwell time) plus 1 sec per passenger over 2. Example, 4 passengers, 1 stop: time = 3 + 2 = 5 sec
Landing Calls	Dwell time 4 sec per stop.
	Transfer time. Dwell time of 4 sec will include transfer of up to six passengers, use 0.6 sec per passenger over six. Example, 8 passengers, 1 stop: time = 4 + 2 × 0.6 = 5.2 sec

$$\text{Average interval} = \frac{\text{round-trip time}}{\text{number of elevators}}$$

The maximum interval for average outgoing traffic conditions should not exceed 40 to 50 sec. This is predicated on the intention of offering all passengers service within 60 sec during the peak outgoing traffic 5-min period.

The intensity of outgoing traffic will vary in all types of buildings. In office buildings a reasonable goal is to provide sufficient elevator service to evacuate the total population in a 25- to 40-min period. The elevators should be able to serve about 18% of the population during an outgoing peak 5-min period. The shorter transfer time, reduced number of stops, increased car loading, as well as an efficient outgoing or down-peak operation will generally make this feasible with the same number of elevators required to serve the in-coming peak traffic. This figure may be modified for particular buildings to a lesser or greater percentage or to provide sufficient capacity to evacuate a building within a time criterion that may be established by a tenant or design requirement.

Probable Stops

The number of stops an elevator makes during the outgoing traffic period depends on how much it fills at each stop. If sufficient passengers are waiting at a floor to fill any car that stops there, it usually descends directly to the lobby. This situation can be determined from information about a particular building: the population of each floor, the nature of population (executive or clerical, for example), the quitting time or reason for leaving that floor. An outgoing traffic schedule can then be established and elevator service calculated accordingly.

When the exact nature of the occupancy cannot be known in advance, as in designing a building for general office use, probable stopping must be estimated. We assume that quitting times are sufficiently staggered so that the peak will not exceed the 5-min average. We also assume that each floor has an equal possibility of originating outgoing passengers. With these assumptions, probable down stopping will be approximately 75% of incoming probable up stopping. As outlined in the discussion of two-way traffic, people tend to leave in groups or with a partner, and experience has shown that average stopping is between 60 and 80% of up-peak or incoming traffic stopping.

Chart 4.1 gives the values for probable incoming up stopping for all conditions of loading. These values should be reduced to 75% as discussed for two-way traffic and the lobby stop accounted for.

Advantage of highest call reversal or travel to the highest down landing call can be taken on the same basis as discussed for incoming traffic, depending on the type of building and the percentage of total population on each floor.

In the following examples, 1.2 times the normal up-peak loading of the elevators will be used to approximate the maximum down-peak loading. As always, judgment must be employed when elevators are related to a particular building, and local practices considered that would warrant using less than that factor.

SAMPLE CALCULATIONS

The following calculations exemplify a few situations to which outgoing traffic studies may apply. Example 6.1A may be considered a typical situation in a diversified office building. Note that if sufficient cars are provided to serve incoming and two-way lobby traffic demands, outgoing traffic is served with comparative ease. If the combination of increased loading and minimum delay in the passenger transfer operation of the elevators were to be applied to Example 6.1A as in Example 6.1C, outgoing capacity could be substantially increased.

Example 6.1B shows an evacuation situation. The elevators described can do a credible job in getting students out of the building. This may not be the prime consideration in elevatoring a classroom building, however; class change periods, with their tremendous interfloor as well as lobby traffic, may be the factor that determines the number of elevators.

Example 6.1C gives one means of solving an extreme outgoing traffic problem and relates actual elevator operation to the traffic served. Stated otherwise, evacuation of a building can be expedited by minimizing the number of stops elevators make, provided that each floor in the building has equal access to vertical transportation. Dependence on performance such as is shown in Example 6.1C would require an elevator operating scheme that is responsive to outgoing traffic and designed to assign each elevator as it leaves the main landing into a zone of the building. Each zone should be so designated to represent equal traffic demand, and the elevators arranged so that each zone has an elevator assigned to it. Assignments can be changed each trip to provide equal service to all floors.

Example 6.1. Outgoing Traffic

A. Given: building population 1000 persons, 15 stories, five 500-fpm elevators, rise 176 ft, up-peak capacity 110 people (see Example 4.1).
Determine: down-peak capacity. Will the elevators accommodate 16.8% of the population during 5 min of down peak?
1. Example 4.1 was developed on the basis of 11 passengers per elevator. Calculating the elevator size needed to meet the (N − 1) rule would be done by finding the four elevators equivalent to five 11-passenger elevators. From Table 4.4, 11-passenger elevator = 2500 lb, 5 × 2500 = 12,500 lb ÷ 4 = 3100 lb = 3500 lb. The recommendation should be five 3500-lb elevators.
2. Nominal capacity of a 3500-lb elevator 16 people times 1.2 gives a maximum down-peak capacity of 19 people.

3. Probable stops, 19 people, 13 upper floors from Chart 4.1
   Probable stops = 10.1 (Chart 4.1) × 0.75 = 7.6

   Time to run down, per stop $\dfrac{176}{7.6}$ = 23.2-ft rise per stop

   (from Chart 4.2) 23.2 ft = 5.4 sec

   Time to run up, $\dfrac{(176 - 23.2) \times 60}{500}$ + 5.4 = 23.6 sec

   Elevator performance calculations:
   Standing time

Lobby time 19 passengers	=	14 sec
Transfer time, down stops (7.6 × 4)	=	30
Door time, down stops (7.6 + 1) × 5.3	=	45.6
Total standing time		89.6 sec
Inefficiency, 10%	=	9.0
Total		98.6 sec

   Running time

Run up	=	23.6
Run down 7.6 × 5.4	=	41.0
Total round-trip time		16.32 sec

   $$HC = \frac{(0 \text{ up } + 19 \text{ down}) \times 300}{163} = 35 \text{ people per elevator}$$

   For 1000 people:
   5-min outgoing capacity with 4 elevators 140 = 14%
   5-min outgoing capacity with 5 elevators 175 = 17.5%
   Five elevators required to meet 16.5% requirement
   Interval: 163/5 = 33 sec vs. 40 to 50 sec desired

B. Given: school, 6 stories, 500 students on floors 2 through 6, 12-ft floor heights
   Determine: size and speed of elevators to evacuate school in 20 min

   Procedure: outgoing 5-min average traffic = $\dfrac{500 \text{ students}}{4 \text{ (5-min periods)}}$ = 125

   Probable stops, maximum 5 (stops at each floor)

   Time to run down, per stop $\dfrac{12 \text{ ft} \times 5}{5}$ = 12-ft rise per stop

   Assume 300 fpm, from Chart 4.2, 12 ft = 5.8 sec

   Time to run up, $\dfrac{(60 - 12) \times 60}{300}$ + 5.8 = 15.4 sec

   Assume 5000-lb elevators 23 × 1.2 = 28 passengers per down trip, 54 in. center-opening doors
   Elevator performance calculations:
   Standing time

Lobby time 28 passengers 6 + (22 × 0.6)	=	19 sec
Transfer time, down stops 5 × 4 = 20, 0.6 × 4 = 2.4	=	22.4 sec
Door time, down stops (5 + 1) × 6	=	36.0
Total standing time		77.4 sec
Inefficiency, 10%	=	7.7
Total		85.1 sec

(Total from page 122, 85.1 sec)

Running time

Run up	=	15.4
Run down 5×5.8	=	29
Total round-trip time		129.5 sec

$$HC = \frac{28 \times 300}{130} = 65 \text{ people per elevator per 5 min}$$

Elevators required $\dfrac{125}{65} = 1.9$, use 2

Interval: $130/2 = 65.0$ sec interval

C. Given: four 4000-lb elevators @ 500 fpm in a 10-story office building, 12-ft floor heights
Determine: greatest number of people that can be moved out of building from all floors in a 5-min period
Procedure: at least one trip to each floor must be made during each 5 min
Divide floors into zones to maximize service:

Zone 1—floors 2,3,4	Zone 2—floors 5,6
Zone 3—floors 7,8	Zone 4—floors 9,10

1. Calculate zone 4.
   Probable stops 2, 1 included in down run to lobby
   Time to run down, per stop 12 ft = 4.4 sec

   Time to run down, floor 9 to lobby $\dfrac{(7 \times 12) \times 60}{500} + 4.4 = 17.5$ sec

   Time to run up, $\dfrac{(8 \times 12) \times 60}{500} + 4.4 = 15.9$ sec

   4000-lb elevator = 19 passengers $\times$ 1.2 = 23 down passengers, assume 48-in. center-opening doors
   Elevator performance calculations:

   Standing time

Lobby time 23 passengers = $6 + (0.6 \times 17)$	=	16.2 sec
Transfer time, down stops $2 \times 4 = 8 + (15 \times 0.6)$	=	17.0
Door time, down stops $(1 + 2) \times 5.3$	=	15.9
Total standing time		49.1 sec
Inefficiency, 10%	=	4.9
	Total	54.0 sec

   Running time

Run up express	=	15.9
Run down per stop	=	4.4
Run down express	=	14.5
Total round-trip time		88.8 sec

   $$HC = \frac{23 \times 300}{89} = 78 \text{ people in 5 min}$$

2. Calculate zone 3.
   Probable stops 2, 1 included in down run to lobby
   Time to run down, per stop 12 = 4.4 sec

   Time to run down, floor 7 to lobby $\dfrac{(5 \times 12) \times 60}{500} + 4.4 = 11.6$ sec

   Time to run up $\dfrac{(7 \times 12) \times 60}{500} + 4.4 = 14.5$ sec

Elevator performance calculations:

Standing time

Lobby time 23 passengers $6 + (0.6 \times 17)$	$=$	16.2 sec
Transfer time, down stops $2 \times 4 = 8 + (15 \times 0.6)$	$=$	17.0
Door time, down stops $(1 + 2) \times 5.3$	$=$	15.9
Total standing time		49.1 sec
Inefficiency, 10%	$=$	4.9
	Total	54.0 sec

Running time

Run up express	$=$	14.5
Run down per stop		4.4
Run down express	$=$	11.6
Total round-trip time		84.5 sec

$$HC = \frac{23 \times 300}{85} = 81 \text{ people in 5 min}$$

3. Calculate zone 2.

Probable stops 2, 1 included in down run to lobby

Time to run down, per stop 12 ft = 4.4 sec

$$\text{Time to run down, floor 5 to lobby } \frac{(3 \times 12) \times 60}{500} + 4.4 = 8.7 \text{ sec}$$

$$\text{Time to run up } \frac{(4 \times 12) \times 60}{500} + 4.4 = 10.2 \text{ sec}$$

Elevator performance calculations:

Standing time

Lobby time 23 passengers $6 + (0.6 \times 17)$	$=$	16.2 sec
Transfer time, down stops $2 \times 4 = 8 + (15 \times 0.6)$	$=$	17.0
Door time, down stops $(1 + 2) \times 5.3$	$=$	15.9
Total standing time		49.1 sec
Inefficiency, 10%	$=$	4.9
	Total	54.0 sec

Running time

Run up express	$=$	10.2
Run down per stop	$=$	4.4
Run down express	$=$	8.7
Total round-trip time		77.3 sec

$$HC = \frac{23 \times 300}{77} = 90 \text{ people in 5 min}$$

4. Calculate zone 1.

Probable stops 3

Time to run down, per stop 12 ft = 4.4 sec

$$\text{Time to run up to floor 4 } \frac{(2 \times 12) \times 60}{500} + 4.4 = 7.3 \text{ sec}$$

Elevator performance calculations:
Standing time
   Lobby time 23 passengers $6 + (0.6 \times 17)$ = 16.2 sec
   Transfer time, down stops $3 \times 4 = 12 + (5 \times 0.6)$ = 15.0
   Door time, down stops $(3 + 1) \times 5.3$ = 21.2
   Total standing time   52.4 sec
Inefficiency, 10%   =   5.3
                            Total 57.6 sec

Running time
   Run up   =   7.3
   Run down 3×4.4 = 13.2
   Total round-trip time   78.1 sec

$$\text{HC} = \frac{23 \times 300}{78} = 88 \text{ people in 5 min}$$

Total HC, four elevators $= (78 + 81 + 90 + 88) = 337$ people per 5 min

$$\text{Average interval} = \frac{(89 + 85 + 77 + 78)}{4} = 82 \text{ sec}$$

Therefore, four 4000-lb elevators @ 500 fpm serving floors 1 through 10 can serve approximately 337 in 5 min, outgoing. Average interval will be approximately 82 sec.

Recalculate Example 6.1C with 4 cars serving all floors. Probable stops, 19 up passengers $\times 1.2 = 23$ down passengers, 9 upper floors $= 8.4 \times 0.75 = 6.3$

$$\text{Time to run down, per stop} \frac{(8 \times 12)}{6.3} = 15.2 \text{ ft} = \text{rise per stop from Chart 4.2} = 4.7 \text{ sec}$$

$$\text{Time to run up} \frac{(96 - 15.2) \times 60}{500} + 4.7 = 14.4 \text{ sec}$$

Elevator performance calculations:
Standing time
   Lobby time 23 passengers $= 6 + (0.6 \times 17)$ = 16.2 sec
   Transfer time, down stops 6.3×4 = 25.2
   Door time, down stops $(6.3 + 1) \times 5.3$ = 38.7
   Total standing time   80.1 sec
Inefficiency, 10%   =   8.0
                     Total 88.1 sec

Running time
   Run up express   = 14.4
   Run down 6.3×4.7 = 29.6
   Total round-trip time   132.1 sec

$$\text{HC: 4 cars} = 4 \times \frac{23 \times 300}{132} = 209 \text{ people in 5 min}$$

Interval: $132/4 = 33$ sec

As can be seen from Example 6.1C, the complete zoning of the elevators provides a substantial increase in handling capacity over having all elevators available to serve every floor. Of course, the increase is not without a penalty. The interval, hence average waiting time, will be extremely long, with many people waiting in each floor lobby. Interestingly, the one car per floor or two-stop shuttle operation would be the expected way an elevator operating system would respond if calls started to wait overlong at each floor.

SPECIAL OPERATIONS

Nature of the Traffic

Heavy outgoing traffic situations can occur in any building and can be generated from various floors. An excellent example is a hotel with a ballroom located two or three floors above the street entrance. Another example is a cafeteria on an upper floor of an office building, where on nice days many people will lunch and then seek the outdoors, creating a heavy down traffic from that floor. In such situations special considerations must be given to the operating program of elevators serving floors that generate heavy traffic.

The nature of the traffic must first be determined: are the passengers all expected to head for a building exit floor or to distribute themselves among that and other floors? Is the traffic from the floor in question all down (or up) with little or no traffic in the opposite direction? In the latter case the loading floor should be considered as an upper "lobby" with all elevators arranged to return to that floor. In addition, the operating system should be designed to allow the loading of more than one elevator at that floor simultaneously, as opposed to normal operating systems that allow only one elevator at a time to stop at an upper floor for a single landing call.

Extra time for inefficiency must be included in situations where passenger discipline and familiarity with elevators is not great. In hotels, hospitals, department stores, and other public places 10 to 20% inefficiency in transfer is to be expected in addition to normal inefficiency. This should be reflected in time considerations and in providing ample platform area as well as sufficient door width.

Escalators should be considered when outgoing traffic is expected to be as great as 200 or more people in a 5-min period. Escalators have capacities of 200 to 500 people in 5 min, depending on width and speed of the steps. This will be further discussed in the chapter on escalators and moving walks.

Escalators can be used effectively to funnel outgoing or incoming pedestrian traffic to an area from which they can distribute themselves to other traffic generators such as checkrooms, lounges, and refreshment stands. When paired escalators are provided to serve two-way traffic, both escalators can be operated in the direction of heavy traffic and other traffic diverted to stairs or elevators. The layout of escalators must be planned for this contingency. As with elevators, sufficient area must be available for persons to enter and leave. It is futile and dangerous to transport a stream of persons to an area faster than they can leave it. An escalator is an unforgiving passenger conveyor—one cannot turn back! Measures of building design and operation such as revolving doors or exit doors should prevent such contingencies, but should they arise the simplest correction is to shut down temporarily or reduce the capacity of the transportation system and have people stay where they are.

Emergency Evacuation

Under certain circumstances it may be necessary to evacuate a building, which can be done with the elevators if a plan is developed and emergency personnel are trained to operate elevators in their most efficient mode. Such circumstances could include a bomb scare, a remote section of the building on fire, or an explosion.

In the event of a fire in the building, the elevators should not initially be used by the occupants and should be automatically returned to a main landing for use by emergency personnel and supervised evacuation. Occupants should be trained by means of fire drills to use stairways.

Modern elevator systems have features for such emergency recall of the elevators, as will be discussed in a later chapter. Once the elevators have been returned to a main landing and while elevators not returned are being accounted for, certain elevators will be commandeered for necessary emergency work. The remainder can be used by emergency personnel to evacuate various floors, one at a time, effectively creating a two-stop elevator operation, loading at one floor and exiting at another. The emergency personnel must give proper attention to disabled personnel on the various floors; an essential prerequisite is for building management to know at what floors such persons are located and for responsible people on that floor to be aware of their presence or absence each day.

Automatic operations to accomplish an emergency evacuation do not appear to be feasible since it is an almost impossible task to program all the contingencies, such as elevators out of service, the location and nature of emergencies, the time when they may occur, and the danger that people requiring special aid may be ignored. Trained police, fire, and building personnel can provide the most important aspect of judgment and supervision.

GENERAL CONCLUSIONS

Consideration should always be given to the outgoing traffic in a building. Expected elevator performance should be calculated and checked against the expected outgoing traffic demand, and necessary adjustment made in elevator capacity and speed.

Outgoing traffic requirements are often effectively met by incorporating suitable features in the elevator operating system. One of the most useful of these procedures is efficient load bypass operation of the elevators as well as means to optimize elevator unloading time at the lobby and to speed their return up into the building to continue the outgoing operation.

Additional operating features include dispatching means so that elevators provide equal service to all floors in the building or to areas of equal demand. To guard against long waits, priority and zoning service should be provided if required by traffic intensity. In extreme down traffic situations such as in schools, "zoning" service, as demonstrated in Example 6.1C, can be employed and automatically effected by clock or operational control. As an alternative, an elevator system with provisions for future addition of these features should be considered. Such an approach is practical for any future expansion within a particular building.

Escalators should be considered for outgoing traffic of substantial magnitude from a limited number of floors. Two-way escalator service allows both escalators to be operated down if their capacity is needed. Clear signs to direct passengers to transportation as well as ample means to leave the exiting area are necessary. Fire drills and other emergency exercises should be practiced to ensure orderly evacuation in the event of an emergency.

7 Elevator Operation and Control*

JON B. HALPERN
LAWRENCE E. PIKE

PART I: GENERAL OVERVIEW OF COMPUTER TECHNOLOGY AS RELATED TO ELEVATOR OPERATION

Many aspects of life have been greatly affected over the past 15 years by the rapid evolution of the computer. Technological development has reached a feverish pace, with no end in sight. Computers have come out of the specially designed, halon-protected, air-conditioned rooms and have entered into coffeepots, thermostats, automobiles, cellular phones, and nearly every other device that can possibly benefit from becoming slightly more "intelligent." The elevator systems of today are no exception.

As computers have become smaller, faster, and less expensive, people have found more and more uses for them. At first, computers were seen as devices that could perform mathematical functions quickly and repetitively, although the very first computers took hours to program to perform simple mathematical functions. Then, as the technology progressed, computers could store vast mountains of information and perform mathematical function on the data. The computer was tireless and could, and usually did, work endless nights processing millions of pieces of information. Many processes could run for days stepping through that ton of information. Out of simple functions evolved algorithms, which were more complex approaches to mathematical problems that required many more steps and a tremendous amount of processing power. An algorithm is a general rule or process which, with a given set of variables and conditions, generates a solution for a given problem.

Because of the limitations of relay logic, both physically in the number of relays you could have, and logically in the growing complexity of large relay logic systems, the complexity of these systems was kept to a minimum. All of these advances in technology brought about elevator systems that could perform better by establishing a set of performance variables and tolerances. But there were some limiting factors in all of this. Systems were fairly static and could be modified only by trained mechanics. Simple changes to a system were not so simple. Accounting for day-to-day changes in traffic patterns was not possible, because the elevator system's only information about the building was designed into this fairly static system. Changes in building occupancy or function could not be

---

*Part I of this chapter was written by Lawrence Pike. Part II was written by Jon Halpern.

The Vertical Transportation Handbook, Third Edition, Edited by George R. Strakosch
ISBN 0-471-16291-4 © 1998 John Wiley & Sons, Inc.

easily accounted for. Basically, the elevator system could not adapt to changes in its environment. It could not *learn!*

With the advent of microprocessor controllers, with phenomenal processing speed and more memory than most early computers had, elevator systems received a "brain." Elevator systems now had the capability to learn from their day-to-day traffic patterns and change their behavior appropriately. Algorithms that had previously been too complex to be implemented in relay logic could now be easily implemented using the plethora of computer languages available to develop such programs.

Technology also evolved in respect to its interface. Computers are no longer ends in and of themselves. They can communicate with each other and share information. They can also communicate with the people responsible for controlling and maintaining an elevator system. An elevator system was traditionally a "black box," closed behind a door that read, "Machine Room." Mechanics have been using the computer as a diagnostic tools for years to "communicate" with their equipment. Only recently has the interface been expanded to include the building management, security personnel, and remote maintenance personnel.

Today building managers can show the performance of their elevator system at peak times in graphical format so that it may be understood by lay persons. They can project what changes in occupancy and function will do to their buildings by taking "real" data and fluctuating it accordingly. Security personnel can lock out floors at any time for any reason. Should an individual be assaulted in an elevator and security personnel see it, they can immediately recall the elevator to the appropriate floor and stop the assault. Maintenance personnel, from their own office, can be alerted to potential problems with the elevator system for which they are responsible. All of these things are possibilities because of the rapid development of technology.

Microprocessor Control: Group Dispatching

The progress made in the dispatching of elevators within a group to improve how quickly hall calls are answered has been very important to the operation of elevator systems. The fewer times passengers have a long wait for an elevator, the happier they are.

Group elevator dispatching systems that operated by using "relay logic" controllers, although successful, were still limited in the way they assigned elevators to answer hall calls. They largely based allocations on distance. The relay-based dispatcher would select the elevator closest to the floor where the registration of the hall call occurred, traveling in the same direction, to answer that call. However, after being assigned to answer the hall call, a passenger in that closest elevator might register a car call, requiring the elevator to stop at another floor before reaching the hall call floor. That might cause it to take longer to answer the call than for another elevator, farther away, to travel directly to answer the call without stopping on the way.

Microprocessor controllers, through their various capabilities, have effected great improvement in the group dispatching of elevators. Most important, they base allocations on how fast it takes an elevator to answer a hall call, rather than on its travel distance. Group dispatchers of the past used programmed relay logic to perform hall call allocations. A group microprocessor supervisory system consists of a computer that uses multitasking/multiprocessing architecture to connect to the individual microprocessor controller of each elevator through high-speed data communication links. This creates a network that can constantly analyze changing building traffic conditions, that is, the number of elevators in service; each elevator's status, loading, position, direction of travel, door opening and closing times; the number of assigned hall calls, hall call demand, and estimated time of arrival, to name a few. The system can then select and dispatch the elevator best suited to answer a given hall call request. The group supervisory system constantly reassesses its allocations by scanning the status of each elevator many times a second. Based on its evaluation of the real-time data, it can reassign hall calls to the elevator that can provide the best service.

Another approach being used by some controller manufacturers is to design a computer system for each individual elevator controller that not only controls all car functions but can provide full dispatching functions. Therefore, each elevator within the group can provide the group dispatching functions by using its microprocessor controller. If an elevator is out of the group operation for service or some other reason, another elevator can take over the dispatching functions without interruption to the building. This method is a form of "distributive processing."

Trade-offs are to be considered regarding these two approaches. Some designers feel that the use of a separate computer to perform the functions of group dispatching offers certain advantages. Because the real-time data being examined within the elevator group's traffic flow is of such short duration, a faster 32-bit system can be used to execute more sophisticated algorithms that produce more efficient dispatching. The disadvantage of this approach is that a dispatcher malfunction or breakdown could result in all of the elevators making stops randomly at all of the floors until the group system can be repaired ("wild card" mode).

In the "distributive processing" method, each elevator in the group can provide the dispatching functions. The advantage is that if an elevator is taken out of the group for service or some other reason, another elevator in the group can take over the full dispatching functions. Current technology typically uses 16-bit processors in distributive processing-type computer controllers for reasons of economics. Although they can run any algorithm, they lack the speed of the more expensive 32-bit processors to run the more sophisticated algorithms used typically in the dedicated group dispatching systems.

Intelligent Dispatching

The microprocessor controller offers further improvement in its dispatching capabilities by "learning" to do "intelligent dispatching." It remembers past experiences and applies them to current decisions.

Artificial Intelligence. Statistical traffic data is continuously monitored and recorded in time segments (usually 15 min), 24 hours a day, 7 days a week. It consists of operational demands and parameters such as the number of hall calls per floor, the time taken to answer each of those calls, the time of day each call occurred, and so forth. It forms a database from which the microprocessor can gather information, enabling the "artificial intelligence" algorithm to predict peak traffic demands by analyzing excessive car and hall call occurrences. The system can then dispatch elevators to meet those needs, even before the calls have been registered, giving passengers the fastest service possible. As traffic patterns and demands change, the statistical traffic data stored within the database changes, along with the dispatching based on the artificial intelligence of the system.

Fuzzy Logic Call Allocation. Fuzzy logic consists of computer-usable algorithms derived from fuzzy set theory. They enable the handling of sets whose members fit only partly into their groupings. A computer that incorporates fuzzy logic can work with imprecise terms, such as "cold," "medium," or "usually right," and with multiple values such as "true," "not true," "very true," "not very true," "more or less true," and so on. This is in contrast to traditional logic, which deals with strict values: "true-false," "either-or," and so forth. Fuzzy logic is best applied in systems that contain ambiguities and uncertainties.

The use of fuzzy logic algorithms in microprocessor controllers enables them to assign a hall call to the elevator that can most efficiently answer the call while handling constantly changing parameters, such as passengers changing their floor destinations, pushing both the "up" and "down" buttons in the hall push-button station, hall calls that remain unanswered owing to "load weighing" sensing a full car, and nuisance calls (six car calls with only one passenger in the car).

A fuzzy logic-enhanced call allocation algorithm coupled with artificial intelligence can provide excellent response to each hall call registered.

Solid State Technology

The use of computers within the various elevator system components provides important benefits. System reliability, accuracy, and flexibility have been enhanced remarkably by solid-state technology.

Reliability. Elevator controllers and all of their various functions, such as dispatching, door operator control, fire fighters service, car and hall call control, and so on, have always used relays, resistors, and other heat-producing components in their circuits. When operating in an elevator machine room environment where hoisting machines and their motors, motor-generator sets, and poor ventilation develop substantial heat loads, relay logic controllers not only add to the heat generation problem, but also suffer from it. Systems become unstable because of too much heat, and their reliability often suffers because of heat-related failures. Replacement of these types of controllers and circuits with controllers that use solid-state components, such as microprocessor chips and transistors, not only results in less heat generation, but enables them to operate more reliably in these environments. Motor-generator sets are being replaced with silicon controlled rectifiers (SCRs), which provide dc voltage to the elevator hoisting machines more efficiently and with practically no measurable heat load generation. The use of solid-state technology within elevator systems has greatly improved reliability because of the small heat loads they produce.

Accuracy. Solid-state technology offers improved accuracy of elevator systems in many ways. For example, the development and use of digital electronic selector/leveling devices have enabled elevator control systems to "digitize" accurately an entire hoistway for determining the exact status of an elevator in which it is traveling. Such a device can monitor an elevator's position, direction, and distance to a predetermined location in a hoistway, such as a slow-down switch, or a final limit switch, and provide the capability to stop the elevator at floor level, with accuracy approaching one-tenth of an inch (0.1 in. or 2.5 mm).

Most of the many components that make up a modern elevator system use solid-state technology that improves their reliability and accuracy. Full curtain door reversal systems feature infrared beams projected across the door opening that fall on solid-state detectors. When the beam is interrupted by someone or something, a signal is sent to the door operator that reverses its closing action.

Flexibility. The flexibility of solid-state technology manifests itself throughout various parts of an elevator system's operation.

The term *relay logic controller* describes a series or array of interconnected relays by which the logic of the circuitry directs the control of the elevator system in accordance with its requirements. In group operation (two or more elevators) the requirements can be so complex as to necessitate a large number of relay devices to accomplish the various tasks. A relay logic group dispatcher is used to communicate direction to each elevator within the group. Thus, the sizes of the individual controllers and dispatchers, because of the number of relays required, are quite large.

A microprocessor can execute the same directions to an individual elevator or to a group of elevators with a simple computer program and interface, used to convert digital information into electrical signals needed to accomplish the same tasks. The result is reduced size, owing to the use of computer chips rather than large numbers of relays.

A huge benefit to be derived from the microprocessor system is its flexibility, allowing it to effect changes in the control of the elevator system by simply making a programming change. This replaces the chore of having to add or delete a large number of relays and make the appropriate wiring changes.

Commands are built into the microprocessor's software program to perform certain tasks. An example is the ability to "lock out" individual car calls or hall calls, preventing access to the elevator from a given floor, or to prevent access to a given floor from the car, or both. In relay logic controllers, it is necessary to install key switches for each car call/hall call push-button to effect the same lockouts.

System Monitoring

Perhaps the most extensive application of computers in the vertical transportation industry is to monitor the operation of elevator systems in real time and by use of its historical database.

Real-time Monitoring. The use of computers to monitor the system operations in real time serves to reduce elevator downtime. Interfacing with all of the controller functions, including the drive system parameters, in real-time provides powerful fault diagnostic capabilities. An elevator technician, using the built-in diagnostics, can diagnose a fault by going directly to the problem area, thereby reducing the time it takes to repair the problem and return the elevator to service. In fact, it is becoming common to be able to diagnose a problem, fix it, and keep the elevator in service before the owner knows the problem exists.

Lobby panels, formerly elaborate displays of lights, indicating the current status of the building elevators by use of an analog-type position indicator and direction arrows are being replaced. State-of-the-art cathode ray tubes (CRT) monitors are being used to display a representation of each elevator as it travels up and down in the hoistway making stops at various floors, the opening and closing of the doors, the registering of hall and car calls, the car's status—Normal, Independent Service, Inspection Service, Fire Service, Out of Service, and so forth. These monitors may be placed in an elevator machine room for use by the technician servicing the elevators, in the Main Lobby of a building for passenger observation of the current position of each elevator, and in the offices of building management personnel for monitoring the elevator system and producing reports to ascertain the overall elevator performance or any potential problem areas.

Monitoring Elevator Operating History. Failure analysis reports provide historical reviews of an elevator's past performance regarding those areas in which failures may have been excessive, such as door operations, leveling, failure to answer car or hall calls, and so forth. By reviewing these reports on a regular basis, an owner and/or the maintenance contractor can be aware of the overall elevator performance and any current or potential problem areas.

For example, an elevator technician making a preventive maintenance call would look first at the Event Log generated by the computer. The computer can usually store historical data for periods of up to one year. In checking the log since the time of his or her last visit, the technician might discover that during a door-close operation, a hoistway door interlock on a given floor had failed to "make up," recycled, failed again, recycled, and finally closed successfully, allowing the elevator to run. Although a shutdown had not yet occurred, the Event Log would show two door-close failures. This would enable the technician to investigate immediately and take whatever action necessary to prevent the interlock from failing, thereby averting a shutdown before it occurred. Most microprocessor controllers can monitor approximately 50 events for diagnostic purposes.

Report Generation

The historical database can also be used to generate reports for use by building management.

Hall Call Waiting Analysis. Every time a hall call is registered, it is logged into the database as to the floor number and the amount of time it took for the elevator system to answer the call. The logs are usually kept in 15-minute increments. An analysis can then be done to study the average waiting time for a hall call to be answered per individual floor, per time period. In other words, how long did passengers have to wait at the main lobby floor for an elevator between the hours of 7:30 A.M. to 9:00 A.M. on a Monday? The study could encompass the average waiting times for all elevators at all floors for any period of time over the last year. This kind of report is a valuable tool for building management in discussing elevator service with existing or potential clients. It is also of great utility in evaluating the elevator maintenance service being provided.

Longest Wait Time Analysis. The same hall call waiting time information can list the longest waiting times encountered for any given floor over any given period of time. This too is a useful tool for management and for maintenance evaluation.

Elevator Downtime Reports. Downtime reports provide a record of elevator downtime and an analysis of a car's time out of service as compared with its time in service. Reports that determine how much downtime is attributed to cleaning, freight/service use, preventive maintenance, or shutdowns requiring maintenance or repair are important management tools. Events that previously required a great amount of record keeping by hand can now be extracted from the historical database and used to provide insights and statistical data for building management through the use of computers.

ABOUT THE AUTHOR

LAWRENCE (LARRY) E. PIKE is a senior designer for J. Martin Associates, an engineering consulting firm specializing primarily in the design of vertical transportation and material handling systems on a worldwide basis. The company presents technical capabilities and proven experience in traffic analysis-system design, equipment survey-modernization, and new construction and inspection-test to establish code compliance.

Mr. Pike earned a B.A. degree, with a double major in mathematics and physics, from Emporia State University in Emporia, Kansas, and has more than thirty-five years of engineering experience. His experience includes seven years of aerospace electronic spacecraft systems engineering; ten years in the fields of computers, electronics, and electro-mechanical systems sales; and twenty years of elevator experience, with five years as an elevator consulting engineer. He is certified by the National Association of Elevator Safety Authorities (NAESA) as a Qualified Elevator Inspector (QEI Cert. #1014.)

PART II: OPERATION AND CONTROL SYSTEMS

Once the number, size, speed, and location of the elevators are determined for a particular building, an adequate operation and motion control system must be provided so that those elevators operate to serve expected traffic. Such systems were alluded to in the previous chapters in discussing various types of traffic. Tied in with serving the predominant traffic period is the ability of the elevators to operate in the best manner to serve that traffic and to perform with a minimum average waiting time and total travel time for passengers. These terms must be translated into specifications so that the supplier of the elevator equipment knows what is expected.

Operations is an inclusive term designating all the electrical decisions designed into an elevator system to control the sequence of movements an elevator or elevator group will make in response to calls for service. For ease of identification, operating systems have been given broad titles that are classified in the A17.1 Elevator Code, as discussed later in this chapter. Individual manufacturers have added their own features and attractions to these operating systems, some providing more than others. The differences between various manufacturers are difficult to define in concise terms, since each has a different philosophy of elevator operation, which will also be discussed later in this chapter. Various suppliers are more than willing to discuss their approaches to solving elevator problems, either by demonstration on previous installations or by verbal and written assurances.

Motion control is the designation for the equipment that determines individual performance characteristics of an elevator; how quickly it can travel from floor to floor, the means and speed of door opening and closing, built-in time factors for passenger transfer, its ability to level swiftly and accurately, and how the elevator displays hall lantern signals are all functions of the control applied to an elevator. This control can and will be modified to a slight extent by the operating system, but in general it is sufficiently distinct to be discussed separately.

Motion Control

An important part of the control system is how power is applied to the elevator to control its starting, acceleration, running, deceleration, leveling, and stopping, which is referred to as motion control.

There are several types of motion control, single- or two-speed alternating current, generator field control utilizing a motor generator, and various types using solid-state electronic devices, both ac and dc, regenerative and nonregenerative. Further distinctions can be made between motion controls using an analog or digital feedback, sometimes referred to as "closed loop" and "open loop," wherein the motion of the elevator is programmed to follow an established pattern. In Chapter 1, the motion control systems applied to various types of elevator machines—hydraulic, geared, and gearless—were indicated. In this chapter, specific applications are discussed. In general, alternating current control is applied to geared and hydraulic elevators; generator field control and solid state is applied to both geared and gearless elevators; and, the latest, variable voltage, variable frequency (VVVF) applied to all types of elevators.

Alternating current resistance control is used to start the pump motor of a hydraulic elevator and is used in a traction elevator to control the starting of the hoist motor in the desired direction. A step or steps of resistance may be inserted between the motor and the line to reduce the in-rush current as the motor is started. The resistance is shorted out when the motor attains normal speed (see Figure 7.1a). In some controls, a three-phase motor may be connected in a Y configuration and switched to a Δ (delta) configuration when up to speed, which will also reduce the starting in-rush current. Single speed alternating current control is used for relatively low-speed elevators of up to 100 fpm (0.5 mps). Stopping is accomplished by removing power from the motor and applying the brake, resulting in floor stopping accuracy that may vary plus or minus 2 in. (50 mm), depending on the loading in the elevator.

Refinements of alternating current motor control include the use of two-speed motors with ratios from 2:1 up to 6:1 between the low-speed and high-speed windings. The elevator runs with the high-speed winding and is switched to the low-speed winding when it is near the floor stop. Final stopping is done with the brake, and stopping accuracy may be improved to plus or minus 1 in. (25 mm). Other approaches use the low-speed winding for releveling; that is, the elevator is switched to the low-speed winding for a floor stop, and if it overshoots floor level, the motor is reversed and the elevator is leveled back to the floor at low speed. Floor stopping accuracy can then be guaranteed within the limits of the dead zone, that is, the distance above and below floor level that the elevator must be out of level before releveling takes place. Normal dead zones are usually ½ to ¾ in. (13 to 19 mm) above and below floor level.

The application of elevators with alternating current motion control was generally limited to apartment houses of up to about six floors, slow-speed freight elevators, and dumbwaiters. It was quite popular from about 1920 to 1960 as the accepted motion control in low-rise apartment houses and is an accepted standard in most low-rise European applications. It must be provided with leveling to meet most handicapped requirements and the new European Union (EU) directives.

Alternating current elevators have been generally superseded by hydraulic elevators for lower-rise applications (five stops or fewer). The hydraulic elevator covers the lower-speed range as do alternating current traction elevators, is universally designed with leveling, and is, in general, more economical to install than an alternating current traction elevator. The alternating current elevator, since it is a counterweighted elevator, requires far less horsepower, hence less energy consumption, than the hydraulic, which requires horsepower sufficient to lift both the dead weight and the carrying load of the elevator. To combat the additional power requirements of hydraulics, manufacturers have added counterweights in various configurations. These systems are referred to as roped hydraulics. The roping can be configured with pulleys (roping $n:1$, $n = 1, 2, 3, 4, 5$) to limit the range of travel of the counterweights so as to save space and achieve the same horsepower savings. The hydraulic elevator usually has a lower initial installation cost since it does not require a car safety device, counterweight, or counterweight rails and it imposes very little load on the building structure.

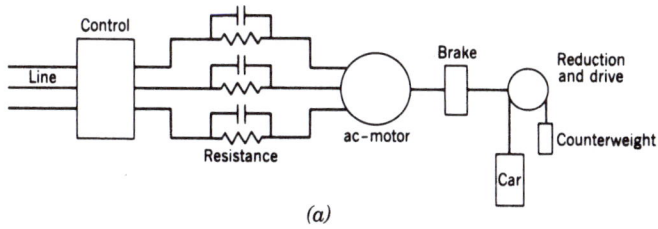

(a)

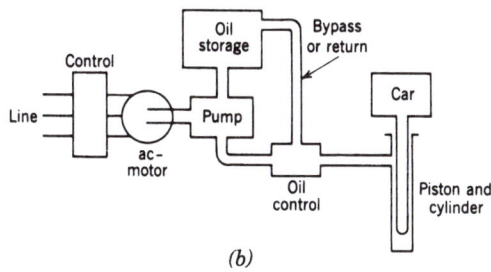

(b)

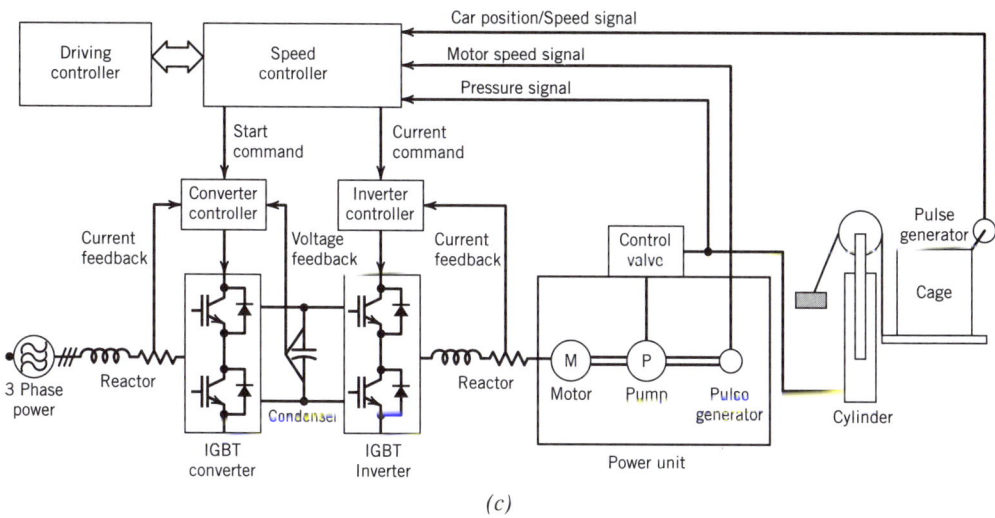

(c)

Figure 7.1. Elevator motion control systems: (a) alternating current; (b) hydraulic; (c) variable speed pump hydraulic (part c courtesy CEMCOLIFT).

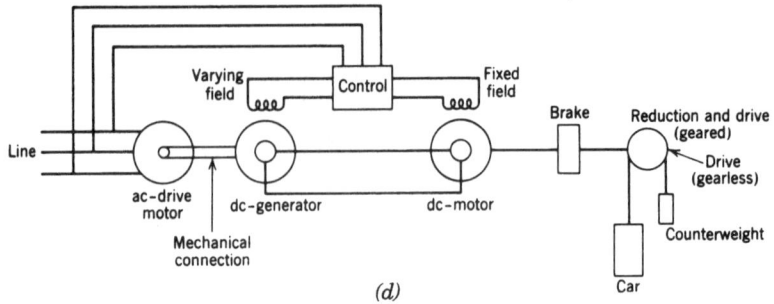

(d)

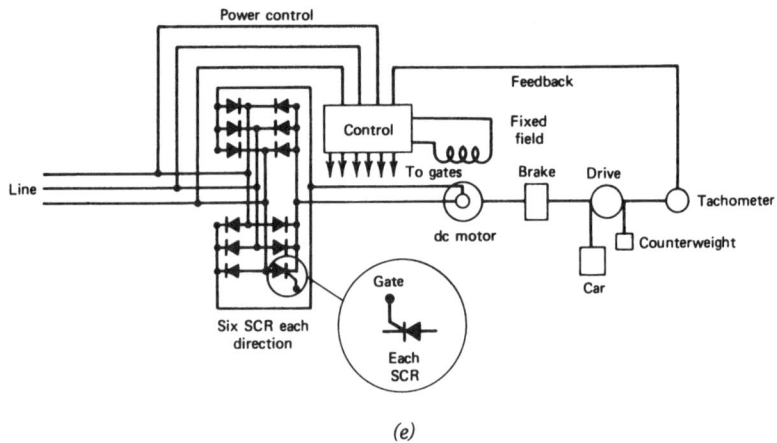

(e)

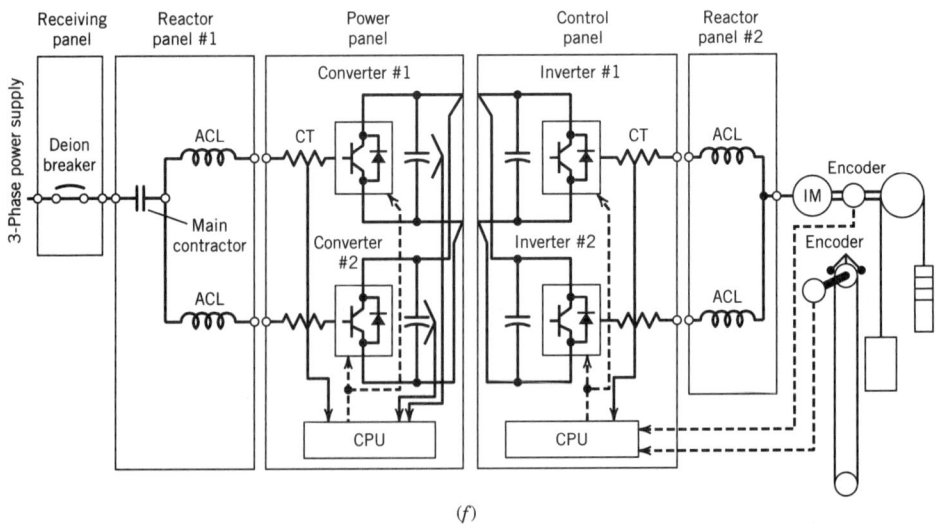

(f)

Figure 7.1. *(Continued)* (*d*) Generator field control; (*e*) SCR (silicon controlled rectifiers) motion control; (*f*) fully regenerative AC VVVF system (part *f* courtesy Mitsubishi).

Hydraulic elevators use resistance (or Y/Δ) for the single-speed motor used to drive the pump supplying oil to the piston and pumping in the up direction only; however, some are beginning to use variable speed pumps. Stopping is accomplished by throttling and then cutting off oil flow (Figure 7.1b). Leveling is accomplished by restricting oil flow or by bypassing a measured quantity of oil. Down control is accomplished by opening a down valve and allowing oil to return to the storage tank.

Hydraulic elevators are restricted in top operating speed to about 200 fpm (1.0 mps). Some new applications in hydraulics include variable speed pump units. Rather than modulating a valve to control the acceleration and deceleration of an elevator, the control system modulates the speed of the motor connected to the pump using variable voltage, variable frequency drive (VVVF). Unlike conventional hydraulics, this control system uses feedback to control the speed of the motor, providing for a smoother, faster, and more consistent ride under all load and temperature conditions. Figure 7.1c shows a system block diagram of the variable speed pump hydraulic. From the diagram it is readily apparent that the variable speed pump hydraulic system is considerably more complex than the simple conventional hydraulic.

Applications for new installations for the midrange geared machines of 200 to 500 fpm (1.0 to 2.5 mps) and high-range gearless machines of 500 fpm (2.5 mps) and above are dominated by variable speed alternating current machine motors and drives. Prior to the 1980s this application range included mostly variable speed direct current (dc) using Ward-Leonard generator field motion control systems and solid-state thyrister (SCR) drives. Today dc applications in this range are almost exclusively for modernization where dc equipment exists and replacement of the dc machinery is, for a multitude of reasons, not economically viable.

Generator field motion control consists of providing a varying voltage to a direct current elevator drive motor. The characteristics of the direct current drive motor are that it has the torque to move the elevator load smoothly up to speed and can absorb the inertia of the moving load by regenerating to stop the elevator with smooth deceleration. Stopping is independent of the brake, with all the energy being absorbed back through the electrical system. The system consists of the dc drive (hoist) motor and the associated direct-connected or geared drive mechanism. The armature of the drive motor is directly connected to a source of controlled dc voltage such as the armature of a dc generator (Ward-Leonard system; see Figure 7.1d) or a bank of silicon controlled rectifiers (SCR; see Figure 7.1e).

With solid-state motion control, the most common type currently being the silicon controlled rectifier (SCR), the source of the varying dc voltage is the bank of SCRs wherein the gate controls the amount of the wave of ac current that can be conducted. This, in turn, limits the voltage applied to the dc drive motor. The control of the firing signal to the gate is done through a feedback loop. The tachometer generator develops a signal that is proportional to the speed of the drive motor. This signal is compared with a reference signal contained in the control circuitry and further modified by signals that indicate that the elevator should either be speeding up, slowing down, or running at constant speed. The gates are then energized to fire at the precise instants that allow the controlled amount of current to flow to the drive motor.

When the elevator is traveling up or down with sufficient car or counterweight load to cause overhauling, the dc hoist motor acts as a generator. This voltage is fed back through the SCR bank through reverse gating and back into the power feeders.

The bank of 12 SCRs shown in Figure 7.1e is arranged so that when the ac power flows toward the machine in the positive wave, only three SCRs conduct; in the negative wave, three others conduct; and when current is flowing away from the machine, one set conducts at each instant to match the incoming power wave characteristics to create ac power from the regenerated dc power.

The SCR system shown in Figure 7.1e is a closed loop; that is, the tachometer feeds back a signal to the control to vary the input to the dc hoist motor armature depending on the speed of the elevator as compared with a desired pattern. The dc generator field control (Ward-Leonard) shown in Figure 7.1d is an open loop; that is, additional control circuitry consisting of a floor selector or switches in the hoistway responds to the position of the elevator and, by contact closure, drops out or pulls in relays to control the direction and amount of dc current being fed to the driving machine.

As the voltage is increased on the drive motor armature, the elevator or load is accelerated up to speed. As the voltage is decreased, speed is reduced until the elevator comes to a complete stop, and the brake is then applied to hold the car at the floor. Releveling requirements owing to changing load at the floor are accomplished by lifting the brake and applying small voltages to the drive motor.

Operating speeds of generator field or SCR-drive elevator-motion control systems are available in any speed up to the present high of 2000 fpm (10.0 mps). For gearless machines, performance is from 4 to 5 sec for a one-floor run of 12 ft (3.6 m) and leveling accuracy under all conditions of loading is from plus or minus $\frac{1}{4}$ to $\frac{1}{2}$ in. (6 to 13 mm) with minimum time penalty incurred for leveling. For lower-performance geared machines, floor-to-floor time is 5 to 6 sec.

Gearless machines are greatly capable of acceleration rates of 4 ft per sec per sec (written as 4 fps^2) (1.2 mps^2) with generator field control and can be made to accelerate faster with SCR motion control. Geared machines are generally capable of acceleration rates of 3 fps^2 (0.9 mps^2) and can also be made to accelerate faster. The limiting factor is not the accelerating rate, but the rate of change of acceleration (sometimes referred to as jerk) that is felt by the riding passenger. This is a matter of personal tolerance, but, in general, an upper limit of 8 ft per sec per sec each second (written 8 fps^3) (2.4 mps^3) is usually the maximum.

Variable speed alternating current control systems are the current standard in new medium- and high-speed elevator applications. Although there are many different ways to attain variable speed alternating current control, I will limit the discussion to variable voltage, variable frequency (VVVF). The most basic principal of ac motor theory and control is that the speed of an ac motor is governed by the number of poles that the motor contains (physical construction) and the frequency of the alternating current applied to the motor. Other factors do play a role in the actual speed of the motor but are eliminated for the sake of this discussion. The basic equation is, Speed is proportional to the frequency divided by the number of poles of the motor. Therefore, if you increase the frequency, you increase the speed, and vice versa. Since the impedance of an ac motor (inductive winding) is proportional to the frequency of the applied voltage as the speed is varied close to zero, the impedance also gets close to zero. Since there is an applied voltage to the ac motor and the current through the motor is inversely proportional to the impedance and proportional to the current, in order to keep the current requirements within reason the applied voltage must also be varied, hence, variable voltage, variable frequency control.

This is a very simplified explanation of a very complex system and should be used only to understand the very basic principles and not to evaluate or design such a system.

The VVVF drive comes in two basic configurations, regenerative and nonregenerative. For most midrange geared applications, a nonregenerative drive is used. This usually means that the braking power is not recovered but usually diverted to a bank of resistors converting the power to heat. The nonregenerative configuration cuts the cost of the equipment considerably; however, it has higher operational power costs. The regenerative configuration is in essence a dual inverter drive. One inverter varies the motor frequency, and the second inverter pumps regenerative power back into the line, almost doubling the basic cost of the drive. A block diagram example of a fully regenerative ac VVVF system is shown in Figure 7.1f. The drive system basically converts the line voltage into direct current and an inverter generates the variable frequency voltage. The devices used in a typical ac drive are MOSFETs (metal oxide semiconductor field effect transistors). These devices are very-high-gain, fast-switching transistors with low saturation on resistance. Even though the resistance is low for large drive applications, the MOSFET devices must be paralleled to provide sufficient output current.

All other operating performance characteristics of ac VVVF systems are similar to those of dc solid-state drive systems.

A chart is shown in Figure 7.2 comparing the energy requirements of different types of drive systems.

As an example of the calculation of the expected running time of an elevator for a given distance, a given top speed, and with an acceleration and deceleration rate of 4 fps^2 (1.2 mps^2), the following derivation is given.

Example 7.1. Expected Running Time of an Elevator

Given: 300 ft (91 m) express run from the ground floor to floor 24, 700 fpm (3.5 mps) elevator speed. What is the time from start to stop? (See Figure 7.3).

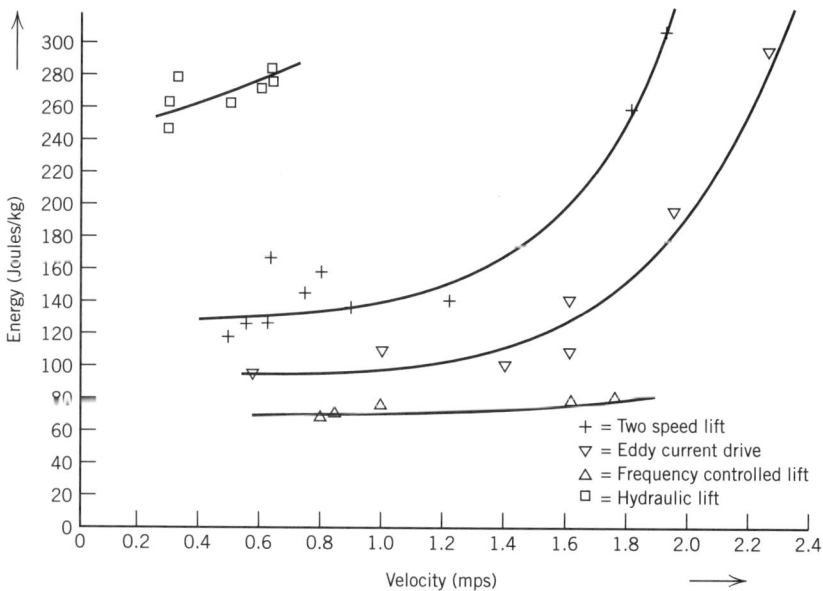

Figure 7.2. Energy requirements of different types of drive systems (Courtesy IAEE).

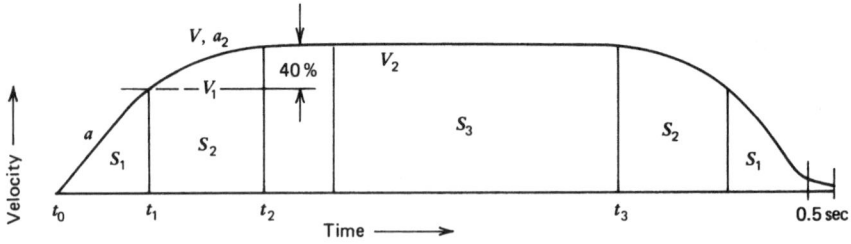

Figure 7.3. Elevator travel—velocity versus time.

Assumptions: Between t_0 and t_1, acceleration = constant at 4 fps². Between t_1 and t_2, the product of the velocity and the acceleration Va = a constant. The time t_1 is when approximately 60% of full speed is attained.

During deceleration the same conditions apply. 0.5 sec is added to allow for leveling into the floor at the end of the slowdown.

$$a = 4 \text{ fps}^2 \quad V_1 = \frac{0.60 \times 700}{60} = 7 \text{ fps}$$

$$V_2 = \frac{700}{60} = 11.67 \text{ fps}$$

$$t_1 = \frac{V_1}{a} = \frac{7}{4} = 1.75 \text{ sec} \quad S_1 = \frac{V_1^2}{2a} = \frac{7^2}{8} = 6.125 \text{ ft}$$

$$t_2 = \frac{V_2^2 - V_1^2}{2V_1 a} + t_1 = \frac{11.67^2 - 7^2}{2 \times 7 \times 4} + 1.75 = \frac{87}{56} + 1.75 = 3.3 \text{ sec}$$

$$S_2 = \frac{1}{3a}\left(\frac{V_2^3}{V_1} - V_1^2\right) = \frac{1}{3 \times 4}\left(\frac{11.67^3}{7} - 7^2\right) = \frac{1}{12}\left(\frac{1589}{7} - 49\right)$$

$$= \frac{1}{12}(227 - 49) = 14.8 \text{ ft}$$

$S_1 + S_2 = 6.125 + 14.8$ = 20.925 ft
$2(S_1 + S_2)$ = 41.85 ft
$S_3 = 300 - 41.85$ = 258 ft
$t_3 - t_2 = \dfrac{258}{11.67}$ = 22.1 sec

Total time = 22.1 + 2(3.3) + 0.5   = 29.2 sec

    Speed regulation between no load and full load with a regulated feedback drive system should be plus or minus 5%. Any tendency to overspeed or underspeed should be governed and corrective measures applied. The A17.1 Elevator Code requires an elevator to be electrically stopped if its running speed exceeds its rated speed by varying percentages of 10% or more, depending on the rated speed of the elevator. A graphic description of the time elevators require at various speeds to get up to full speed from a stop, and immediately slow down to a stop, is shown in Figure 7.4. These times include 0.5 sec for leveling.

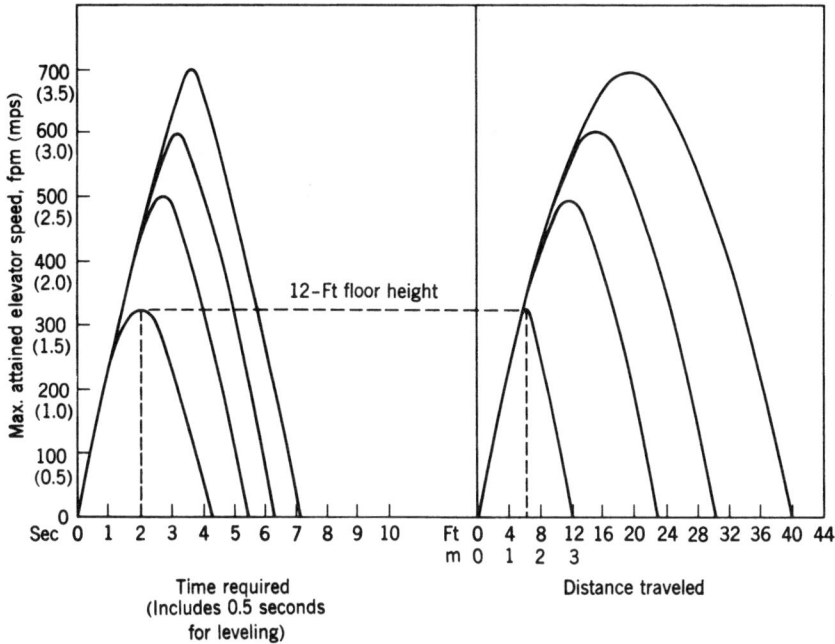

Figure 7.4. Elevator speed versus time and distance.

Door Control

Operation from floor to floor or getting the elevator up to speed and stopping is the major function of the motion control system. A secondary, similar motion control system is the control of door operation. Like the elevator, the doors must be accelerated, decelerated, and stopped during both their open and close cycles. The A17.1 Elevator Code limits the force of the closing doors to 7 ft pounds (in the European code, CEN81, this force is stated as 0.29 joules) and the thrust of the closing doors to 30 lb (13.6 kg); that is, 30 lb are required to prevent the doors from closing. These forces are for doors with door protective devices, such as a safety edge, which is a mechanical device that, when displaced, activates a switch to reopen the doors; photoelectric devices, which will reverse the doors when closing, or other devices, such as ultrasonic, infrared, radar, or electronic proximity devices, which will sense a presence in the path of a closing door and reverse the door to minimize impact.

Closed-loop door control in various forms has improved door performance considerably. Door controllers can use voltage velocity position, acceleration, and current feedback to improve the operational reliability and safety of door systems (see Figure 7.5). They can provide faster and smoother performance while ensuring that the maximum force and kinetic energy requirements dictated by code are met. The force and kinetic energy requirements can at times be the limiting performance factor, as the kinetic energy requirements dictate the maximum speed of the doors (KE = 0.5 * mass * velocity^2, where mass is the mass of the car door, landing door, and all the movable connections, including the door operation; velocity is the velocity of the door at its maximum speed in closing).

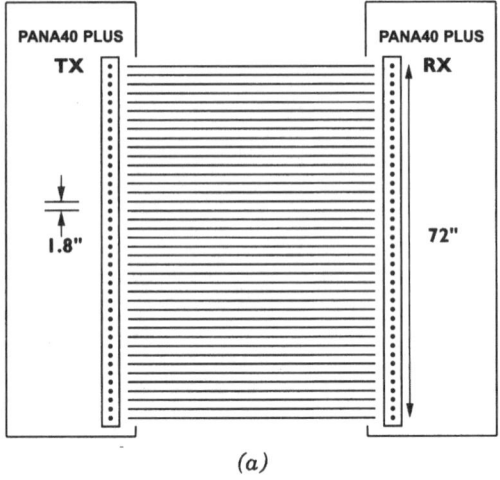

(a)

(b)

Figure 7.5. (*a*) The "PANA 40" superdense beam curtain (Courtesy Janus Elevator Products, Inc.). (*b*) The Progard R radar proximity detector (Courtesy Schindler Elevator Co.). (*c*) The "intensive mirror" principle as applied to the protection of both the car and the hoistway doors.

Second mirror
at 136°
to first mirror

Beam emerging at 90° to incident beam

First mirror at 22.5°
to incoming beam

Incident beam

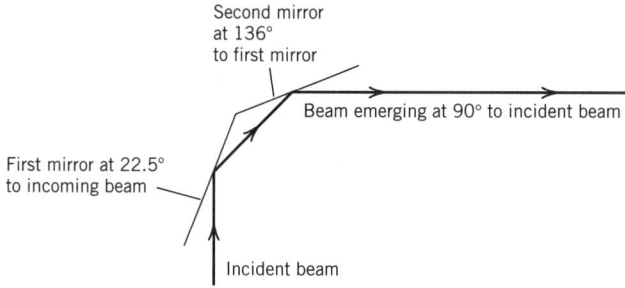

The "Insensitive Mirror" Principle

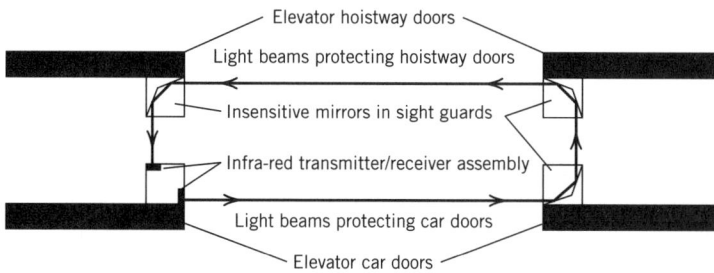

Elevator hoistway doors

Light beams protecting hoistway doors

Insensitive mirrors in sight guards

Infra-red transmitter/receiver assembly

Light beams protecting car doors

Elevator car doors

(c)

Figure 7.5. *(Continued)*

If no door edge protective device is used or if the protective device fails, the closing force of the door is required by the A17.1 Code to be 2.5 ft poundal (0.115 joules) or less. This is accomplished by reducing closing speed and is featured by "nudging" an electrical circuit that operates if the doors are held open excessively long (15 to 20 sec) in addition to a reasonable time for passenger transfer of 2 to 5 sec. Nudging can be found in two forms, one that continues to try to close the doors with a slight forward movement and pressure, and the other that stops the doors until the obstruction is removed.

The operating system should establish the length of time that doors remain open at a floor, which is called "dwell-time." Such times are usually about 2 to 3 sec for a stop made in response to a car call, and from 3 to 5 sec for a stop in response to a landing call, depending on the relation between the location of the elevator and the location of the landing button fixture. A further, separate, adjustable time should be established for loading or unloading at a main terminal floor. The time for loading at a main terminal should be varied, depending on traffic. For example, during heavy traffic periods the elevator doors should start to close about 5 to 15 sec after a car call is registered, depending on the expected loading of the elevator, or sooner if the load sensing device detects a predetermined load.

Additional refinements of the adjustment of dwell-time include shortening the time the doors remain open after a passenger transfers into an elevator after being detected by a photoelectric device in the entranceway, and extending dwell-time for a car call if the elevator is filled to a predetermined percentage of capacity load to allow additional time for passengers to leave. Advanced systems have an electronic proximity device or an ultrasonic detector on the elevator doors, which allows the doors to be closed within the minimum transfer time.

The time the doors remain open, in an installation designed for heavy traffic, should be adjusted to the load in the elevator. Transfer to and from a lightly loaded car can be faster than to and from a filled car. The adjustment of the dwell-time should permit this distinction. Similarly, transfer time for loading or unloading at the lobby will be longer and should be recognized. Any of the aforementioned transfer times should be independently adjustable to suit the requirements of the building in which the elevators are installed, as well as the characteristics of the traffic they are expected to serve. Separate car-stop and landing-stop dwell-times should be specified for any building to match expected passenger transfer time with the operation of the elevator.

In addition to the time required to open and close the doors plus the dwell-time, there is usually a slight delay between the time the doors are fully closed and the elevator starts. This time, referred to as "lock time" or "build-up time," is usually about 0.5 to 0.8 sec and consists of the time necessary to ensure that the door is closed and locked and to start the mass of the elevator.

Doors can be started to open as an elevator levels to the floor, provided the various aspects of control can ensure that the doors will not be open wide enough for passenger transfer prior to the time the elevator stops. This slight preopening can improve elevator performance time and can be employed with motion control systems designed to ensure that such preopening can be done within a minimum distance of floor level.

The combination of all the times discussed to this point adds up to a performance time for an elevator on a one-floor run or trip (Figure 7.6). One of the means of assuring a quality elevator installation is to specify and to measure the performance time of elevators as installed. The performance time, or the time from the start of door closing at one floor until the doors are fully opened at another floor, is a measure of quality and can easily be determined. The measurements should be consistent under all conditions of loading, from full load to no load in the car, and in both the up and down direction.

Leveling

An extremely important aspect of an elevator's performance is the accuracy with which it levels to the floor and maintains floor level as passengers enter and leave. Leveling opera-

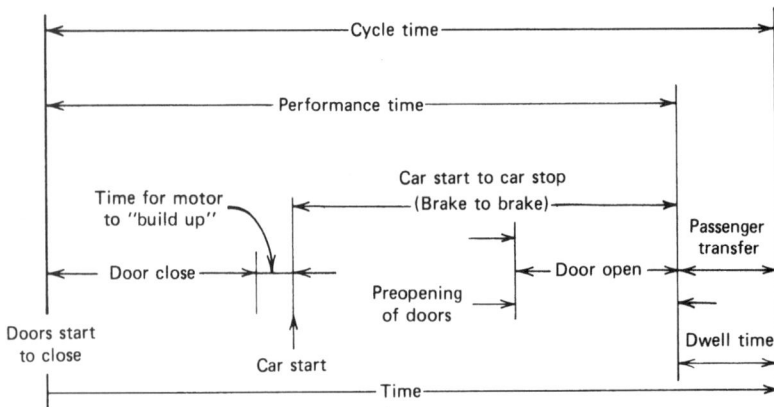

Figure 7.6. Elevator time elements—floor-to-floor performance.

tion should be direct, that is, the elevator should come to a stop at a floor within the prescribed leveling tolerance of plus or minus ¼ in. (7 mm) for a quality installation, without overshooting and leveling back or stopping short of floor level and leveling into the floor (often called spotting). Once the elevator is at floor level and the passenger load enters or leaves, the floor level should be maintained with minimum perceptible movement, which should always be toward floor level, never away. Suitable provisions should be made in the motion control system to maintain an elevator stopped at floor level under any adverse conditions, which may include building sway, to be discussed in a later chapter.

When lanterns are provided at the landings, they should light in advance of a car's arrival at the floor and in the direction the car is expected to leave that floor. Adequate advance lighting of a lantern will give prospective passengers a chance to approach the elevator and board it promptly, thus saving valuable seconds in transfer time. Recent legislation suggests the combination of advanced landing lantern operation and door-hold open time to meet designated requirements to provide ample time for the mobility impaired to board the elevator.

There are many other things an individual elevator should do. They may be best summed up by requiring any elevator to "inform" prospective passengers. If they observe the lantern signals, board the right elevator, and operate the car buttons promptly, the elevator should take them swiftly and smoothly to their destination.

OPERATION SYSTEMS

The ability of elevators to move in response to demands for service is provided by the operating system. In early elevators, hydraulic or steam-driven, the operating device was a continuous rope that ran the length of the hoistway and actuated a valve in the basement. To go up, the rope was pulled down to admit steam or water to the elevator driving means. To go down, the rope was pulled up and the driving machine was reversed. The pull in the opposite direction of travel had a distinct advantage; by sending the rope through a restricted hole in the car platform, a stop ball on the rope could be placed at the extremes of elevator travel. The continued upward or downward movement of the car caused the rope to be pulled by the car and would stop the elevator (Fig. 7.7).

Handrope or "shipper rope" operation, as it was called, had many advantages and continued in vogue for many years; early, electric elevators employed a handrope to open and close contactors to start, stop, and reverse the elevator. Some elevators still have a handrope, although it has been generally outlawed by local safety codes. In the early days when hoistways were open or enclosed by waist-high gates, one could reach in, pull the rope, and summon the elevator—an "automatic" albeit very unsafe operation.

Car-switch Operation

With the introduction of electric elevators, the natural step was to put a switch or lever in the car which when operated would electrically start the car in the up or down direction, and if centered would stop the elevator. The electric impulses to the motor room from the moving elevator were transmitted by a traveling cable. This was an electrical cable that hung from some point near the middle of the hoistway and was attached to the underside of the car. With electric elevators and increased concentration on the safety of passengers, such things as hoistway door interlocks and car gates were introduced and subsequently

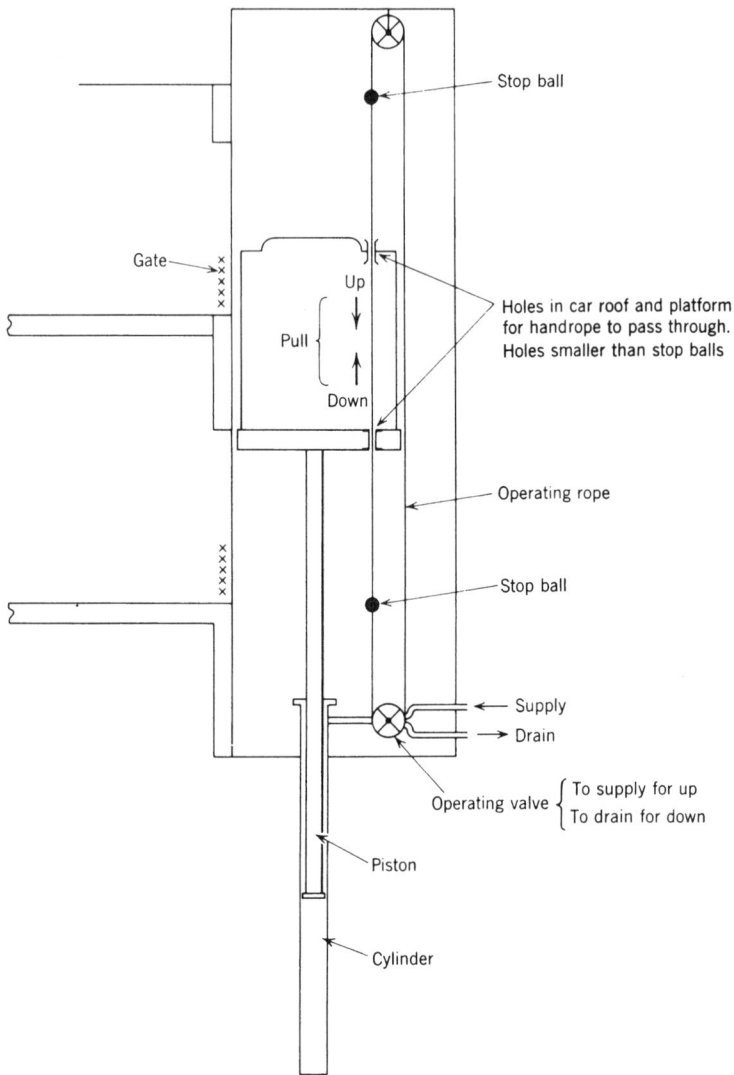

Figure 7.7. Hydraulic elevator—handrope operation.

required by safety codes. The interlock prevents the elevator from being operated unless the door is closed and locked. Electrical signaling devices such as annunciators and call bells were also introduced as hoistways became more fully enclosed and an efficient means to signal the elevator operator was required.

Car-switch operation, as it was known, was the chief operation system available in office buildings until the early 1920s. Also developed were elaborate signaling systems to inform the operator where he was to stop, as well as buzzer systems so that the elevator starter at the main floor could direct the elevators. To aid the starter in the main lobby in this task, supervisory or "starter's" panels with lights were added to indicate where landing calls were waiting, as were either dials or lights to show the position of the elevators in the hoistways.

Automatic Operation

Parallel with the development and refinement of car-switch operation for office buildings came the introduction of the simple automatic electrical operating system for single elevators. This was primarily intended to make the elevator acceptable to the small residential apartment house where the traffic did not warrant a full-time elevator attendant. Safety was established by the use of interlocks on the hoistway doors and contacts of the car gate to ensure that both were closed before the elevator could be run. An automatic operation system known as Single Automatic was introduced and became the accepted apartment house operating system in the United States for the period from about 1890 to 1920.

Single Automatic Push-Button operation, or Single Automatic operation as it is presently known, consists of single buttons at each landing and a button for each floor in a car operating panel. The car can be called from any floor, provided the door and gate are closed and no one is operating the car. Once the car is intercepted, entering it, closing the hoistway door and car gate, and operating a car button ensures the riding passenger exclusive use for that trip. This is a light-service operation; the elevator can serve only one call at a time and the next passenger must wait until it is free before he or she can use it. To indicate availability, an "in use" light is generally placed in the landing call fixture. When the light is out, a passenger can call the car. To make Single Automatic operation work properly, sequence switching and timing must be provided so that the passenger who enters will have a chance to register the call before another passenger can call the car to another floor.

Single Automatic operation remains a valuable operation in light-traffic buildings where exclusive use of an elevator is desired. It is ideal for garages and factories where only one vehicle can fit on an elevator at one time. Single Automatic operation is the preferred operation for apartment houses in Europe where elevators are small and people prefer to wait to gain exclusive use.

Signal Operation

As elevator speeds were increased past 500 fpm (2.5 mps), the skill of the elevator operator on a car-switch elevator had to increase so that the operating handle could be dropped precisely and the next stop made with fair accuracy. Various schemes to overcome this difficult precision were tried.

A light would flash just before the next required stop, and the operator had to respond with split-second accuracy. Leveling devices were applied to eliminate jockeying at the floor.

These approaches were satisfactory until taller buildings required speeds in excess of 700 fpm (3.5 mps). No operator could recognize when to stop when floors were passing at about a second apart, and even then the decision to stop had to be made at least 20 to 30 ft from the desired floor.

The need for an improved operating system for fast elevators became apparent, and signal control was introduced. Signal control is a push-button operating system in which the elevator operator registers the call for the floors at which the passengers wish to stop and operates a button or lever to start the car. Acceleration, response to floor stops, deceleration, and leveling are all done automatically. Response to landing calls is done in the same manner; the operator does not know what landing calls the elevator will respond to until the car starts to slow down.

Signal control, or signal operation as it is presently known, removed the speed limitation from elevators. Speeds of 1000, 1200, and 1400 fpm (5, 6, and 7 mps) were attained, and all the notable buildings of the late 1920s and the 1930s had signal-control elevators. These include the Empire State and Chrysler Buildings and most of Rockefeller Center, as well as hundreds of others. Most buildings of that era have been modernized, and Group Automatic Operation has now been provided.

Technology also came to the aid of the elevator starter. In addition to indicator lights to show waiting landing calls and the position or motion of each elevator, electrical scheduling systems were introduced. These systems attempted to operate cars on schedules maintained by the operator's response to signal lights in each elevator. The light indicated when to start and was timed from the departure of the previous car to maintain a time spacing of elevators on both the up and down trips.

The objective of spacing elevators throughout a building to render the best possible service to waiting passengers was only imperfectly realized. Elevator operators saw little point in rushing when traffic seemed light and could not resist the opportunity to stop at a floor and enjoy a smoke or conversation. Electrical dispatching gave way to human dispatching, and many such systems were abandoned.

Collective Operation

While signal operation was being introduced for the taller office buildings, demand was increasing for automatic operation for apartments, hospitals, small office buildings, and any building that required elevators but had only limited traffic. The need for better service than the Single Automatic elevator could provide led to the introduction of the collective elevator and improved means of automatic elevator door operation.

Collective operation, as the name implies, is a means to collect or remember and answer all the calls in one direction, reverse the elevator, and then "collect" and answer all the calls in the opposite direction.

The common collective operation is Selective Collective Operation. With full selective collective, both up and down landing call buttons are provided. In addition to car calls, the elevator will stop only for up landing calls in the up direction and down calls in the down direction, all calls being remembered until answered. A variation of selective collective is Down Collective, in which only down landing buttons are provided at each upper floor. This is acceptable in most apartment houses where people generally want service up to their apartments from the lobby (up car call) and down to the lobby from their floors (down landing call). Another variation is Single-Button Collective, whereby the landing call button will stop either an up- or down-traveling elevator (sometimes referred to as interceptive collective).

Automatic operation of elevators is enhanced by the power operation of the doors and ample protection to keep the closing doors from striking the passengers. Earlier automatic elevators had manual doors and gates, usually equipped with spring or weighted closers. The passenger had to open both, once an elevator answered a landing call. With power operation, the doors open automatically, remain open for a time interval, and automatically close.

If the leading edge of the door touches or is about to touch a person or object, a sensing device on the door edge should actuate the reopening of the door and recycle the closing operation. A photoelectric device projecting across the entrance can be used in the sensing device to ensure that the door will remain completely open until the entrance is clear. With provisions for disabled persons, many elevator codes are requiring photoelectric devices.

Selective collective is the accepted operation for single elevators in any type of building and for service elevators in major office buildings. Its utility has been enhanced by added features such as home landing, so that the elevator is always returned to a given floor on completion of any trip; load sensing devices, which prevent the car from stopping for additional landing calls if the weight in the car exceeds a predetermined percentage of capacity; independent service operation, wherein the car can be used to handle special loads and not respond to landing calls; and attendant operation, so that operation is manually controlled for security and freight-handling requirements.

Two-car selective collective operation is commonly known as Duplex Collective. This operation can be either full selective or down collective and consists of the following operations.

One of the two cars is designated as the home-landing car. It can be either car and may change after each trip. The other car is called "free" and may park at its last call or some designated landing (middle free car parking) or in a designated upper area of the building. The home-landing car is designated to respond to any landing call at the lobby or above it and below the free car. The free car is designated to respond to any up landing call above it or to any down landing call.

If the home-landing car is taken to an upper floor by a car call, the free car becomes the home-landing car. If the free car is set in motion by a car or landing call, and a call occurs behind the free car (either an up or down landing call below an up-traveling free car, or an up or down landing call above a down-traveling free car), the home-landing car can start to "help out" the free car. In addition, the home-landing car should start if any landing call, behind or not, remains unanswered beyond a predetermined time or if the free car does not move within a predetermined time. This latter feature, known as the "milkman circuit," was developed because the milkman would block the elevator with his carrying rack while he delivered to the various apartments. Since tenants could be registering calls ahead of the free car—for example, if the milkman's car was at the second floor—the home-landing car might never start; hence, the timing logic had to be developed.

Duplex Collective will adequately serve the normal expected traffic in an apartment house if the elevators have sufficient capacity to serve the needs of the building. Duplex Collective operation has no means to separate elevators if they are both traveling in the same direction, and the cars will "bunch."

If heavier traffic persists for any period of time because of, say, an underelevatored situation, Duplex Collective operation is undesirable as a system for heavier-trafficked buildings.

As we stated, with duplex collective a call behind the free car starts the home-landing car. In a busy situation, one or many calls behind may appear immediately. The former home-landing car is traveling right behind the free car and in the same direction. With many calls, the elevators are "leap-frogging," that is, both will arrive at the lobby about the same time, will leave again, and start the cycle over—a condition known as "bunching." Instead of "two-car coordinated operation," the operation is "two cars bunched," or not much better than a single large elevator.

A number of steps must be taken to improve two-car operation as well as any group operation for heavier traffic. Some of these steps have been described in the discussions of the various types of traffic in previous chapters. The application of various traffic-handling features to elevators results in another class of operation known as Group Automatic Operation.

Group Automatic Operation

In viewing the history of elevator operations, the almost parallel development of signal operation with attendant for the larger, busier building, and the collective (both single and duplex) nonattended operation for residential and lighter-traffic buildings can be observed. In about 1949, a number of developments occurred that started the rapid elimination of elevator attendants in any type of building.

At that time the completely automatic, operatorless elevator was introduced. This innovation consisted of a number of steps such as the development of an operating system that caused the elevators to respond to changing traffic demands, the refinement of the door protective devices, and the establishment of timing and scheduling systems logic circuits that maintained elevators on operating schedules, in response to traffic requirements and without the necessity of human interference.

The demand for good automatic operation for larger groups of elevators was apparent for some time. Various attempts to meet this demand were made using three- and four-car selective collective operations, which were generally unsatisfactory for the changing heavier-traffic situations such as incoming, two-way, and outgoing traffic. With a collective system and with increased traffic, serious bunching can and does occur. However, the use of automatic collective operations, plus the fact that people had been operating elevators by themselves in various types of buildings for years, allayed any fear that they would be unwilling to do so in a large office building.

All the developments came together in 1949 when the Otis Elevator Company introduced, through the efforts of their chief development engineer, William Bruns, a fully automatic elevator system for an office building in Texas. In addition, there was a citywide strike of elevator operators in Chicago, which provided the marketing awareness. The total system provided the necessary scheduling system, which made the decisions to start elevators on schedule, introduced a proximity system of door protection, and developed the various load-sensing and protective devices to replace the other functions of an elevator attendant.

The closing of the car and landing doors was the major task of elevator attendants. It was their duty to ensure that no one was coming and to close the doors to start the elevator. Reliable electric power door operators were available, and to provide good automatic operation, it became essential to ensure protection against the doors' striking a person and a means to reverse such doors swiftly if a person or object was encountered during closing. This task was aided by a reduction in the speed, and thus the kinetic energy of the closing doors, which resulted in only a slight time penalty over the speed at which they were being closed by an attendant. Otis introduced an electronic proximity device as a door protection device, whereas other elevator companies improved the movable door safety edge and photoelectric devices by making the edge retractable and altering dwell time after the light beam was interrupted by passenger transfer.

The proximity-type of door detection was developed based on the natural phenomenon that all people and objects have a degree of electrical capacitance to ground. An electronic field generated by a device at the edge of the elevator car door reacts to this capacitance to create an electrical signal that reverses the closing doors. The electronic field can be adjusted in sensitivity to include the edge of the hoistway door in addition to the car door so protection against both the hoistway and car doors could be provided. Unfortunately, if the sensitivity is increased too greatly, false reversal can occur, particularly during periods of increased humidity. The further development of ultrasonic detectors may provide more reliable and less sensitive door protection. Today modern door detection devices use infrared beams in various configurations as shown in Figure 7.5.

A safety edge is simply a movable strip on the leading edge of the car door, which is deflected when it touches a person or object. This deflection actuates a switch to cause electrical reversal of the door. The photoelectric device projects a beam of light across the entrance which, when intercepted, also provides a signal to reverse the closing doors. The photoelectric device can also provide an electrical signal to decrease the amount of time the doors remain open (dwell-time). This signal, plus additional operational logic to determine whether the passenger is moving out of the elevator (stop in response to a car call) or boarding the elevator (stop in response to a landing call), can be helpful in minimizing the time spent at a floor and improve elevator service. A further refinement of logic is to relate the time the doors remain open to the number of people in the elevator through load sensing. If the car is filled, additional time can be allowed over the dwell time allowed for a lightly loaded elevator.

The elevator attendant had one additional task in relation to door operation. If the elevator was delayed, the attendant's job was to enforce the closing of the doors. With automatic elevators this is accomplished by electrical logic, which can initiate door closing after a certain elapsed time. If someone is holding the door an excessive length of time, about 15 or 20 sec, nudging, as described earlier in this chapter, will occur.

A further task of the operator was to count the number of people in an elevator to limit loading and to leave the loading floor when sufficient passengers were on board. If the car was filled at an upper floor, the attendant could also bypass additional landing calls. These functions are replaced by load-sensing devices. To reduce the danger of overcrowding, loading restrictions were developed by limiting the area of the car in relation to its capacity in pounds or kilograms to limit the number of people in the car. Another function of the attendant was to register the passengers' calls, which they could easily do themselves.

The final task of the operator was to pay attention to the elevator starter or to an electrical signal. This function was replaced by initiating door closing and car starting by electrical signals and by relying on the door safety edges to protect the passengers if transfer was still in progress.

The elevator starter in a main lobby generally provided the signals and the schedule of elevator operation in relation to the various demands for service as noted from a landing call and car position indicator in the lobby. Specialized electrical computers were developed to weigh all the factors a starter must consider, plus many more that human limitations did not allow. Based on these factors, the various electrical signals for starting, stopping, and reversing elevators are given, and the system operates completely unattended.

Group Automatic Operating systems were developed by the various elevator manufacturers, based on their differing philosophies of elevator operation and logic. During the period from about 1950 to 1982, the Otis Elevator Company introduced the Six Program Autotronic, Four Program Autotronic, Basic Autotronic with Multiple Zoning, VIP 260, Elevonic 101, and, in 1982, the Elevonic 401. Westinghouse's systems were the Selectomatic Six Pattern, Selectomatic Four Pattern, and Selectomatic Mark IV and Mark V. Haughton, called its systems Auto Signamatic, 1090, 10921C, and Aconic. Both Haughton and Westinghouse were acquired by the Schindler Elevator Company, who currently call their systems "Miconic." Dover's system is called the Traflomatic, and Montgomery's is the Miprom. Each represented some degree of change or development, and most systems in the 1990s employ some form of microprocessor using programmable computer and solid-state electronic circuitry.

These group automatic elevator operating systems were all designed to accommodate the various differences between incoming, two-way, and outgoing traffic in busy buildings. Success of these efforts to gain public acceptance of automatic elevators in any type of building is evidenced by the fact that every elevator installed today is completely automatic with attendant features only for special services.

Group Automatic Operation is merely the means to provide the most effective operation of a fixed number of elevators to serve the varying traffic requirements encountered as a building functions during its intended use. The results are most important and can be simply stated as providing vertical transportation with the minimum waiting and riding time that optimizes the traffic-handling capability of the system. Stated otherwise, a grouping of elevators in a building has a maximum people-handling "quantity." The group operations system provides this quantity at optimum "quality" of service.

Different elevator manufacturers use different philosophies in approaching this task, and the following paragraphs will be a critical look and comparison of some of the logic (software) of the decisions used to effect elevator operation for various traffic situations, as well as some background on the electronic (hardware) systems.

Group Automatic Supervisory Systems

Perhaps the most dynamic aspect of the changes occurring in modern elevator operating systems has been the introduction and use of microprocessors with programmable features. The elevator operational controllers need no longer be massive and hard-wired with a myriad of relays to translate the built-in design logic required to respond to many calls for service as a result of changing conditions. The microprocessor affords the ability to establish a basic design, and the programmable feature allows relatively easily made changes to adapt that basic logic to the particular requirements of a building.

Computer buzzwords such as ROM, CPU, RAM, PROM, REPROM, LSI, *burn in, algorithm,* and others are abounding among elevator technicians, but through it all, the same basic requirements remain: making elevators provide their optimum handling capacity with the optimum quality of service. The microprocessor allows the collection of more information at an extremely high speed, so more meaningful and timely directions can be given to the individual elevators to best deploy them for the immediate and constantly changing tasks of serving people.

The inputs to the microprocessor can be divided into three classes. The primary inputs are those generated by people who are either the prospective or riding passengers. There are car calls, landing calls, and the loading of each elevator. The loading is represented by the passenger weight on the car platform. The secondary inputs are the fixed and readily recognizable aspects of individual elevators and the elevator system: the location and current operation of each elevator, whether it is stopped, with or without passengers (parked), with its doors open or closed; its direction of travel; the number of floors served; the elevator's capacity and speed and the time factors for door operation; dwell time; and performance from one floor to the next floor, or two or more floors away. The tertiary inputs are programmable and can consist of such factors as the time of day, day of the week, basic strategies such as the monitoring of the time individual calls wait, selected parking positions of the elevators, decisions as to whether a car will stop at or bypass a particular landing call, the predicted time to reach the highest call for reversal, the time to reach the main lobby, or other fixed decisions that represent the philosophy of the manufacturer as to how best to serve particular traffic situations.

These inputs must be transmitted either on an individual wire from the source to the microprocessor or by multiplexing; that is, each group of inputs is converted to digital signals and sent along a single wire and again converted at the receiving source. Multiplexing can serve to reduce the number of wires in the traveling cable (hence, cost and weight) substantially.

Since there is no obvious movement of relays with microprocessors, a means of checking proper operation must be provided for maintenance personnel. This may be done by indicator lights on the controller, by displaying the sequence of events on a printer or monitor, or by special test kits or tools. The intricacies of solid-state circuitry require that the entire electronic system be sectionalized and each part placed on a separate circuit board so that it can be readily replaced while the defective board is being repaired.

The ASME Safety Code for Elevators A17.1 defines group automatic operation as follows:

> . . . automatic operation of two or more nonattendant elevators equipped with power operated car and hoistway doors. The operation of the cars is coordinated by a supervisory control system including automatic dispatching means whereby selected cars at designated dispatching points automatically close their doors and proceed on their trips in a regulated manner. It includes one button in each car for each floor served and up and down buttons at each landing (single buttons at the terminal landings). The stops set up by the momentary actuation of the car buttons are made automatically in succession as a car reaches the corresponding landing irrespective of its direction of travel or the sequence in which the buttons are actuated. The stops set up by the momentary actuation of the landing buttons may be accomplished by any elevator in the group, and are made automatically by the first available car that approaches the landing in the corresponding direction.

The key word in the definition is *regulated,* which implies that the elevator is directed to perform certain operations based on data gathered by the group supervisory system at the instant the elevator is started (dispatched). Dramatic changes can take place from instant to instant as people are operating landing and car buttons and the elevators are changing positions. For this reason, the initial directions given an elevator as it starts may change before the elevator makes its first stop. A view of some traffic situations will attempt to show some of these changes.

Starting at a point in time when there is no traffic, which can be some time during the course of the working day, but usually before starting time and after quitting time, elevators are parked. The Otis philosophy was to park them throughout the building in zones, each of which consisted of a floor or number of floors either with equalized population or because of some importance, such as an executive floor. The lobby was a separate zone, and the number of zones was generally equal to the number of elevators in the group so each elevator could be stationed in a zone. The section of the floors making up each zone provided a basis for later operational logic decisions.

The Westinghouse approach was to park elevators at the place where they completed their last call and to move them from that point to where an elevator would be needed for a later requirement. Westinghouse also divided the building into zones, but based on measuring demand rather than designated parking. With Westinghouse, the floors the elevators served were divided into one, two, or three or more floor zones for down service and usually two zones for up service above the lobby.

With both Otis and Westinghouse, one car was usually parked at the lobby. When a demand for service appeared, which had to be a landing call if the elevators were inactive, an available car proceeded to answer it. With Otis, an available car was one that was in the zone where the call was placed, and with Westinghouse, it could be either the closest elevator or one that was designated as next available.

If traffic remained such that no more than the number of elevators available were needed to serve the demand at one time, there would be little need for sophisticated operations. It was when traffic exceeded that amount or when the traffic occurred in a recognizable pattern such as incoming, outgoing, or two-way, that the logic of the group supervisory system became most important.

A group automatic system should be capable of being programmed to suit the expected traffic activity in any type of building. For example, if the number, speed, and capacity of the elevators are chosen to serve a specific type of peak traffic, the group operations must initially include the necessary features to serve that traffic and have the capability of being adjusted or reprogrammed to accommodate future changes in traffic patterns or requirements.

TRAFFIC DEFINITIONS AND OPERATIONS

In the discussion of two-way traffic in Chapter 5, traffic was categorized into light, moderate, and heavy traffic, and preliminary definitions were given. Those definitions are modified as follows.

Intensity of Traffic

Light traffic occurs when the number of people riding or requiring elevator service at one time does not exceed the number of elevators in the group. Traffic is light, for example, if there are four cars in a group and no more than four calls, either car or landing, are expected to be registered at one time.

Moderate traffic occurs when the number of people riding or seeking elevator service at a given time is such that the available elevators in a group must be shared by more than one person and the average loading is not expected to fill any elevator beyond 50% of its capacity.

Heavy traffic occurs when the demands on the elevator system are such that the available capacity must be equalized among many passengers and priority of service may have to be given to passengers riding in one direction over those seeking to travel in the opposite direction.

Various traffic conditions require various operations, as described in the following paragraphs.

Incoming Traffic. Light incoming traffic periods require at least one elevator parked at the loading lobby to receive passengers. Operation should be to respond to the highest required call and immediately return to the lobby if no other elevator is at the lobby to receive passengers.

Moderate incoming traffic periods require that all the elevators should be at the loading lobby or returning to the lobby to receive passengers. The elevator travels to the highest required call and returns immediately to the lobby. More than one elevator should be able to be loaded at the same time, and any filled car should be dispatched.

Heavy incoming traffic periods require all the aforementioned operations plus a system to give priority to lobby traffic during designated periods of working days. Service to upper-floor up and down landing calls can be temporarily denied if no car is available at the lobby for loading.

Two-way Traffic. Light two-way traffic periods require the elevators to be available throughout the building, either at rest or in motion, so that a car is either at or traveling toward the next expected call. One approach is to station elevators at various zones of the floors served and to operate them as individual units during periods of minimum demand.

Moderate two-way traffic periods require operation of all elevators in a predetermined regulated manner to minimize the waiting time for landing calls and to provide concentrated elevator service in the direction of heaviest traffic.

Heavy two-way traffic periods require that in addition to the predetermined operation to minimize landing call waiting time, priority operations for longer waits and for potentially long-wait calls should be instituted. Means should be provided to move elevators from areas of light demand into areas of heavy directional demand. For intense situations a system that bypasses some calls to deploy elevators to floors with potential long-wait landing time should be employed.

Interfloor traffic operations include all the actions required for two-way traffic with the elevators concentrating on floors of heaviest demand. Trips to terminal floors should be restricted unless definite demands are registered for travel to those floors.

Outgoing Traffic. Light outgoing traffic periods should reduce time spent by any elevator at the unloading terminal. This is true for any outgoing situation. Elevators should return up into the building to serve down demand with travel no higher than necessary.

Moderate outgoing traffic periods require elevators to minimize time spent at the unloading landing, and bypass landing calls if the elevators are filled up to or beyond a predetermined loading.

Heavy outgoing traffic periods require all of the foregoing plus a system to restrict service to up landing calls if down landing calls are being bypassed during a predetermined time each day. In addition, some means should be provided to divide the available elevator service to minimize the possibility of lower floors being bypassed by cars filling at upper floors. Priority should be given to outgoing traffic limited to a predesignated time at the end of a working day. A system to automatically restrict up landing call service when passengers only wishing to go down operate both up and down landing calls is often required.

Nighttime and Traffic Lulls (Off-peak). With any group automatic operating system it should be decided how the elevators will be stationed during nighttime and periods of traffic lulls, which can occur at any time during the day. As previously mentioned, this may be done by parking the elevators in zones or at other strategic locations

Operational Strategy

The Otis Elevator Company used the zone parking approach to provide the starting point for the operation of elevators during various traffic patterns. Both Otis and Westinghouse, now Schindler, as well as most manufacturers use load-sensing devices in the cars and landing call button information to change the operation of the elevator group during certain distinct traffic patterns. For example, an absence of substantial landing call demand

and filled elevators leaving the lobby is a definite indication of incoming traffic activity. Conversely, a heavy down landing call demand and filled cars traveling in the down direction indicate heavy outgoing traffic.

Continued down traffic with elevators bypassing down landing calls and down landing calls waiting beyond a predetermined time was a signal that Westinghouse used to cause elevators to travel directly without intermediate stopping to down landing zones (predesignated zones consisting of two or three floors), with down landing call or calls exceeding a predetermined time. Otis formerly employed similar strategy of distributing the elevators among the various zones or concentrating service in the zone with the heaviest demand.

A microprocessor and the integration of many data inputs is now used for additional strategies of matching car and landing calls. By matching a car call for a particular floor with a landing call at the same floor in the direction the car is traveling, improved elevator efficiency will result. As can be appreciated, rapid changes and innovations are constantly introduced, and the ability to reprogram elevator operating systems affords building owners a substantial management opportunity to maintain their elevators at their optimum traffic-handling ability.

Providing for Contingencies

A group automatic operating system should consider all contingencies. If, for example, a car is delayed at a floor beyond a predetermined period of time, some means to disengage that car from group operation should be provided. This will ensure that the delayed elevator does not interfere with other elevators in the group or prevent other elevators from stopping at that floor in response to landing calls. A signal at some central location should communicate the delay or operating failure.

If for any reason the group operating system fails to function and elevators are not moving as required, auxiliary means should be provided. This auxiliary means should operate the elevators in a random fashion and provide a signal to inform a responsible person of the failure.

A third possible serious failure could occur if for any reason the landing buttons fail to function. Passengers would be stranded on floors, unable to summon elevator service. A means should be provided to detect such failure and operate the elevators to make predetermined stops so that at least one car stops at each floor. A warning signal should occur in some central location so that necessary corrective action can be taken. If an emergency occurs, such as a fire, as indicated by a detecting system, such random stopping should be aborted and all elevators returned to a designated floor for use by emergency personnel.

Elevators are complex combinations of electrical and mechanical elements. There may be thousands of electrical signals interchanged during the course of a single trip. Elevators have been proven extremely reliable with minimum downtime. The probability of a passenger being trapped is extremely remote and if it does occur, adequate means to inform someone are required by elevator safety codes. The common information system is an alarm bell button located in the car operating panel. Many buildings have either a two-way communication system in each elevator or a telephone connected to the building office. Others have alarm systems connected to central protection agency offices. In any building where it is possible for only one person to be in the building and riding an elevator, this latter consideration is essential.

Passengers in a stalled elevator are safest staying where they are. If they try to get out without help, the elevator may move and they may be injured. Elevator safety codes require immediate emergency lighting in the car. A low-voltage electric fan operated from the same emergency batteries can provide some air circulation. The alarm bell should also be connected to the emergency power source. In buildings with standby generators, the car lights and fan should be connected to the standby power source when the generator is started. Adequate switching should be provided since it is desirable to turn off the car lights when the elevator is parked by normal means. Operation of elevators during a power failure when standby power is available will be discussed in a later chapter.

SPECIAL OPERATING FEATURES

Many of the operations and features described in the preceding pages may be applied to any type of elevator operation and are not restricted solely to group automatic operation. When elevators are specified for a particular building, the desired features should be described in detail and the elevator supplier required to supply them or offer an acceptable substitute.

In addition to normal operating features, it is desirable to take any elevator out of a group and use it for special service. This is necessary in an office building during moving or in a hospital to transfer a patient. At that time it is required to have a system of operation from the car buttons only, bypassing landing calls. A key switch should be located in the car operating panel, the actuation of which will cancel all existing car calls and cause the car to bypass landing calls. Direction can be established by operating the car button for the desired floor, and the car is started by operating an appropriate car start button. A key switch, master keyed to the building key system, is preferred to avoid abuse by unauthorized persons, and making the key removable only in the off position ensures that the car will be restored to normal service when the special operation is complete. Such a system of operation is generally referred to as independent service, emergency service, or hospital service.

In certain buildings it is desired to summon an elevator to a particular floor to give priority service to an executive, to emergency personnel, or for other reasons. A key switch can be located at that floor which, when operated, will call an available car to that floor. The question often to be answered is, Should the car be allowed to complete its existing calls or should the passengers who are on that car be made to travel to the priority call? For emergency personnel or in a hospital the answer is obvious, but what would happen if a vice president called the car and the president happened to be riding? Nevertheless, once the question about existing car calls is answered, the key switch will call the car to the floor of call, and from that floor the car may be operated as described for independent service. Quite obviously, such special service must be limited or conflicts could arise. Elevator operation during emergencies will be discussed in a later chapter.

A third form of special service is to operate one or more of the elevators in a group from a separate or independent riser of landing buttons. When on this service, the elevator responds only to the separate button riser and the other cars respond only to the normal landing buttons. Independent riser service can be abused if its use is unsupervised. One solution is to have the separate riser operate an annunciator so that an attendant on the separated car can control its use. In addition, a lighted sign in the landing button faceplate can inform the users of the elevator's availability for this special operation.

How to serve basements and special floors can often be a problem. Extra time allowances must be made for special stops if they are to be served during any traffic situation. In addition, all the elevators in the group should serve the special floors or basements unless they are to be served only during independent service operations. Failure to follow this procedure impairs accessibility to and from those floors.

With a single basement, a suggested criterion is that if less than 25% of the critical traffic on the elevators is expected to originate from the basement, the basement should be served only on call. Any car not selected for dispatching or a down traveling car should travel to the basement only if a car or landing call for basement service exists. If more than one basement is served, the car should not travel any lower than the lowest up landing call in the basement area and return up.

If more than about 33% of the critical traffic is expected from the basement, consideration should be given to making the basement the lower dispatching landing. Application engineering; that is, establishing the number and size of the elevators for a building, needs to include additional time factors for stops at both the basement and a second entry floor. This needs to be done since it is expected that some cars will be filled at the basement and will not stop for additional passengers at the second entry floor. If the basement is a cafeteria and open only at lunchtime, the dispatching landing for the group of elevators should then be switched from the lobby to the basement by either a clock-controlled switch or automatic means.

Floors above the main group of floors the elevators serve should be treated similarly, in effect like upside down basements.

Many other situations could arise in applying the proper operations to a group of elevators, the foregoing being some of the most common. Odd situations should be recognized early in building design and sufficient additional operations engineered into the group automatic operating system to meet the expected problems.

ADJUSTING OPERATION TO TRAFFIC

Initiating the proper operations of the elevators at the proper time is a task that must be accomplished automatically. In earlier systems an attendant was required to change the mode of operation or program, as it was known. Too often the system was set for one operation and not changed, so that when it was time to go home, the passengers had to fight elevators that had been left on "up-peak" operation. Modern systems should be completely responsive to the traffic situation. Sophisticated systems that utilize programmable microprocessors to determine where elevators can be directed to best serve the prevailing traffic are available and offer immeasurable opportunity.

With the increasing availability of advanced computer circuitry and microprocessors that can process more information, greater utilization of elevator capacity should become practicable and elevators will be able to provide service of the highest quality. In preceding chapters the basis of calculating required elevator capacity and quality of service was shown. The elevators chosen for a given building and the operating scheme furnished should be capable of providing that required performance.

ADVANCES IN OPERATIONAL STRATEGY—ARTIFICIAL INTELLIGENCE

Artificial Intelligence in Group Dispatchers

Artificial intelligence (AI) is defined as machines that exhibit humanlike intelligence such as learning, reasoning, and solving problems.

Automatic elevator group supervisory systems have for many years met the criterion for artificial intelligence. Advanced strategies, such as divide and conquer algorithms, used by even the last generation of relay supervisory systems provided for strategies that could easily outperform humans in dispatching in real time. They learned using relays as memory devices and solved a reasonably complex problem implementing a complex strategy. Therefore elevator systems for the last 20 years have met the criterion for using artificial intelligence.

Advances in modern computers and computational methods have made it easier to provide higher levels of artificial intelligence at much lower costs, and with much less hardware, taking up much less physical space. Methods such as minimal-cost algorithms, linear predictive methods, knowledge-based inference engines, fuzzy logic, neural nets, and the ability to compute faster, and learn and remember more, has all contributed to the advances.

The technical details of these algorithms and the method used in artificial intelligence is well beyond the scope of this book, but I would like to describe some of their uses.

Minimal-cost algorithms have been the central engine for almost all modern day dispatching systems. Cost is one parameter that we choose to minimize that will, in effect, maximize performance. Other such parameters are hall call waiting time, travel time in the elevator, total trip time, or any other parameter one chooses as important. These parameters can be weighted differently by placing levels of importance to each of them. Supervisory systems then take these cost functions and, in real time, try to compute a set of call assignments that minimize the function, and then distribute the assignments to the elevators. The difficulty in solving the minimal cost solution is that these decisions must be made in real time, which usually means that the optimal solution is never actually found, as it takes computational time to find the optimal solution and the time is usually limited to less than one second. What occurs most often is that some very good suboptimal solution is found within the time limit and is used.

Methods for predicting the future are also used in modern supervisory systems. Linear predictive methods provide a tool for memorizing the past for use in predicting the future. In a large number of elevator applications it can be shown that, in a broad-based way, what happened yesterday will most probably happen today and what happen a minute ago will also happen in the next minute. If this is not taken too literally and it is agreed that it has some statistical validity then the information can be used to predict the future and improve dispatching ability. If we can predict the future correctly more than 50% of the time we then can then probably improve our dispatching algorithm. If we predict incorrectly we pay a penalty as we are basing our decision in anticipation of an event occurring, and when the event does not occur the system is out of position and has wasted resources. An analogy is a goal keeper who, based on some criterion, can predict to which side of the goal his opponent will place the ball or puck. If he is a good predictor he will perform better than without prediction and if he is a poor predictor he will perform worse.

This statistical prediction method can be used in a variety of instances. We can predict passenger arrivals, peak periods, car calls, traffic trends, and more. We can utilize this information in our cost algorithms and provide features such as smart parking/zoning as well as many other features that improve passenger service and comfort.

Advanced supervisory systems have provided for widely improved group performance. Unfortunately we are approaching the limit of performance because the performance improvement curve is not a linear progression. It becomes increasingly more difficult to improve waiting or travel time to its pure optimal point. Dealing with the physical constraints of an elevator system as we know it today—fixed cabins, fixed maximum speed, fixed number of elevators in a group—improvements in elevator supervisory systems will only provide for minimal overall performance improvements.

Destination-Based Group Supervisory Systems

Destination-based supervisory systems are by no means a new idea. Such systems were initially introduced in about 1960, using a relay-type controller, which proved too costly and complex. However, until the introduction of the microprocessor and additional technologies such as entry terminals and voice announcement, the destination-based system was difficult to introduce as a practical, usable product.

In destination-based group systems a passenger enters a desired final destination in the hall prior to entering the elevator, rather than pressing an up or down hall call button. By means of an illuminated sign over the designated elevator, the supervisory system must then respond to the passenger to indicate which elevator is assigned for that trip. This assignment process must be virtually instantaneous and fixed. The passenger can then travel over to the assigned elevator and wait for its arrival. This eliminates the need for hall gongs and lanterns. Upon the arrival of the elevator, the passenger boards the cabin and needs only to wait for the elevator to travel to the selected destination. The passenger does not need to press any car call buttons; in fact, there are no car call buttons in the car. Only position indicators and destination confirmation indicators are included in the cabin for the comfort of the riders.

The elimination of the car calls in the car creates a tremendous psychological barrier for the user to get used to. It is a major paradigm shift. Once boarding an elevator, the passenger may no longer change destination. The point of no return is moved back into the hallway. Yet this point of no return is common to almost every other form of transportation. Passengers do not get to change their minds after boarding a plane or a bus; they get on or off the transportation device only at predetermined stops and may reboard other devices to actually get to the new destination. The same holds true for the elevator system. A passenger desiring another destination must leave the cabin at the next stop, rebook the new destination with the system, and take the newly assigned elevator.

Schindler Elevator markets and sells a destination-based group system under the name Miconic 10. The "10 stands for the decade keyboard (10-button telephone-style) used to enter the passenger destination in the hallway, as shown in Figure 7.8. The keyboard has a display integrated that confirms the destination selection and replies to the passenger with the assigned elevator within a few seconds. Voice synthesis is used to communicate to blind persons.

One of the major advantages of destination-based group systems is that the dispatch algorithm knows all of the final destinations of its passengers and how many passengers are waiting on each floor. This information can be used to assign elevators much more efficiently, maximizing the number of coincident stops and thereby minimizing the number of stops (see Figure 7.9). The result is a system that has considerably more handling capacity than a conventional group system. In fact, it is improved to a level that an entirely new set of handling capacity equations are required, as shown in Table 7.1. In effect, the destination-based group system takes the channeling system to the limit, dynamically splitting the bank of elevators into n partitions, where n is the number of elevators in the group.

Figure 7.8. Miconic 10 decade keyboard (Courtesy Schindler Elevator Co.).

The handling capacity of this system is 1.5 times that of a conventional system if there are six cars in the group. Such increased handling capacity can mean that in designing a high-rise building, fewer elevators may be used to handle a given population. This can result in the elevators taking up less core space of the building, thereby increasing rentable space. The only caveat is the paradigm shift a passenger must make in entering her destination. There are currently a dozen or so installations of this sort in the world today, with good passenger feedback. Only time will tell whether the system will become widely accepted.

Channeling

The most efficient elevator is one that makes only two stops—one for all the passengers to get on and one to allow all the passengers to exit. This was recognized in the early days of elevator operation when passengers were required to line up in front of a designated elevator and that elevator took them all to the same floor. It was a common operation in highly populated buildings, such as major insurance firms, and was often extended to limiting the stops of manually operated elevators to a limited number of floors.

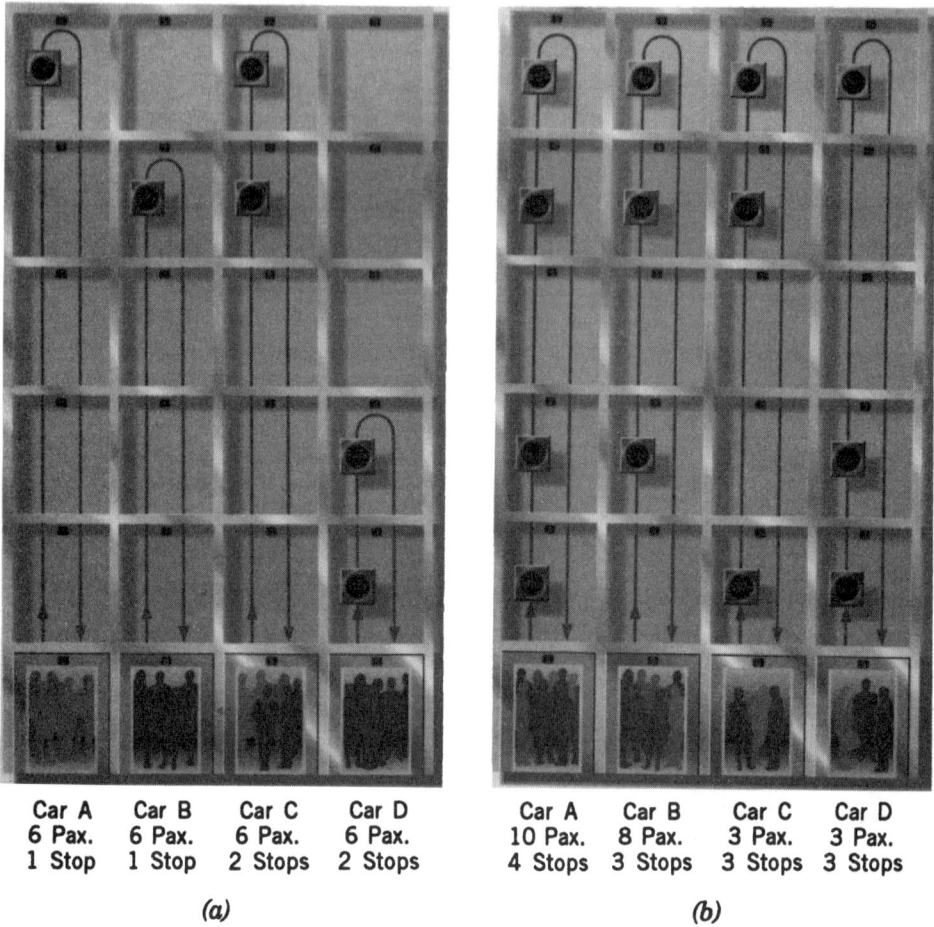

Car A   Car B   Car C   Car D
6 Pax. 6 Pax. 6 Pax. 6 Pax.
1 Stop 1 Stop 2 Stops 2 Stops

(a)

Car A   Car B   Car C   Car D
10 Pax. 8 Pax. 3 Pax. 3 Pax.
4 Stops 3 Stops 3 Stops 3 Stops

(b)

Figure 7.9. (*a*) Efficiency Miconic 10; (*b*) Conventional (Courtesy Schindler Elevator Co.).

TABLE 7.1. Probabilities, Waiting Time (Conventional System vs. Miconic 10).

Conventional:	$X = n \left(1 - [(n-1)/n]^P\right)$
M10:	$X = (2\,n/c)\left(1 - [(2 \times n - 1)/2 \times n]^{cP}\right)$
Conventional:	$Y = n - \sum\limits_{i=1}^{n-1} (i/n)^P$
M10:	$Y = n - \sum\limits_{i=1}^{n-1} (i/n)^X$
Conventional:	$I = RTT/c;\; W = mw\,(I/2)$
M10:	$I = RTT/c;\; W = (Y/X)\,(I/2)$

X = probable stops
Y = probable reversal floor
n = upper floors served
P = passenger load, departing
c = cars/group
I = interval (sec)
W = waiting time (sec)
RTT = round-trip time (sec)
mw = waiting time multiplier: $mw = 2$ at up-peak; $mw \geq 1.2$ at light loading

Attempts were made to incorporate such operation into automatic elevators by arranging the system to cause certain elevators to stop at particular floors during designated times of the day, and by requiring passengers to que in front of the elevators designated by a lighted sign over each car. The elevators and floor designation were fixed and any change required extensive rewiring.

A modern version called "channeling" has been introduced and, by means of computer circuits, the system learns which floors are popular during certain times of the day. Lighted signs in the main lobby direct passengers to the elevator that will serve their floor and, unlike the earlier versions of spotting, there are no fixed elevators for particular floors and the floor destinations can vary depending upon traffic and the availability of the elevators in the system. Channeling requires the prospective passengers to pay attention to the elevator they board and has found favor in some headquarter-type office buildings where the population is less transient than in a hotel or multipurpose building.

Multicar Single Hoistway

Another way to increase the handling capacity of a group of elevators is to place multiple independent cabins in a single hoistway. Unlike a double-deck elevator, as described in Chapter 14, one cabin is not attached to the second; however, this freedom must be managed rather carefully and brings forth an entire new set of both mechanical and operational control problems.

The concept of more than one elevator in a single hoistway was introduced during the 1930s when the Westinghouse Elevator Company prototyped an installation in a building in Pittsburgh. Economics and technical problems aborted the continued development of the concept and the approach has lain dormant until recently.

Otis recently announced the release of a system named Odyssey, which can place multiple glass-enclosed cabins, called transitors, in a single shaft. These transitor modules not only move vertically in the shaft but are also capable of moving horizontally so that multiple cars in a single shaft can be shuffled. Cars can be loading and unloading at predetermined locations such as sky lobbies, and other transitor modules can pass.

The idea of off-loading an elevator car and moving it horizontally is growing, and some practical applications have been developed, although not incorporated in a commercial structure to date. One prominent amusement ride uses this idea to provide a unique experience for the elevator passengers. Automated people movers are being perfected and increasingly applied, so the idea of moving passengers horizontally is becoming an accepted practice and little imagination is needed to conceive of bringing such a vehicle to a stop, moving it onto a platform, and lifting it to a higher destination where it can be off-loaded. This proposal was presented in the second edition of *Vertical Transportation* and is repeated in the last chapter of this edition. One can see the increase of hoistway utilization, as well as elevator handling capacity and availability, by viewing the charts accompanying that proposal. We would expect that such a concept will be commercially applied in the near future, especially on a super-high building.

Because this product is new, there are no practical applications installed to date. However, Odyssey seems to be a product that has true potential. A conceptual rendering of Odyssey's multitransitor design is depicted in Figure 7.10.

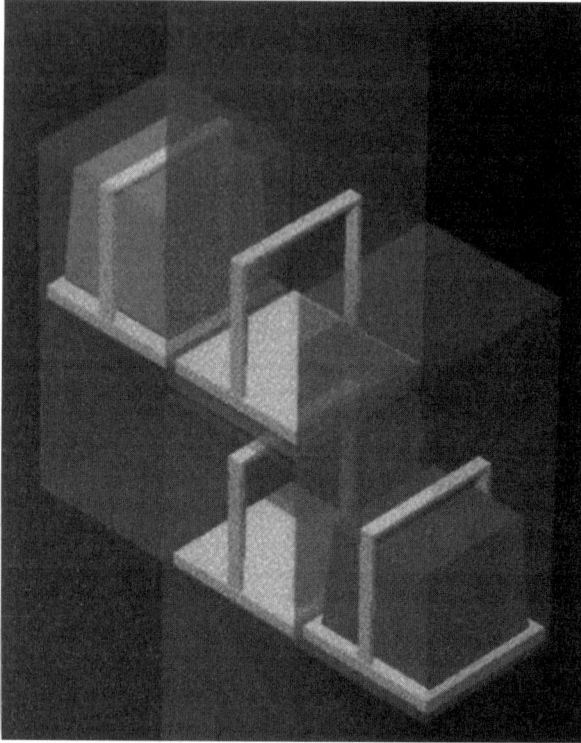

Figure 7.10. Odyssey Multitransitor Rendering (Courtesy Otis Elevator Co.).

Machine Vision

Another area for research and development is a means to determine how many people are waiting for elevator service at a lobby or on a particular floor. One landing call can represent one person or ten people. Elaborate systems have been tried to provide buttons for each elevator stop at each floor and to provide elaborate illuminating signs over each elevator to indicate the floors at which they will stop. As can be surmised, the confusion of finding and standing in front of the right elevator, where six or eight elevators stop at a common lobby, can be overwhelming. A reliable electronic crowd counter could be a great asset to improve elevator service.

The era of the microprocessor and programmable control applied to elevators is just beginning. The key will be to ensure the traditional reliability that has made the elevator an accepted part of everyone's life with minimum concern about safety and practically no concern about availability of service.

NEEDS FOR FUTURE DEVELOPMENT

As sophisticated as elevator operating systems are becoming, there is still vast opportunity for new development and improvement. No elevator company at present has the means to determine whether an elevator is full in volume, although devices that measure weight in the elevator car are common. A means to determine whether the space within an elevator is fully occupied would be especially valuable in a hospital, where one cart or stretcher can fill up an elevator, making it inefficient to stop for additional landing calls until that load is removed.

Elevator companies should recognize the difference between operation and motion control. It should be possible to obtain the most sophisticated group operation with a low-rise hydraulic elevator installation as with a system in a high-rise downtown building. Yet this is not possible with present technology, which is an injustice to the owner of a suburban office building. It is simple to separate operation from motion control and amazing that it has not been practiced by elevator manufacturers.

ABOUT THE AUTHOR

JON B. HALPERN is the executive vice president of Millar Elevator Industries, Inc., a wholly owned subsidiary of the Schindler Elevator Company. Mr. Halpern has a master of science degree and a professional degree in electrical engineering from Columbia University, New York, and a bachelor of science degree in electrical engineering from The George Washington University in Washington, D.C. In his 20 years at Millar, Mr. Halpern has held various positions ranging from a design and development engineer of elevator control and dispatch systems to various management positions. He also recently returned from an overseas assignment as program manager of modernization for Schindler Europe. Mr. Halpern has published numerous articles on elevator control and dispatch systems, and is a member of the International Association of Elevator Engineers (IAEE).

8 Space and Physical Requirements

ROBERT S. CAPORALE

SPACE FOR ELEVATORS

Good internal and external pedestrian circulation is an essential feature of a well-designed building. Individual floor plans may be readily changed to suit specific tenant requirements, but the facilities for pedestrian circulation between floors in the building are fixed. Good vertical transportation is one of the first aspects of good building design.

All aspects of the circulation within a building must be considered—the lobbies and corridors must be properly sized, with sufficient elevators of proper size and adequate stairs provided. Elevator and escalator planning must consider both the space allocated for the equipment and the space provided for people to use that equipment. Such arrangements become essential on the lobby floor, where people must be able to access elevators or escalators easily and must also be able to enter and leave the building with comparative ease. Lobby amenities such as stores and information areas must be properly located so as not to interfere with pedestrian circulation. Ample queuing space for escalators and waiting space for elevators are both important elements in building design, as are adequate exits to accommodate the crowds that elevators may discharge.

Chapter 2 discussed elevator lobby space and its essential contribution to efficient elevator utilization. Elevator traffic handling emphasizes the importance of the sufficient number, size, and speed of elevators, as well as the proper door opening width and arrangement. This chapter focuses on the building space required to accommodate elevators.

REQUIRED SPACE

To make an elevator platform travel up and down smoothly and safely, certain structural requirements must be met. Each elevator must have guides or rails that keep it in a substantially vertical path and provide the columns from which an elevator may be supported if, for any reason, its continued operation becomes unsafe. Such support is accomplished by means of safety application, which consists of a device below the elevator platform, that is applied to stop and hold the elevator on the rails if it should happen to overspeed in the down direction. During safety application, the entire weight of the elevator is arrested and transferred to the building structure through the rails. Stopping an elevator that has a structural weight of 3 or 4 tons (3600 kg) plus a passenger load of 2000 to 4000 or more pounds (1800 kg), which is traveling at hundreds of feet (1 mps or more) per minute, by

The Vertical Transportation Handbook, Third Edition, Edited by George R. Strakosch
ISBN 0-471-16291-4 © 1998 John Wiley & Sons, Inc.

clamping it on the building structure through the guide rails will require sizable rails and rail supports. Space for rails, therefore, is a necessity (Figure 8.1*a*).

Rail Support

In addition to the rails, adequate rail supports or brackets must be provided. The brackets are fastened to the building steel by the elevator contractor or, in a concrete building, by means of inserts set in the concrete by the concrete contractor or by the elevator contractor (Figure 8.1*b*). The vertical space between brackets is critical, because the rail has a design stiffness and can be allowed to deflect only about ⅛ in. (3 mm). Elevator rails are designed for fastening at floor levels up to about 14 ft (4.2 m) apart. If longer spans are expected, additional intermediate supporting beams should be provided in the structure, heavier rails should be used, or additional horizontal space to reinforce the elevator guide rails and provide the necessary stiffness should be provided. Heavier rails with a larger

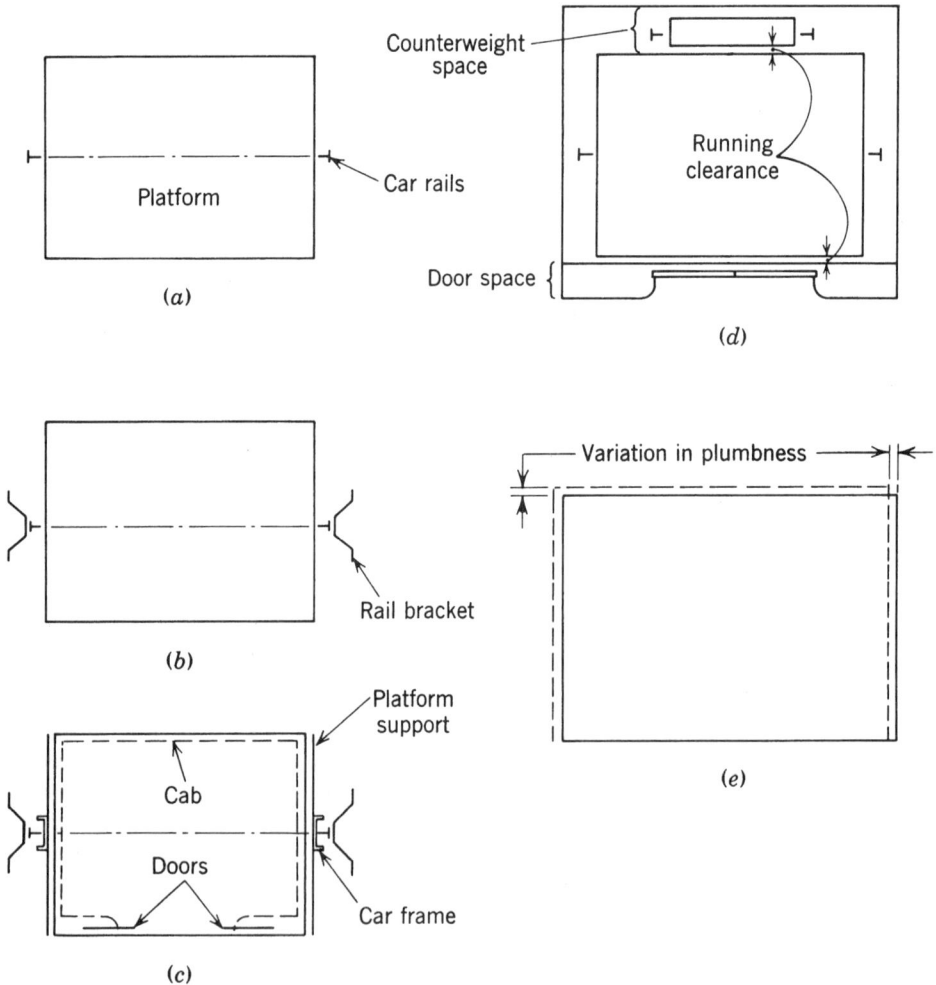

Figure 8.1. Elevator layout—plan.

cross section are stiffer and, depending on the total load of the elevator, can allow a greater bracket spacing.

The maximum vertical distance between brackets, related to the size of elevator rails and the expected load on those rails, is established in the A17.1 Elevator Code. The latest edition of the code should be checked for the current requirements. Additional considerations must include the installation's seismic risk and any A17.1 and local requirements that have to be followed.

Noise

The noise an elevator causes, operating with the normal ambient noise in a building, is usually unobjectionable. Elevator noise may be noticed and objected to under quiet conditions, however, such as in a residential building or hospital during quiet periods. The counterweight or elevator passing sleeping rooms adjacent to the elevator hoistway can cause disturbing noise, and any vibration that may be present can be transmitted through the walls and structure. This situation can be avoided by providing sound and vibration isolators on the elevator brackets where they are fastened to the building structure (Figure 8.2). The best approach is to architecturally place such rooms away from the hoistways.

Elevators operating during the night create a certain amount of noise when they stop at a floor and doors open and close. Operating systems that park elevators at upper floors should be designed so that such parking is done without door operation. Door operations should occur only in response to a car or landing call.

Structure-borne elevator noises and vibrations may be transmitted to places remote from the hoistway through the building structure. The elevator machine should be isolated from its support, and the secondary or deflecting sheave attached to the machine rather than to

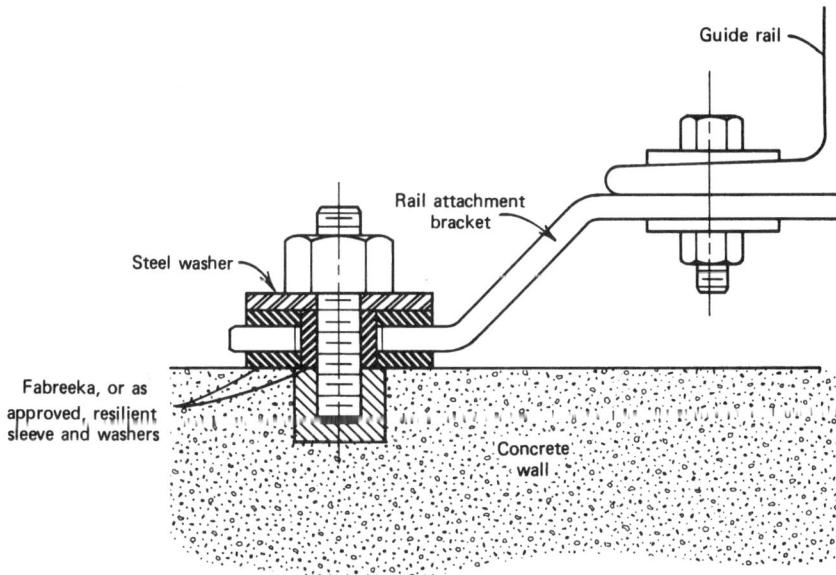

Figure 8.2. Structure-borne noise isolation arrangement of elevator rail brackets—necessary if sleeping rooms or sensitive equipment is located adjacent to an elevator hoistway (Courtesy Cerami Acoustical Associates).

the supporting structure. Electrical equipment such as isolation transformers used with solid-state motor drives should be isolated from the building structure to break up sound paths and guard against the transmission of objectionable structure-borne frequencies.

As a high-speed elevator travels through the hoistway, various noises are generated. A "puff" may occur when the elevator passes each floor as the air currents eddy at landing sills, and various pockets and structural members in the hoistway can create additional "puffs." These can be overcome by aerodynamic design and structural considerations.

Wind created around hoistway doors where stack effect is present will create whistling noises. These can best be overcome by minimizing the stack effect or, as a last resort, weather-stripping the hoistway doors, which usually requires additional maintenance considerations.

Each noise problem must be individually approached. Sensible initial design can avoid the expense of later correction.

Platform and Enclosure

The elevator must be enclosed. Because the interior should create a comfortable surrounding and is seen by everyone entering and leaving a building, it merits the best architectural design and finest craftsmanship. Providing a comfortable interior requires space and creates weight. In addition, the car enclosure must be completely isolated from both sound and vibration, which may be transmitted to and from the structure that supports it. Such isolation, usually accomplished by elastomer pads, may require additional rail space. The supporting structure, called the car frame, must be constructed to withstand the force of a safety application and the lifting forces on the elevator, as well as the weight of the mechanism that will operate the doors (door operator), switches, cams, and so on. All this weight requires a structure that may become quite substantial and takes up space (Figure 8.1c).

This space, if measured from the side of the hoistway to the side of the car platform, amounts to 8 in. (200 mm) for elevators up to 500 fpm (2.5 mps) and to about 10 to 12 in. (250 mm) for higher-speed elevators, double-deck elevators, and elevators with capacities of 5000 lb (2300 kg) or more.

Access to the elevator from each floor is guarded by a hoistway door. This door and entrance assembly is required by local building codes to be of fire-labeled construction and must be certified for the type of wall used, whether masonry, drywall, or poured concrete. In addition, a mechanism must be provided to lock the door safely when the elevator is not at a floor and automatically open it when the car is leveling to the floor. Space for both the doors and its mechanism is required (Figure 8.1d).

The required minimum clearance between moving pieces of equipment in hoistways as prescribed by most codes is ¾ in. (19 mm); therefore, the distance between car and landing doorsills must be, at a minimum, ¾ of an inch (19 mm). Owing to necessary construction tolerances that must be allowed, this distance is usually set as 1¼ in. (31 mm). In order to set the landing sills true and plumb, they are usually set to overhang the front edge of the hoistway by 1 in. (25 mm), thereby establishing the minimum space between the car doorsill and the edge of the hoistway structure as 2¼ in. (57 mm). The doors on each floor will require front space in addition to the running clearance and sill overhang.

In a traction-type elevator the structural load and part of the passenger load, from 40 to 50% of the rated load, is counterweighted, and the counterweights travel up or down in the opposite direction from the elevator car. Counterweights and their guides are usually located behind the car, although some installations, such as elevators in hospitals, have counterweights at the side. Horizontal space for the necessary cast iron or steel counterweight is required (Figure 8.1d).

The total space for counterweights behind the elevator, as measured from the back hoistway wall to the back edge of the car platform, is approximately 15 in. (380 mm) for counterweights without safety devices and 17 in. (430 mm) for counterweights with safety devices as would be required if space located below the elevator hoistway could be occupied. This requirement should be determined before space is allocated. A counterweight with a safety device will also necessitate a heavier rail. If seismic requirements must be fulfilled, additional bracketing and support space of about 4 in. (100 mm) should be provided. If a hydraulic elevator is being considered, all the foregoing requirements except the space for the counterweight at the rear must be provided. In lieu of the rear counterweight space, a minimum allowance of 2 in. (50 mm) is provided between the rear of the platform and the back wall of the hoistway.

Doors

The width of the clear opening of hoistway doors is directly related to the available clear width of the hoistway. Single-slide and center-opening door space requirements are established by multiplying the desired width of the opening by 2 and adding 6 to 8 in. (150 to 200 mm) for structural requirements behind the open doors.

For example, if 4-ft 0-in. (1200-mm) center-opening doors are desired, the clear hoistway must be 2 times 4 ft (2400 mm), plus 8 in. (200 mm), or 8 ft 8 in. (2600 mm) wide.

For two-speed doors, the hoistway width required is 1.5 times the desired width of opening, plus 4 to 6 in. (100 to 150 mm) and 11 in. (275 mm) for the strike jamb. Therefore, the clear hoistway width for a 4-ft (1200 mm) two-speed door is 1.5 × 4 ft plus 15 in., or 7 ft 3 in. (2200 mm), preferably 7 ft 5 in. (2260 mm). The strike jamb side of the two-speed opening must be at least 3 in. (75 mm) in from the platform edge. Two-speed center-opening doors are established similarly to center-opening doors; however, the door space in this case is 1.5 times the opening width.

In Example 8.1, the 7-ft 0-in.- (2140-mm)-wide platform given can accommodate, under ideal conditions, a 4-ft 0-in- (1200-mm)-wide center-opening door. More practically, a 3-ft 10-in.- (1150-mm)-wide center-opening entrance should be employed because building tolerances must be considered. If the platform width is increased to 7 ft 2 in. (2200 mm), per Table 4.4 in Chapter 4, the 4-ft 0-in. (1200-mm) center-opening entrance can be easily accommodated.

It is important to take that architectural or structural infringements such as mail chutes or columns are located so as to avoid interference with the door opening.

As previously discussed, the front-to-back space required for various doors from the edge of the hoistway doorsill to the back of the hoistway wall is 5 in. (127 mm) for center-opening doors and 6½ in. (165 mm) for two-speed doors. If flush transom panels are used over center-opening doors, this space must be increased about 1½ in. (38 mm).

The front wall may be of masonry or of gypsum board construction and must preserve the fire integrity of the entrance assembly as well as the fire rating of the hoistway. Entrance assemblies are required to be tested under fire conditions and certified by an independent laboratory. Details of such a test can be found in the American Society for Testing and Materials publication ASTM 152.

Elevator hoistways in buildings are required to be in fire-rated enclosures, and the elevator entrance assemblies should be designed, tested, and installed so that their fire resistance is comparable to the fire rating of the hoistway enclosures. This was seldom a consideration when hoistway enclosures were constructed of masonry; the entrance was bricked-in solid and secure. With drywall construction using gypsum board, the interface between the entrance assembly and the adjacent wall is a critical connection, which must be made in accordance with the way the entrance was fire tested. Figure 8.3 shows a typical interface with an entrance frame: the important consideration is that the connecting "J" stud be fastened to the entrance and the long face of the stud be at the shaft side.

Once the plan of the various elevators in a building is established, the planner must give attention to providing the necessary vertical space. At this point, the discussion focuses on conventional traction elevators. Details on hydraulics, basement machine arrangements, and underslung elevators are discussed later in this chapter.

Construction Tolerance

An elevator must run up and down as plumb as possible. If it does not, it may experience extra wear on the elevator guides. Buildings are often not plumb; tolerance may vary from a fraction of an inch (25 mm or less) to an inch or more. Elevator contractors ask for 1 in. (25 mm) plus or minus plumbness in an elevator shaft, which means that a difference of 2 in. (50 mm) may exist between the top and bottom in either the front-to-back or side-to-side direction or both. Higher-rise buildings may be built with a slight lean to compensate for prevailing winds or may be designed to sway with a calculated deflection and period because of wind loading. Such contingencies should be ascertained and the necessary provisions made in the design of the elevator system (Figure 8.1*e*).

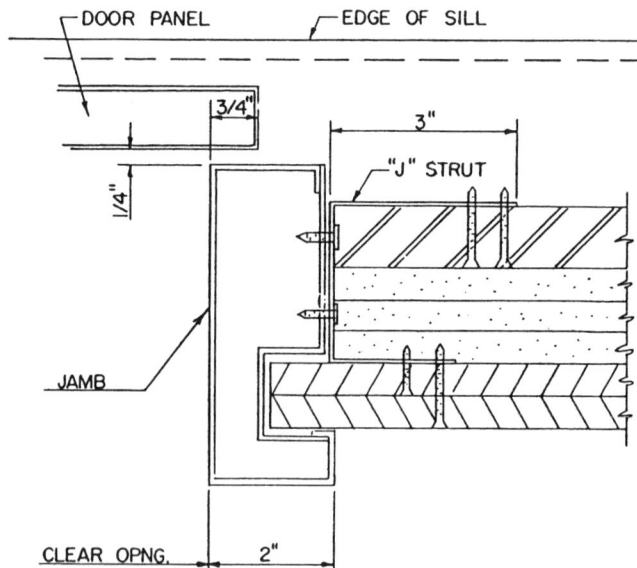

Figure 8.3. Plan of center-opening entrance as fire tested in a nonmasonry wall (dry wall)—maximum door height 8 ft 6 in. (2600 mm) (Courtesy Jaros, Baum & Bolles).

With all the foregoing specifications, the net usable platform area available in any elevator layout is 60 to 75% of the hoistway required for the elevator. Details of the reason for allocating this space are shown in Figure 8.4. Industry's effort to reduce this nonusable space is continuous, including use of stronger steel, new approaches to door arrangement, and new fastenings. Since the safe operation of an elevator is predominant and most of the space required at present is to ensure maximum safety, the prescribed space allowances must be provided; however, it is important also to design for the optimum-size car.

To "rough out" the required elevator hoistway for an elevator of a given platform size or, conversely, to determine what platform size can be accommodated in a given hoistway, the following steps can be used. All dimensions are approximate.

1. Allow 8 in. (200 mm) on either side of the car platform for rails (see Figure 8.5a). The total clear hoistway width will therefore be equal to the car platform width plus 16 in. (450 mm).
2. Allow 15 in. (310 mm) at the rear of the platform for the counterweight (17 in. (430 mm) if counterweight safeties are required) and 2¼ in. (57 mm) from the front of the platform to the edge of the farthest-protruding landing door ledge. The 2¼ in. (57 mm) is composed of the 1¼-in running clearance and the 1-in (25-mm) landing doorsill overhang (see Figure 8.5c).

Rated Load (mass) (kg)	Maximum Available Car Area (see note) (m²)	Maximum Number of Passengers	Rated Load (mass) (kg)	Maximum Available Car area (see note) (m²)	Maximum Number of Passengers
100	0.40	1	975	2.35	13
180	0.50	2	1000	2.40	13
225	0.70	3	1050	2.50	14
300	0.90	4	1125	2.65	15
375	1.10	5	1200	2.80	16
400	1.17	5	1250	2.90	16
450	1.30	6	1275	2.95	17
525	1.45	7	1350	3.10	18
600	1.60	8	1425	3.25	19
630	1.66	8	1500	3.40	20
675	1.75	9	1600	3.56	21
750	1.90	10	1800	3.88	24
800	2.00	10	2100	4.36	28
825	2.05	11	2500	5.00	33
900	2.20	12			
Beyond 2500 kg, add 0.16 m² for each 100 kg extra.					

(a)

Figure 8.4. Elevator capacity vs. inside car area. (a) European CEN81 Code.

Maximum Inside Net Platform Areas for the Various Rated Loads

Rated Load (lb)	Inside Net Platform Area (ft²)	Rated Load (lb)	Inside Net Platform Area (ft²)
500	7.0	4,500	46.2
600	8.3	5,000	50.0
700	9.6	6,000	57.7
1000	13.25	7,000	65.3
1200	15.6	8,000	72.9
1500	18.9	9,000	80.5
1800	22.1	10,000	88.0
2000	24.2	12,000	103.0
2500	29.1	15,000	125.1
3000	33.7	18,000	146.9
3500	38.0	20,000	161.2
4000	42.2	25,000	196.5
		30,000	231.0

To allow for variations in cab designs, an increase in the maximum inside net platform area not exceeding 5%, shall be permitted for the various rated loads.

1 lb = 0.454 kg
1 ft² = 0.0929 m²   *(b)*

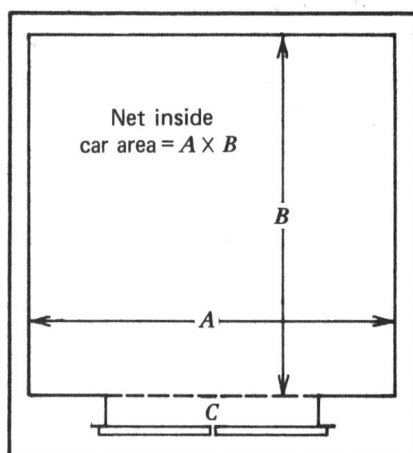
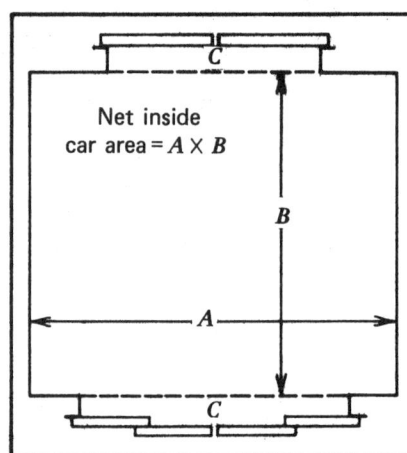

(c)

Figure 8.4. *(Continued)* *(b)* North America A17.1 and Canadian B44 Codes. *(c)* Measurement diagram (A × B) = A17.1 and B44 areas, (A × B) + area C = CEN81 area.

3. Door space in front of the hoistway consisting of the total sill and space for doors will vary with the door type and is calculated as follows:

   a. From edge of sill to back of door, 2¼ in. (57 mm).

   b. Each door panel is 1¼ in. (32 mm) thick.

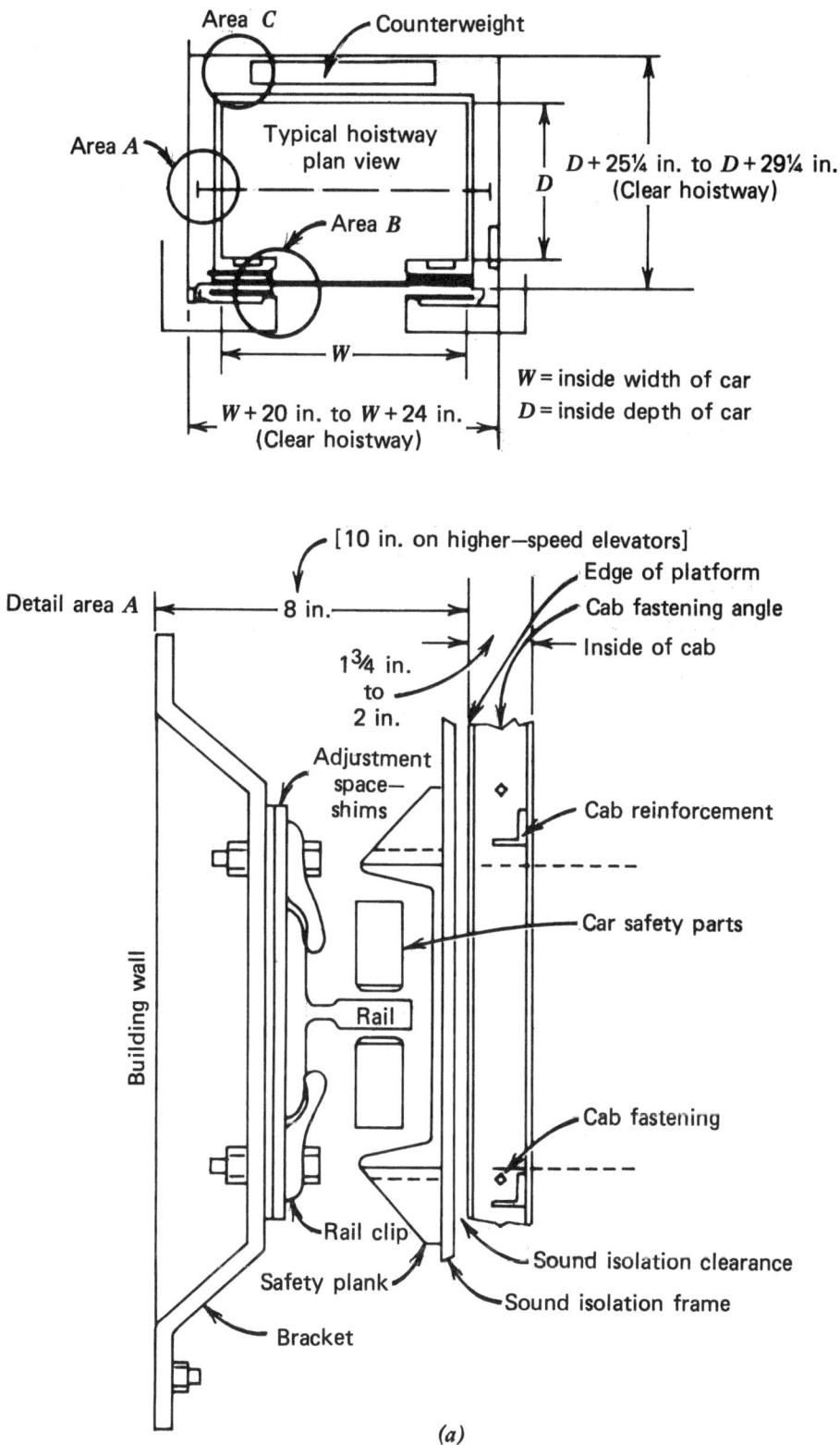

Figure 8.5 Elevator construction. (*a*) Detail of Area *A*.

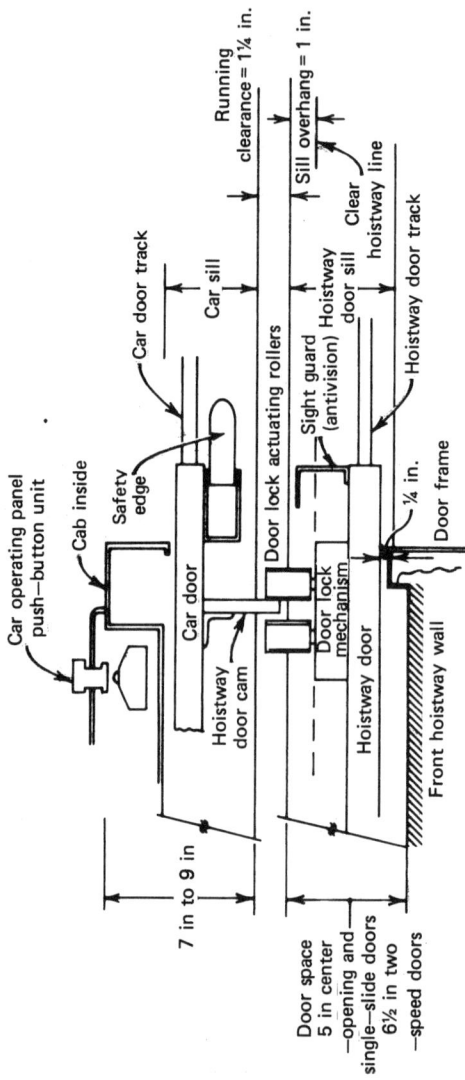

(b)

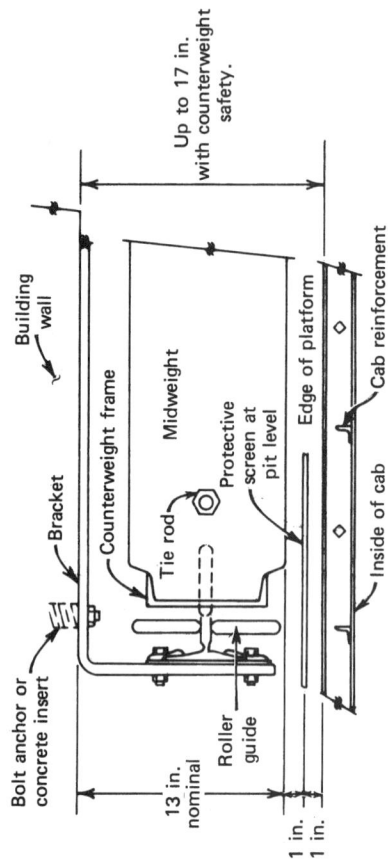

(c)

Figure 8.5 (*Continued*) (*b*) Detail of Area *B*. (*c*) Detail of Area *C*.

c. One-quarter inch (6 mm) is allowed between each door panel and the door frame. To this is added 1¼ in. (32 mm) for protrusion of entrance frame construction into the hoistway, for a total of 1½ in. (38 mm) door, clearance, and frame space.

Therefore, with single-panel center-opening or single-slide doors a total door space of 5 in. (127 mm) is required, and with two-speed doors the required space is 6½ in. (165 mm) (see Figure 8.5b).

To estimate the clear inside car dimensions for a given platform size, the following steps are required:

1. From the platform width subtract 2 in. (50 mm) on each side, for a total of 4 in. (100 mm). The 2-in. (50-mm) space is needed for structure and fastening, as shown in Figure 8.4.
2. From the front-to-back dimension of the platform subtract 2 in. (50 mm) on the back and 6 in. (150 mm) if single-slide or center-opening doors are used, or 8 in. (200 mm) if two-speed doors are used, at the front, for a total of 8 or 10 in. (200 or 250 mm), to obtain the net front-to-back inside car dimension.

The A17.1 Elevator Code allows a net area related to the elevator capacity in pounds, as shown in Figure 8.4c. This area is measured at a point 36 in. (915 mm) above the car floor and does not include the space required by the car doors nor the framing space around the doors. This framing space can be 2 to 4 in. (50 to 100 mm) deep, and to obtain the net front-to-back dimension, the gross depth is further reduced by 2 or more inches (50 mm), depending on the cab design. The A17.1 Elevator Code allows the inside area to vary with a tolerance of +5% to accommodate nonconventionally shaped cabs.

Example 8.1. Elevator Capacity

Inch-pound: Given a 7-ft-wide by 5-ft-deep platform, what is the elevator capacity? Assume center-opening doors.

Width: 7 ft   = 84 in. − 4 in. = 80 in.
Depth: 5 ft   = 60 in. − 9 in. = 51 in.
Area: 80 × 51 = 4080 in.^{2} 144 = 28.4 ft^2
Allowable A17.1 Code area 2500 lb = 29.1 ft^2
(See Figure 8.4b.)
Therefore, rate elevator 2500 lb.

Metric: The European CEN81 (see Chapter 16) inside area requirements are shown in Figure 8.4a. Given a 2140-mm-wide by 1525-mm-deep platform, what is the elevator capacity? Assume center-opening doors.

Width: 2140 mm − 100 mm = 2040 mm
Depth: 1525 mm − 180 mm = 1345 mm
Area: 2040 mm × 1345 mm = 2.74 m^2
Add back space between door jamb inside car and depth to face of door.
28 mm × 1200 mm = 0.03 m^2
Total inside area: 2.77 m^2
Allowable area for 1200 kg per EN81 Code 2.8 m^2 (see Figure 8.4b).
Therefore, rate elevator 1200 kg.

The A17.1 Code does not include the area between the inside door jambs to the face of the doors, whereas the EN81 Code does; therefore the difference in calculation.

OVERHEAD AND PIT REQUIREMENTS

Pit Space

When an elevator stops at a floor of a building, certain parts of the structure of the elevator are either above or below the cab as it is seen at that floor. Below the cab floor is the platform, a structural base composed either of wood and angle iron or all steel. This platform should be cushioned on elastomer pads, on a sound isolation subframe. A common method of weighing the load in an elevator car is by measuring the deflection of the elastomer pads and by actuation of microswitches or a transducer for the various degrees of car loading (Figure 8.6).

An alternative method used for weighing loads consists of strain gauges mounted on the car frame crosshead. Transducers react to crosshead stress caused as the car loading varies, and at predetermined set points initiate car dispatching, floor bypassing, or antinuisance car call canceling operations. Both load-weighing methods require the car and platform to be vibration and sound isolated from the car frame.

The platform and its sound isolation frame rest on the safety plank, which also supports the elevator safety device, clamps that will stop and hold the elevator on the rails if it overspeeds in the down direction. The safety plank must also be designed to absorb the impact from the elevator buffers located in the pit (Figure 8.7a).

When the elevator is stopped at the lowest floor, there will be a few inches overtravel before it strikes the buffer. The buffer is designed to stop the elevator if, for some reason, it travels at its operating speed past the lowest floor. The buffer absorbs the kinetic energy of the moving car and brings it to a stop within the limit of the buffer stroke. The stop is not gentle, but it is within safe limits. The buffer is not required by safety codes to bring a free-falling elevator to a stop; this is the specific function of the car safety mechanism.

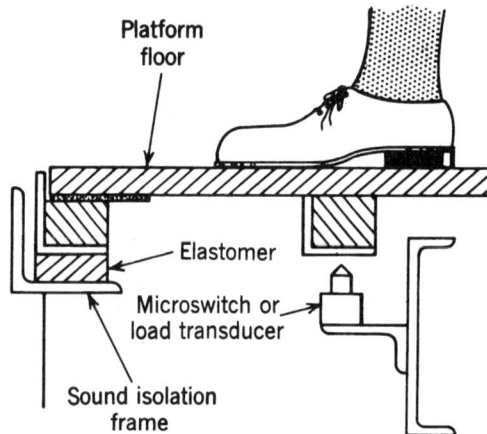

Figure 8.6. One form of load weighing by measuring platform deflection.

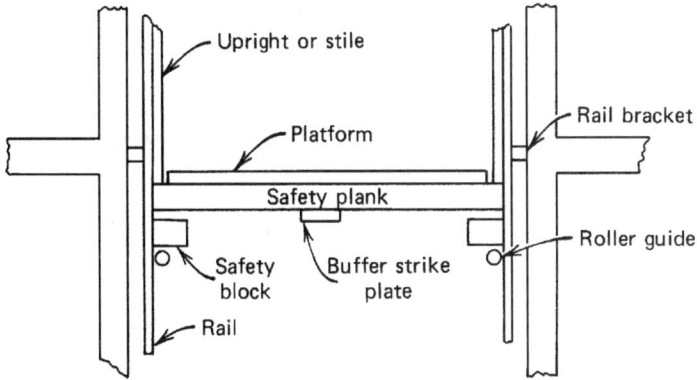

(a)

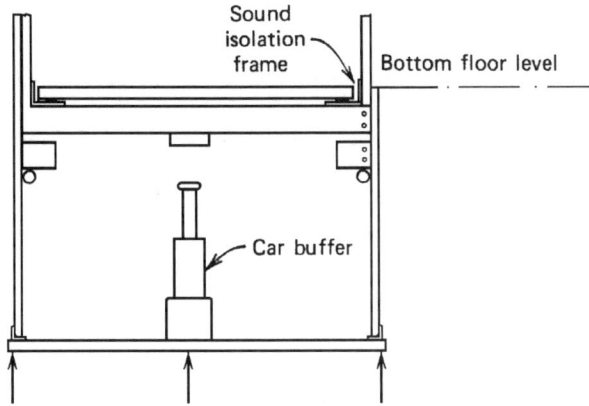

Buffer
impact

Pit detail

Rail impact
upon safety
application

(b)

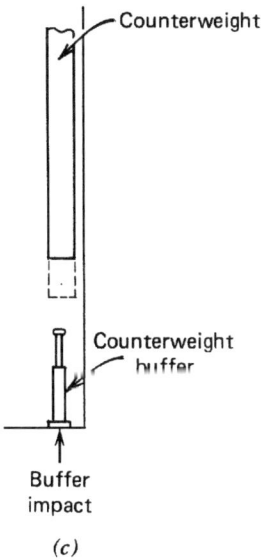

Buffer
impact

(c)

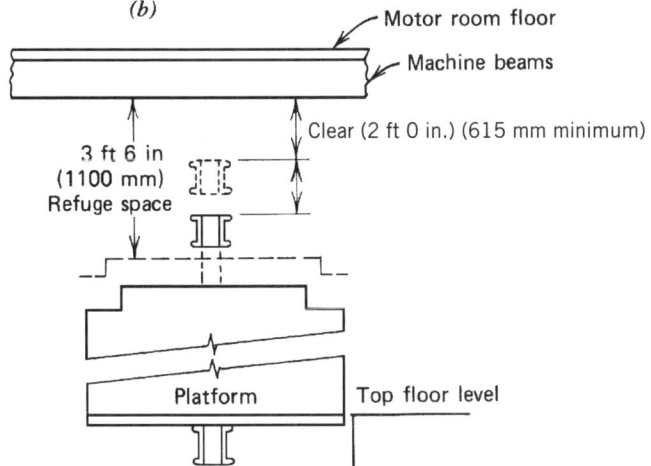

(d)

Figure 8.7. Elevator layout procedure—elevation.

The depth of the pit must include the depth of the platform and support space required for the elevator car (Figure 8.7), plus operating clearance, in addition to the buffer standing and stroke space (Figure 8.7c). For elevator speeds up to and including 200 fpm (1.0 mps) spring buffers are used. For speeds over 200 fpm gradual buffers must be used, which are usually hydraulic buffers commonly referred to as "oil buffers." The stroke of the buffer, defined by the A17.1 Elevator Code, is a function of car speed; see Table 8.1. The depths required for typical pits range from 5 ft (1.5 m) for low-speed cars to 16 ft (5 m) or more for very high speeds. Table 8.1 shows depths for the more common speeds.

For any special elevators the buffer stroke may be reduced if certain other precautions are taken. One of these is a means to ensure that the elevator brake is applied if the car is traveling at high speed when it is within a determined distance of the pit. If, for example, it normally takes about 20 ft (6 m) to slow down and stop a 700-fpm (3.5-mps) elevator, and the elevator is traveling that fast within, say, 15 ft (4.5 m) of the pit, applying the brake at that point should slow the car sufficiently so that a shorter buffer will stop the car.

When the elevator car lands on the buffer and the buffer is fully compressed, the counterweight is at its highest point. The counterweight will also jump a short distance, less than one-half the buffer stroke, if it is traveling up at full speed and is abruptly stopped. Sufficient clearance above the counterweight must be maintained so that there is no danger of its striking the overhead.

The counterweight will land in the pit if the car passes the topmost landing. The counterweight will have a buffer of equal stroke to the car buffer. In addition, space under the counterweight must be provided for normal running clearance [Figure 8.8b, (B)] plus sufficient additional space to accommodate the normal stretching that is experienced with elevator ropes and to avoid causing the counterweights to land [Figure 8.8b, (C)]. This rope stretching can result from temperature changes, aging, or rope strands settling against the core after manufacture owing to rope loading or various other causes. About 0.25% of rope length is a nominal allowance for this contingency after the initial installation stretch of 0.5 to 1% of the rope length.

In addition to the buffers, the pit must often contain provisions for guiding compensating ropes. These are ropes that are attached to the bottom of the car and run down the hoistway into the pit, where they turn back up (often around a sheave) for attachment to the bottom of the counterweight. In taller buildings this pit-mounted compensating rope guide consists of a weighted sheave riding on short tracks in the pit. In very tall buildings where elevator speeds exceed 500 fpm (2.5 mps), if seismic shock is a possibility or if building sway is anticipated, the guiding mechanism must be arranged, often by code requirements, to link the car and counterweight together in such a way that when the car safety applies it acts on the entire mass of the system. This avoids excessive jump of the up-traveling weights when the high kinetic energy of the down-traveling elevator is arrested. Such a system, known as tied-down compensation or monomass safety, creates a substantial up-pull where it is attached to the pit floor.

For slower-speed elevators, special ropes, traveling cables, or chains interwoven with sash cord or jacketed with plastic may be used as compensation. The sash cord or plastic jacket is used to overcome the rattling noise made by moving chains. Chain compensation is usually confined to elevators of 500 fpm (2.5 mps) or less where an increased noise level is acceptable. Traveling cable compensation may be used up to 700 fpm (3.5 mps).

In determining the pit depth, consideration of other equipment is often necessary. Electrical cables that hang from the bottom of the elevator to a point near the middle of the hoistway to carry all the power and signals to the elevator car must travel into the pit

TABLE 8.1. Pit Depths. Traction Elevators—Overhead Machines

| Speed (fpm) | 100 | 200 | 300 | 400 | 500 | 600 | 700 | 800 |
(mps)	0.5	1	1.5	2	2.5	3	3.5	4
Depths								
a. With restrained rope compensation	—	—		8 ft 0 in. 1.6 m	8 ft 6 in. 2.6 m	9 ft 2 in. 2.8 m	9 ft 10 in. 3.0 m	10 ft 6 in. 3.2 m
b. With chain, free rope, or traveling cable compensation	5 ft 0 in. 1.5 m	5 ft 0 in. 1.5 m	5 ft 4 in. 1.6 m	7 ft 10 in. 2.4 m	8 ft 4 in. 2.5 m	—	—	—
c. With reduced stroke buffer and either restrained rope, chain, traveling cable, or free rope compensation	—	—	5 ft 0 in. 1.5 m	5 ft 4 in. 1.6 m	8 ft 0 in. 2.4 m	8 ft 6 in. 2.6 m	8 ft 6 in. 2.6 m	9 ft 2 in. 2.8 m
Buffer type	Spring	Spring	Oil	Oil	Oil	Oil	Oil	Oil

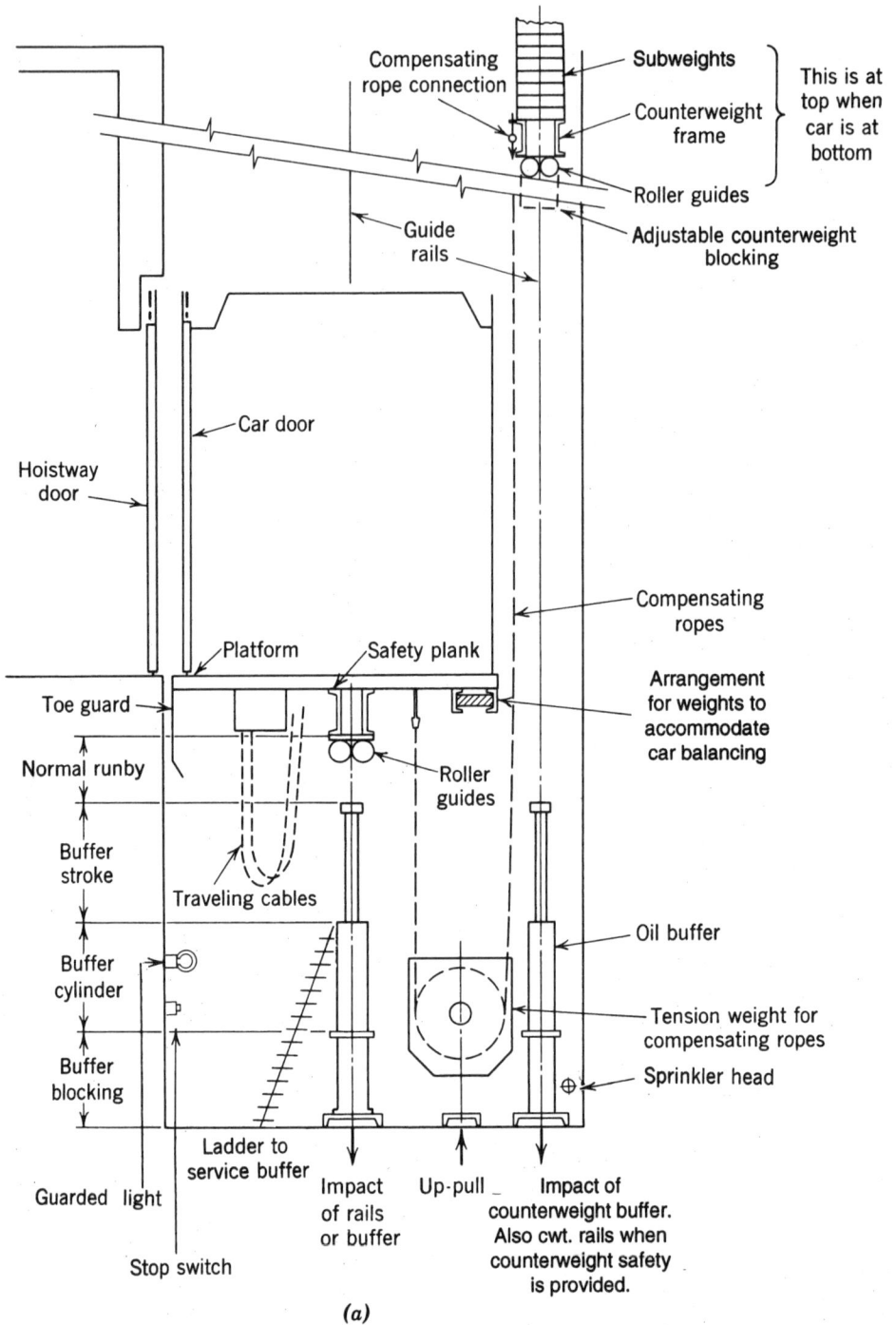

Figure 8.8. (*a*) Pit Construction and equipment—with rope compensation. (*b*) Pit and overhead space with chain or traveling cable compensation.

184

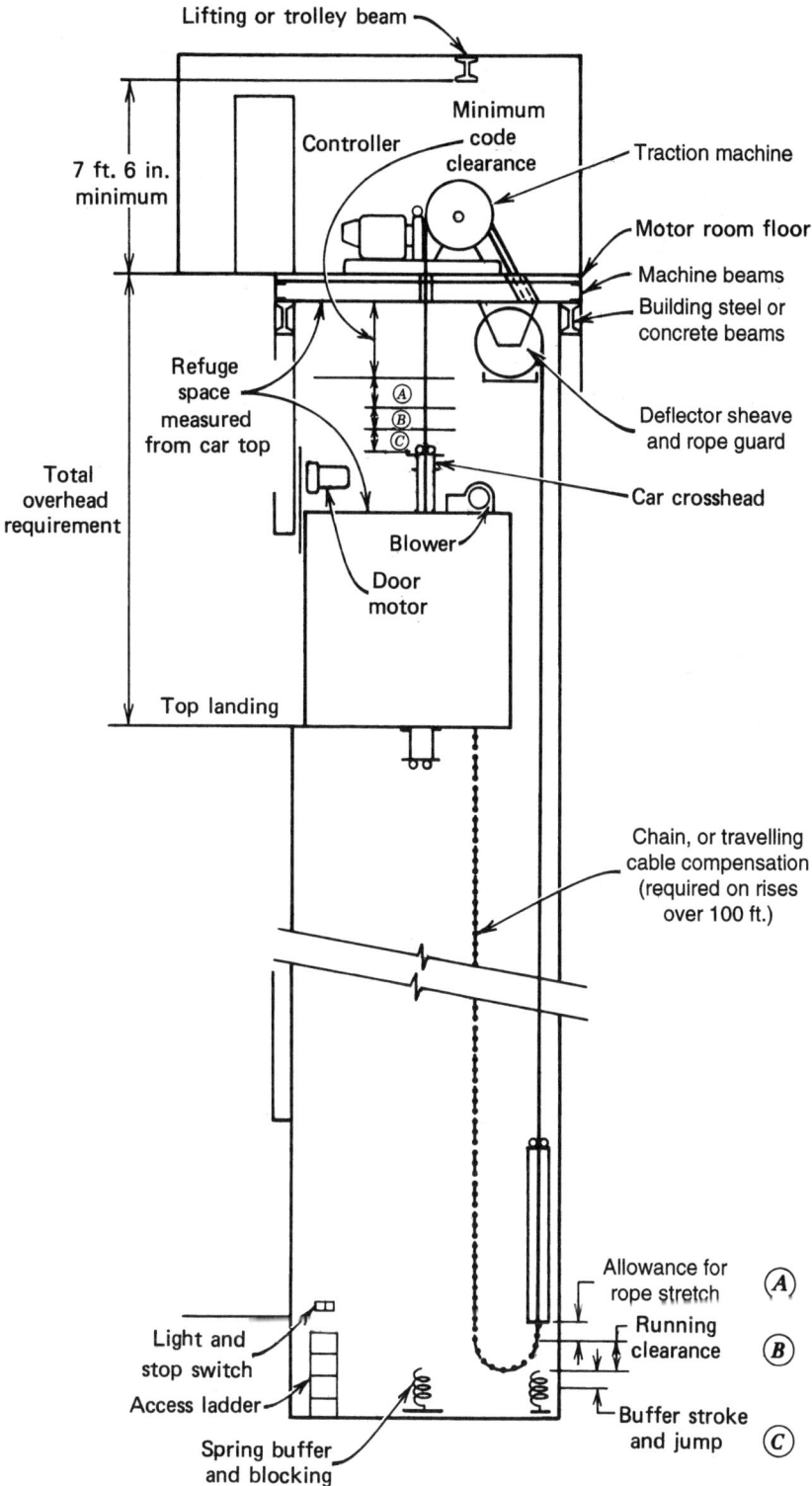

Figure 8.8. *(Continued)*

without coming in contact with other pit equipment. A tensioning device for the elevator governor or safety rope, which controls the application of the car safety, must also be located in the pit.

The total design of the pit (Figure 8.8) must include provisions for all the necessary elevator equipment, lighting, an elevator stop switch, access ladders, and, in some jurisdictions, sprinklers to guard against possible fire, as well as sufficient depth for the possible run-by of the elevator. In addition, a safety clearance imposed by elevator codes requires at least 2 ft (0.6 m) between the bottom of the car platform and the pit floor when the elevator is on a fully compressed buffer. Sufficient support must be provided for the various impacts that may occur. These impacts include those of the rail on safety application, those on the buffers if they are called on to perform, and, in the event of safety application, the up-pulls on the compensating arrangement. All these reactions are shown on the elevator layout provided by the consulting engineer or the elevator manufacturer. Architects and structural engineers must design the pit structure to accommodate these loads on the bottom of the hoistway.

Overhead Space

The overhead space required to accommodate the elevator when it is stopped at the top landing must be allocated in a manner similar to the way pit space is allocated. On top of the elevator car is the mechanism that operates the elevator doors and the elevator lifting structure or crosshead. In addition, space must be provided for the car blower, various car-top-mounted elevator operating devices, plus refuge space for a person on the car top. Appropriate clearances must be maintained between car tops and overhead structures. Close points may be located at the bottom of the beams that support the elevator machinery, or a sheave used to deflect the hoisting ropes back toward the counterweight.

If the elevator passes the top landing, a number of events occur. First, the counterweight lands and starts to compress the counterweight buffer. If the elevator is traveling at full speed, the car may have a slight jump. To accommodate a person working on top of the car, an area with enough clearance for someone to crouch safely will be required. The A17.1 Code requires a refuge space for a person who may be on top of an elevator car. All this adds up to the space required for run-by. The space required from the top landing to the top of the motor room floor may be 20 to 30 or more ft (6 to 9 m).

Figure 8.8 shows this space and how it is allocated. Considerable variation may be expected with various car speeds, sizes, and arrangements of elevator equipment.

Calculating Pit and Overhead Space

The interior height of the elevator cab is one of the main factors in calculating the required overhead height of an elevator hoistway as measured from the top floor served to the underside of the machine room floor. If we assume a desired clear inside car ceiling height of 8 ft (2400 mm), the height of car lighting, which can vary from 12 to 18 in. (300 to 450 mm) depending on the lighting fixtures selected, must be added. This establishes the overall height of the cab. On top of the car, space is needed between the bottom of the crosshead and top of the cab for either rope shackles or a 2 to 1 sheave mounted within the crosshead. This space ranges from 12 to 18 in. (300 to 450 mm). The structural crosshead is 10 to 12 in. (250 to 300 mm) high. The overtravel of the elevator is about 6 in. (150 mm), and the elevator car will move upward the length of the buffer stroke plus one-half of the stroke for possible jump when the counterweight lands on its buffer. When the counterweight is resting on its fully compressed buffer, the elevator code requires a minimum of 2 ft (600 mm) from the highest point of car structure (usually from the top of the crosshead) to the underside of the top of the hoistway or the closest possible striking point.

Summing up the foregoing, the factors shown in Table 8.2 are required.

The A17.1 Safety Code for elevators also requires a 3-ft 6-in. (1050-mm) refuge space on top of the elevator car enclosure to accommodate a maintenance person who may be required to work on top. This must be provided over an area of 650 in.^{2} (0.4 m^2) with no side less than 16 in. (400 mm) when the elevator is at its upper extreme of travel. With large high-speed elevators there is usually sufficient space on the car top and overhead space is generally ample to provide the required refuge space. Refuge space usually requires additional overhead for slower-speed elevators, especially for hydraulic elevators where there is little need for extensive structure on top of the elevator car.

Machine Room Space

The preferred location for traction elevator machinery is directly above the elevator hoistway. For traction elevator applications this location can also be below, at the side or at the rear (a front location will interfere with doors) and adjacent to the hoistway. The machine must be sufficiently low in the hoistway so that the hoist rope lead from the driving machine sheave to the sheaves in the overhead does not cause excessive sidewise stray of the ropes. The machine room space for hydraulic elevators may be remote, with piping carrying the oil to and from the elevator and the pumping unit. Machine-below arrangements require special study, and additional discussion may be found later in this chapter.

TABLE 8.2. Factors Required for Calculating Pit and Overhead Space

Requirement	Example
1. Height of cab ceiling*	8 ft 0 in.
2. Space for lighting	1 ft 3 in.
3. Space between top of enclosure and underside of crosshead	1 ft 0 in.
4. Depth of crosshead	0 ft 10 in.
5. Overtravel	0 ft 6 in.
6. Counterweight buffer stroke at 500 fpm	1 ft 5 in.
7. Counterweight buffer jump	0 ft 8 in.
8. Rope stretch and counterweight buffer clearance (assume)	0 ft 8 in.
9. 2-ft clearance	2 ft 0 in.
	13 ft 40 in.
Distance from top landing served to underside of the top of the hoistway (slab or structure)	16 ft 4 in. (5000 mm)

*Cab ceiling height is also related to landing door height. For example, if the landing doors are 7 ft 0 in. high, the cab ceiling should be a minimum of 7 ft 6 in. high.

Equipment in the machine room varies with the load the elevator must carry and its speed. The following items are usually included: the hoisting machine and the electric elevator controller, a governor for safety application, a motor generator with any elevator of the generator-field-control type (replaced with a solid-state motor drive unit in more modern control systems), a floor-selecting device on the larger elevators or those that serve many stops, and, for a group of three or more elevators, a group-dispatching controller. Some elevator manufacturers may include a group controller as part of each elevator controller.

In addition to elevator equipment (which, by most elevator safety codes, is the only equipment allowed in the elevator machine room), a main power disconnect switch is required for each elevator. Lighting and ventilation are also necessary. The ventilation must be capable of removing the heat released by the elevator equipment to maintain a maximum temperature in the motor room of 104°F (40°C). (For some solid-state controls, 95°F (35°C) is required.) Heating should be provided in colder climates to keep the machine room no less than 50°F (10°C).

Reasonable access to and from the machine room should be provided. Mechanics must maintain the equipment and occasionally remove and replace parts. Trapdoors and trolley beams provide for this contingency in gearless installations. The repairman's ingenuity must often suffice in the smaller, single-car installations, but prior consideration of the problem and provisions for adequate machine room access will save the owner time by expediting repair work should it become necessary. Space should be provided around each piece of electrical equipment, and requirements have been established by the National Electrical Code® (ANSI/NFPA 70) or by local codes. This space usually consists of 36 to 48 in. (900 to 1200 mm) in front of electrical controllers, depending on their location in relation to walls or other equipment. Space to remove the armature of a motor should be provided in front of each hoisting motor or generator (Figure 8.9).

Secondary Levels

All of the aforementioned items must be considered in the allocation of space for the machine room. On average, the plan of the motor room for each elevator usually requires the space over the elevator hoistway plus an equivalent area in front of the hoistway. In larger installations, elevator equipment may advantageously be placed on two levels.

In smaller or slower installations all equipment is usually located on one level as shown in Figure 8.9, which reflects clearances as required by the National Electrical Code C1, 1996 edition. The latest edition of these codes or local codes should be used as a reference. In larger installations a secondary sheave located below the driving machine is often necessary, and because there must be access to this sheave, a sub-machine room or secondary level is required. This level can be a half floor below the elevator machine room floor and provided with suitable access for personnel (Figure 8.10). The lobby space between elevators at the floor below the elevator machine room can often be used for additional elevator equipment.

Venting

Elevator hoistways usually extend to the highest point in the building and can become filled with smoke as a result of a building fire. Many building codes require venting of the hoistway to the atmosphere. An open vent can cause wind and weather problems; therefore, most codes allow dampers that open automatically in the event that a smoke detector

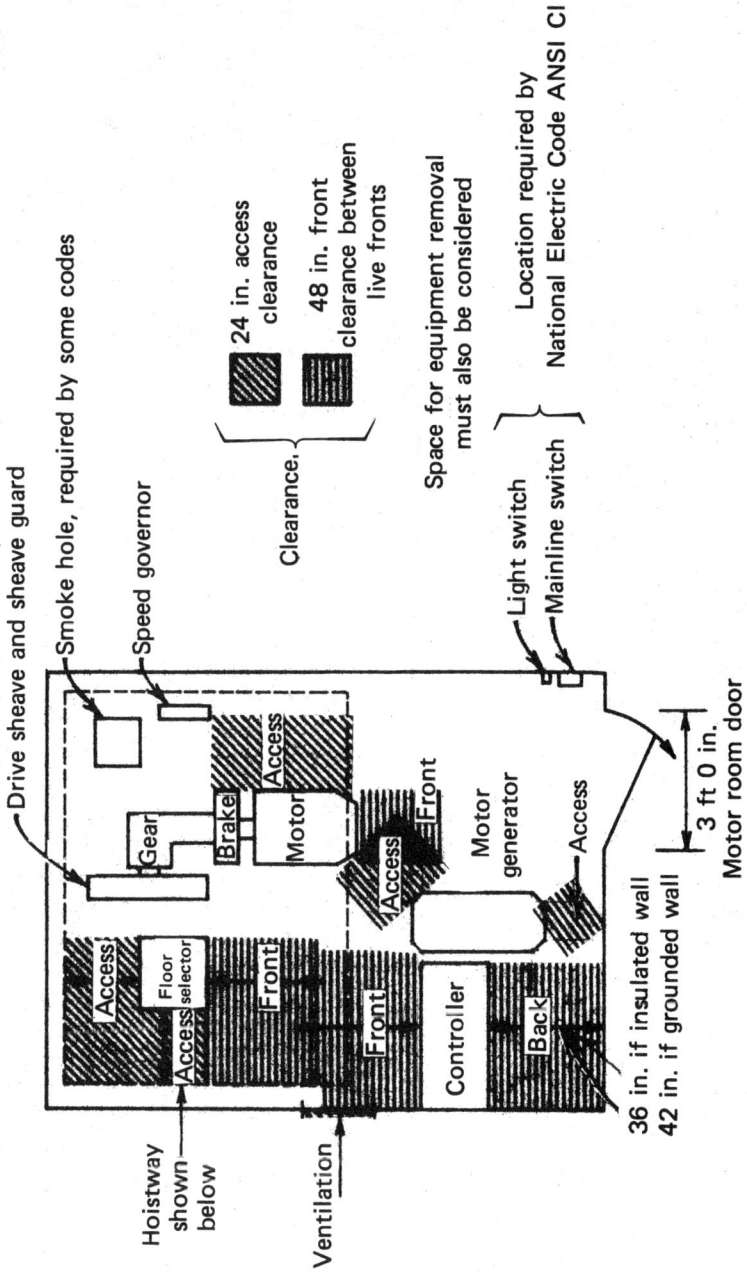

Figure 8.9. National Electric Code® (ANSI/NFPA 70) clearance requirements.

Drive sheave and sheave guard

Smoke hole, required by some codes

Speed governor

24 in. access clearance

48 in. front clearance between live fronts

Clearance.

Space for equipment removal must also be considered

Location required by National Electric Code ANSI Cl

Light switch

Mainline switch

Hoistway shown below

Floor selector

Access

Front

Front

Gear

Brake

Access

Motor

Access

Front

Motor generator

Access

Ventilation

Controller

Back

36 in. if insulated wall
42 in. if grounded wall

3 ft 0 in.

Motor room door

189

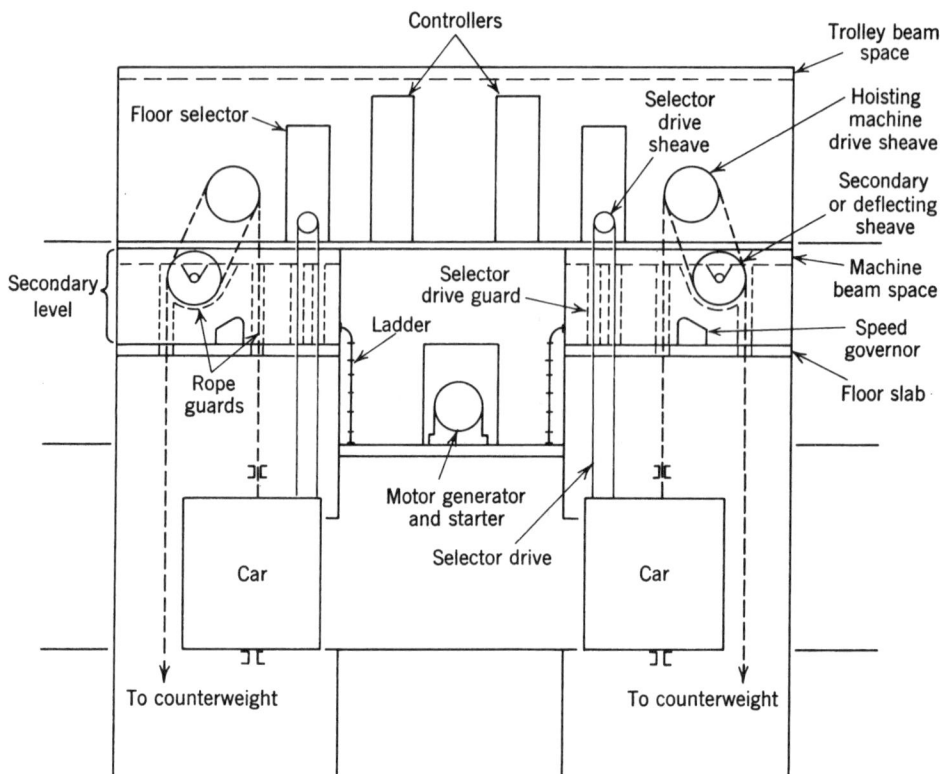

Figure 8.10. Cross section—elevator machine room.

in the hoistway operates. Local requirements and applicable building codes must be consulted for specific requirements regarding venting, vent opening location, and vent sizes.

Particularly in high-rise buildings, hoistway vents, if left opened directly to the outside, can induce vertical air movement throughout the building. Building heating and air-conditioning systems create positive air pressures in areas surrounding the elevator hoistways and, depending on the outside air temperature and atmospheric pressure, vertical air movement in hoistways can be quite substantial if building air is allowed a direct path to the outside. This phenomenon, known as stack effect, can be so great as to restrict elevator door operation at some floors, particularly in winter months when inside and outside air temperature differentials can be extreme.

Building entries are often provided with vestibules and revolving doors to minimize the infiltration of outside air into the building; however, permanently opened hoistway vents can exacerbate the problem and nullify the effects of elevator lobby and building entry enclosures.

Although hoistway vents are necessary to allow smoke—which may infiltrate into hoistways during a fire—to be extracted from the building, they must be carefully designed and controlled to open only when needed—during fire emergencies.

Close coordination between the elevator system designer, the architect who designs the hoistways, and the heating, ventilating, and air-conditioning (HVAC) system designer is

essential. The location and size of vent openings must be considered in regard to anticipated hoistway and machine room equipment. Vent openings must not be restricted, and
space requirements must be established and maintained in these areas to ensure adequate
air flow out of the hoistway during a fire emergency. Proper placement of hoistway and
machine room equipment must be accommodated and necessary access to equipment provided—unrestricted by vent damper enclosures or controls—for equipment maintenance,
inspection, and testing.

ELEVATOR REACTIONS

Concurrent with the space allocations in the machine room are the loads that the equipment
imposes on the structure. The primary load for a traction elevator is due to the elevator-
hauling load that is transmitted through the driving machine sheave shaft to the elevator
machine beams. These beams are installed by the elevator contractor, and the structure
must be designed to provide for the loads imposed. Two or three beams are placed across
the hoistway, and the reactions at the end or support point of each beam are calculated and
indicated on the elevator layout. This load includes the suspended load and an allowance
for impact on the structure, which is usually double the suspended load. In addition, the
weight and location of each piece of elevator equipment is shown on the layout, and the
floor or supporting structure must be designed to accommodate these loads. Connecting
conduits for electrical wiring can be buried in the floor, or the wiring can be installed in
overhead troughs. Trolley beams for moving equipment should be provided. Loading and
location, which will vary with each installation, are functions of elevator capacity, roping
arrangement, and speed.

Typical Layouts and Manufacturers' Standards

Most of the larger elevator manufacturers have preprepared layout information for a wide
variety of elevator applications. These layouts are available on request and often represent
the particular manufacturer's best arrangement for a particular elevator size, capacity, and
speed. There may be variations among manufacturers for the same situation. One may use
a machine with 2:1 roping, whereas another may use 1:1 roping for the same load and
speed. If competitive bidding is expected, it is necessary to ensure that any bidder's
equipment will fit and the space allocated will suit the greatest requirements.

Many typical layouts can be varied in some dimensions, and the manufacturer's representative's advice should be sought. In some typical layouts, which may be based on a
particular manufacturer's "model" elevator, no variation is permitted because all parts
have been preengineered.

In addition to dimensional and structural loading requirements, the typical layout
should include information regarding electrical power requirements and the expected
amount of heat to be released by the operating elevator. Any additional items, such as access ladders, lighting, power for intercommunication systems, life safety systems, or special requirements, should also be provided with the layout data.

For a larger, more complex installation a competent elevator consulting engineer
should be engaged because numerous unique problems of space allocation and complex
electrical requirements can arise. In a taller building there are no standard column spac-
ings and the available land area may not allow use of an acceptable standard module.

Building structural elements such as columns and beams are often integrated into the elevator hoistways, and the elevators may be of unique sizes because of space allocations. Electrical requirements become complex when split feeders and standby power components are considered, and special building operating, security, and life safety requirements may involve unusual elevator interfaces.

Changing a typical layout to suit a particular situation should be attempted only with expert advice. As may be seen from the earlier discussion of space requirements, an assumed minor change in any dimension can have a chain effect and render the standard layout useless. In European countries many elevators are completely standardized and sizes are established by the International Standards Organization (ISO), so any deviation may increase the cost of the installation. In the United States, if a manufacturer's "model" is intended to be used, such changes can impose a substantial price penalty.

Samples of typical layouts that have been developed by the National Elevator Industries Inc. (the U.S. elevator manufacturers trade association) are shown in Figures 8.11a–h. Note that all the necessary dimensions are provided.

ADDITIONAL MACHINE ROOM CONSIDERATIONS

Power and ventilation requirements were mentioned in the discussion of machine room requirements. These items concern the efficient operation of elevators, and particular attention should be given to their installation.

Energy—Absorbed and Regenerated

Electrical energy is required primarily to start and run an elevator (traction-type) and only partially to lift the load. Except for energy consumed in opening and closing the doors, much of the energy a traction elevator requires to lift loads is returned to the line when such a load is lowered. Stated otherwise, elevators require energy to lift the uncounterbalanced portion of the load when people come to work; and those same people, as they leave, theoretically cause the elevator to pump back an equal amount of energy. In practice, friction, impedance, and other factors introduce losses, and some energy will always be used by an elevator in excess of energy returned. An approximate guide is that regeneration caused by a fully loaded elevator traveling down is about 75% of the energy required to lift that same load.

Efficiency of traction elevator machinery varies from 50 to 70% for the geared-type elevator with roller guides, and from 75 to 85% for gearless elevators. Energy losses are generally changed to heat, which must be dissipated by the building ventilation system.

The power required by any fully or partly loaded elevator may be readily derived from the following formula:

Horsepower (hp) =

$$\frac{\text{load in car (lb)} \times \text{percentage of load that is unbalanced} \times \text{velocity (fpm)}}{33,000 \times \text{efficiency of the hoisting machine}}$$

Example: Traction elevator, 2500 lb, 500 fpm, gearless machine, 40% counterbalance

$$\text{hp} = \frac{2500 \times 0.60 \times 500}{33,000 \times 0.80} = 28 \text{ hp approximate}$$

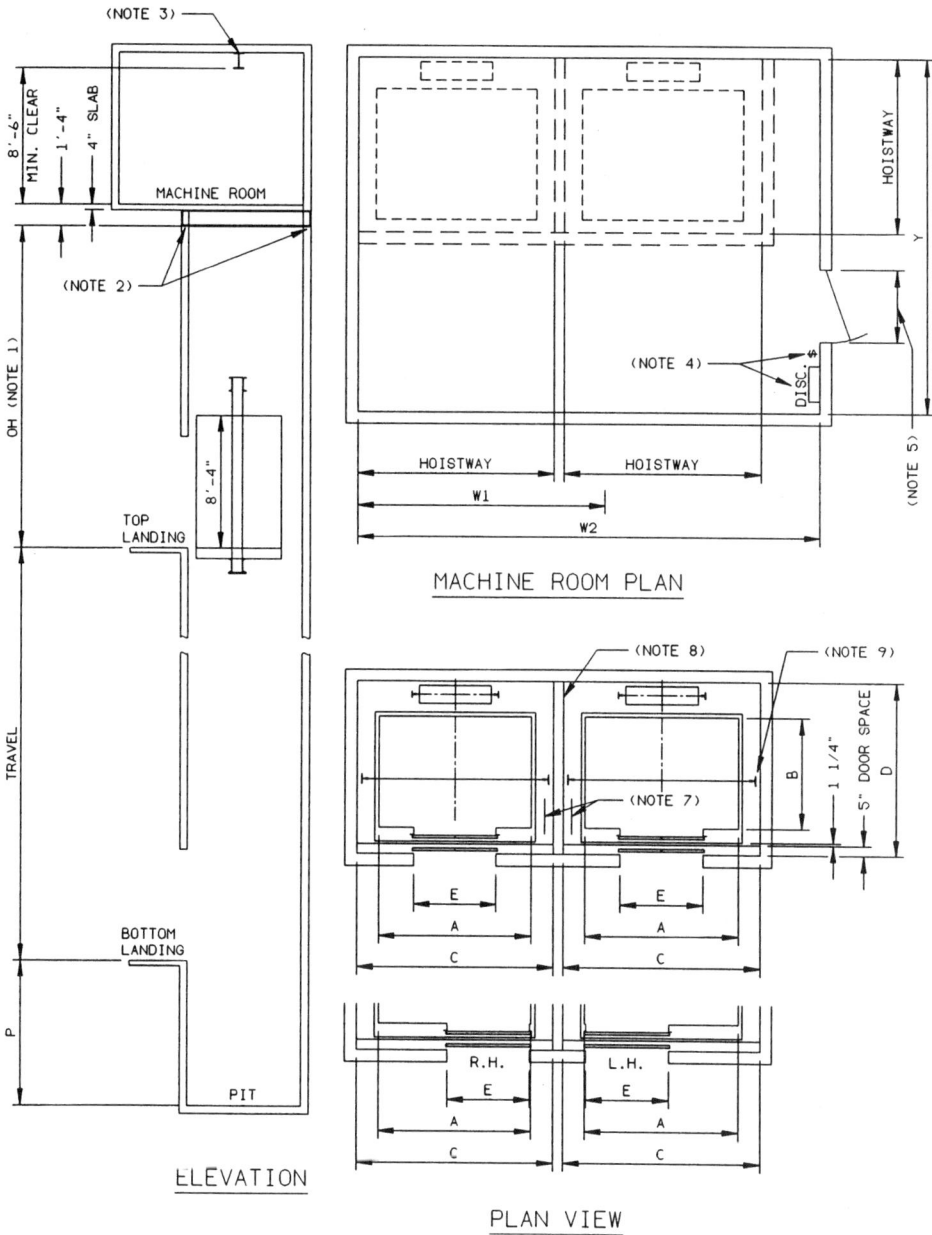

Figure 8.11. Sample typical layouts. (*a–b*) Electric passenger elevators with rated speeds of 500–700 fpm. (Reprinted with permission from *NEII Vertical Transportation Standards* 7th Edition, including 1994 Supplement. Copyrighted © 1992 and 1994, National Elevator Industry, Inc., Fort Lee, New Jersey. This reprinted material is not the complete and official position of National Elevator Industry, Inc. on the referenced subject which is represented only by the standard in its entirety.)

"OH" BOTTOM OF BEAM (Note 1) ft-in.					
SPEED (fpm)	RATED LOAD (lb)				
	2000	2500	3000	3500	4000
100	14-6	14-6	14-6	14-6	14-6
150	14-6	14-6	14-6	14-6	14-6
• 200	14-6	14-6	14-6	14-6	14-6
250	15-0	15-0	15-0	15-0	15-0
300	15-0	15-0	15-0	15-0	15-0
• 350	15-0	15-0	15-0	15-0	15-0

CAR & HOISTWAY								
RATED LOAD (lb)	AREA ■	ft-in. (Note 6)					ENTRANCE (Note 10)	
		A	B	C	D	E	TYPE	
2000	24.2	5-8	4-3	7-4	6-10	3-0	SSSO ▲	SSCO
2500	29.1	6-8	4-3	8-4	6-10	3-6	SSSO ✱	SSCO ▲
3000	33.7	6-8	4-7	8-4	7-4	3-6	SSSO ✱	SSCO ▲
3500	38.0	6-8	5-3	8-4	8-0	3-6	SSSO ✱	SSCO ▲
4000	42.2	7-8	5-3	9-6	8-0	4-0	SSCO ✱	

"P" PIT DEPTH ft-in.					
SPEED (fpm)	RATED LOAD (lb)				
	2000	2500	3000	3500	4000
100	5-6	5-6	5-6	5-6	5-6
150	5-6	5-6	5-6	5-6	5-6
• 200	5-0	5-0	5-0	5-0	5-0
250	5-0	5-0	5-0	5-0	5-0
300	5-0	5-0	5-0	5-0	5-0
• 350	5-0	5-0	5-0	5-0	5-0

MACHINE ROOM ft-in.								
RATED LOAD (lb)	W1 WIDTH	W2 WIDTH	"Y" DEPTH FOR RATED SPEED (fpm)					
			100	150	200 •	250	300	350 •
2000	7-4	15-0	16-0	16-0	16-0	16-0	16-0	16-0
2500	8-4	17-0	16-0	16-0	16-0	16-0	16-0	16-0
3000	8-4	17-0	16-0	16-0	16-0	16-0	16-0	16-0
3500	8-4	17-0	16-0	16-0	16-0	16-0	16-0	16-0
4000	9-6	19-5	16-0	16-0	16-0	16-0	16-0	16-0

Notes

1. "OH" Dimensions are based on a 8'-4" overall car height.

2. Supports for elevator machine beams not by elevator supplier. Refer to page 12 details.

3. Hoisting beams not by elevator supplier.

4. Location of the main power disconnecting means, the car light disconnecting means and lighting switch shall be in accordance with requirements of ASME A17.1 & ANSI/NFPA No. 70.

5. 3'-6" x 7'-0" Recommended.

6. Refer to PREFACE to determine approximate platform dimensions.

7. Pit ladder not by elevator supplier.

8. Dividing beams, not by elevator supplier, to be designed to sustain rail forces. Consult elevator supplier.

9. Refer to page 12 for typical guide rail bracket details.

10. Refer to page 11 for hoistway and entrance details, information on required hoistway door sill supports, etc.

11. When compliance with seismic risk zone 2 or greater requirements is anticipated, see page 10 for additional hoistway space.

Notes

■ Maximum allowable inside car area in ft^2 per ASME A17.1, Rule 207.1.

✱ These car dimensions and entrance types provide wheelchair accessibility and accommodate an ambulance type stretcher (76 in. x 24 in.) in the horizontal position.

▲ These car dimensions and entrance types provide wheelchair accessibility.

• These are the most commonly used speeds.

(b)

Figure 8.11. *(Continued)* (Reprinted with permission from *NEII Vertical Transportation Standards* 7th Edition, including 1994 Supplement. Copyrighted © 1992 and 1994, National Elevator Industry, Inc., Fort Lee, New Jersey. This reprinted material is not the complete and official position of National Elevator Industry, Inc. on the referenced subject which is represented only by the standard in its entirety.)

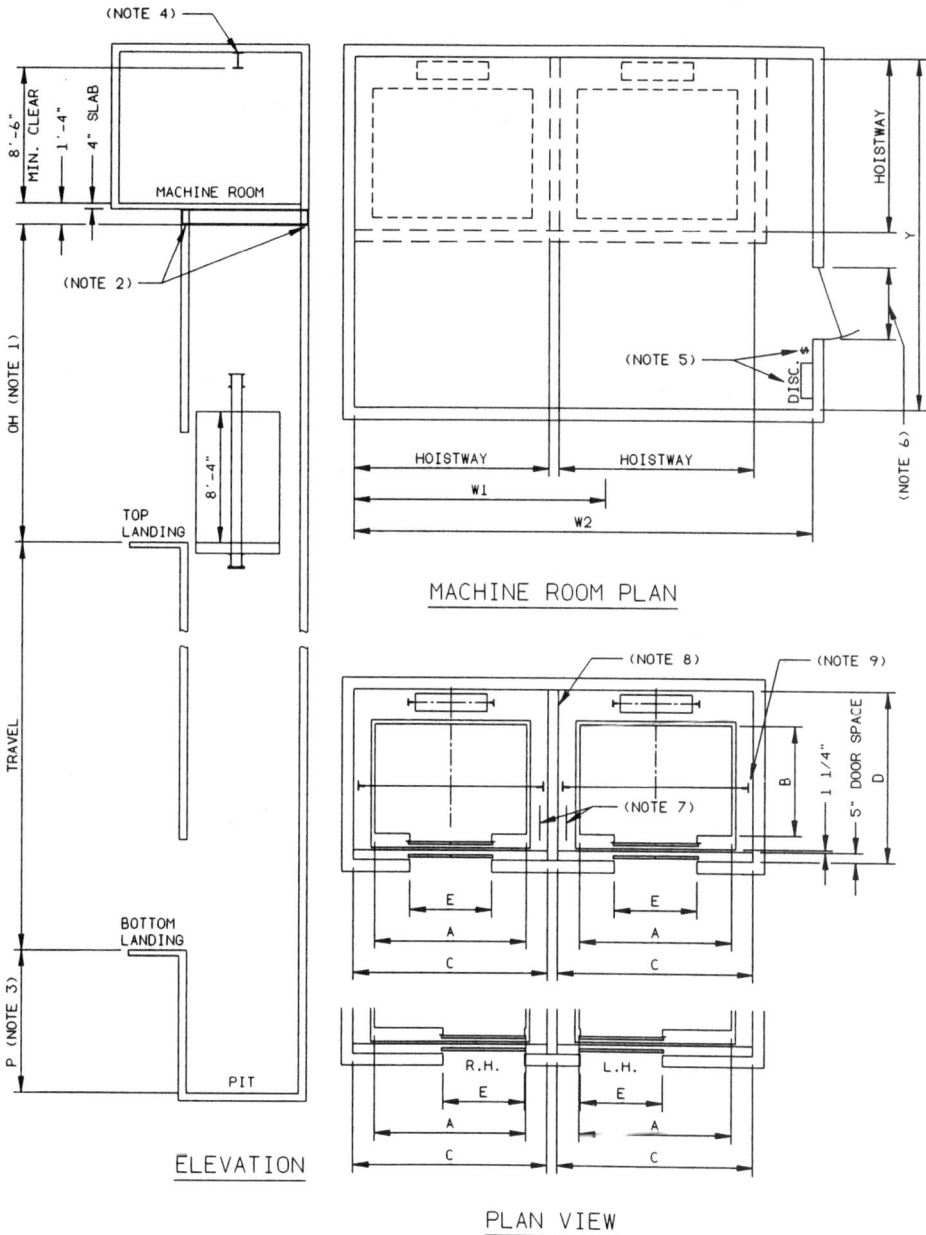

Figure 8.11. *(Continued)* (*c–d*) Electric passenger elevators with rated speeds of 100–350 fpm. (Reprinted with permission from *NEII Vertical Transportation Standards* 7th Edition, including 1994 Supplement. Copyrighted © 1992 and 1994, National Elevator Industry, Inc., Fort Lee, New Jersey. This reprinted material is not the complete and official position of National Elevator Industry, Inc. on the referenced subject which is represented only by the standard in its entirety.)

"OH" BOTTOM OF BEAM (Note 1) ft-in.					
SPEED (fpm)	RATED LOAD (lb)				
	2000	2500	3000	3500	4000
500	17-6	17-6	18-3	18-3	17-6
600	18-6	18-6	18-6	18-6	18-6
700	20-6	20-6	20-6	20-6	20-6

CAR & HOISTWAY								
RATED LOAD (lb)	AREA ■	ft-in. (Note 11)					ENTRANCE (Note 10)	
		A	B	C	D	E	TYPE	
2000	24.2	5-8	4-3	7-4	6-11	3-0	SSSO ▲	SSCO ▲
2500	29.1	6-8	4-3	8-4	6-11	3-6	SSSO ✱	SSCO ▲
3000	33.7	6-8	4-7	8-4	7-5	3-6	SSSO ✱	SSCO ▲
3500	38.0	6-8	5-3	8-4	8-1	3-6	SSSO ✱	SSCO ▲
4000	42.2	7-8	5-3	9-6	8-1	4-0	SSCO ✱	

"P" PIT DEPTH (Note 3) ft-in.					
SPEED (fpm)	RATED LOAD (lb)				
	2000	2500	3000	3500	4000
500	10-1	10-1	10-1	10-1	10-1
600	11-5	11-5	11-5	11-5	11-5
700	11-5	11-5	11-5	11-5	11-5

MACHINE ROOM ft-in.					
RATED LOAD (lb)	W1 WIDTH	W2 WIDTH	"Y" DEPTH FOR RATED SPEED (fpm)		
			500	600	700
2000	7-4	15-0	18-6	18-6	18-6
2500	8-4	17-0	18-6	18-6	18-6
3000	8-4	17-1	18-6	18-6	18-6
3500	8-4	17-1	18-6	18-6	18-6
4000	9-6	19-5	18-6	18-6	18-6

Notes

1. "OH" Dimensions are based on a 8'-4" overall car height.

2. Supports for elevator machine beams not by elevator supplier. Refer to page 12 details.

3. For travel greater than 400'-0" increased pit depth may be required. Consult elevator supplier.

4. Hoisting beams not by elevator supplier.

5. Location of the main power disconnecting means, the car light disconnecting means and lighting switch shall be in accordance with requirements of ASME A17.1 & ANSI/NFPA No. 70.

6. 3'-6" x 7'-0" Recommended.

7. Pit ladder not by elevator supplier.

8. Dividing beams, not by elevator supplier, to be designed to sustain rail forces. Consult elevator supplier.

9. Refer to page 12 for typical guide rail bracket details.

10. Refer to page 11 for hoistway and entrance details, information on required hoistway door sill supports, etc.

11. Refer to PREFACE to determine approximate platform dimensions.

12. When compliance with seismic risk zone 2 or greater requirements is anticipated, see page 10 for additional hoistway space.

Notes

■ Maximum allowable inside car area in ft^2 per ASME A17.1, Rule 207.1.

✱ These car dimensions and entrance types provide wheelchair accessibility and accommodate an ambulance type stretcher (76 in. x 24 in.) in the horizontal position.

▲ These car dimensions and entrance types provide wheelchair accessibility.

(d)

Figure 8.11. *(Continued)* (Reprinted with permission from *NEII Vertical Transportation Standards* 7th Edition, including 1994 Supplement. Copyrighted © 1992 and 1994, National Elevator Industry, Inc., Fort Lee, New Jersey. This reprinted material is not the complete and official position of National Elevator Industry, Inc. on the referenced subject which is represented only by the standard in its entirety.)

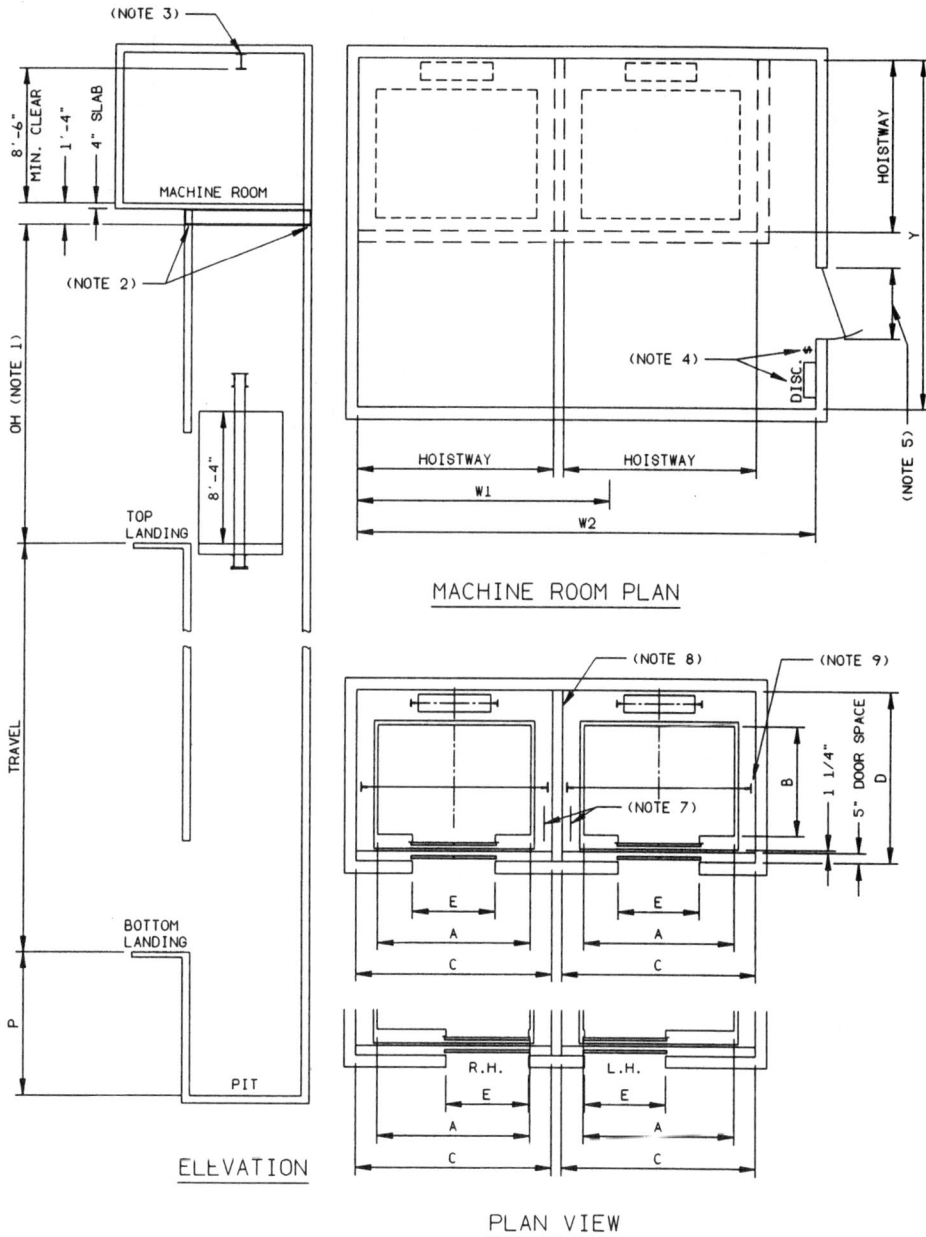

Figure 8.11. *(Continued)* *(e–f)* Electric passenger elevators with rated speeds of 800–1200 fpm. (Reprinted with permission from *NEII Vertical Transportation Standards* 7th Edition, including 1994 Supplement. Copyrighted © 1992 and 1994, National Elevator Industry, Inc., Fort Lee, New Jersey. This reprinted material is not the complete and official position of National Elevator Industry, Inc. on the referenced subject which is represented only by the standard in its entirety.)

"OH" BOTTOM OF BEAM (Note 1) ft-in.				
SPEED (fpm)	RATED LOAD (lb)			
	2500	3000	3500	4000
800	20-8	20-8	20-8	20-8
1000	24-6	24-6	24-6	24-6
1200	24-6	24-6	24-6	24-6

CAR & HOISTWAY								
RATED LOAD (lb)	AREA ■	ft-in. (Note 6)					ENTRANCE (Note 10)	
		A	B	C	D	E	TYPE	
2500	29.1	6-8	4-3	8-6	7-0	3-6	SSSO *	SSCO ▲
3000	33.7	6-8	4-7	8-6	7-6	3-6	SSSO *	SSCO ▲
3500	38.0	6-8	5-3	8-6	8-2	3-6	SSSO *	SSCO ▲
4000	42.2	7-8	5-3	9-6	8-2	4-0	SSCO *	

"P" PIT DEPTH (Note 12) ft-in.				
SPEED (fpm)	RATED LOAD (lb)			
	2500	3000	3500	4000
800	14-2	14-2	14-2	14-2
1000	17-0	17-0	17-0	17-0
1200	20-0	20-0	20-0	20-0

MACHINE ROOM ft-in.					
RATED LOAD (lb)	W1 WIDTH	W2 WIDTH	"Y" DEPTH FOR RATED SPEED (fpm)		
			800	1000	1200
2500	9-6	19-0	19-0	19-0	19-0
3000	9-6	19-0	19-0	19-0	19-0
3500	9-6	19-0	19-0	19-0	19-0
4000	9-6	19-5	19-0	19-0	19-0

Notes
1. "OH" Dimensions are based on a 8'-4" overall car height.

2. Supports for elevator machine beams not by elevator supplier. Refer to page 12 details.

3. Hoisting beams not by elevator supplier.

4. Location of the main power disconnecting means, the car light disconnecting means and lighting switch shall be in accordance with requirements of ASME A17.1 & ANSI/NFPA No. 70.

5. 3'-6" x 7'-0" Recommended.

6. Refer to PREFACE to determine approximate platform dimensions.

7. Pit ladder not by elevator supplier.

8. Dividing beams, not by elevator supplier, to be designed to sustain rail forces. Consult elevator supplier.

9. Refer to page 12 for typical guide rail bracket details.

10. Refer to page 11 for hoistway and entrance details, information on required hoistway door sill supports, etc.

11. When compliance with seismic risk zone 2 or greater requirements is anticipated, see page 10 for additional hoistway space.

12. If full stroke buffers are to be used, consult elevator supplier for increased pit dimension.

Notes
■ Maximum allowable inside car area in ft^2 per ASME A17.1, Rule 207.1.

* These car dimensions and entrance types provide wheelchair accessibility and accommodate an ambulance type stretcher (76 in. x 24 in.) in the horizontal position.

▲ These car dimensions and entrance types provide wheelchair accessibility.

(f)

Figure 8.11. *(Continued)* (Reprinted with permission from *NEII Vertical Transportation Standards* 7th Edition, including 1994 Supplement. Copyrighted © 1992 and 1994, National Elevator Industry, Inc., Fort Lee, New Jersey. This reprinted material is not the complete and official position of National Elevator Industry, Inc. on the referenced subject which is represented only by the standard in its entirety.)

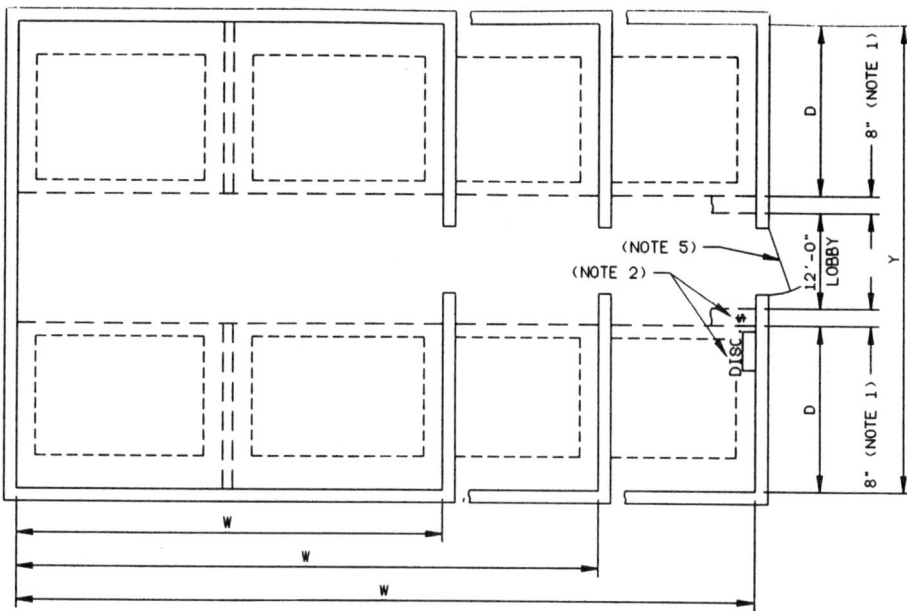

MACHINE ROOM PLAN (OVERHEAD)
PASSENGER ELEVATOR

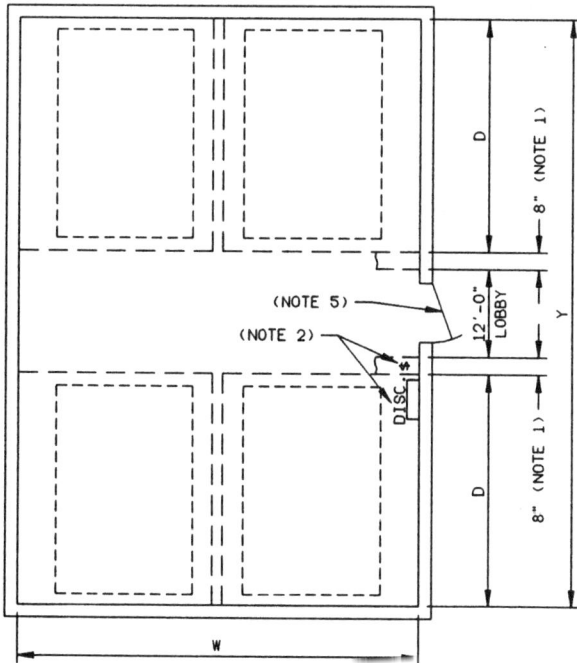

MACHINE ROOM PLAN (OVERHEAD)
HOSPITAL ELEVATOR

(g)

Figure 8.11. *(Continued)* *(g–h)* Electric passenger and hospital elevators. (Reprinted with permission from *NEII Vertical Transportation Standards* 7th Edition, including 1994 Supplement. Copyrighted © 1992 and 1994, National Elevator Industry, Inc., Fort Lee, New Jersey. This reprinted material is not the complete and official position of National Elevator Industry, Inc. on the referenced subject which is represented only by the standard in its entirety.)

Notes
1. Inside hoistway to inside hoistway
   dimension is 13'–4" minimum. When
   wall thickness varies the total of
   lobby and two hoistway walls must
   be adjusted to the minimum 13'–4".

2. Location of the main power disconnecting
   means, the car light disconnecting means
   and lighting switch shall be in accordance
   with requirements of ASME A17.1 &
   ANSI/NFPA No. 70.

3. The "Y" dimension requires a minimum
   of twice the "D" dimension of the
   appropriate set of elevators plus
   13'–4".

4. The "W" dimension for groups of
   elevators facing across lobby from
   each other requires the "W1" from
   the appropriate set of elevators
   plus the addition of the "C"
   hoistway width dimension plus
   the width of one divider beam
   per each additional elevator
   added to the width.

5. 3'–6" x 7'–0" Recommended.

6. Should any of machine room space
   shown require reduction in areas
   shown due to ductwork, equipment
   access doors in floor, etc. in the
   area, or if columns and/or windbracing
   are in this area, consult elevator
   supplier for additional space
   requirements.

(h)

Figure 8.11. *(Continued)* (Reprinted with permission from *NEII Vertical Transportation Standards* 7th Edition, including 1994 Supplement. Copyrighted © 1992 and 1994, National Elevator Industry, Inc., Fort Lee, New Jersey. This reprinted material is not the complete and official position of National Elevator Industry, Inc. on the referenced subject which is represented only by the standard in its entirety.)

The energy consumption of a group of elevators operating over a period of time requires a number of calculations, including determination of full-, partial-, and no-load trips, direction, and number of car stops. The task for a particular large building requires the help of a special computer program.

The energy required by an automatic elevator varies with its size, speed, and usage. It can be easily determined by a recording watt-hour meter after the building is completed and has an established population.

Estimates can be based on known examples. If we take a typical office building as an example and either measure or calculate the energy requirements during the course of a 10-hr day, the following results can be expected.

Example 8.2. Energy Requirements for an Automatic Elevator

Given: Six low-rise elevators, 3500 lb (1600 kg) at 500 fpm (2.5 mps), serving floors 1 to 10

Average energy required during up-peak traffic   48 kW
Average energy required during heavy two-way traffic   36 kW
Average energy required during light two-way traffic 30 kW
Average energy required during down-peak traffic   30 kW

Assume the following distribution of traffic during a 10-hr day.

Up-peak	1 hr × 48 kW =	48 kWh
Heavy two-way	4 hr × 36 kW =	144 kWh
Light two-way	4.5 hr × 30 kW =	135 kWh
Down-peak	0.5 hr × 30 kW =	15 kWh
		342 kWh

Total energy consumption for the day is 342 kWh, which divided by 6 is 57 kWh per elevator. At 6 cents a kWh, this amounts to about $21.00 for the six elevators per day, or for a 250-working-day year, $5130.00 per year.

Example 8.2 is based on elevators with motor generator sets. If solid-state motor drives are used, the motor generator idling losses are eliminated and the energy requirements are reduced by about 30 to 35%. Therefore, the total energy per elevator would be about 40 kWh per day. The annual savings for the six elevators would be $1750.

With higher rises and higher speeds, values as high as 100 kWh per day are possible. As a general rule, elevators in commercial and institutional buildings require 50 to 100 kWh per day, depending on load, speed, and rise, and elevators in residential buildings from 10 to 25 kWh per day for lower-rise (up to about 20 floors) and 25 to 50 kWh per day for higher buildings.

As compared with heating, air-conditioning, and lighting, elevators require a very small percentage of the total building energy needs, amounting to about 2 or 3% in a large office building. (See Table 8.3.)

This is, to some extent, a result of the regenerative nature of most traction elevator systems. Because of the 40 to 50% overbalancing of counterweights, traction elevators traveling down full (car heavier than counterweight) or up empty (counterweight heavier than car) can, under some conditions, actually be generating power into, rather than drawing it from, the electrical power supply system. This regenerative energy is to some extent released as heat in the machine room, and to a lesser degree transferred back into the building electrical power supply system where it is absorbed by other building systems—

TABLE 8.3. Starting and Running Current: Typical Generator Field Control Elevators @ 460 V, 3φ, 60 Hz

Geared	2500 lb @ 200 fpm 1200 kg @ 1.0 mps	3000 lb @ 300 fpm 1400 kg @ 1.5 mps	3500 lb @ 350 fpm 1600 kg @ 1.6 mps
M.G.* starting from rest	40 A	60 A	60 A
Elevator start full load up	45 A	70A	85 A
Elevator run full load up	30 A	45 A	55 A
Gearless	2500 lb @ 500 fpm 1200 kg @ 2.5 mps	3000 lb @ 700 fpm 1400 kg @ 3.5 mps	3500 lb @ 1000 fpm 1600 kg @ 5.0 mps
M.G. starting from rest	75 A	120 A	250 A
Elevator start full load up	70 A	120 A	180 A
Elevator run full load up	45 A	70 A	110 A

*M.G. = motor generator (values will vary with various manufacturers).

With solid-state motor drive, the M.G. start from rest current is eliminated and the other values remain approximately the same.

lighting and motors of building mechanical equipment. Solid-state control systems do not as readily accommodate the transfer of regenerative power, which is therefore converted mostly to heat, which is released by elevator machinery into the machine room environment. This physical property of solid-state control systems, although having little effect on the overall power consumption of an elevator system, is of great concern regarding the design of machine room ventilation systems and is covered in more detail in the following section on ventilation.

The demand for power created by an elevator system is of concern. The feeders that supply the elevator motor must be of sufficient size to serve that demand with a minimum of voltage drop. Elevator motor generators are driven by ac induction motors, as are the pumps on hydraulic elevators of about 10 to 75 or more horsepower. The greatest power demand is usually created when the motor starts from rest. An equal or higher-power demand for a generator field control elevator or an elevator with an SCR (silicon controlled rectifier) motion controller may occur when the elevator itself is started up with a full load in the car. Normal running current is generally much lower than that used for either the starting of the motor generator or the starting of the elevator with full load up. Elevator power requirements are usually stated with three values: (1) motor generator starting from rest, (2) elevator starting full load up, and (3) elevator running full load up. The motor current for other conditions is lower.

With any induction motor, resistance starting (or reconnecting motor windings from Y to Δ) is a means of reducing the peak starting current of the motor. There is a mathematical probability that not all the elevators in a group will start up with a full load at one time. Therefore, a diversity factor can be calculated for any group of two or more cars. Typical values run from 0.87 for two cars to 0.75 for six cars, the value indicating the probable percentage of the total current required to operate all the elevators in the group at one time.

A set of typical starting and running current values appears in Table 8.3. The building designer is usually required to confirm with the local electric company the voltage characteristics and availability of power before the elevator manufacturer undertakes production of the equipment.

Electric current requirements for ac resistance traction and hydraulic elevators are easier to calculate. In such a case, the elevator is either running or stopped, unlike one using generator field control whereby a generator is running whether or not the elevator is running. With ac traction or hydraulic elevators, starting is either across the line or through a resistance step. The horsepower of the elevator motor establishes the electrical power required. Typical values are given in Table 8.4.

Power Failure

Buildings that have a vital function, such as hospitals, should have standby power systems so that essential elevator service is maintained in the event of utility power failure. In practice, usually two sources of outside utility power are supplied if possible, and means to switch from one to another provided. The elevator system is designed to operate from two power feeders, some of the elevators on one feeder and the rest on a second feeder run from the main building power panel to the machine rooms. If one source of power fails or if a feeder is destroyed, at least half the elevators will be available.

When a power failure does occur, a number of events should be designed to take place. First, car lighting and a small circulating fan should go on in each elevator car immedi-

TABLE 8.4. Starting and Running Currents: Typical ac Traction Machine (Counterweighted) and Hydraulic Elevators (Uncounterweighted), 460 V, 3φ, 60 Hz (Amperes Starting/Amperes Running)

	2500 lb @ 100 fpm 1200 kg @ 0.5 mps	2500 lb @ 125 fpm 1200 kg @ 0.6 mps	2500 lb @ 150 fpm 1200 kg @ 0.75 mps
ac traction across the line	25/15	35/20	45/25
Hydraulic across the line	175/30	200/35	275/65

	3500 lb @ 100 fpm 1600 kg @ 0.5 mps	3500 lb @ 125 fpm 1600 kg @ 0.6 mps	3500 lb @ 150 fpm 1600 kg @ 0.75 mps
ac traction across the line	50/25	65/30	75/35
Hydraulic across the line	200/35	225/40	300/70

ately, supplied by a battery pack on the car itself. This lighting is required by the A17.1 Elevator Code. Ventilation can be provided by a fan with a permanent magnet motor that draws minimum battery current. The standby generator should start automatically, and, when it is up to speed, each elevator should automatically return to a main floor and park with its doors open. The standby power is then switched to the next elevator. When all elevators are returned, emergency personnel should be able to manually operate a selector switch to choose the elevator they wish to use.

In a high-rise building it is desirable to size the standby generator to run at least two elevators in the building. A service elevator serving all floors is dedicated to the standby power system for use by emergency personnel, and the remaining elevators are evacuated one at a time until all are accounted for. Figure 8.12 shows an arrangement for standby power using the split-feeder approach.

Hydraulic elevators can be arranged to operate on standby power utilizing a valving arrangement that allows the hydraulic elevator to be lowered without causing the pump to start. Standby power is used to operate the down valve, run the elevator to the lower landing, and open the doors. The same arrangement can be achieved manually by operating the down valves and using an emergency unlocking device to open the doors. Alternatively, a battery pack may be wired into the hydraulic elevator operating systems to perform the lowering and door opening automatically.

With any standby power system, the elevators must be provided with a signal to the elevator system indicating that the source of power is a standby power source and is limited. Elevator controllers must be arranged to operate so that only one or a designated minimum number of elevators can operate at one time to provide the mode of operation desired. The specifications for the elevators must include all the necessary details for the guidance of the elevator design engineer.

Ventilation

The heat released by ac resistance traction or hydraulic elevators is relatively easy to estimate. Most of the energy supplied to the motor will be changed to heat through braking

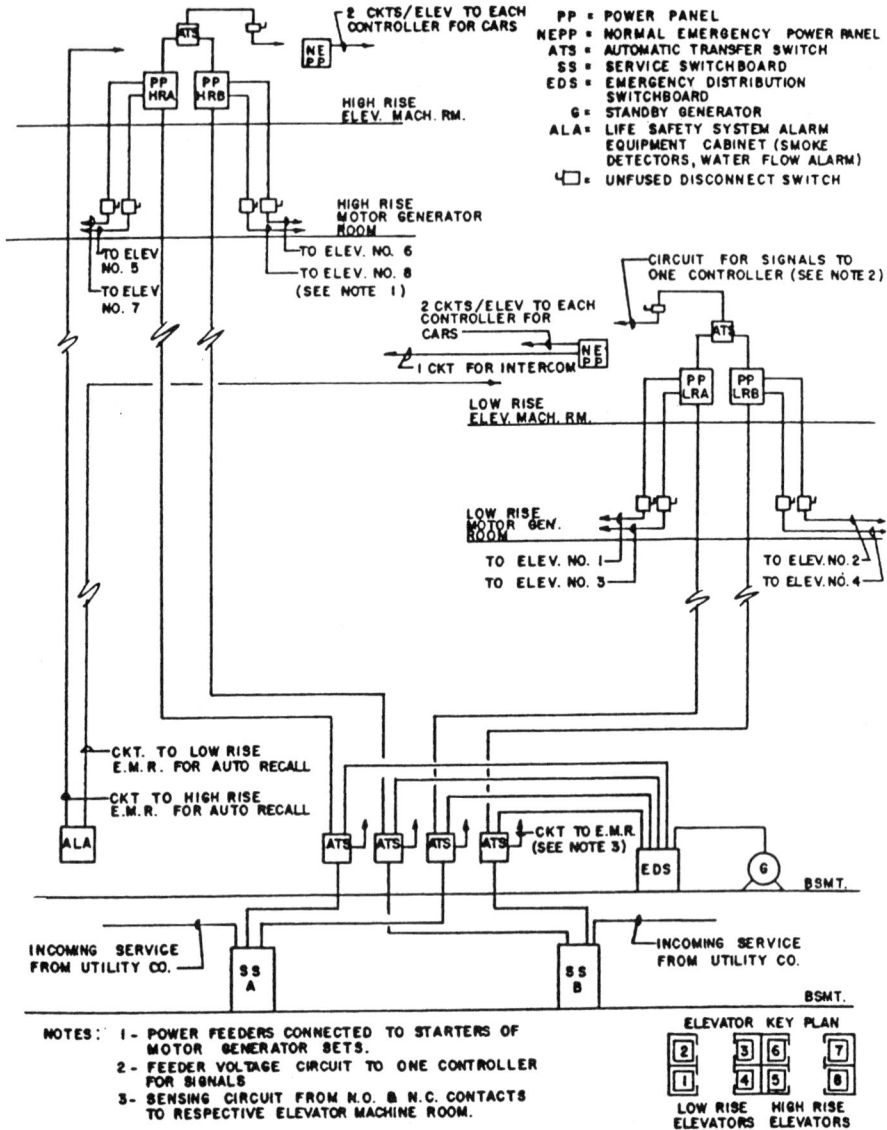

Figure 8.12. Circuitry for standby power where split elevator feeders are provided. Split feeders are provided to a group of elevators so that only one-half of the elevators will be out of service if one feeder fails (Courtesy Jaros, Baum & Bolles).

when stops are made, or in a hydraulic elevator, through the temperature rise of the oil. An elevator typically runs up less than half the time, whereas a substantial portion of its time is spent stopped. Inasmuch as most ac resistance traction or hydraulic elevators are found in low-rise buildings, it is fair to estimate a maximum of 20% of time will be spent running full-load up, the worst condition. Converting this percentage of elevator motor horsepower to Btu's gives a good approximation of the heating expected in the elevator machine room. (One horsepower per hour equals 2544 Btu.) For efficient elevator operation, this heat must be dissipated.

Dissipation of machine room heat is important in any elevator installation. If the machine room gets too hot, electrical wiring insulation deteriorates, oil loses its viscosity, and erratic operation such as poor leveling, abrupt starts, and poor brake action can be experienced.

The heat generated by an elevator may be considerable during certain periods of the day. For example, during the morning in-rush in an office building, elevators are leaving the lobby with full loads, so energy demands are at their maximum for a half-hour or so. At that time a typical office building elevator may generate 25,000 to 40,000 Btu/hr.

Adequate ventilation is required in any elevator machine room to dissipate the heat and provide a temperature of no more than 104°F (40°C) (95°F (35°C) for solid-state controllers) or no less than 50°F (10°C) and the machine room may require cooling or heating. In smaller, single-car installations a thermostatically controlled exhaust fan may accomplish this. In larger installations the air-conditioning system should be designed to maintain these temperature limits. Table 8.5 shows a sample of typical heat amounts generated by elevator equipment with a regenerative system. If the generator is remote from the elevator drive motor, one-half of the heat emanates from the generator or SCR controller and one-half from the hoist motor.

Nonregenerative systems, that is, solid-state motor drives (SSMD), variable voltage, variable frequency (VVVF), and ac variable voltage (ACVV) systems, do not utilize a rotating machine such as a motor generator to control hoist machine voltage and current characteristics. These static control systems therefore do not readily allow regenerative energy to be passed back into the building electrical power supply system; hence, they are referred to as nonregenerative systems. Systems that are designed using these nonregenerative devices release the regenerative energy that is developed by elevators operating in the overhauling mode into the machine room in the form of heat. Machine room air-conditioning or natural ventilation systems must be designed to maintain the machine room ambient temperature ranges previously indicated, based on the higher equipment heat release. Heat release levels 25% higher than those regenerative systems (Table 8.4) can be expected. HVAC designers must closely coordinate machine room temperature requirements and machine room air-conditioning system design with elevator equipment designers.

TABLE 8.5. Sample Heat Release Parameters; Typical Gearless Installations Under Busy Conditions (mj = mega-joule)

1. 2000 lb @	500 fpm	30-story apartment (250-ft rise)
900 kg @	2.5 mps	18,000 Btu/hr per elevator (19 mj)
2. 3000 lb @	500	10-story office building (100-ft rise)
1400 kg	2.5 mps	22,000 Btu/hr per elevator (23 mj)
3. 3500 lb @	700	Serving floors 1, 10 to 20 in an office building
1600 kg	3.5 mps	35,000 Btu/hr per elevator (37 mj) (240-ft rise)
4. 4000 lb @	500	10-story hospital (100-ft rise)
1800 kg	2.5 mps	25,000 Btu/hr per elevator (26 mj)
5. 3500 lb @	1000 fpm	Serving floors 1, 30 to 45 in an office building
1600 kg	5.0 mps	40,000 Btu/hr per elevator (42 mj) (550-ft rise)
6. 6500 lb @	400 fpm	Service elevator in an office building serving floors 1 to 20 (240-ft rise)
3000 kg	2.0 mps	37,000 Btu/hr per elevator (39 mj)

An estimate of machine room heating should be given by the elevator equipment manufacturer. This estimate can be calculated from the number of floors an elevator serves, its rise, the speed, and the duty load, plus information as to its expected use. The percentage of full-load running time is then calculated from probable-stop data and the heating losses expressed in Btu or joules. (One Btu = 1055 joules.)

Table 8.5 gives some typical values for expected heat release.

Emergency Operation

In all hospitals and in many other buildings it is desirable to provide a standby power supply to operate the elevators. In all buildings it is necessary to have a separate power source for the car lights and fan to reassure passengers if, for any reason, car lighting fails. This should be a constantly charging battery pack located on the car top so it can be serviced and used to energize car lights when needed. A battery light is necessary since standby generators require time to start. Car lighting depends on cables traveling to the car, which may fail. When elevators are essential, such as in a hospital, they should be capable of operation during a power failure.

The amount of standby power available determines how elevators can be operated. If the supply is limited, the consulting engineer can specify that the elevator equipment manufacturer arrange the installation so that only one elevator operates at one time automatically to evacuate passengers, and then a choice can be made manually by use of a selector switch for one elevator to remain in service. The electrical contractor provides the wiring and switch-over to provide power at the machine room. The elevator manufacturer provides interlocking for each elevator controller to ensure that only one elevator is operated at one time, and the selecting switch to determine which elevator will operate after all are evacuated.

The building's standby generator system should supply sufficient power to operate any one elevator (or more if the standby plant is large enough) as well as sufficient provisions to absorb the current regenerated by the elevator if it should travel down with a full load. This absorption of regenerated power is essential to prevent overspeeding of the elevator. The value of absorption needed is given by the elevator equipment manufacturer and must be accommodated by the standby power system.

Hydraulic elevators have a manual lowering valve that can be used to move a car to a next-lower floor, or they can be arranged so that the standby power source or a battery pack can be used to cause them to travel to the lowest landing and open their doors. Smaller-geared elevators can be cranked to a landing. Any elevator that stops for other than normal reasons is potentially dangerous. No attempt to move a car or remove the passengers in that car should be made without adequate precautions. The passengers' condition should be ascertained, and they should be advised of imminent movement. Intercommunication systems connected to a standby power source are a necessity.

ELEVATOR LAYOUTS

Attention has been given to establishing the space, ventilation, and power requirements of passenger overhead traction elevators in particular. Layouts for hydraulic elevators, elevators with basement machines, freight elevators, and other types will follow, in general, the steps that have been outlined. Based on car size, the hoistway plan is established, the pit and overhead determined, the machine room laid out, supports for equipment and impacts on the structure determined, and power and ventilation established. For other than overhead traction elevators, there are additional considerations, which are discussed in the following paragraphs.

Hydraulic Elevators

Hydraulic elevators require a hoistway space for only the car since they usually do not have a counterweight. Sufficient space must be provided for the pump, control equipment, and piping to and from the elevator shaft from a machine room, which may be remote. Adequate pit space for the plunger and cylinder supports are necessary, and impacts on buffers must be considered. In areas where there is groundwater, consideration must be given to tying down the cylinder and well casing lest it float up, as well as protecting it against corrosion by electrolytic action. In the elevator hoistway, adequate run-by space above the top landing must be provided, as well as sufficient support for fastening rail brackets at established distances (usually at floor levels). The vertical loads on the building are minimum, which makes the hydraulic elevator attractive for low-rise structures and, especially, for heavy-duty freight and truck elevators (see Figure 8.13).

Freight Elevators (see Figure 8.14)

For both electric and hydraulic heavy-duty freight elevators, adequate rail supports must be provided, usually at shorter distances than at floor levels, depending on the expected loading on the elevator. These supports can be the building steel at each floor and at intermediate levels, or a solid concrete hoistway into which the elevator contractor can anchor supporting brackets. For elevators loaded by hand trucks or forklifts, fabricated steel supports may be erected at each side of the hoistway (Figure 8.15) to provide for the additional intermediate rail brackets required.

The supports for elevator guides and brackets must be designed to absorb the impact of the truck as it loads on the elevator. The impact consist of the braking load on the truck wheels, the twisting motion of unbalanced loading, and opposing forces on the top and bottom elevator guide shoes transmitted to the rails. Figure 8.16 shows the directions some of these forces will take.

If an elevator is to be loaded by an industrial truck, it must be designed to hold both its capacity load and some of the weight of the truck. Depending on the type of load, either all or part of the truck may be on the elevator as the last portion of the load is deposited. The elevator machine brake (or hydraulic cylinder), car frame, and platform must all be designed to withstand this extra "static" load. The elevator need not lift this load but must be able to level, that is, move the platform level with the floor with the extra load on board. An extensive discussion on freight elevators is included in Chapter 13.

Basement Traction Machines

It is often desirable to install an elevator without a rooftop penthouse. The basement machine in Figure 8.17 requires a minimum penthouse, but a hydraulic elevator usually does not need a penthouse and is limited in speed to about 150 fpm (0.75 mps) and in rise to about 70 ft (21 m). Hydraulic elevators have higher electrical horsepower requirements than counterweighted electrical elevators, and the high power demand associated with

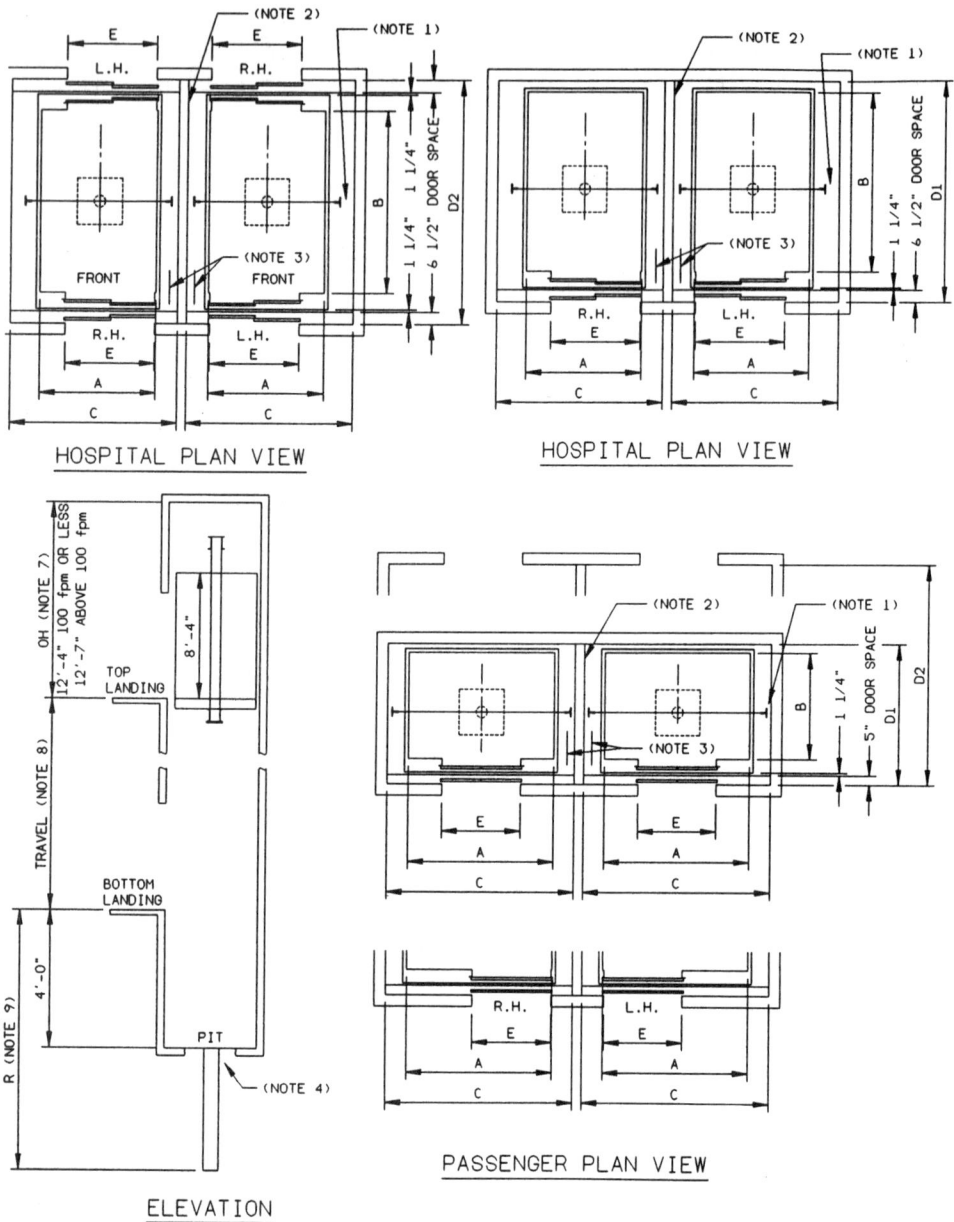

Figure 8.13. Hydraulic passenger and hospital elevators; front entrance or front and rear entrance. Rated speeds 75–200 fpm. (Reprinted with permission from *NEII Vertical Transportation Standards* 7th Edition, including 1994 Supplement. Copyrighted © 1992 and 1994, National Elevator Industry, Inc., Fort Lee, New Jersey. This reprinted material is not the complete and official position of National Elevator Industry, Inc. on the referenced subject which is represented only by the standard in its entirety.)

HOSPITAL CAR & HOISTWAY							
RATED LOAD lb	AREA ■	ft-in. (Note 5)					ENTRANCE (Note 6)
		A	B	C	D1	D2	E	TYPE
4000	42.2	5-8	7-4	7-5	9-3	10-4	4-0	
4500	46.2	5-8	7-10	7-5	10-0	11-0	4-0	2SSO *
5000	50.0	5-8	8-6	7-5	10-5	11-5	4-0	

PASSENGER CAR & HOISTWAY							
RATED LOAD lb	AREA ■	ft-in. (Note 5)					ENTRANCE (Note 6)
		A	B	C	D1	D2	E	TYPE
2000	24.2	5-8	4-3	7-4	5-11	6-10	3-0	SSSO ▲	SSCO
2500	29.1	6-8	4-3	8-4	5-11	6-10	3-6	SSSO *	SSCO ▲
3000	33.7	6-8	4-7	8-4	6-3	7-2	3-6	SSSO *	SSCO ▲
3500	38.0	6-8	5-3	8-4	6-11	7-10	3-6	SSSO *	SSCO ▲
4000	42.2	7-8	5-3	9-6	6-11	7-10	4-0	SSCO *

Notes
1. Refer to page 12 for typical guide rail bracket details.
2. Dividing beams, not by elevator supplier, to be designed to sustain rail forces. Consult elevator supplier.
3. Pit ladder not by elevator supplier.
4. If cylinder hole is not drilled before pit floor is poured a 3'-0" x 3'-0" square opening must be provided in the pit floor.
5. Refer to PREFACE to determine approximate platform dimensions.
6. Refer to page 11 for hoistway and entrance details, information on required hoistway door sill supports, etc.
7. "OH" Dimensions are based on a 8'-4" overall car height.
8. Consult elevator supplier for limitations for maximum travel.
9. Provisions for hydraulic cylinder requires a well hole with dimension "R" = approximately the travel plus 7'-0".
10. When compliance with seismic risk zone 2 or greater requirements is anticipated, see page 10 for additional hoistway space.

Notes
■ Maximum allowable inside car area in ft² per ASME A17.1, Rule 207.1.
* These car dimensions and entrance types provide wheelchair accessibility and accommodate an ambulance type stretcher (76 in. x 24 in.) in the horizontal position.
▲ These car dimensions and entrance types provide wheelchair accessibility.

MACHINE ROOM REQUIREMENTS

A. Minimum machine room size for a single elevator is 7'-0" x 11'-0" x 7'-9" high.
B. Minimum machine room size for a duplex elevator is 17'-0" x 8'-2" x 7'-9" high.
C. Recommended machine room door size is 3'-8" x 7'-0".
D. It is recommended that the machine room be located adjacent to the hoistway and at or near the bottom terminal landing. Consult elevator supplier for exact size and location.
E. Location of the main power disconnecting means, the car light disconnecting means and lighting switch shall be in accordance with requirements of ASME A17.1 & ANSI/NFPA No. 70.
F. Location of switches are similar to those shown on electric elevator machine room plans.

Figure 8.13. (*Continued*) Reprinted with permission from *NEII Vertical Transportation Standards* 7th Edition, including 1994 Supplement. Copyrighted © 1992 and 1994, National Elevator Industry, Inc., Fort Lee, New Jersey. This reprinted material is not the complete and official position of National Elevator Industry, Inc. on the referenced subject which is represented only by the standard in its entirety.

multiple hydraulic elevators may suggest electric traction as the preferred equipment. Installing an electric traction machine in the basement or adjacent to the hoistway at an upper floor below the top landing served has the advantage of the counterweighted elevator with its lower horsepower demand and higher speeds. These traction elevator arrangements are called "machine below."

Figure 8.14. Freight elevators (Courtesy Montgomery KONE).

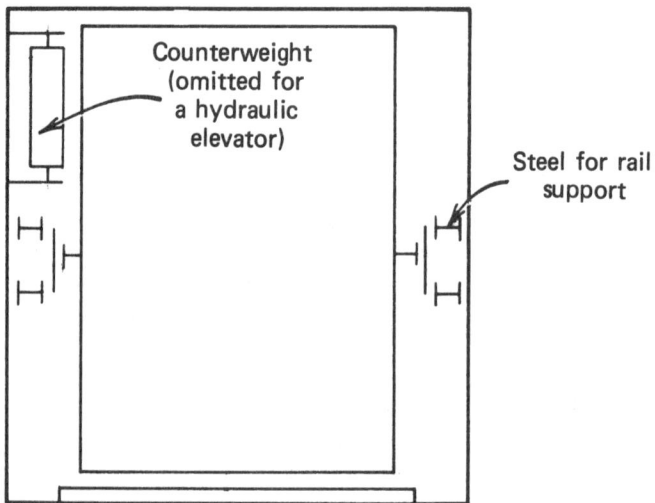

Figure 8.15. Heavy-duty freight elevator—schematic.

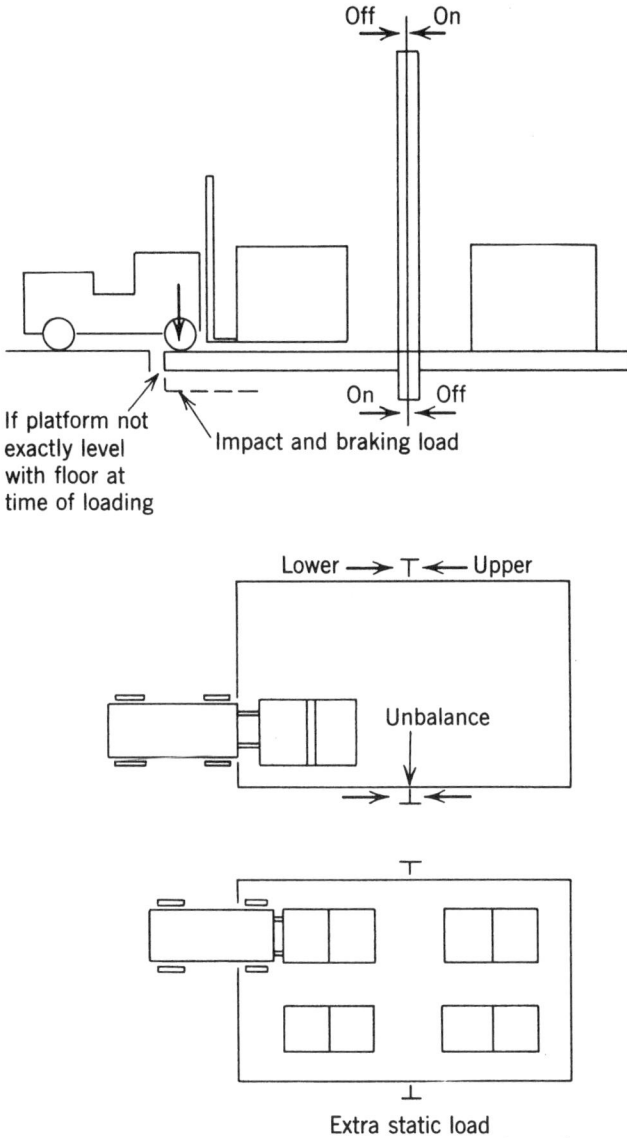

Figure 8.16. Loads produced by an industrial truck.

This class of traction machine below can be of two types, underslung or overslung (also called "direct pickup"). The underslung arrangement consists of lifting sheaves located under the car platform and requires somewhat more pit depth and hoistway space than the overslung system. An overslung is lifted at the crosshead like any conventional traction elevator. Either arrangement requires overhead space for rope sheaves and a machine room located either to the rear or side of the hoistway. With more than one elevator in a line, the rear location is preferred, and with three or more elevators in a line, provides the best layout (Figure 8.18).

With any machine-below arrangement, upthrust forces on the machine are equal to the entire weight of the elevator and counterweight plus impact allowances, which require sufficient tie-down, in the form of anchors, to the building steel or a concrete foundation block.

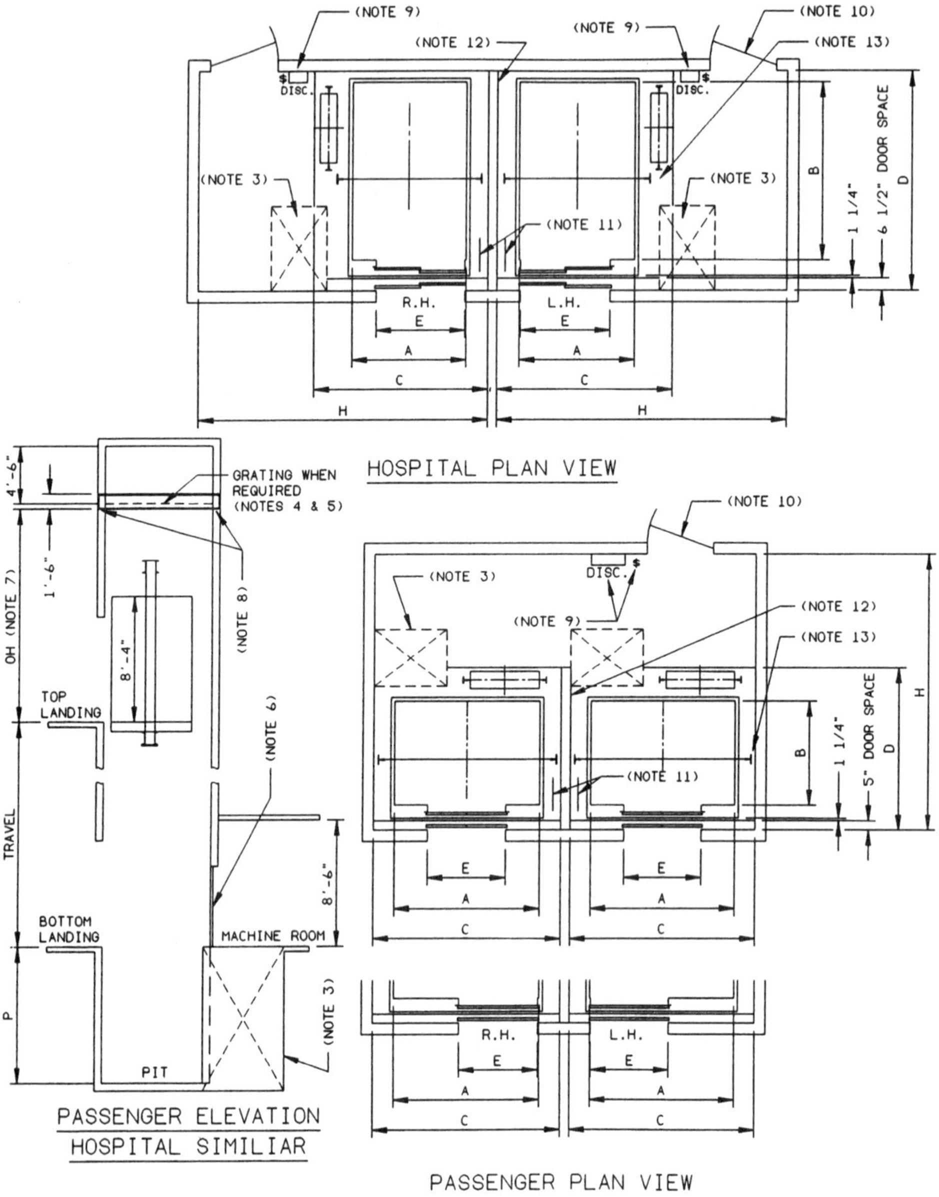

Figure 8.17. Electric passenger and hospital elevators—basement machines. Rated speeds of 100–350 fpm. (Reprinted with permission from *NEII Vertical Transportation Standards* 7th Edition, including 1994 Supplement. Copyrighted © 1992 and 1994, National Elevator Industry, Inc., Fort Lee, New Jersey. This reprinted material is not the complete and official position of National Elevator Industry, Inc. on the referenced subject which is represented only by the standard in its entirety.)

<table>
<tr><th colspan="6">"OH" BOTTOM OF BEAM (Note 7)
ft-in.
(FOR HOSPITAL ADD 1'-6")</th></tr>
</table>

SPEED (fpm)	RATED LOAD (lb)				
	2000	2500	3000	3500	4000
100	15-0	15-0	15-0	15-0	15-0
150	15-0	15-0	15-0	15-0	15-0
200	15-0	15-0	15-0	15-0	15-0
250	15-0	15-0	15-0	15-0	15-0
300	15-0	15-0	15-0	15-0	15-0
350	15-0	15-0	15-0	15-0	15-0

HOSPITAL CAR & HOISTWAY								
RATED LOAD (lb)	AREA ■	ft-in. (Note 15)						ENTRANCE (Note 14)
		A	B	C	D	E	H	TYPE
4000	42.2	5-8	7-4	8-2	9-3	4-0	22-6	2SSO *
4500	46.2	5-8	7-10	8-2	10-0	4-0	22-6	
5000	50.0	5-8	8-6	8-2	10-5	4-0	22-6	

"P" PIT DEPTH ft-in. (FOR HOSPITAL ADD 6")					

SPEED (fpm)	RATED LOAD (lb)				
	2000	2500	3000	3500	4000
100	5-6	5-6	5-6	5-6	5-6
150	5-6	5-6	5-6	5-6	5-6
200	5-0	5-0	5-0	5-0	5-0
250	5-0	5-0	5-1	5-1	5-1
300	5-0	5-0	5-1	5-1	5-1
350	5-0	5-0	5-1	5-1	5-1

PASSENGER CAR & HOISTWAY									
RATED LOAD (lb)	AREA ■	ft-in. (Note 15)						ENTRANCE (Note 14)	
		A	B	C	D	E	H	TYPE	
2000	24.2	5-8	4-3	7-4	6-10	3-0	22-0	SSSO ▲	SSCO
2500	29.1	6-8	4-3	8-4	6-10	3-6	22-0	SSSO *	SSCO ▲
3000	33.7	6-8	4-7	8-4	7-4	3-6	22-0	SSSO *	SSCO ▲
3500	38.0	6-8	5-3	8-4	8-0	3-6	23-0	SSSO *	SSCO ▲
4000	42.2	7-8	5-3	9-6	8-0	4-0	23-0	SSCO *	

Notes

1. A basement machine is often desirable to eliminate rooftop penthouse. It is also desirable if future travel is required.

2. For machine located at other than bottom landing or other desired location, consult elevator supplier for requirements.

3. Sufficient tie down of machine is required. Consult elevator supplier for up–pull reaction to properly size machine foundation. Machine foundation bolts must be set before concrete foundation is poured. Elevator supplier will provide size, number of bolts and location.

4. A governor access door is required above grating level. Elevator supplier does not provide this door but will provide information regarding size and location.

5. In overhead space a floor (grating) will have to be provided when governor cannot be serviced by reaching thru the access door. Floor not by elevator supplier.

6. Partition between machine room and hoistway required in accordance with ASME A17.1, Rule 100.1a. Not to be installed until elevator is in place.

7. "OH" Dimensions are based on a 8'-4" overall car height.

8. Support for elevator overhead beams not by elevator supplier.

9. Location of the main power disconnecting means, the car light disconnecting means and lighting switch shall be in accordance with requirements of ASME A17.1 & ANSI/NFPA No. 70.

10. 3'-6" x 7'-0" Recommended.

11. Pit ladder not by elevator supplier.

12. Dividing beams, not by elevator supplier, to be designed to sustain rail forces. Consult elevator supplier.

13. Refer to page 12 for typical guide rail bracket details.

14. Refer to page 11 for hoistway and entrance details, information on required hoistway door sill supports, etc.

15. Refer to PREFACE to determine approximate platform dimensions.

16. When compliance with seismic risk zone 2 or greater requirements is anticipated, see page 10 for additional hoistway space.

■ Maximum allowable inside car area in ft^2 per ASME A17.1, Rule 207.1.

✱ These car dimensions and entrance types provide wheelchair accessibility and accommodate an ambulance type stretcher (76 in. x 24 in.) in the horizontal position.

▲ These car dimensions and entrance types provide wheelchair accessibility.

Figure 8.17. (*Continued*) Reprinted with permission from *NEII Vertical Transportation Standards* 7th Edition, including 1994 Supplement. Copyrighted © 1992 and 1994, National Elevator Industry, Inc., Fort Lee, New Jersey. This reprinted material is not the complete and official position of National Elevator Industry, Inc. on the referenced subject which is represented only by the standard in its entirety.

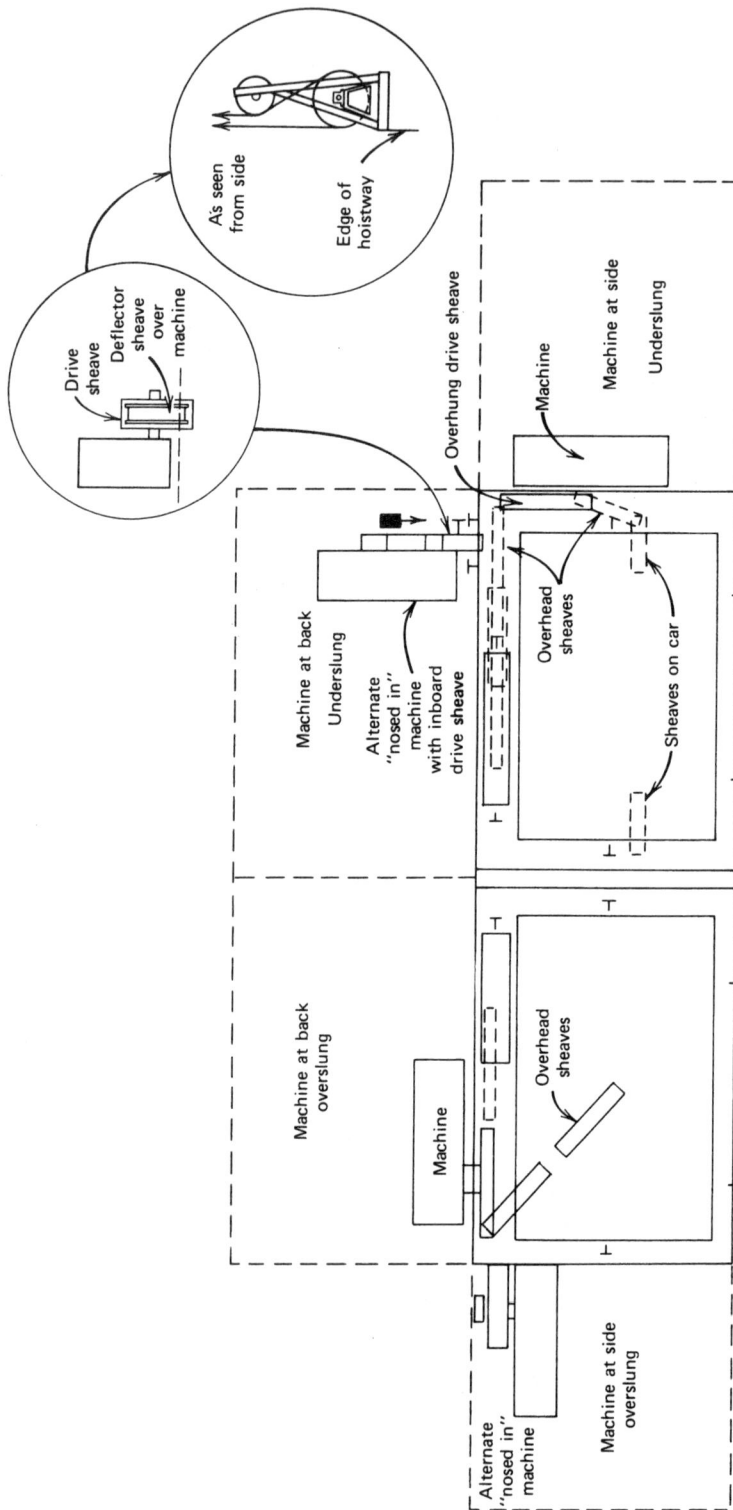

Figure 8.18. Variations on machine room arrangements with "machine below."

Labels within figure:

As seen from side

Edge of hoistway

Drive sheave

Deflector sheave over machine

Overhung drive sheave

Machine

Machine at side
Underslung

Machine at back
Underslung

Alternate "nosed in" machine with inboard drive **sheave**

Overhead sheaves

Sheaves on car

Machine at back
overslung

Machine

Overhead sheaves

Alternate "nosed in" machine

Machine at side
overslung

214

Underslung or overslung elevators can be designed for a full range of lifting capacity and speed; however, speeds above the geared elevator range are quite special and costly. Front and rear or front and side entrances on an elevator with a machine-below arrangement are extremely special. Layout space requirements for underslung elevators necessitate space of up to 14 in. (350 mm) on each side of the car for the sheaves under the car. With the overslung arrangement, space requirements are the same as for an overhead traction machine.

Dumbwaiters

Dumbwaiters are of either the traction or drum type. The size of a dumbwaiter is limited to 9 ft² (0.8 m²) of platform area and a height of no more than 4 ft (1200 mm). Anything over that size must be classified as an elevator or a special lift and must comply with the codes governing elevators.

Dumbwaiters are operated from a landing and are equipped with doors and electrical locks or contacts to prevent their operation if a door is opened. Typical operations include (1) call and send for a two-stop dumbwaiter, (2) multifloor buttons at one floor to send the dumbwaiter to any designated floor with return, call, or multifloor buttons at other than the main floor, and (3) multifloor button stations at all floors for complete flexibility. Loading can be at counter height, under counter, or arranged so carts can be rolled on at floor level. The latter arrangement requires automatic leveling and carts with wheels of sufficient diameter to bridge the running clearance or the installation of an automatic drawbridge.

Dumbwaiters are manual loading and unloading devices. Automatic loading and unloading can be provided in conjunction with power-operated doors to create an automatic material-handling system, which is detailed in a later chapter. Most dumbwaiters are completely standardized and can be arranged with front entrances only, front and rear entrances, or entrances at the front and side. Both power-operated doors and manually operated doors are available.

SPECIAL LAYOUT REQUIREMENTS

Counterweight Safeties

Occupied space must occasionally be located under an elevator or groups of elevators as they serve the upper floors in a building, or if a garage or other basement space is used under the elevators. In some cities, railroad or subway tracks run beneath buildings. In such cases, safety demands provisions against the contingency of a falling elevator or counterweight. The elevator is protected against falling by the required car safety. The counterweight does not normally have a safety, but one can be installed and is required if space below the elevator can be occupied. The counterweight rails are made heavy enough to withstand the safety application load, and the elevator pit must also be designed to withstand this possible impact.

The counterweight with a safety may require as much as additional 2 to 4 in. (50 to 100 mm) of hoistway depth and space for a counterweight governor in the elevator machine room.

A recently developed possible alternative to providing counterweight safeties is the use of a hoist rope brake device. Mounted in the machine room, these devices consist of pneumatic or spring-applied mechanisms that are arranged to clamp the hoist ropes should the car overspeed in either direction. The use of a rope brake still requires the provision of car safeties, but in some jurisdictions where the hoistway is located over occupied space, the use of a rope brake may mitigate the use of counterweight safeties. The additional hoistway space required for the installation of counterweight safeties will not be required if they are replaced with a rope brake. Appropriate machine room space must be provided around the rope brake for proper installation, maintenance, and periodic inspection and testing of this overspeed safety device.

Rear Openings

An elevator car with a rear opening in addition to the normal front entrance requires special layout consideration. The elevator counterweight should be located at the side rather than in the rear, and the machine room must be designed to locate the machine so that the side counterweight may be accommodated. The common hoistway arrangement for front and rear openings is known in the elevator trade as "No. 4" construction (from an early standard layout). This arrangement consists of a support beam for the car rails across the shaft, with counterweight space behind it (Figure 8.19). These provisions require hoistway space for front and rear entrances, calculated similarly to conventional front-only entrances, plus the additional counterweight space. Another use for No. 4 construction is to provide a deeper car if front-to-back space is limited and additional space is available at the side. Extreme care must be taken if two elevators having No. 4 construction are placed side by side. Machine and deflector sheave interference may occur, necessitating that one machine be placed at a higher level than the adjacent machine.

Corner-post Arrangement

A variation to providing entrances on an elevator at the front and at the side is a corner-post arrangement. The car frame is specially built, the rails are placed in opposite corners

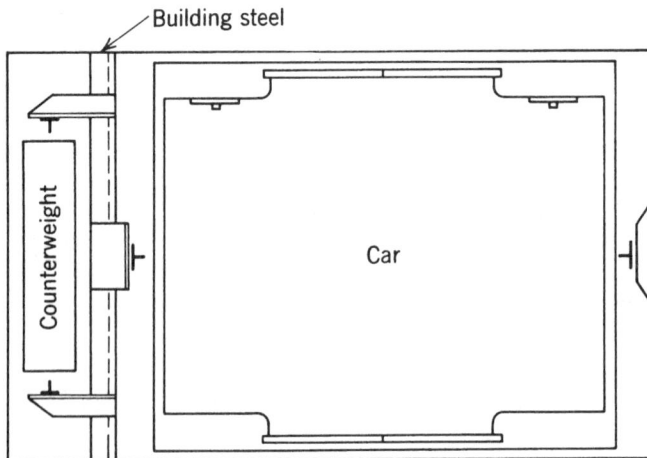

Figure 8.19. "No. 4" construction.

of the hoistway, the counterweight is properly located, and the machine room is specially arranged. There is usually only enough space for two-speed doors on both the front and side openings (Figure 8.20).

A corner-post elevator is difficult to lay out and expensive to erect. It should be avoided, and elevators with front and rear entrances provided, if at all possible.

Special Arrangements

There are many other possible elevator layouts: observation-type elevators on the outside of buildings, wall climbers, explosion-proof elevators in chemical processing plants, shipboard elevators, special elevators to automatically load and unload rolls of paper or pallets, and elevators with revolving platforms. These all require some special considerations and early collaboration with elevator engineers. Similar problems may have already been solved, and a specialist's knowledge is needed to aid in the new application. A later chapter on special installations will discuss many unique applications.

Considerations During a Modernization

Space conditions are a major consideration during a modernization. Unlike the design of a new installation, the replacement and upgrade of existing equipment must be performed within the available space, which is often restricted by existing construction. Pit depths and overhead clearances cannot be extended without costly structural and architectural modifications to the building. Hoistway clearances must also be carefully checked throughout the entire car travel to ensure that upgraded equipment can be accommodated.

If the existing speed and capacity of an elevator is maintained, retaining the pit and overhead clearances will usually obviate the need for major building modifications. Often,

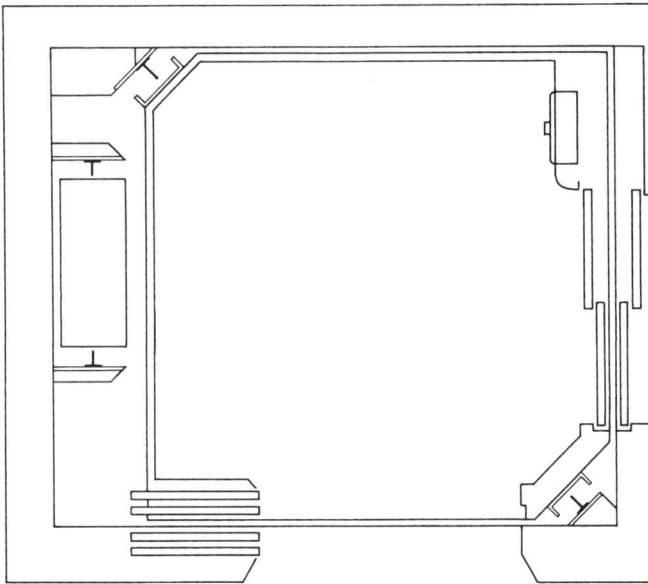

Figure 8.20. Corner-post arrangement (note necessity to use two-speed doors).

however, car and counterweight buffers and safeties must be replaced and careful coordination of new equipment space requirements with existing conditions must be undertaken.

Although elevator car size and capacity are usually retained in a modernization, the upgrade of car interior finishes and design must include a careful analysis of the total car weight resulting from the new design. The maximum allowable hoist machine sheave shaft load and existing counterweight frame design must be checked to determine the allowable car weight that can be accommodated within the existing hoistway and machine room space.

Quite often the inside area of a car is substantially less than that allowed by the code (see Figure 8.4). If such is the case, it may be possible to decrease the pound capacity to a lower rating, which can have a positive effect on the various loads that have to be considered.

Hoistway door clearance, sill and wall conditions, and door hanger pockets for tracks must be checked to ensure that new equipment will fit into these available spaces. Plumb lines should be used to measure overall side-to-side and front-to-back hoistway dimensions throughout the entire vertical run of the hoistway.

In addition to close scrutiny of elevator system components, guide rail attachments and brackets must be checked to be sure that they are in sound condition and reusable with new car and counterweight guides and safeties that may have to be installed.

New machine room equipment—controllers and solid-state motor drives—generally require less space than their existing relay-based counterparts. Additional machine room considerations can include adding air-conditioning, new power panels, improved lighting, various fire suppression devices, and a communications system.

Elevator modernizations constitute a major segment of the elevator business. Successful modernizations require precise planning and expert analysis of existing equipment and space conditions. Modernization should be undertaken by specialists who have a good deal of experience in performing equipment assessment, installation supervision, and project management.

ABOUT THE AUTHOR

ROBERT S. CAPORALE's 33 years of experience in the construction industry started at the engineering firm of Jaros, Baum and Bolles, where he advanced to the position of associate and was the principal designer, field engineer, and inspector on some of world's largest vertical transportation and materials-handling projects. In 1991, he joined Syska and Hennessy Engineers as vice president and director of the Transport System Group and, in 1993, started at *Elevator World Magazine* as an associate editor. He was appointed editor in 1997. Educated at the State University of New York (SUNY) where he earned an AAS degree in Electronics, Bob is a National Association of Elevator Safety Authorities International (NAESA) certified elevator inspector (QEI), a member of the American Society of Mechanical Engineers (ASME), and a founding member of the National Association of Vertical Transportation Professionals (NAVTP).

9 Escalators and Moving Walks

DAVIS L. TURNER

ESCALATORS VERSUS MOVING WALKS

In this chapter we emphasize the moving stairway and moving walkway types of vertical and horizontal transportation. Of the two, the escalator (or moving stairs) is the more important since it is used more frequently and has been in use for almost 100 years. The moving walkway, either inclined or horizontal, was introduced in its modern form in the 1960s. Earlier versions date back to the Columbian Exposition in Chicago in 1893.

In the 1970s developmental work was begun on an accelerating moving walkway. This device has an entering speed of about 100 fpm (0.5 mps) and will accelerate the pedestrian to about 500 fpm (2.5 mps). At the end of the travel, deceleration will take place and exit speed will be about 100 fpm (0.5 mps). Numerous measures have to be perfected, one of the most important being a handrail capable of matching the acceleration, speed, and position of the steps or pallets. When perfected, the accelerating moving walkway is intended to provide convenient horizontal transportation for distances in the 200 to 2000 ft (60 to 600 m) range. Beyond that distance, an automated moving vehicle seems to be the optimum approach. Such automated moving vehicles will be discussed in a later chapter. Although this chapter emphasizes escalators, the rules and suggestions for escalator application may be extended to include moving walkways.

The Importance of Escalators

There is no better way to guide people in a given path in a building than by providing an escalator. Department store owners discovered this years ago, and the most successful stores have their escalators as centers of attraction. The most desirable space in the store is located in line with or next to the escalators. Major expositions have used escalators to direct people to desirable sights and have used moving walkways to keep people moving past exhibits to gain maximum exposure (Figure 9.1).

Transportation terminals, subway stations, and other areas in which large groups of people must be moved from one level to another in surges are ideal applications for escalators and moving walks to speed circulation and avoid congestion. Everyone can be moved at a constant speed, and people are carried efficiently from one place to another. When people are walking, some are slow, others are fast, some have baggage, others are accompanied by children, so that walking is often slowed to the speed of the slowest pedestrian. With a moving device, the velocity is established and constant.

The Vertical Transportation Handbook, Third Edition, Edited by George R. Strakosch
ISBN 0-471-16291-4 © 1998 John Wiley & Sons, Inc.

Figure 9.1. Modern escalators in a dramatic setting. (Courtesy Westinghouse Elevator.)

Escalators provide an effective means to make the second floor or basement space as attractive as street-floor space. In a commercial building this increases revenue. In an institutional building service performance is enhanced, horizontal walking distance is shortened, and a greater concentration of service rendered can be attained.

An effective security barrier and checkpoint can be created in a building by using escalators to a second-floor lobby and having the elevators start from that floor. All people entering must use the escalator to gain access to the second-floor lobby; this expedites checking of identification by a security guard.

Escalators are essential in buildings with double-deck elevators. Since people must separate into those going to the odd-numbered floors and those going to the even-numbered floors, escalators are the only effective means of providing convenience during this separation process. This will be discussed more fully in the chapter on special elevators.

Escalators are found in many places besides their initial field of applications in stores and transportation facilities. Today, schools, hospitals, factories, office buildings, and restaurants have escalators. So do hotels, motels, museums, theaters, convention halls, sports arenas, and other buildings that must accommodate large groups of people within a short period of time.

Brief History of Escalators

Among the earliest known applications of moving stairs is that dating back to 1859, when Nathan Ames was granted patent No. 25,076 for "revolving stairs." The design comprised an equilateral triangle of moving steps and carried passengers up one side and down the other (see Figure 9.2*a* and *b*). The impracticality of jumping on one side and jumping off on the other probably caused the demise of Ames's design.

The modern escalator, as we know it today, is a result of two inventions and extensive development. About 1892, almost a third of a century after Ames's revolving stairs was patented, Jesse Reno designed and patented a moving inclined ramp featuring cleated triangular platforms on a continuously operating belt, which were "combed" at the top and bottom (see Figure 9.3*a, b,* and *c*). An early installation was on the Third Avenue elevated

N. AMES
REVOLVING STAIRS

No. 25,076   Patented Aug. 9, 1859

WITNESSES:

(a)

N. AMES
REVOLVING STAIRS

No. 25,076   Patented Aug. 9, 1859

WITNESSES:

(b)

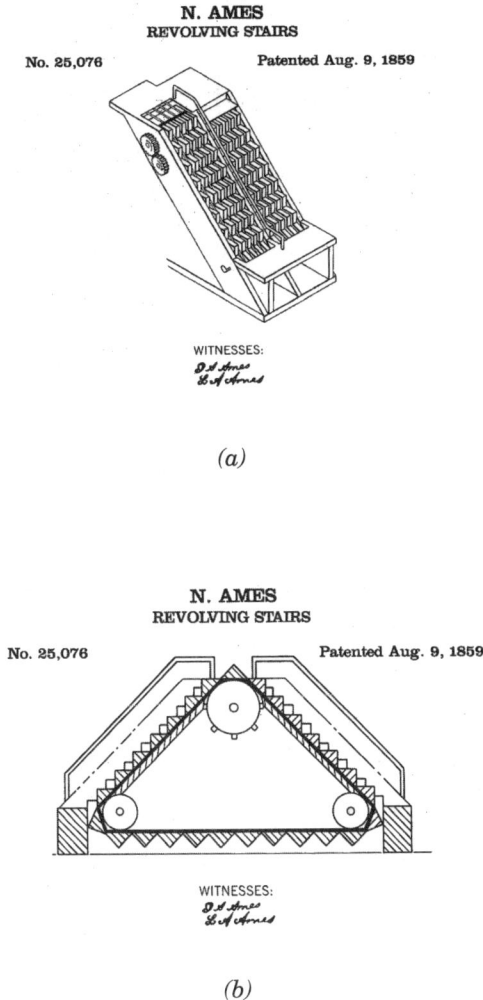

Figure 9.2. N. Ames revolving stairs, patented August 9, 1859.

line at the 59th Street station on the New York City transit system in about 1900, where it was in operation until the line was torn down in 1955. Concurrently and independently, in about 1892 George H. Wheeler invented and patented a flat-step "inclined elevator" with a handrail (see Figure 9.4a and b). This invention was further developed by Charles D. Seeberger and the Otis Elevator Company. The newly developed flat-step escalator was exhibited at the Paris Exposition in 1900 (Figure 9.5). Otis also created the word "Escalator," which was a registered trademark of Otis until it was declared to be in the public domain in the 1930s, when it was used as a title in the A17.1 Safety Code.

Wheeler's inclined elevator had flat steps and a triangular diverting baffle, or "shunt," at the top and bottom, where people had to sidestep on and off the escalator.

Both the Reno and the Seeberger/Wheeler types of escalators were manufactured by Otis as separate products. In 1922 further developments in the design of the escalator

(a)

J. W. RENO
ENDLESS CONVEYOR OR ELEVATOR

No. 470,918   Patented Mar. 15, 1892

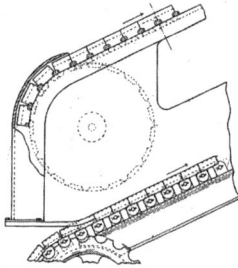

WITNESSES: INVENTOR

BY

ATTORNEYS

(b)

J. W. RENO
INCLINED ELEVATOR

No. 637,526   Patented Nov. 21, 1899

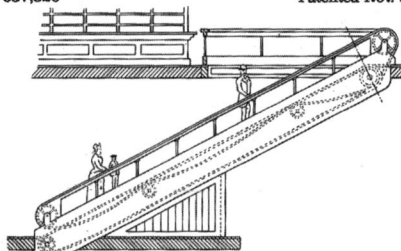

WITNESSES: INVENTOR

BY

ATTORNEY

(c)

222

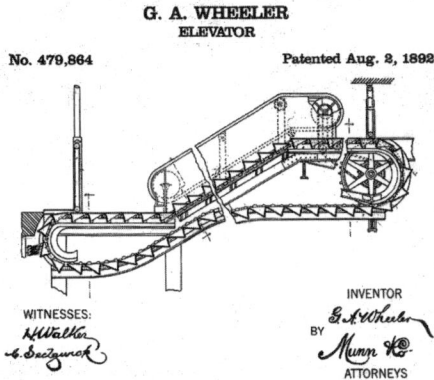

(a)

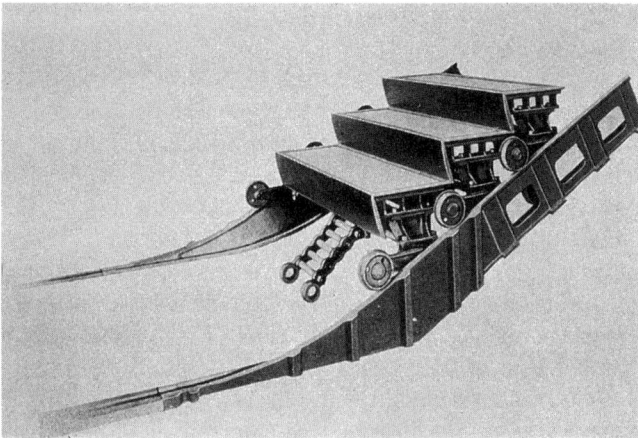

(b)

Figure 9.4. G. A. Wheeler elevator, patented August 2, 1892: (*a*) flat step design; (*b*) flat step detail.

resulted in the forerunner of today's modern escalator, which combined Reno's cleated steps and combs with Seeberger's flat steps. The shunt was eliminated at the landings, so passengers entered and departed in line with the direction of travel of the escalator with the help of the flat, cleated steps and combs at the top and bottom.

EFFECTIVE APPLICATION

Escalators can be advantageously applied to any building if certain requirements are met. Equipment should be located so that most people entering the building can see it. Access

---

Figure 9.3 (opposite). Reno moving inclined ramp: (*a*) side view; (*b*) handrail detail; (*c*) "inclined elevator" overall view.

Figure 9.5. Seeberger flat step escalator with passenger diverter. Passengers stepped on and off escalator to the side as the steps continued under diverter (Courtesy Otis Elevator).

to an escalator must be attractive and in the path of the heaviest expected traffic. Evaluation of expected traffic volume, which can range from a few people continuously to hundreds in a peak 5-min period, depends on the type of building facility and its use.

Escalators are most effectively used as continuously running, unidirectional conveyors. The efficiency of the escalator derives from its continuous availability to passengers. There is no waiting for an escalator as there is with elevators. Escalators can be started and stopped on demand, but this requires additional special considerations, which will be discussed later. Starting, stopping, or reversal is best done by an attendant and with the assurance that no one is riding at the time. This is the current ASME A17.1 Code requirement. Pairs of escalators are necessary for two-way service.

Ample space for people must be provided at the entry and exit landings of an escalator. The escalator can feed people into an area much more rapidly than they can climb a stairway or walk through a restricted opening to leave that area. If an unloading area is restricted, people could be crowded into it with possibly dangerous results, because the escalator is an unforgiving conveyance. Such restrictions as doors or gates should be interlocked with the escalator or ramp to ensure that the restriction is removed before the escalator can be run. Where escalators feed a restricted area such as a subway platform, security personnel must be alerted to the possibility of platform overcrowding if subway service is interrupted, and instructed to stop the escalators.

In many localities a building can have fewer stairs if fire-protective enclosures are provided around the escalator. This enclosure must be equipped with sufficient doors and space at the landings for the doors to swing with the traffic without impeding prompt passenger transfer. When escalator traffic of any magnitude is expected, its volume may reach the capacity limit of the unit and ample loading area must be provided. Means should be provided to automatically stop escalators that may travel into an area of a building endangered by a fire. This will be discussed later in this chapter.

One of the benefits of an escalator is its continuous motion, providing service with zero interval in elevator terms. Normally, people need not wait but may enjoy service the moment they reach the entry level. If the capacity of the escalator is exceeded, a wait may be necessary; however, the waiting time is readily apparent. If more people are expected to arrive than the escalator system can handle, additional facilities, higher speed, or adequate alternative routes should be offered. Stairs adjacent to pairs of escalators are an absolute necessity if the escalators are the primary means of entering or leaving a building lobby. Escalator handling capacity is discussed later in this chapter.

If escalators are the primary means of vertical transportation they must be supplemented by one or more elevators. Some physically impaired persons can usually negotiate an escalator, but it is almost impossible—and actively discouraged—for a person in a wheelchair to do so. Similarly, a person with impaired vision would have difficulty picking a proper tread to step on, as would many older people with motion difficulties.

Mothers with baby carriages or strollers should be encouraged to use elevators rather than escalators or moving walks. Escalators are almost impossible to negotiate with a carriage, and the incline of a walk may make a loaded carriage difficult to hold back. A loaded food cart, which may weigh from 50 to 100 lb (20 to 40 kg) can be dangerous on an inclined walk unless the wheels are locked. Locking must be done automatically, since it would be an impossible task to ensure that all individuals using a walk will secure their carts. Inclined elevators as an adjunct to escalators are discussed in a later chapter.

TRAFFIC-HANDLING ABILITY

Escalators and moving walkways are rated by step width and by speed in feet per minute of meters per second. Because escalators and walkways are usually driven by ac induction motors, operating speed is constant under varying load conditions and rating is at a single speed. Historically, escalators were offered with speeds of either 90 fpm (0.45 mps) or 120 fpm (0.6 mps) along the incline. More recently escalator speeds have been standardized at 100 fpm (0.5 mps). Faster escalators have been provided in some areas, but their use is not common and a factor of diminishing return can result; the steps may move too fast for people to use them. Recent editions of the ASME-A17.1 Safety Code limit the rated speed of escalators to 125 fpm (0.64 mps).

Very high-speed escalators are used in the subway stations of Moscow, Kiev, and Leningrad. This is due to the exceedingly high rises in many of their stations, up to a record of 214 ft (65 m) in Kiev. Speeds as high as 200 fpm (1.01 mps) are used to overcome the extremely long riding time, and handling capacity is of secondary importance. Many escalators are equipped for two-speed operation by manual switching; these escalators can be run at 120 fpm (0.6 mps) for high traffic periods and at 90 fpm (0.45 mps) during the rest of the day, with a consequent reduction in operating mileage and wear and tear on the equipment.

Although the normal angle of incline of an escalator is 30° by design, the angle is permitted to vary to 31° owing to building conditions, as established by the ASME-A17.1 Safety Code. In European and Asian countries escalators with inclines of 27.3° or 35° are also common. The 27.3° incline is designed to run parallel to and in line with stairwells in public transportation facilities. The 35° escalator is designed to take up less space but is usually limited to 6 meters of rise (approximately 20 feet) and 100 fpm (0.5 mps).

The number of flat steps a passenger encounters upon entering an escalator is extremely important. A flat step is a step that is level with the preceding step prior to the step rising or depressing along the incline. Observations have shown that the greater the

TABLE 9.1. Escalator Capacities (30° Incline)

Step Width	Speed	Maximum Capacity Theoretical[1]	Nominal Capacity Observed[2]
24 in. (600 mm)	90 fpm (0.45 mps)	422/5 min 5063/hr	168/5 min 2025/hr
	100 fpm (0.5 mps)	469/5 min 5626/hr	187/5 min 2250/hr
	120 fpm (0.6 mps)	562/5 min 6751/hr	225/5 min 2700/hr
32 in. (800 mm)	90 fpm (0.45 mps)	506/5 min 6075/hr	Same as 24 in (600 mm)
	100 fpm (0.5 mps)	562/5 min 6751/hr	Same as 24 in (600 mm)
	120 fpm (0.6 mps)	675/5 min 8102/hr	Same as 24 in (600 mm)
40 in. (1000 mm)	90 fpm (0.45 mps)	675/5 min 8102/hr	337/5 min 4051/hr
	100 fpm (0.5 mps)	750/5 min 9002/hr	375/5 min 4501/hr
	120 fpm (0.6 mps)	900/5 min 10800/hr	450/5 min 5401/hr

[1]Based on the following:

1.25 persons per step or 5 persons every 4 steps for 24-in. (600-mm) escalator

1.5 persons per step or 3 persons every 2 steps for 32-in. (800-mm) escalator

2.0 persons per step for 40-in. (1000-mm) escalator

[2]Based on one person every other step for 24-in. (600-mm) wide and 32-in. (800-mm) wide, and one person per step for 40-in. (1000-mm) wide escalator

number of flat steps, the more easily passengers adjust to the moving escalator and traffic handling ability is expedited, within certain limitations. This is to be expected, since a person's stride is about 30 in. (762 mm) and two flat steps are about 32 in. (812 mm). The ASME-A17.1 Safety Code prescribes a minimum of two flat steps and a maximum of four flat steps for escalators installed in jurisdictions that subscribe to that code. The greater number of flat steps is generally applied to high-speed and high-rise escalators.

Escalator handling capacities are developed for the purpose of determining the proper number of escalators required for a given application. The escalator loading used to determine these handling capacities are not the same as the loads used to determine structural rated loads, machinery rated loads and brake rated loads by the applicable codes. Table 9.1 gives theoretical handling capacities for various escalator widths and speeds.

Because moving walkways can be installed at any angle from 0° (a moving walkway) to 12° (inclined walk), operating speed varies with the angle of inclination. At any speed with an inclined walk, the entering and exiting area should not exceed 3° for boarding or exiting, and make a smooth transition to inclined motion. With level boarding and exiting, operating speed can be higher than if the passenger must board at an incline. Operating speeds, angles, and walk widths are established by the ASME A17.1 Code and are briefly shown in Tables 9.2 and 9.3. Latest editions of the code and local codes should be consulted.

The nominal widths of escalators are either 24, 32, or 40 in. For European escalators the corresponding step widths are 600 mm (24 in.), 800 mm (32 in.), and 1000 mm (40 in.).

The 24-in. is wide enough for one person per step, and the 40-in. allows a person with baggage or two adults to ride side by side. In actual observed practice, one person is on every other step of a 24- or 32-in. escalator and one person on each step of a 40-in. escalator. A further advantage of the 40-in. escalator is that people in a hurry may pass a standing rider if all the riders stand to one side. If one were available, an escalator with a 46-in.-step width would easily allow passing or two people per step.

TABLE 9.2. Operating Speeds of Moving Walkway [Based on 40-in. (1000-mm) Nominal Trend Width*]

Incline of Ramp on Slope	Maximum Speed with Level Entrance and Exit
0 to 8°	180 fpm (0.9 mps)
Over 8 to 12°	140 fpm (0.7 mps)

*Speed, angles, and capacities will vary with width. See A17.1 Code.

TABLE 9.3. Moving Walkway Capacities, 40-in. (1000-m) Nominal Width[a]

	Treadway Speed	Maximum Capacity Theoretical[b]	Nominal Capacity Observed[c]
0° incline	180 fpm (0.9 mps)	1200/5 min, 14,400/hr	600/5 min, 7200/hr
5° incline	140 fpm (0.7 mps)	932/5 min, 11,180/hr	466/5 min, 5600/hr
10° incline	130 fpm (0.65 mps)	867/5 min, 10,400/hr	434/5 min, 5200/hr
12° incline	125 fpm (0.63 mps)	833/5 min, 10,000/hr	416/5 min, 5000/hr

[a]Speed, angles, and capacities will vary with width. See A17.1 Code.
[b]2.5 ft^2 (0.23 m^2) of treadway per person.
[c]5 ft^2 (0.46 m^2) of treadway per person.

Escalator capacities are generally expressed in passengers per hour. Capacities expressed by escalator manufacturers are theoretical and assume that each step carries either 1¼ or 2 passengers, depending on the width. A reasonable estimate of actual output would be about 50% of theoretical output, as shown in Table 9.1. The values of Nominal Capacity, Observed, given in Table 9.1 are based on averages over time and normal queuing conditions. Certain applications, such as in mass transit facilities or stadiums and arenas, impose extreme loading or queuing conditions on escalators over relatively short periods of time, such as at the conclusion of an event in a stadium or the arrival of a train at a station during an urban rush hour. With these queuing conditions "crush loads" can be imposed on the escalator, and it is advisable to station attendants at the entries of escalators to control the loading. As mentioned earlier, this condition also imposes demands on the entry and exit areas immediately in front of and at the end of the escalators, and security personnel must be alerted to these possibilities.

Walkways are rated in much the same way. Nominal width is expressed as the pallet or treadway width, and the A17.1 Code limits width depending on the incline of the walkway. Escalators and moving walkways should have moving handrails at both sides of the steps. A listing of nominal inclines and ratings is shown in Table 9.3.

ARRANGEMENT AND LOCATION

Two general arrangements of escalators are descriptively named *parallel* and *crisscross*. Both arrangements may have up and down equipment side by side or separated by a distance. A third possible arrangement, which could be called *multiple parallel,* consists of more than two escalators side by side between the same exiting and entering levels, primarily to serve more traffic than a single escalator could handle. Flexibility is provided by operating all the units but one in the direction of heavy traffic.

The various arrangements are sketched in Figure 9.6. The crisscross arrangement is the most popular in department stores because it uses floor space effectively, structural requirements are minimized since escalators can be stacked above each other, and it achieves maximum exposure of passengers to merchandise on the various floors. Separating crisscross escalators increases exposure to the various floors and eases the intermin-

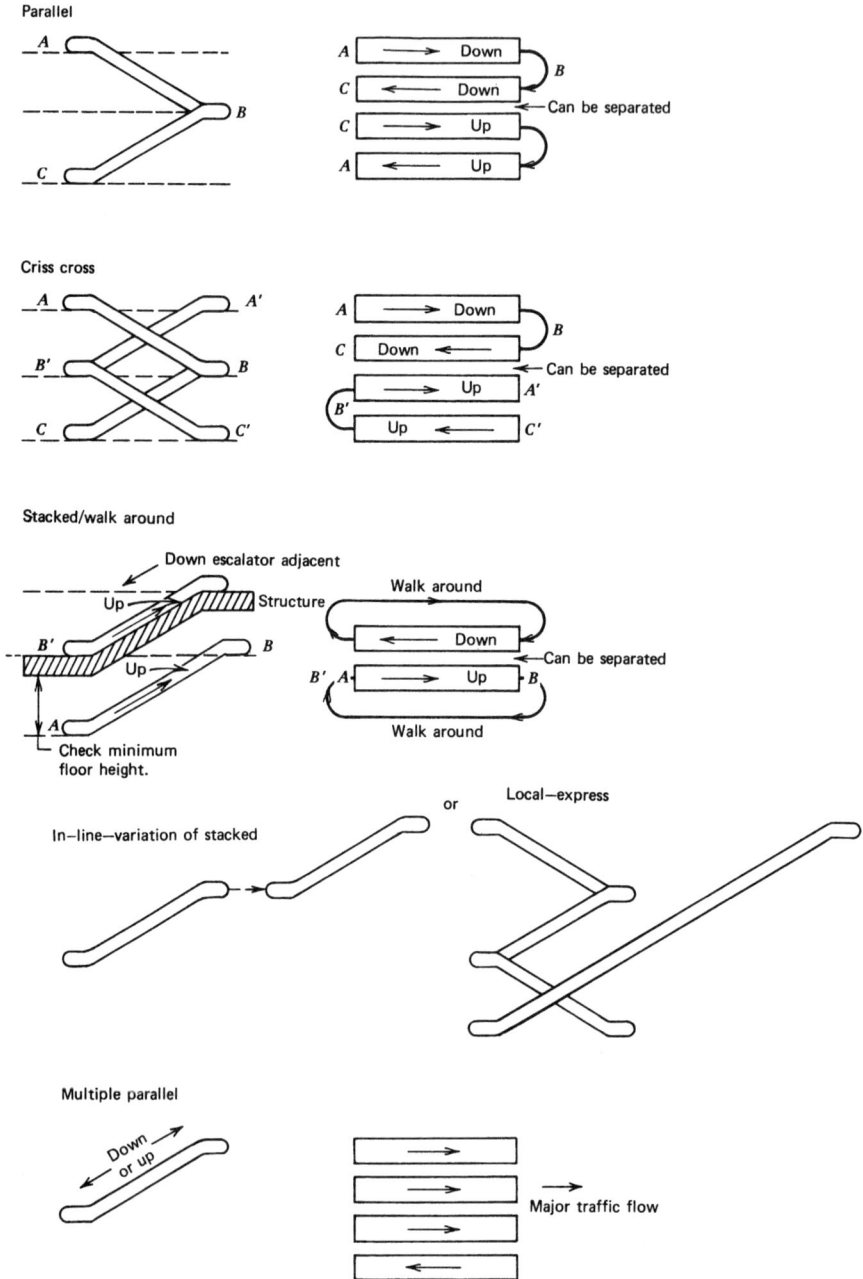

Figure 9.6. Escalator and moving walkway arrangements.

gling of riding passengers and people wishing to board. The separated crisscross arrangement is considered the safest by many users because only one escalator is presented to the riding passenger and there is minimum confusion about whether it is going up or down.

The parallel arrangement provides the least congested flow with the greatest traffic-handling ability and the most impressive appearance to the prospective passenger. Massing at the escalator or ramp entrances immediately attracts people to that area. The open appearance at upper landings provides space for decoration or a high-traffic selling area, as well as additional open space for intermingling traffic. The arrangement is limited if it is intended to board many people at each floor simultaneously, unless a barrier is created to form a queuing line (Figure 9.7).

The multiple parallel arrangement is provided when many people must be transported to another level in minimum time, such as in a busy commuter terminal. Passenger demands may be served by three of a four-unit installation, operated in the direction of the morning and evening traffic and reverted to two units in each direction during the rest of the day. Frequent applications are in buildings where the main lobby must be located above the street because of a subway or railroad beneath the building. Here, again, direction of operation is changed to conform with traffic. Units would also be reversed in sports arenas, exhibit halls, and theaters at the start and end of performances.

SPACE REQUIREMENTS

An escalator or moving walk can be thought of as being constructed in three component units: an upper portion, a lower portion, and a midsection.

The lower unit includes the newel (where the handrail is reversed), the lower step return and step tensioning device, and the landing plate and step entry. The upper unit consists of the upper newel, the upper landing, and the driving mechanism (motor and control) for steps and handrail (some manufacturers place their drive motors in the

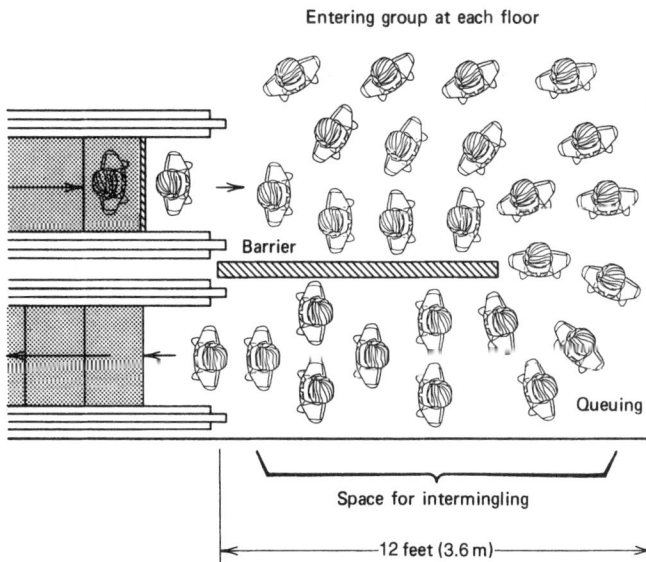

Figure 9.7 Exiting and queuing space.

midsections). The midsection can be of indefinite length (within limits) and consists of balustrading, steps, step tracks, supports, and so on.

Normal support points for escalators and short-run inclined walks are at the top and bottom and are established by a distance from the working points. A working point, or more accurately, a design working point, is an imaginary point located at the intersection of the finished floor line, at the upper and lower landings, and the line formed by the nose line of the steps along the incline. While these points are easily located on drawings and layouts, they do not actually exist on the escalator. For the purpose of developing a layout of an escalator and locating it within the building structure, all distances to escalator supports are measured from these working points.

The escalator or walk layout procedure begins with determination of the vertical rise and approximate location of the escalator and the upper and lower pedestrian access space, to establish the upper and lower working points (see Figure 9.8).

The horizontal projection of the distance between the design working points is dependent on the angle of inclination, α, of the escalator and is given as:

$$L = \text{rise } (H) \div \tan \alpha$$

α	$\tan \alpha$	$1/\tan \alpha$
27.3°	0.52688	1.935
30°	0.57735	1.732
35°	0.70021	1.428

It is interesting to note that on a 30° escalator the length of the escalator, actual distance between the design working points, is twice the rise of the escalator.

Once both working points are established, the particular manufacturer's standard space requirements must be met to locate the necessary supports that must be built into the building structure. The reactions on these supports depend on the length and width of the escalator.

For the nominal 24-in (600-mm)-wide escalator serving a 12-ft (3.6-m) rise, the lower end of the truss imposes a load of about 19.1 kips (8.7 t), and the upper end 20.4 kips (9.3 t). These loads include the weight equivalent of plaster facing on the sides of the es-

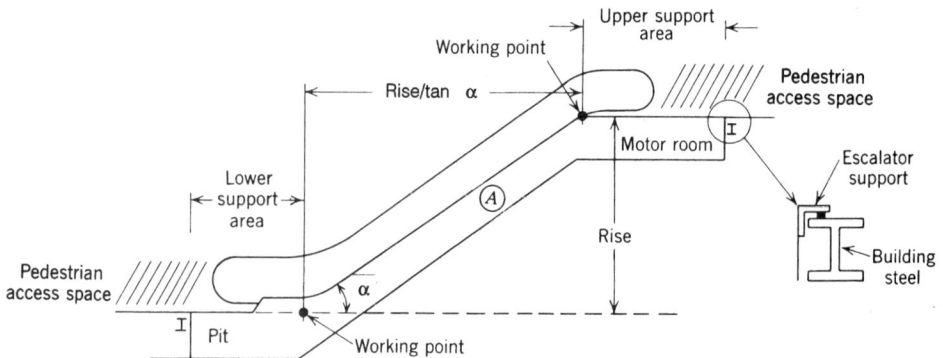

Figure 9.8. Escalator space requirements. Some escalators have the motor located at Ⓐ.

calator and normal balustrade treatment. The loads are, for the most part, uniformly distributed over the width of the escalator. Escalators are mounted on the building steel by means of an angle, so the load given is for the escalator width. Rises more than about 18 ft (5.5 m) for the 48-in. (1000-mm) and more than 23 ft (7 m) for the 32-in. (600-mm) escalators, as well as for most moving walks, usually require intermediate supports. A table of the highest loads imposed by U.S. manufactured escalators is given in Table 9.4.

The loads of moving walkways will vary considerably with the angle of incline and the length of the walkway. When a desired application of a walkway is established, it is best to work with a manufacturer's representative or consulting engineer to determine necessary space requirements and the loads imposed on the structure.

The values given in Table 9.4 are only examples. These reactions can vary greatly from one model to another from any manufacturer, and from manufacturer to manufacturer, owing to machine location and weight, truss structure, the number of flat steps at the upper or lower landing, handrail drive type and location, and many other factors. The manufacturer's literature should be used as a reference in preparing layouts or determining structural support size and locations for various buildings and applications.

Rises in excess of 20 ft (6.0 m) generally require an intermediate support located somewhere between the two design working points. An intermediate support will result in a three-support or, if the rise is high enough, a four-support escalator truss. Complex engineering calculations are needed to determine the reactions on the building and the required support structure. The exact location of the support and the resultant building reactions vary from manufacturer to manufacturer. In some cases a manufacturer can apply special truss designs to withstand the loads imposed by extra-high-rise escalators without the use of an intermediate support and still comply with the structural requirements of various codes. Therefore, it is recommended that the manufacturer be consulted for design considerations in high-rise escalators or extra-long moving walkways.

The plan of an escalator includes the width of the steps, the width of the balustrading plus space for the truss support, and external decoration. A 24-in (600-mm) escalator typically has an overall truss width of about 4 ft 0 in. (1219 mm); a 32-in. (800-mm) escalator, about 4 ft 8 in. (1422 mm), and a 40-in. (1000-mm) escalator, 5 ft 4 in. (1625 mm) (Figure 9.9). Since most escalators are side-by-side, the overall width of two escalators is shown. Widths for moving walkways vary with the width of the tread and, to some degree, the angle of incline.

The overall length of an escalator is much shorter than a moving walkway for the same rise because of the greater angle of inclination of the escalator. An average 24-in. (600-mm), 30° escalator for a 12-ft rise requires about 36 ft (10.9 m) between supports, excluding upper and lower access areas. A 10° walk serving the same rise will require about 82 ft (25 m), excluding upper and lower access areas. An average escalator truss depth including soffit is about 3 ft 6 in. (110 mm), whereas the truss for a walk can be of very limited depth, depending on the rise. Escalator trusses are deeper because they must provide space for the returning steps, whereas the walk truss need accommodate only returning pallets. A listing of escalator space requirements for various rises is given in Table 9.5.

Escalator Layouts

The first step in developing an escalator layout is to determine the distance between the working points based on the floor-to-floor height, which is the rise of the escalator. From the working points the data given in Table 9.5 and the dimensions shown in Figure 9.10

TABLE 9.4. Maximum Load per Escalator on a Building Structure in Kips (1000-lb Units) or Metric Tons (t) (1000-kg Units)

Width/Rise		10 ft (3 m)	12 ft (3.6 m)	14 ft (4.2 m)	16 ft (4.8 m)	18 ft (5.5 m)	20 ft (6.0 m)
24 in. (600 mm)	Upper	16.6 kips (7.5 t)	17.7 kips (8.0 t)	18.8 kips (8.5 t)	19.9 kips (9.0 t)	21.0 kips (9.5 t)	22.1 kips (10.0 t)
	Lower	15.5 kips (7.0 t)	16.6 kips (7.5 t)	17.7 kips (8.0 t)	18.8 kips (8.5 t)	19.9 kips (9.0 t)	21.0 kips (9.5 t)
32 in. (800 mm)	Upper	17.7 kips (8.0 t)	18.9 kips (8.6 t)	20.1 kips (9.1 t)	21.3 kips (9.7 t)	22.5 kips (10.3 t)	23.8 kips (10.8 t)
	Lower	16.2 kips (7.5 t)	17.8 kips (8.1 t)	19.0 kips (8.6 t)	20.2 kips (9.2 t)	21.5 kips (9.7 t)	22.7 kips (10.3 t)
40-in. (1000 mm)	Upper	19.3 kips (8.8 t)	20.7 kips (9.4 t)	22.0 kips (10.0 t)	23.3 kips (10.6 t)	24.6 kips (11.2 t)	25.9 kips (11.8 t)
	Lower	18.2 kips (8.3 t)	19.6 kips (8.9 t)	20.9 kips (9.5 t)	22.2 kips (10.0 t)	23.5 kips (10.7 t)	24.8 kips (11.3 t)

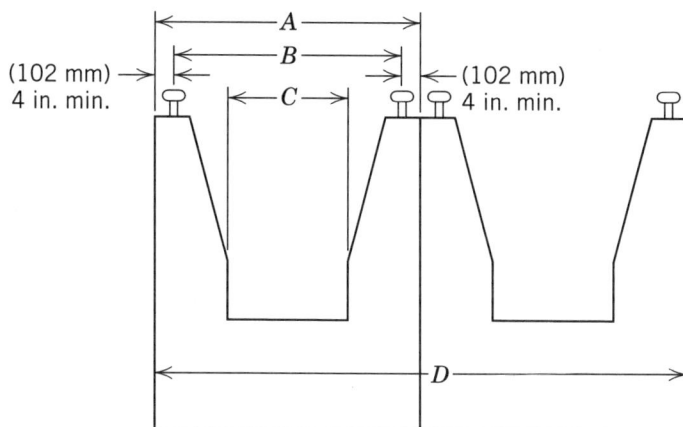

Dimension	A	B	C	D
24 in. (600 mm)	48 in. (1219 mm)	36 in. (914 mm)	24 in. (609 mm)	96 in. (2438 mm)
32 in. (800 mm)	56 in. (1422 mm)	44 in. (1117 mm)	32 in. (813 mm)	112 in. (2845 mm)
40 in. (1000 mm)	64 in. (1626 mm)	52 in. (1321 mm)	40 in. (1016 mm)	128 in. (3251 mm)

Figure 9.9. Escalator cross section. Dimension A is the overall escalator width measured across the decking profiles. The decking generally overhangs the truss by approximately 1" (25.4 mm) on each side.

are used to determine the approximate distance for the upper support point to the lower support point, depending on the number of flat steps required. At this point the designs of various manufacturers should be checked to ensure that sufficient space is allowed for each manufacturer to install the equipment. Alternatively, the greatest dimensional requirement should be developed with the understanding that adjustment can be made once the escalator contractor is chosen.

When the support points are established, vertical and horizontal spaces for the truss, upper structure, and pit are determined. If escalators are to be located one above the other, it is important to have floor-to-floor heights of at least 11 ft 6 in. (3.5 m) to obtain sufficient headroom at the entrance to the lower escalator.

TABLE 9.5. Horizontal Space Required, 30° Escalator

Rise	Distance Working Point to Working Point	Distance Edge of Support to Edge of Support	
		2 Flat Steps	3 Flat Steps
10 ft (3.0 m)	17 ft 4 in (5.3 m)	33 ft 10½ in (10.31 m)	36 ft 6½ in (11.12 m)
12 ft (3.6 m)	20 ft 10 in (6.3 m)	37 ft 4 in (11.37 m)	40 ft 0 in (12.18 m)
14 ft (4.2 m)	24 ft 3 in (7.4 m)	40 ft 9½ in (12.43 m)	43 ft 5½ in (13.24 m)
16 ft (4.8 m)	27 ft 9 in (8.5 m)	44 ft 3⅛ in (13.5 m)	46 ft 11⅛ in (14.3 m)
18 ft (5.5 m)	31 ft 2 in (9.5 m)	47 ft 8¾ in (14.54 m)	50 ft 4¾ in (15.35 m)
20 ft (6.0 m)	34 ft 8 in (10.6 m)	51 ft 2¼ in (15.6 m)	53 ft 10¼ in (16.4 m)

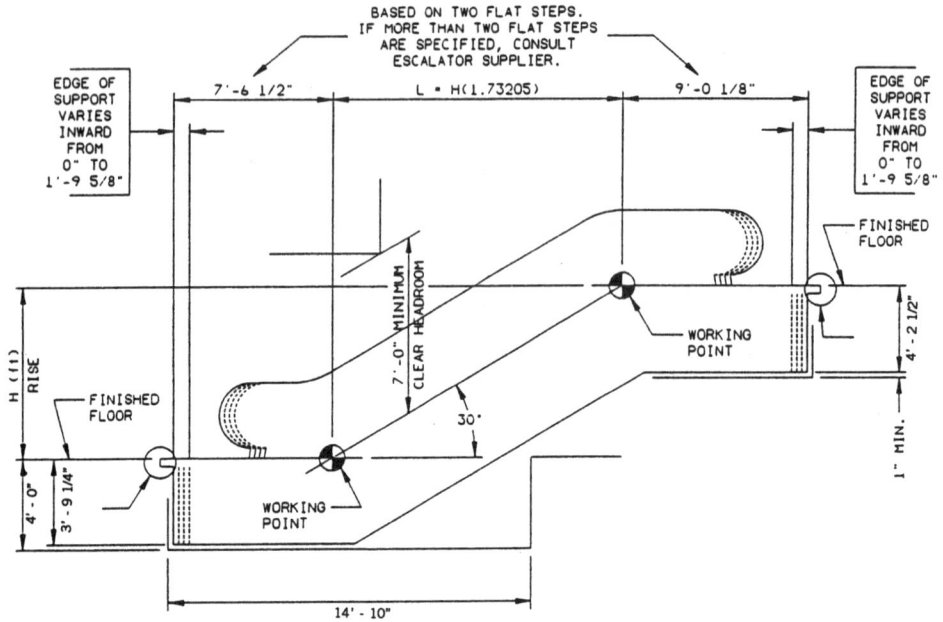

Figure 9.10. Escalator pit and overhead requirements. (Reprinted with permission from *NEII Vertical Transportation Standards* 7th Edition, including 1994 Supplement. Copyrighted © 1992 and 1994, National Elevator Industry, Inc., Fort Lee, New Jersey. This reprinted material is not the complete and official position of National Elevator Industry, Inc. on the referenced subject which is represented only by the standard in its entirety.)

After the truss is located, the newels can be established based on the location of the working points and the number of flat steps. Access area to the escalator is determined from the end of the newel to the nearest obstruction and should be a minimum of 12 ft (3.6 m).

Additional Requirements

Access to equipment is required to properly service an escalator or walk. Normal access includes the removal of the top and bottom landing plates. The bottom landing plate covers the step or belt tensioning device, and in some escalators the top landing plate covers the escalator or walkway machine and upper drive; in others the machine is located in the truss. With this latter design, a side access door may be desirable. Alternatively, steps must be removed to service the machine.

Power must be provided to the escalator or walkway controller. The power supply must meet the requirements of the particular manufacturer, and the respective motor horsepowers of escalators are usually 10 hp for the 24-in. (600-m) and 15 hp for the 40-in. (1000-mm) for rises up to about 22 ft (6.7 m). Walkway power requirements vary with different widths and angles of incline, and equipment suppliers should be consulted.

To dissipate the heat released when an escalator or ramp is operated for long periods under load conditions, ample ventilation for the motor area should be provided. This can be grillwork if the machine room is in an open area, or forced ventilation if the machine room is confined. The approximate heat release of escalators is about 10,000 Btu/hr for the 10-hp, 24-in (600-mm) escalator and 15,000 Btu/hr for the 15-hp, 40-in. (1000-mm) escalator. The heat dissipation requirements of moving walkway machinery will vary with the horsepower.

FEATURES OF ESCALATORS AND MOVING WALKS

Because escalators and walks should serve people of all ages and abilities, they must be inherently safe. It has been determined that escalators are generally safer than stairs, but comparable statistics for moving walkways have not been compiled. Escalator safety has been the result of experience, with much research and development. Early escalators, for example, had wide step cleats with grooves of about ½ in. (13 mm). Well-designed modern escalators have cleats no more than ¼ in. (6 mm) wide and grooves not less than ⅜ in. (9 mm) deep, as required by the A17.1 Code. These cleats and grooves are "combed" as they move at the top or lower landing to dislodge soft shoe soles and debris so as to prevent accidents (Figure 9.11).

The A17.1 Code requires the same type of combing action on moving walks. Escalator step risers are also cleated so that people who ride with their toes against the riser will not have their soft shoe soles drawn between the steps as the steps straighten out (Figure 9.12). The footwear shown in the illustration may be dated, but the hazard is amply illustrated. As the lower step rises to meet the upper step at the exit of an up escalator, the friction of the material against the moving step riser will cause the shoe material to be drawn into the gap between the steps. The combing action will oppose this downward motion. The same hazard exists as the side of the steps travels along the skirt panel.

Some people ride with their feet pressed against the side of an escalator. As the steps flatten, there is a possibility of a soft shoe such as a sneaker or rainwear being drawn between the step and the side. If this occurs, a switch actuated by the deflection of the side should stop the escalator (Figure 9.13). Escalator and walk brakes must be designed to stop the fully loaded treadway as quickly as possible to minimize personal injury. The ASME A17.1 Code requires that a stop be made at a rate of deceleration not to exceed 3 ft per sec per sec (0.9 mps^2). A promising development consisting of raised treads adjacent to the skirts discourages riders from pressing their feet against the skirt. The treads need

Figure 9.11. Escalator step with cleated risers.

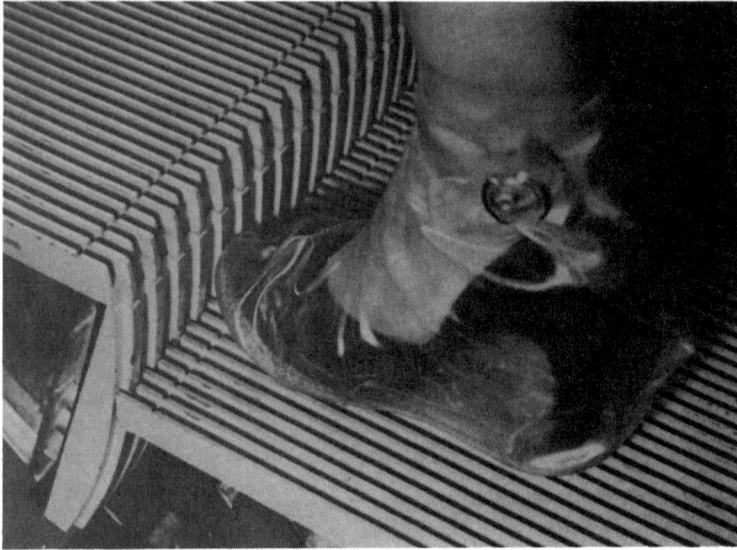

Figure 9.12. Combing action of cleated risers.

only be raised about ⅛ in. (3 mm). In addition, the treads adjacent to the skirt and at the leading and trailing edges of each step can be marked in a high-visibility color such as yellow or orange so that riders will avoid these areas.

Other approaches addressing this problem include the use of an antifriction coating on the skirt panel, applied lubricants, brushes mounted on the skirt, or filler plates to minimize clearance between the skirt and step.

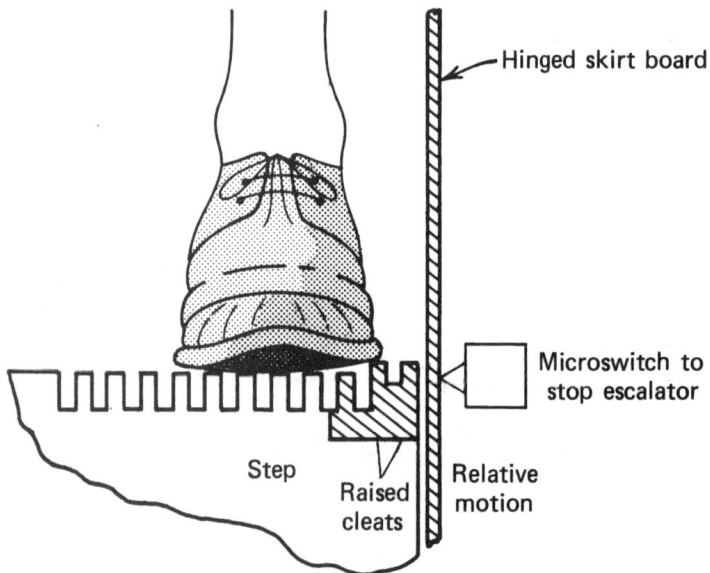

Figure 9.13. Skirt panel and step arrangement to avoid pinching soft shoes.

All escalators and walkways should be reversible so that their capacity can be utilized in either direction. Reversing switches are key operated and generally located at the top or bottom landing in the newel post. An emergency stop switch is required for escalators and walkways, and its location is prescribed by A17.1 to be mounted in sight of the passengers. To discourage mischief, a hinged plastic cover is used over the stop switch. Lifting the cover sets off an alarm, and the stop switch can then be operated. The alarm can also be monitored at a security station so that emergency personnel can be alerted.

The loading and exiting newels of escalators and walks should be extended so that passengers can grasp the handrail and become adjusted to the speed of the steps as they board. Two flat steps should be provided as a minimum before the incline begins. Illumination under the steps, which shines through the gap between steps, is an effective safety feature to provide tread delineation at the entrance and exit.

Balustrade treatments have undergone radical changes since the early days of escalators. A wide variety of materials is available for the balustrading of both escalators and walkways, including stainless steel, bronze, glass (also called crystal balustrading), aluminum in various colors, laminated plastics, tinted glass, lighted glass panels, and fiberglass material. Handrails are now being made in a variety of colors. The basic oval section of the handrail has remained unchanged, as this shape seems to provide the firmest and most comfortable gripping surface. The entry of the handrail into the newel is made as inaccessible as possible to minimize hazard, especially to curious children (Figure 9.14).

Escalator Operation

Escalators are generally started and stopped manually by an attendant using a keyed switch located at the top or bottom of an escalator. The attendant can ensure that no one is on the escalator and then start it in the desired direction. Stopping is also done by an attendant who can ensure that no one is riding at the time. This stop can be somewhat harsh

Figure 9.14. Extended newels (Courtesy Montgomery KONE).

because it is the same as that used for emergency stopping when the various safety switches are actuated.

It is possible to equip escalators with two levels of stopping deceleration, the emergency stop and a controlled stop that extends the stopping distance of the steps and is somewhat gradual. The controlled stop is limited to a deceleration of about 1.5 fps^2 (0.46 mps^2), whereas an emergency stop deceleration is about 3 fps^2 (0.9 mps^2). The controlled stop was introduced in the United States by the consulting engineering firm of Jaros, Baum & Bolles. William S. Lewis, the partner in charge of vertical transportation, became concerned about escalators in an office building during a fire emergency whereby the escalators, if not stopped, could bring people into a smoke-involved floor. By specifying and encouraging the development of a controlled stop, it became possible to stop the escalators if a smoke detector is actuated, or stop them from a remote control if a fire emergency occurs. The controlled stop has been applied on a number of escalators in various buildings with excellent results.

With a controlled stop and with future development of a gradual, controlled start, it becomes feasible to consider escalators that can be automatically started and stopped and to have a single escalator serve traffic in both the up and down directions. This is presently done without controlled starting and stopping and is accomplished by means of floor treadle or mat switches at the top and bottom landings of escalators. However, this situation may not be in compliance with local safety codes.

The operation of a stopped escalator is initiated, for example, by a person stepping on the top treadle and starting the escalator in the down direction. When the escalator starts, the passenger can travel down, and once he or she steps off and passes the lower treadle, the escalator will stop a preestablished time after the lower treadle switch is actuated. Similarly, a passenger wishing to go up steps on the lower treadle and starts the escalator in the up direction, and it will stop a short time after the upper treadle switch is actuated.

The hazard is that any person who may have bypassed one of the treadles may be walking on an escalator, and the abrupt start or stop will upset that person. With both a controlled start and stop, that hazard can be minimized.

A single reversible escalator can provide a cost-effective vertical transportation solution to an application where heavy traffic in one direction is expected at any time and in the opposite direction at another time, such as in a train station or a commuter parking garage.

Special Applications

Escalators and walkways can be installed practically anywhere. Outdoor applications should be designed to be weatherproof and include heaters to prevent icing or excessive water accumulation. The operation of escalators under exposed conditions is not recommended since wet or icy steps can be hazardous. Suitable shelter over the escalator or ramp must be provided as required by the A17.1 Code.

Escalators have been installed aboard ships to operate while a ship is experiencing motion: roll, pitch, and yaw. Shipboard applications have included access to and from the flight deck on aircraft carriers, between the restaurant and the kitchen on cruise liners (for waiters carrying trays), and for general passenger use.

Many industrial plants have escalators to serve their employees during shift changes and in the normal course of their duties. The highest and fastest escalators in the world is reported to be in the Russian subway system and to have a vertical rise of up to 214 ft (65 m) and operate at 200 fpm (1 mps).

ESCALATORS VERSUS ELEVATORS

Many office buildings and schools have escalators as the primary means of vertical transportation. Escalators can often be more cost-effective than elevators for a given building condition or for solving a particular vertical transportation problem.

One of the best examples is a high school or college in a single building. The major transportation demand occurs when classes change and may include as much as 40 to 50% of the student population moving during a 5-min period. For a six-story school of about 2500 students, 8 to 10 elevators would be required to give everyone floor-to-floor service. This number of elevators is seldom provided, and the accepted approach is to have the elevators stop at every other floor and have about half the students walk a floor. With the skip-stop arrangements, only about six elevators are required. With a pair of escalators servicing each floor, everyone can ride and the average trip requires no more time than the average elevator trip. The cost of the escalator arrangement plus two elevators for disabled passengers and freight service may amount to about the same as or less than equivalent elevator service.

Another example of the effectiveness of escalators was shown in an eight-story industrial research building with a population of about 2200 persons. Parking was in four areas on the ground, each with good access to main highways, so the arrival rate at the building amounted to about 25% of the population in a 5-min period. Sufficient vertical transportation would have required a group of seven elevators, with the elevators operated so that one elevator was designated to serve each floor. People would have been directed to line up in front of the elevator for their floor and wait. Once on their floor, they would either have to travel back to the lobby to get to another floor or wait until the peak was over.

A pair of 24-in. (600 mm) escalators, both operated in the same direction during the incoming and outgoing periods, provided the necessary short-time handling capacity. Evaluating all the factors, including the usable space required by the elevators versus the escalators, capital investment, maintenance cost, and cost of elevator pit and penthouse structures, is required to make the best recommendation.

Escalators in combination with elevators can solve some intricate vertical transportation problems. An example is a merchandise mart where buyers converge during show times once or twice a year. A number of suppliers hire space to display their products on the various floors. The average buyer starts at the top of the building and moves down from floor to floor. For this traffic, elevators can be used for travel solely between the top and bottom and escalators used in the down direction only, to take the buyers from floor to floor. Because it may take a buyer more than a day to travel the entire building, elevators could be arranged to stop at some middle floor, in addition to the top, to serve people who may want to return to the lobby during the noontime.

Escalators can be used to serve extra heavy incoming, outgoing, and two-way traffic. Travel time should be considered, and based on average floor heights of 12 ft (3.6 m) and an escalator speed of 100 fpm (0.5 mps), each floor would take about 26 sec to traverse, including the time required for a passenger to turn around and board the next up escalator. This time consists of 6 sec for entering and exiting, 14 sec for traveling, and 6 sec to walk around to the next rise of escalators. If maximum riding time of about 3 min is desired, effective application of escalators would be limited to buildings of no more than seven floors.

CALCULATING ESCALATOR OR MOVING WALK REQUIREMENTS

As with any vertical transportation application, the initial step is to qualify the peak traffic demand and the characteristics of that demand. In office buildings, the peak demands may be incoming and outgoing traffic. In a department store, peak demand may be two-way traffic during holiday seasons. In a transportation terminal, peak escalator travel requirements may occur as trains or planes arrive and discharge their passengers. In sports arenas, peak demand can be related to how quickly people can enter the facilities or, perhaps, how quickly the parking lots can be emptied after an event is over.

Examples 9.1 and 9.2 illustrate the application of escalators to an office building that has large floor areas and is of limited height. Before the escalators are considered, a number of additional considerations should be made. A building that can provide space for 500 persons per floor will cover an area of at least 60,000 to 75,000 ft^2 (5600 to 7000 m^2), based on a density of 100 ft^2 (9 m^2) of net area per person. A building of this size could be 650 ft (200 m) long and 100 ft (30 m) deep, or 300 ft (90 m) long by 225 ft (70 m) deep. If only a single pair of escalators is provided, they should be in the center of the square building so that internal walking distance is minimized. If the building is long and narrow, a number of groups of escalators are necessary at points that limit walking distance to a maximum of 200 ft (60 m). This latter consideration may change the entire economic aspect of the solution to favor elevators. With any vertical transportation arrangement where more than one group of elevators or escalators serve the same floors, each group must have about 20% excess capacity to compensate for unequal demand.

Assuming that the building is square and that a single center core of escalators can be used, the next step is to estimate the peak demand. If the discharge points of the parking or local transportation system are equally accessible to the entrances of the building, a high rate of arrival can be expected. Assuming further that this will be a single-purpose office building with all employees expected to start and quit at the same time, an arrival rate of up to 20% of the population in 5 minutes can be expected. The calculations would be as in Example 9.1.

If it is quitting time, full capacity will probably be used plus stairways, so the time will probably be less than 30 minutes. Because the example is based on a single rise of escalators, both up and down escalators should be provided. Using both up and down escalators operating down at 100 fpm (0.5 mps), the time will be less than 15 minutes.

If the building is evacuated in 15 minutes, it is highly unlikely that local transportation or the exits from the parking lots can accommodate that type of traffic peak. On that basis, only the single down rise of escalators needs to be used.

In addition to escalators, service elevators for vehicular traffic, moving furniture, deliveries, mail carts, and other movements necessary for office activity must be provided. In most areas, passenger elevators must be provided that comply with the Americans with Disabilities Act (ADA) and the Americans with Disabilities Act Accessibility Guidelines (ADAAG), which mandate the provision of vertical transportation service for physically impaired people qualitatively equal to that provided for all others. Service elevators for materials movement must be provided in addition to the passenger elevators.

Example 9.3 is a typical store situation in which the interest is in providing transportation to turn over the customer attendance on each floor within some given time period. A customer density of 20 ft^2 (1.9 m^2) of net selling area per person was used. This will, of course, vary with different types of stores and for different floors within the same store and will vary with the expected clientele. The requirement is to provide sufficient vertical transportation so that people can be carried to a floor to replace those already there who must have transportation exiting. This continuous replacement of patrons is called "turnover."

Example 9.1. Incoming Traffic*

Given: five-story building, 500 people per floor, arrival rate 20% of population in 5 min. Assume: equal attraction per floor; therefore 100 people must be carried to each floor.

Floor	Population	5-min Demand	Escalator Must Carry
5	500	100	100
4	500	100	200
3	500	100	300
2	500	100	400
1	500	100	500

Choose escalator to carry 400 persons in 5 min from Table 9.1

Floors 1 to 2   40 in. @ 100 fpm nominal capacity 375/5 min vs. 400 demand
Floors 2 to 3   40 in. @ 100 fpm nominal capacity 375/5 min vs. 300 demand
Floors 3 to 4   40 in. @ 100 fpm nominal capacity 375/5 min vs. 200 demand
Floors 4 to 5   24 in. @ 100 fpm nominal capacity 187/5 min vs. 100 demand

It can be noted that a 40-in. (1000-mm) escalator at 100 fpm (0.5 mps) is used to meet the expected demand from floors 1 to 2 (400 people in 5 min). Although the initial handling capacity of this escalator may appear to be insufficient to handle the expected demand, the estimated handling capacity of this escalator is calculated using 1 passenger per step loading, Nominal Handling Capacity from Table 9.1. If the calculated load of the escalator assumes 1.25 passengers per step (5 passengers for each four steps), the estimated handling capacity increases to 468 passengers per 5 minute period, adequate to accommodate the expected demand. Because speeds are constant and uniform from one escalator to the next, 100 fpm (0.5 mps) traffic moves smoothly with little congestion. If we wished to determine how quickly people could leave the building, an additional calculation to determine the outgoing traffic capacity of the escalators has to be made, as in Example 9.2.

---

*Note: Examples 9.1, 9.2, and 9.3 are based on Imperial units. To convert to metric, the following factors may be used:

12 ft = 3.6 m
100 fpm = 0.5 mps
60,000 sq ft = 5,581 sq m
75,000 sq ft = 6,976 sq m
650 ft = 198 m
100 ft = 30.5 m
300 ft = 91.5 m
225 ft = 68.6 m
200 ft = 61 m

Example 9.2. Outgoing Traffic

Floors 4 to 5   24 in (600 mm) @ 100 fpm (0.5 mps) nominal capacity 187/5 min, 37/min
Floors 3 to 4   40 in (1000 mm) @ 100 fpm (0.5 mps) nominal capacity 375/5 min, 75/min
Floors 2 to 3   40 in (1000 mm) @ 100 fpm (0.5 mps) nominal capacity 375/5 min, 75/min
Floors 1 to 2   40 in (1000 mm) @ 100 fpm (0.5 mps) nominal capacity 375/5 min, 75/min

500 persons on the fifth floor will require
13.4 min + 1.6 min (riding time at 26.5 sec per floor)   =   15.0 min
500 persons on the fourth floor will require
6.6 min + 1.2 min (riding time at 26.5 sec per floor)   =   7.8 min
500 persons on the third floor require
6.6 min + 0.78 min (riding time at 26.5 sec per floor)   =   7.38 min
500 persons on the second floor
6.6 min + 0.34 min (riding time at 26.5 sec per floor)   =   6.94 min

Maximum time to evacuate   =   37.12 min

Example 9.3. Two-way Traffic

Given: six-story store building, 20,000 ft^2 (1900 m^2) net selling area per floor, turnover 1 person per 20 ft^2 (1.9 m^2) per floor per hour.
Demand each floor 20,000/20 = 1000 persons per floor per hour × 2 (up and down)
= 2000 persons per floor per hour.

Floor	Demand	Demand on Escalators per Hour		Floors	Equipment Required	Capacity
		Up	Down			
6	2000	1000	1000			
				5 to 6	One pair 24-in. @ 100 fpm (600 mm @ 0.5 mps)	2250/hr/escalator
5	2000	2000	2000			
				4 to 5	One pair 24-in. @ 100 fpm (600 mm @ 0.5 mps)	2250/hr/escalator
4	2000	3000	3000			
				3 to 4	Two pairs 24-in. @ 100 fpm (600 mm @ 0.5 mps)	2250/hr/escalator
3	2000	4000	4000			
				2 to 3	Two pairs 24-in. @ 100 fpm (600 mm @ 0.5 mps)	2250/hr/escalator
2	2000	5000	5000			
				1 to 2	Two pairs 40-in. @ 100 fpm (1000 mm @ 0.5 mps)	4500/hr/escalator
1	2000					

The demand from the first to second and from second to third floors requires two pairs of escalators from floor to floor. These would be put in different locations in relation to the entry to the store and the merchandise features on each floor. A judgment must be made as to whether 24-in. (600-mm) or 40-in. (1000-mm) escalators should be provided above the second floor. Because of the merchandising plan of the store, it may be desirable to have the wider escalators to encourage greater patronage of the floors above the second floor and to accommodate two "socially related" people per step.

The complete vertical transportation system of any store must include passenger elevators for one-stop shoppers, shoppers with strollers, and physically impaired patrons. In fact, any building with escalators as a primary means of vertical transportation should have an elevator for this contingency. Stocking the various floors in a department store requires service elevators, which can combine passenger and freight functions.

Further discussions of stores are given in the chapter on commercial buildings.

APPLICATION OF ESCALATORS AND MOVING WALKS

Proper application of escalators or moving walkways requires determination of the expected demand and the nature of the demand on the system. This is true of any vertical transportation system and must be part of the study undertaken when a facility is planned. An estimate as to how many people will be seeking vertical transportation in a period of time must be made. This is an operational and management problem and may depend on many considerations in addition to those listed in the examples. In a sports arena the rate at which tickets can be sold or collected influences vertical transportation requirements. In a store, whether mass transportation is a factor, the nature of the business, and the price of the merchandise, must all consider the availability of parking space.

If both elevators and escalators serve the same floors of a building and people are given their choice to use either, it has been demonstrated that the following approximate division of traffic will occur (assuming the elevators provide reasonably good service):

	Division of Traffic	
Floors Traveled	Escalator	Elevators
1	90%	10%
2	97%	25%
3	50%	50%
4	25%	75%
5	10%	90%

Obviously, if minimum elevator service is provided, the percentages will favor the escalators. At about six floors, with both elevator and escalator service available to the sixth floor, people will ride elevators to the sixth floor and travel one or two floors down to a destination by escalator.

Location is all-important in the application of escalators and walkways. They have a known through-put of passengers which, once begun, must be accommodated, and suitable provisions must be made. By proper location, the use of an escalator or moving walkway may be enhanced. If the entrance area is restricted, fewer people will be able to use an escalator or walkway than if the entrance is wide open. Convenience is another factor; if the facility will save people walking or climbing, they will go out of their way to use it. If a building is on a hillside and an escalator provides ready transportation from one level to the other, all the people in the neighborhood will use that escalator if allowed to do so. Providing community transportation can contribute to success of commercial facilities: people will use escalators if it will save them time, and stores can obtain exposure to potential patrons.

Application opportunities vary with each location and each building. The foregoing discussion is designed to create the awareness of what a walkway or escalator can do. The successful application will depend on its placement and its convenience as it is being used.

ABOUT THE AUTHOR

DAVIS L. TURNER graduated from Cornell University with a degree in electrical engineering in 1968 and is a 35-year veteran of the elevator industry, having begun his career with Otis Elevator Company in 1962 in New York in the construction department. Throughout his career with Otis, he assumed positions of increasing responsibility and gained experience in sales, marketing, construction, service, modernization, and engineering.

In 1988, Mr. Turner joined Mitsubishi Elevator Company, a subsidiary of Mitsubishi Electronics America, as its vice president and general manager. In 1989 he was appointed president of the company. Under his direction Mitsubishi's elevator business in the United States expanded into three states and established a reputation for the highest quality of equipment and service available in the industry.

In 1996, Mr. Turner formed Davis L. Turner & Associates, an independent vertical transportation consulting firm located in Mission Viejo, California. The firm provides elevator and escalator technical expertise to a wide range of clients.

Mr. Turner is a member of the following industry associations: American Society of Mechanical Engineers, Board of Safety Codes and Standards, alternate member of the A17 Main Committee, member of the Mechanical Design Committee, member of the Escalator and Moving Walk Committee, member of the National Interest Review Committee; National Association of Elevator Safety Authorities, member of the advisory board; member of the Elevator Industry Group of Southern California, chairman of the Membership Committee; and member of the city of Los Angeles Board of Examiners for Elevator Constructors.

Mr. Turner is also active in community, civic, and cultural organizations and holds the following positions: member of the Advisory Board of the Japan America Society of Southern California and president of the Cornell Alumni Association of Orange County (California) from 1995 to 1997. He served as the campaign chairman for the Elevator Escalator Safety Foundation (EESF) in 1997.

10 Elevatoring Commercial Buildings

DEFINITION

Commercial buildings are buildings in which space is rented or used for a definite commercial purpose. This would include all types of businesses, professional office buildings, stores, industrial buildings, self-parking garages, and so forth. Apartments and hotels will be considered as residential buildings and schools and hospitals as institutional buildings.

Commercial buildings have definite vertical transportation requirements, because the arrival and departure of their populations are usually concentrated within certain periods of the working day. Traffic patterns vary with the use and location of the building, and some of the major variations are discussed in this chapter.

Although commercial buildings can be located anywhere and a trend to decentralization has been evident, most of them are still in the central business districts of cities and are usually concentrated in areas with reasonably good horizontal transportation. The efficiency of transportation to and from buildings will greatly influence pedestrian circulation patterns within any building and affect its elevator traffic.

High-density horizontal transportation may be provided by a transit terminal near a building. As the trains arrive, groups of people enter the building, most people timing their transit trip to arrive almost at the time they must start to work. With a train capable of discharging hundreds of people in a short time, the building's vertical transportation system is subject to severe incoming peak demands.

The other extreme of horizontal traffic affecting a building's elevators is evident in a suburban office building with remote parking. Demand on the building's elevators may be directly related to the time required for people to park their cars and walk to the building. If there are local coffee shops, the potential elevator passenger may arrive early, have breakfast, then enter the building. If eating facilities are provided in the building, the people may go to their desks and immediately return to the coffee shop.

How people arrive, when they go to lunch, what they do when they are at work, and how they leave are all factors in elevatoring any commercial building. A systematic consideration of these factors follows.

POPULATION

As important as the arrival rate of traffic is the number of people who will occupy the building. To some extent, tradition governs allocating space within a building. To perform a job, a person requires a certain minimum space, which can be as little as 10 or 15 ft^2 (1 or 1.5 m^2). People must get to and from their desks, which requires another 10 ft^2

The Vertical Transportation Handbook, Third Edition, Edited by George R. Strakosch
ISBN 0-471-126291-4   © 1998 John Wiley & Sons, Inc.

(1 m^2). If files or records are used, about another 10 ft^2 (1 m^2) must be allocated for that function. The minimum space per person is therefore about 30 to 40 ft^2 (3 to 4 m^2). If dealing with visitors is necessary, additional room to transact business is required, which is about 25 ft^2 (2.5 m^2). Size is also status, and office workers expect a minimum of about 100 ft^2 (10 m^2) to perform their tasks. The average manager gets about 200 ft^2 (20 m^2), and the executive may have more than 300 ft^2 (30 m^2).

The nature of the task to be performed greatly influences commercial building population. A law office, with its necessary reference files and library, requires more square feet per occupant than a drafting office, in which total working area is within reach of the drawing board. Computer departments with their large machines must have more space per occupant than the word processing department, where a person can operate at a single console.

In professional buildings the examining room may require specialized equipment and an area of 200 ft^2 (20 m^2) or more. Technicians' shops—eyeglass fitting, denture making, photo laboratories, and so on—are compact and may average less than 50 ft^2 (0.5 m^2).

Where there are many tenants on a single floor, a considerable amount of space is required to provide the necessary anterooms and passageways. With single large tenants, more of the space is usable since passageways can be minimized and public space is often limited to elevator lobbies and a corridor around the building core.

This latter aspect of larger open floor areas is being refined with a trend toward open space planning and landscaped offices. The traditional partitions are eliminated, and modularized workstations, "cubicles," consisting of desks, tables, and files integrated as a semimovable unit are utilized. In this way, grouping of people with associated functions can be accomplished and aisle space minimized. The open space plan is more applicable to larger areas of 5000 ft^2 (500 m^2) or more and may result in population densities approaching 100 ft^2 (10 m^2) or less without the sensation of crowding. The vertical transportation engineer must pay close attention to space utilization, experience, and trends in the area where a new building is planned so that accurate population estimates can be made.

Office buildings can be classed as diversified, single-purpose, or combined single-purpose-diversified. The completely diversified office building can be one in which no more than one tenant occupies more than a single floor and less than 25% of the tenants are in a similar line of work. This last qualification is important, because if all tenants are in the same business, competition may cause them to start work at the same time, have the same luncheon habits, and have similar patterns of visitor traffic. The tenants may be diversified, but their impact on the building will be the same as that of one large firm. We shall refer to this as single-purpose diversified occupancy.

The single-purpose building is exactly that, one firm occupying an entire building or a substantial portion of a building. The notable difference between a pure single-purpose building and a single-purpose diversified building is that the first provides opportunities to control traffic by staggering employee working times. In buildings with multiple groups of elevators—low-rise, mid-rise, high-rise—it is quite possible for one section to be single-purpose and another diversified if one tenant occupies the floors served by a single group of elevators.

In determining how many elevators are required for a given building, it is necessary to quantify and qualify its population. It is seldom known at the time the building is planned exactly who will occupy each floor, and the quantity of population must be averaged for each floor based on the type of tenancy expected. Table 10.1 gives typical values of popu-

TABLE 10.1. Population Factors, Commercial Buildings

Diversified	150 ft^2 (15 m^2)	net usable area per person*
Diversified single-purpose	135 ft^2 (13.5 m^2)	net usable area per person
Single-purpose	120 ft^2 (12 m^2)	net usable area per person

*Net usable area = gross area less elevator shaft and lobby space, mechanical space, stairways, janitorial, columns, toilets, corridor around core, and convector space.

lation related to the net usable square feet in each building. This net usable square feet should not be confused with net rentable area, which often includes columns, toilets, elevator lobbies, radiator or convector space, and a portion of an air-conditioning equipment room that may be on another floor. If net rentable area is used as a basis for population, an allowance must be made.

The thoroughness with which the building program is planned has a great influence on the population of the building. If a building is being built on an investment basis, tenant rental to follow, population should be established conservatively, based on the experience of comparable buildings (Tables 10.2 and 10.3). If the building is for a specific tenant who has planned and allocated space requirements, advantage should be taken of that planning in establishing population, with allowance for expansion.

Because elevators must be planned to serve the needs of the building population, the importance of correctly estimating population cannot be overemphasized. It can be costly to underestimate population to make undue allowances for absent employees. These variables can depend on current business conditions, which may change and affect the entire basis of elevatoring. The essential consideration is that space is available that can be used by personnel whether or not it is occupied at any particular time. It is also extremely important to estimate both resident and visitor traffic.

This latter point can be emphasized when the arrival traffic in an existing building is observed. Traffic will arrive at the building over about a 1- to 2-hr time period, and the total count should be close to the building population on that particular day. During that time period, a certain 5 min will represent the peak of the arriving traffic. In some buildings there may be two or three five-min periods when the peak traffic is almost identical

TABLE 10.2. Approximate Net Usable Area,* Various Height Buildings [15 to 20,000 Gross ft^2 (1500 to 2000 m^2) per Floor]

0 to 10 floors	Approximately 80% gross
0 to 20 floors	Floors 1 to 10 approximately 75% gross
	11 to 20 approximately 80% gross
0 to 30 floors	Floors 1 to 10 approximately 70% gross
	11 to 20 approximately 75% gross
	21 to 30 approximately 80% gross
0 to 40 floors	Floors 1 to 10 approximately 70% gross
	11 to 20 approximately 75% gross
	21 to 30 approximately 80% gross
	31 to 40 approximately 85% gross

*Net usable area is approximately 85% of full-floor standard rentable area.

TABLE 10.3. Suggested Building Population Factors Related to Building Height—Based on Net Usable Area

0 to 10 floors	125 ft^2 (12.5 m^2) per person
0 to 20 floors	Floors 1 to 10 125 ft^2 (12.5 m^2) per person
	11 to 20 130 ft^2 (13 m^2) per person
0 to 30 floors	Floors 1 to 10 125 ft^2 (12.5 m^2) per person
	11 to 20 130 ft^2 (13 m^2) per person
	21 to 30 140 ft^2 (14 m^2) per person
0 to 40 floors	Floors 1 to 10 125 ft^2 (12.5 m^2) per person
	11 to 20 130 ft^2 (13 m^2) per person
	21 to 30 140 ft^2 (14 m^2) per person
	31 to 40 150 ft^2 (15 m^2) per person

Other Commercial Space

Professional buildings	200 ft^2 (20 m^2) per doctor's office
Self-parking garages	300 ft^2 (30 m^2)[a] 1.2 persons per auto
Stores	Customer density of 10 to 40 ft^2 (1 to 4 m^2) of net selling area[b]
Industrial buildings	
Factories	Depends on manufacturing layout and product
Drafting	80 to 100 ft^2 (8 to 10 m^2) per draftsman

[a]300 ft^2 is an average for small and large autos.
[b]Net selling area is area open to the public.

owing to the variances in horizontal external traffic to the building or starting times of firms occupying the space.

If the traffic were to be observed over many days, a certain pattern would emerge, and the average peak 5-min traffic based on those many observations is what the elevators should be designed to serve. The peak 5 min also has a direct relation to the resident population. The percentage may remain the same, but the quantity of people will vary as the population changes. The following section describes the characteristics of peak traffic.

ELEVATOR TRAFFIC IN COMMERCIAL BUILDINGS

Once the population that requires vertical transportation is established, the next step is to determine the quantity and characteristics of vertical transportation required. As discussed in the chapter on elevator traffic, we are seeking a critical 5-min traffic period on which to base handling capacity and against which to check all other active traffic periods.

The up-peak or start-to-work period has traditionally been the basis for establishing the quantity of elevator-handling capacity for an office building. This is usually a critical period, as surface transportation and subways discharge passengers at the building and arrivals are at their highest level. It is essential to clear the lobby and get people to their desks so they can begin work.

During recent times, starting about 1970, a number of changes have been evolving. Many firms have recognized the difficulties of commuting and are allowing employees flexibility in starting and quitting times. In some localities, local governments, in an effort to reduce traffic congestion, have encouraged staggered starting and quitting times. Progressive firms have recognized that there is no need for all departments to start work at a fixed time and have instituted a staggered work schedule. The net result of these efforts has been a remarkable change in elevator peak traffic, wherein a considerable down traffic exists during heavy periods of up traffic. No longer is the up-peak a pure up-peak, but rather an up-peak with approximately 10% of the total up traffic traveling down—up-peak with 10% down traffic.

The traffic will usually peak during a 5-min period before the time most people start work. The intensity of this peak, consisting primarily of up traffic with some down traffic, is stated as a percentage of building population and forms a means of relating one building to another and the basis of establishing the elevatoring of each (Figure 10.1).

The other periods of the day when elevator traffic may be critical are at lunchtime and quitting time. Because it is general practice to stagger lunchtimes, these traffic peaks are usually not as severe as the incoming peak. This is not always true, however, because lunchtime may also be a very critical traffic period for a single-tenanted building of a full-service cafeteria is provided. Quitting time is an intense period of elevator traffic, yet elevator efficiency is greater during an outgoing rush. People will crowd elevators more than they will during arrival periods because they are usually more anxious to leave work than they are to get to work. With this crowding, and because the passengers are distributed over many floors rather than waiting in the lobby as during up peak, the elevators tend to make fewer stops, hence more trips in a given period of time. The net result is that, for a given number of elevators, outgoing capacity is a substantial percentage greater than incoming capacity.

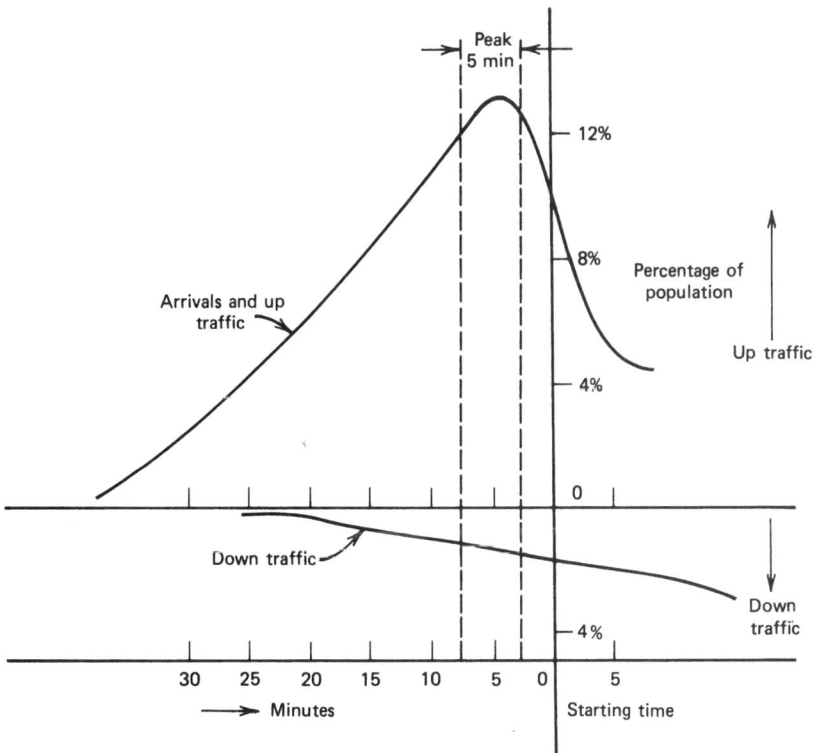

Figure 10.1. Typical arrival rate and elevator traffic distribution in an office building.

If good horizontal transportation exists, the incoming peak traffic with or without opposing down traffic will remain the critical traffic period. If an extended arrival rate is expected, the noontime period of two-way traffic may become critical. In professional office buildings, stores, and industrial buildings with shift changes the two-way traffic period is the most critical.

Sample percentages of the population that must be served during a critical 5-min period in various building types are given in Table 10.4.

Trading Floors

A development in the late 1980s into the 1990s has had a profound influence on building population and elevator traffic in that time span. Many office buildings in downtown areas have allocated space to trading floors where brokers are stationed and conduct transfers of securities and other financial instruments. Each has a desk, a computer, and multiple phones and many remain at his or her station for endless hours, with messengers coming and going, delivering buy and sell orders as well as other paper. It is not uncommon for meals to be catered and for support staff to travel to and from these floors to other areas of the building.

Population density is high, often in the area of 40 to 60 ft^2 (4 to 6 m^2) per occupant. The concern is not the resident population but, rather, the traffic that is generated by their staff. As an arbitrary estimate, it can be considered that each will have 4 to 6 people come and go to visit them each working hour. For example, 100 brokers or traders will account for 1200 elevator trips per hour to and from their floors, or 100 persons, two-way traffic, each 5 min. Because these floors must be intensively planned in a building program, either new or modernized, early information on the impact on the building can be gained.

TABLE 10.4. Expected Peak Traffic Periods—Various Commercial Buildings

	Percentage of Population in a 5-min Period		
	Peak Arrivals	Up-peak with 10% Down Traffic	Noontime or Two-way
Office Buildings			
Diversified offices	10 to 11%	11 to 12%	10 to 12%
Diversified single-purpose	11 to 13%	12 to 15%	12 to 15%
Single-purpose	12 to 18%	13 to 20%	13 to 17%
Other Building Types			
Characteristic Peak Traffic:			
Professional buildings	Peak traffic	Two-way, based on 1 to 2 visitors per doctor each 15 min coming and going	
Garages—self-parking (assume sufficient ramps to fill or empty garage in 1 hr)	Peak traffic commuter garage	10% to 15%, one-way traffic	
	Peak traffic, store, transportation terminal garage	10 to 15%, one-way traffic	
Stores	Population to be turned over, i.e., up and down (two-way) in 1 hr		
Industrial buildings	Peak traffic, 15 to 20% (up-peak or two-way).		

The result may be that a separate group of elevators designed to serve this function will have to be installed.

Interval

The quality of service given by any elevator system is also reflected by the interval or frequency of that service. Because the reputation of commercial buildings and the rentals they can command are based on the quality of service a building offers, the best quality of elevator service is a necessity. A building may offer the best space layout and services, but if people must wait too long for elevator service the value of other advantages can be lost. Excessive waiting times have been determined by analyzing service in buildings where complaints were minimal, and experience has shown that an up-peak without down traffic loading interval of between 25 and 30 sec produces excellent elevator service in any office building. Loading intervals of 30 to 35 sec are considered good where some degree of down traffic exists. Loading intervals of more than 35 sec are certain to lead to complaints in office buildings of any type and should be avoided.

Two-way traffic operating intervals of 30 to 40 sec are generally acceptable provided that the elevator operating system is designed to maintain that range or better. With a two-way operating interval of 40 sec designed and poor elevator operation, waits of more than 40 sec will be frequent and complaints will be received. In general, 30 sec is the maximum wait an average person will accept without complaint in a busy commercial atmosphere.[1]

In professional buildings, stores, or industrial buildings, the maximum two-way operating interval should never exceed 50 sec. The preferable safe maximum should be 40 sec for quality service. Again, the elevator operating system should be designed to maintain that maximum or better.

The use of Table 10.5, on suggested intervals, is predicated on providing sufficient traffic-handling capacity for the five-minute peaks as shown in Table 10.4. Multiple entrances to the building, upper-floor cafeterias, roof or basement stops on the elevators, and any odd openings will adversely affect the operating interval and must be considered in calculations.

Capacity and Speed

The combination of capacity and speed (often referred to as duty) of elevators for commercial buildings should be selected to provide service of the highest quality. The typical office building tenant is continually using elevators in his or her own and other buildings and soon learns what to expect in the way of service standards.

The minimum recommended size of an elevator for any building is the 2500-lb (1200-kg), 82 in. (2100 mm) wide by 51 in. (1300 mm) deep car interior. This size allows the architect to take advantage of superior inside decoration and allows the use of 48-in. (1200-mm) center-opening doors.

For most office buildings the 3000-lb (1400-kg) car should be the minimum. Office buildings that are prominent and heavily traveled should have 3500-lb (1600-kg) and 4000-lb (1800-kg) elevators, and monumental buildings such as the headquarters of large corporations should always have 4000-lb (1800-kg) or larger elevators.

---

1. Refer to Chapter 5 for a discussion of interval related to waiting time.

TABLE 10.5. Suggested Intervals

Office Buildings	Up-peak[a]	Up-peak[a] with 10% Down Traffic	Two-way[b]
Diversified	25 to 30 sec	30 to 35 sec	35 to 45 sec
Diversified single-purpose	23 to 28 sec	28 to 33 sec	33 to 43 sec
Single-purpose	20 to 25 sec	25 to 30 sec	30 to 40 sec
Professional buildings	—	—	30 to 50 sec
Self-parking garages	40 to 50 sec	—	40 to 60 sec
Stores	—	—	30 to 50 sec
Industrial buildings	25 to 30 sec	—	30 to 40 sec

[a]Loading interval—time between elevators departing from a main terminal.
[b]Operating interval—frequency of elevators passing an upper floor.

For commercial buildings other than offices the minimum recommended size of elevator varies with the traffic demand and use of the building, as shown in Table 10.6.

Elevator speed may be of secondary importance versus prompt floor-to-floor travel. If an elevator cannot attain a floor-to-floor performance time of 9 to 10 sec including prompt and efficient door-opening and closing speed, it will seem to be a slow, sluggish elevator. This performance is a function of the elevator motion control system and should be of the highest quality in any prestigious building. Speed becomes very important where express runs are provided, such as in the high-rise portions of an office building of 15 or more floors (see Table 10.8). Speed is also of considerable importance if the eleva-

TABLE 10.6. Suggested Elevator Capacity (pounds)—Commercial Buildings

Type of Building	Class of Building		
	Small	Average	Large or Prestige
Offices, suburban	2500	3000	3500
Service elevator[b]	4000	4000	4000
Offices, downtown	3000	3500	4000[f]
Service elevator[b]	4000	4500	6500
Professional offices			
Passenger[a]	2500	3500	4000
Service[b]	4000	4000	5000
Stores			
Passenger[c]	3500	3500	4000
Service[d]	4000	4000 to 6000	6000 to 8000
Garages	2500	3000	3500
Industrial[e]	4000	4000	4000

[a]As a practical consideration, 4-ft center-opening doors on a 3500-lb car will allow a mobile stretcher to enter.
[b]A hospital-shaped car should be provided.
[c]Wide, shallow cars with widest possible center-opening door desirable.
[d]Capacity is primarily to obtain largest-size elevator possible.
[e]Consideration should be given to combination passenger-freight elevators.
[f]Special single-purpose occupancy may require 5000 lbs.
Metric equivalents, 2500 lb = 1200 kg   3000 lb = 1400 kg
                   3500 lb = 1600 kg   4000 lb = 1800 kg
                   6500 lb = 3000 kg   5000 lb = 2300 kg

tor must make many stops, such as in a tall, slim building of 20 or more floors where only one group of elevators may be provided to serve many floors.

For the low-rise, suburban-type office building of three or four floors, hydraulic elevators with speeds of up to 150 fpm (0.75 mps) are often acceptable. Consideration should always be given to underslung traction elevators, which are capable of higher speed without the necessity of extensive overhead structure. Table 10.7 shows recommended elevator speeds for various types of commercial buildings.

When buildings have both high- and low-rise elevators, the high-speed elevators are required to traverse the express run in the shortest possible time. The limitation is the rate of change of acceleration, as described in Chapter 7. This results in a certain minimum number of floors being required for the express run with various speeds of elevators. This variation is shown in Table 10.8. An example of low-rise–high-rise elevatoring is given later in this chapter.

LAYOUT AND GROUPING OF ELEVATORS

In any multistory commercial building the vertical transportation system should visually dominate the lobby. Since the system is, in effect, the main entrance to the upper floors, people should be directed to the elevators or escalators both physically and visually. Signs, clearly visible from each building entrance, should plainly indicate each system and the floors it serves.

TABLE 10.7. Suggested Elevator Speeds

	Class of Building			
	Small	Average	Large or Prestige	Service
	Office buildings (including professional offices)			
Up to 5 floors	200 fpm[a]	300 to 400 fpm	400 fpm	200 fpm[a]
5 to 10 floors	400 fpm	400 fpm	500 fpm	300 fpm
10 to 15 floors	400 fpm	500 fpm	500 fpm	400 fpm
15 to 25 floors	500 fpm	700 fpm	700 fpm	500 fpm
25 to 35 floors	—	1000 fpm	1000 fpm	500 fpm
35 to 45 floors[b]	—	1000 to 1200 fpm	1200 fpm	700 fpm
45 to 60 floors[b]	—	1200 to 1400 fpm	1400 to 1600 fpm	800 fpm
over 60 floors[b]	—	—	1800 fpm	800 fpm
Stores				
Up 2 to 5 floors	150 fpm[a]	200 fpm	300 fpm	200 fpm[a]
5 to 10 floors	400 fpm	400 fpm	500 fpm	400 fpm
10 to 15 floors	500 fpm	500 fpm	500 to 700 fpm	400 fpm
Garages				
2 to 5 floors	200 fpm[a]	200 fpm[a]	200 fpm[a]	
5 to 10 floors	200 fpm	300 fpm	400 fpm	
10 to 15 floors	300 fpm	400 fpm	500 fpm	

[a]150-fpm hydraulic acceptable.
[b]Sky lobby design should be considered for this height.
Metric equivalents, 150 fpm = 0.75 mps, 200 fpm = 1 mps
$\qquad$ 300 fpm = 1.5 mps, etc.
$\qquad$ mps (approx.) = fpm/200 = mps

TABLE 10.8. Time and Distance Required to Attain Full Speed or to Slow Down from Full Speed (Approximate)

Ultimate Elevator Speed, fpm (mps)	Time (sec)	Distance, ft (m)	Minimum Number of Floors* Required for Express Run
500 (2.5)	2.9	12 (3.6)	4
700 (3.5)	3.6	20 (6)	8
1000 (5.0)	4.7	38 (11.5)	12
1200 (6.0)	5.5	53 (16)	20
1400 (7.0)	6.2	70 (21.2)	24
1600 (8.0)	6.9	80 (24.2)	28

*Based on 12-ft (3.6-m) floor heights, 1 m = 3.3 ft (approx.).

Vertical transportation should be grouped in one area, either the central "utility core" of the building or a service tower along one wall. Long corridors from the main entrance to the elevator lobbies should be avoided. The main entrance will be the one closest to the main horizontal transportation. People will not take an indirect route to a main entrance if there is a secondary entrance next to the transit station.

Elevators in office buildings are commonly installed in groups of either four, six, or eight cars. In many buildings the high-rise shaft space should be conserved for service elevators, building services, stairs, or smoke shafts. The highest-rise elevators may be in a five- or seven-car group. This allows the sixth or eighth shaft of a six- or eight-car core to be used for a service elevator.

As pointed out in the section on grouping of elevators (Chapter 2), six- and eight-car groups should have open-ended lobbies to leave space for people to wait before they board elevators. The space adjacent to the elevator lobby is a necessary reservoir for people entering at the main entrance, and its extent is based on the ability of the elevators to serve the incoming traffic. It should be equal, as a minimum, to the elevator lobby area.

Some typical core arrangements for high-rise buildings are shown in Figure 10.2. The elements included in the core are usually two stairways, electrical closets, telephone closets, air distribution shafts, toilets, pipe shafts, passenger elevators, and a service elevator. The balancing of required core space and gross area per floor is often the result of repeated alternative arrangements until the most favorable net-to-gross area is accomplished.

Once determined for a particular building size, the layout and space requirements of elevators is relatively inflexible. The design of the building should always proceed by designing the elevators and core first if a satisfactory elevator plant is expected and the cost of redesigning is to be avoided. The elevator design must be resolved early because most other aspects of the building design depend on the elevator design, and it is often necessary to contract for elevators long before contracts are awarded for other aspects of the building.

TRANSFER FLOORS

Local and express or multiple rises of elevators in a building should always be provided with a transfer floor. Any tenant who occupies floors served by more than one group of el-

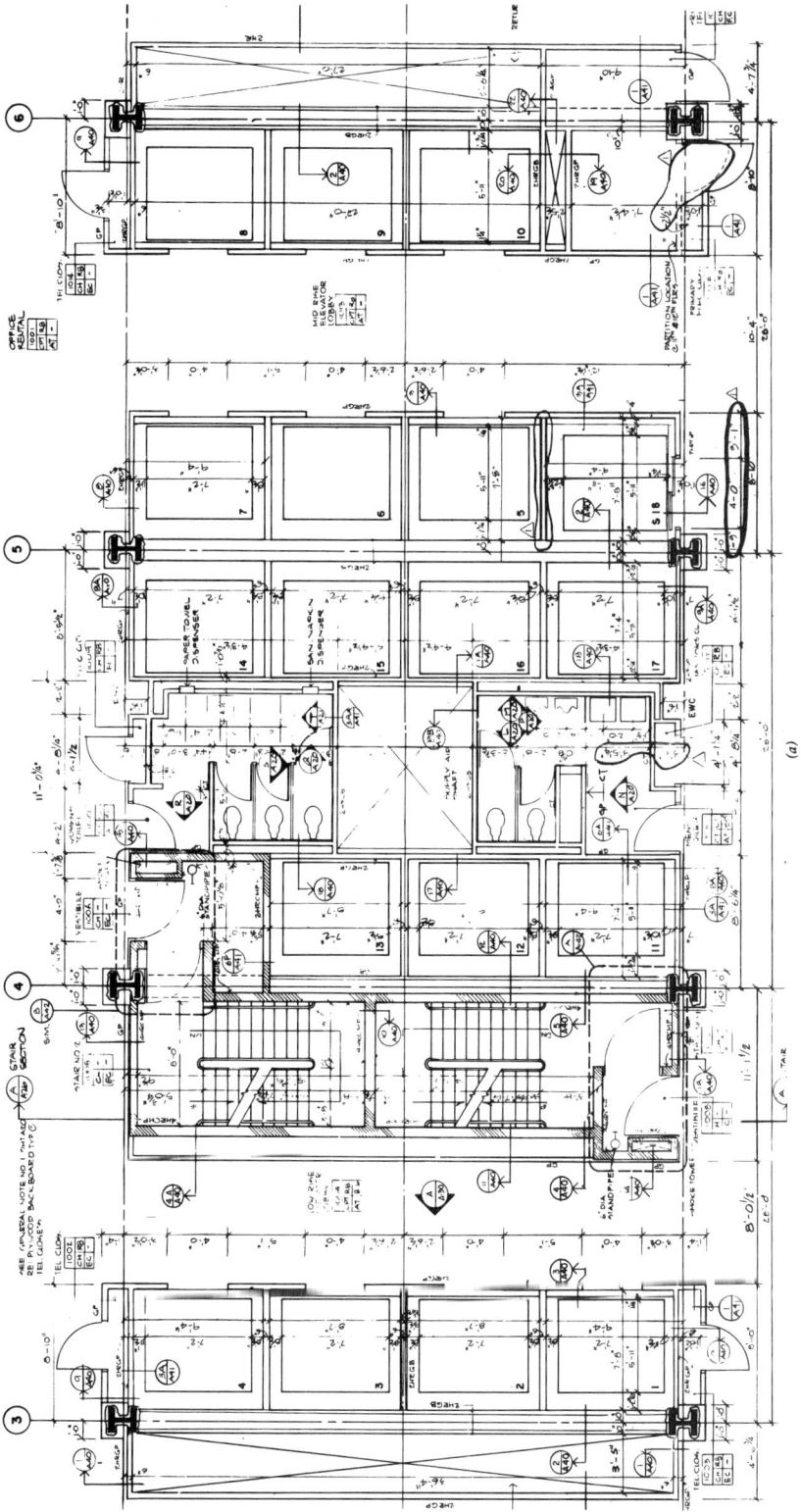

Figure 10.2. Typical office building elevator and utility cores. (*a*) Three groups of elevators in a 40-story building (Courtesy Skidmore, Owings, and Merrill).

255

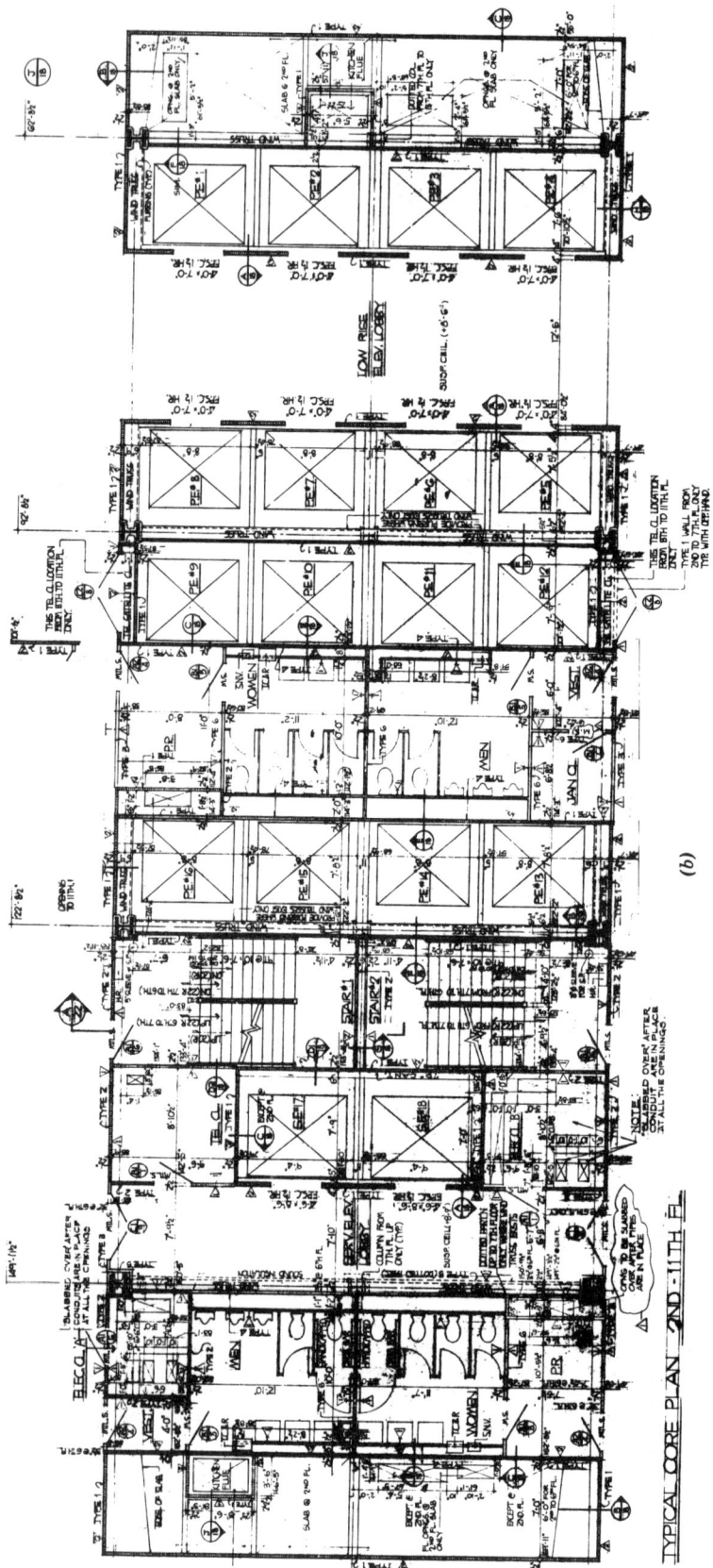

Figure 10.2. (Continued) (b) Two groups of elevators in a 28-story building (Courtesy Emery Roth and Sons).

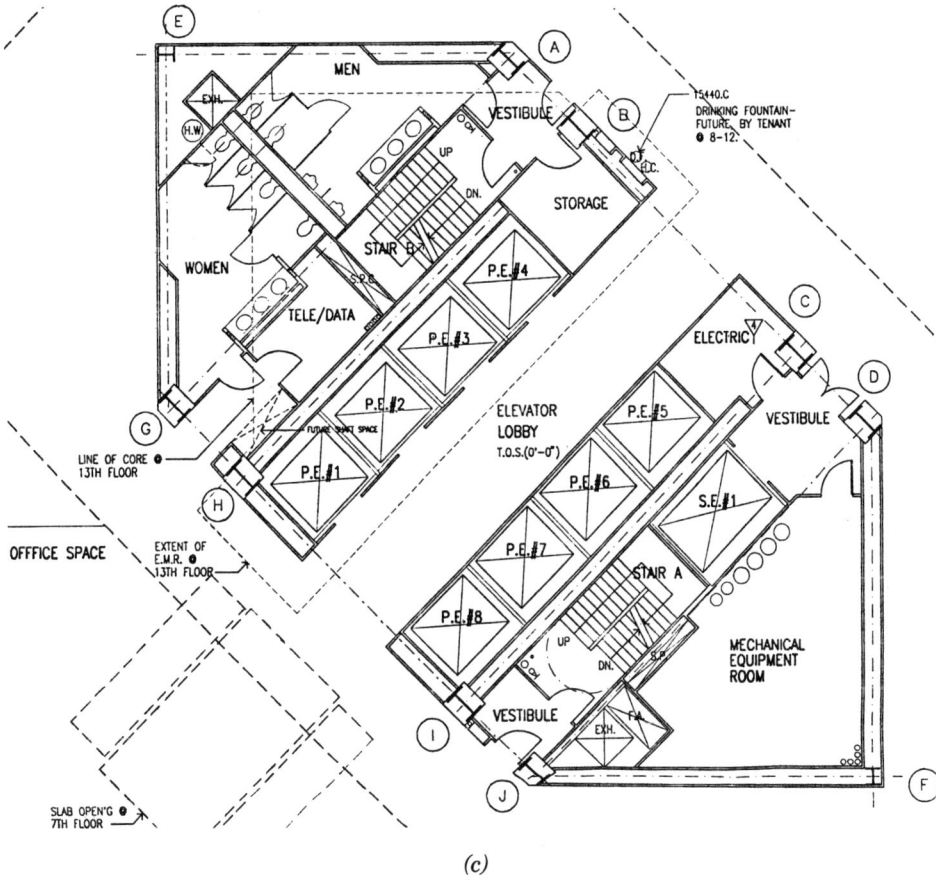

MEN

VESTIBULE

15440.C
DRINKING FOUNTAIN–
FUTURE, BY TENANT
● 8-12.

STORAGE

STAIR B

P.E.#4

WOMEN

TELE/DATA

P.E.#3

ELECTRIC

VESTIBULE

P.E.#2

P.E.#5

ELEVATOR
LOBBY
T.O.S.(0'-0')

P.E.#6

LINE OF CORE ●
13TH FLOOR

P.E.#1

S.E.#1

OFFFICE SPACE

EXTENT OF
E.M.R. ●
13TH FLOOR

P.E.#7

STAIR A

MECHANICAL
EQUIPMENT
ROOM

P.E.#8

UP

DN.

VESTIBULE

EXH.

SLAB OPEN'G ●
7TH FLOOR

(c)

Figure 10.2. (*Continued*) (*c*) Single group of elevators in a low-rise building (Courtesy Fox & Fowle Architects).

evators requires a transfer floor. Otherwise, people would have to travel to the lobby and change to the next group of elevators. At the transfer floor people will also have to change elevators but will not be backtracking. The time saved can be substantial and depends on the relative interval in each group of elevators and the traffic at that moment.

At the transfer floor the elevator operating system should be arranged to allow the higher-rise elevator to stop only for an up landing call or down car call. This avoids the possibility of a tenant using the high-rise group to reach the transfer floor in the morning, when it is the first stop after the express run, so employees do not have to wait for intermediate stops, and the lower-rise elevators in the evening, when the tenant floor is the first stop in the down direction. If the employees were to fill the car, they would get priority down service. Response to only up landing calls and down car calls at the higher-rise transfer stop avoids this possibility. This operation is referred to as limited transfer floor service.

A second alternative to transfer floor and elevator flexibility requirements is to provide for future entrances on the high-rise elevators at overlapping floors. The entrances could be blocked closed door frames, and as the building matures and changes in elevator requirements are apparent, these entrances could be put into use and necessary adjustments made to the elevator operation.

Transfer between a high-, intermediate-, and low-rise elevators can be a requirement in a single-purpose office building. Normally, a single transfer floor on each group of elevators should be sufficient; however, if the nature of the business is such that a great deal of interfloor traffic will take place, openings can be installed in a high-rise elevator hoistway at the low-rise transfer floor. The opening that serves the low rise should be designed to operate only in response to an up landing call or down car call. Such an arrangement is often referred to as a "crossover" floor.

ELEVATOR OPERATION IN COMMERCIAL BUILDINGS

Based on an estimate of the expected traffic in a building, the elevators proposed, and the handling capacity and interval calculated for various periods of the day, a traffic flowchart can be prepared. This flowchart will indicate the complexity of the elevator traffic and the varieties of elevator operations required to handle that traffic. A rough chart based on observations made in similar existing buildings can be used as a start, and additional refinements added as requirements are analyzed in greater detail. An example of such planning is shown in Figure 10.3.

Figure 10.3 shows the expected traffic in a typical diversified office building, a pattern characteristic of most such buildings. The intensity will vary inversely with the degree of tenant diversity.

In planning elevator operations the service expected of the elevators must be considered. As an example, if much visitor traffic is expected during the day, the system may have to provide for varying degrees of two-way traffic in the heavy up, heavy down, and balanced directions. If considerable traffic between floors is expected, additional provisions for traffic of that nature may be needed.

Each building has its unique requirements depending on location, expected tenancy, the nature of the tenants' businesses, and visitor traffic, as well as external transportation to and from the building. Typical traffic flow diagrams for other types of commercial buildings (see Figure 10.4), supplemented by information on operation given in Chapter 7, may be used as a guide for a particular building in selecting a suitable operating system.

Special Requirements

Commercial buildings are, in general, public buildings and are expected to have many visitors. The elevator layout and design should be such that it will be attractive to the visitor and offer service. Tenants, as well as visitors, appreciate attractive elevator design in addition to prompt and efficient service with minimum effort.

To fulfill this requirement the architect must employ an informational system that is clear and distinct. Signs that indicate the floors served by each elevator group must be clearly visible from anywhere in the building lobby. Landing lanterns for each elevator should be easily discerned from every point in the elevator lobby. The elevator call button must also be plainly identifiable and readily accessible from routes of passenger approach and located so that passengers will wait near the elevators. Floor numerals applied to the edges of hoistway doors at each floor help the rider identify his or her floor. Codes enacted to provide service to disabled persons require such an approach.

Communication between elevator passengers and building staff through a central telephone system or intercom is required in many localities and a rule in the A17.1 Elevator

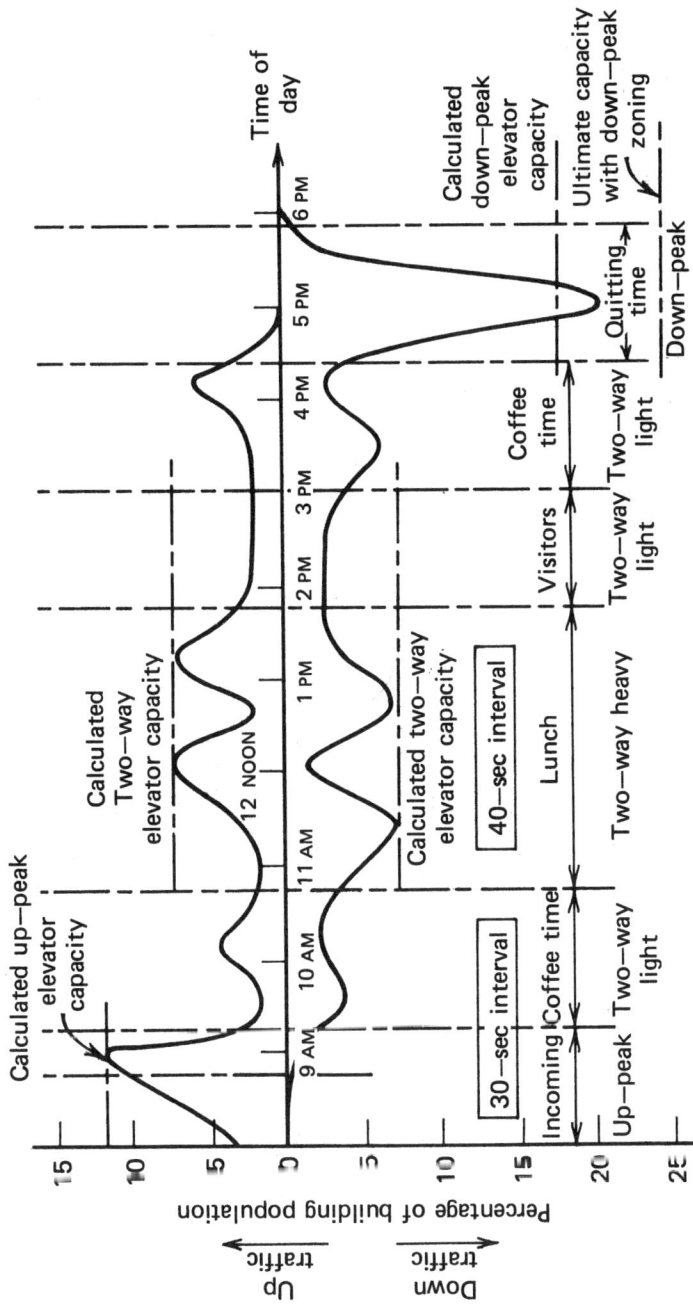

Figure 10.3. Typical traffic flow—diversified office building traffic observed on and off elevators at the lobby.

259

Traffic	Operating Requirements
Up-peak	Requires full capacity of the elevators.
	Requires forcing of elevators back to the lobby.
	Usually definite peak at a predetermined time, clock control enforced.
	Consideration needed for down and interfloor traffic.
Two-way	Light traffic—requires deployment of elevators to anticipate landing calls.
	Heavy traffic—requires spacing and strategic relocations of elevators to equalize service to landing calls.
Down-peak	Minimize time spent at the lower terminal.
	Maintain elevators in motion.
	Deploy elevators to traffic centers to equalize service.
	Down-peak can be anticipated by a clock.

Figure 10.3. *(Continued)*

Code. If the building has a receptionist or lobby floor attendant, the attendant should be able to communicate with the elevators so that action can be taken if necessary. In many buildings a lobby indicator panel shows the position of the various elevators and can be equipped with flickering indicators to call attention to a car delayed beyond a predetermined time. Communication by intercom can establish the reason, and necessary help can be summoned. Building personnel can also observe possible elevator system failures at the lobby station by means of indicator lights and take appropriate action. Such lobby stations are often located at a security desk together with other building monitoring functions. In some localities, a fire command center is developed so that emergency operations, including elevator operations, can be directed from a central location.

In smaller buildings the indication and communication equipment can be located in the building office, security room, or at any place where surveillance is available. In buildings where only part-time help is in attendance, an emergency call system to a central protective agency can be provided. Many local safety codes as well as the A17.1 Elevator Code require communication outside the building if the building is not manned 24 hr each day.

Operations to call a particular car to the lobby for cleaning or some other purpose are desirable. One such arrangement is a number of key switches, one for each car, in a control panel. Operating the switch can call the car so it may be placed on special service, such as operating from its car buttons independently of the group, or so it may be shut down for necessary cleaning or maintenance. The switch should also serve to close elevator doors when a car is locked out of service for, say, nighttime security.

Modern elevator systems employ microprocessors that have the capability of providing a video display of car position and function, as well as indicating delays and malfunctions. Special commands can be inputted via a keyboard located in the lobby or at a designated location. Such inputs can be the ability to recall elevators, lock out selected floors, restrict cars to serve certain floors, and other functions, depending on building requirements.

Extended programs in such computers may include diagnostics, that is, indicating the location and nature of troubles plus performance monitoring. Performance monitoring includes periodic displays of how the elevators are responding to traffic, one such indication being the presence of long-wait landing calls. Extended discussion of these functions can be found in the chapter on traffic studies later in this book.

Figure 10.4. (*a*) Professional building—elevator traffic in and out at lobby. (Scale will depend on the number of professional offices in the facility.) (*b*) Commuter garage.

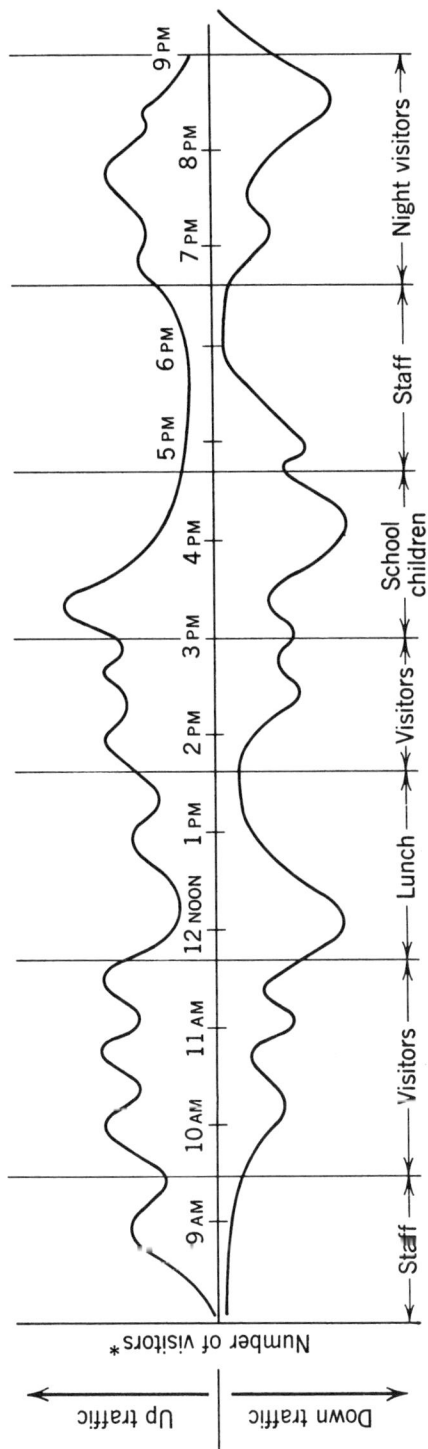

(*a*) Professional building

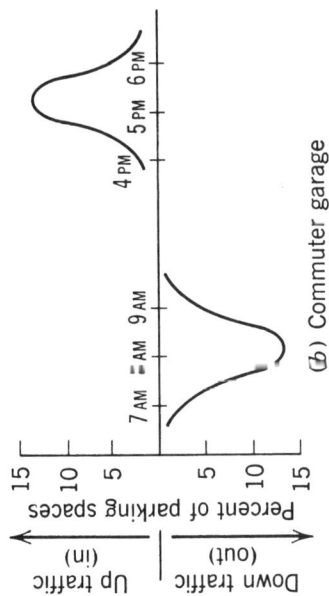

(*b*) Commuter garage

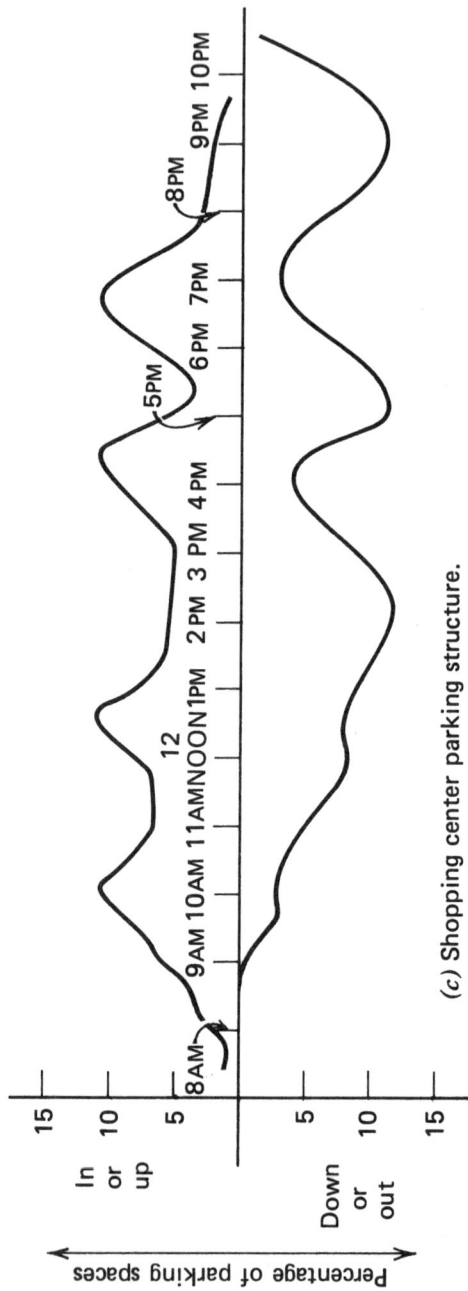

Figure 10.4. (*Continued*) (*c*) Shopping center parking structure. (Traffic is based on automobiles and occupants in and out of garage entrance. If garage is a basement or above-ground facility, location of pedestrian exit and entrance floor assumed at ground level.)

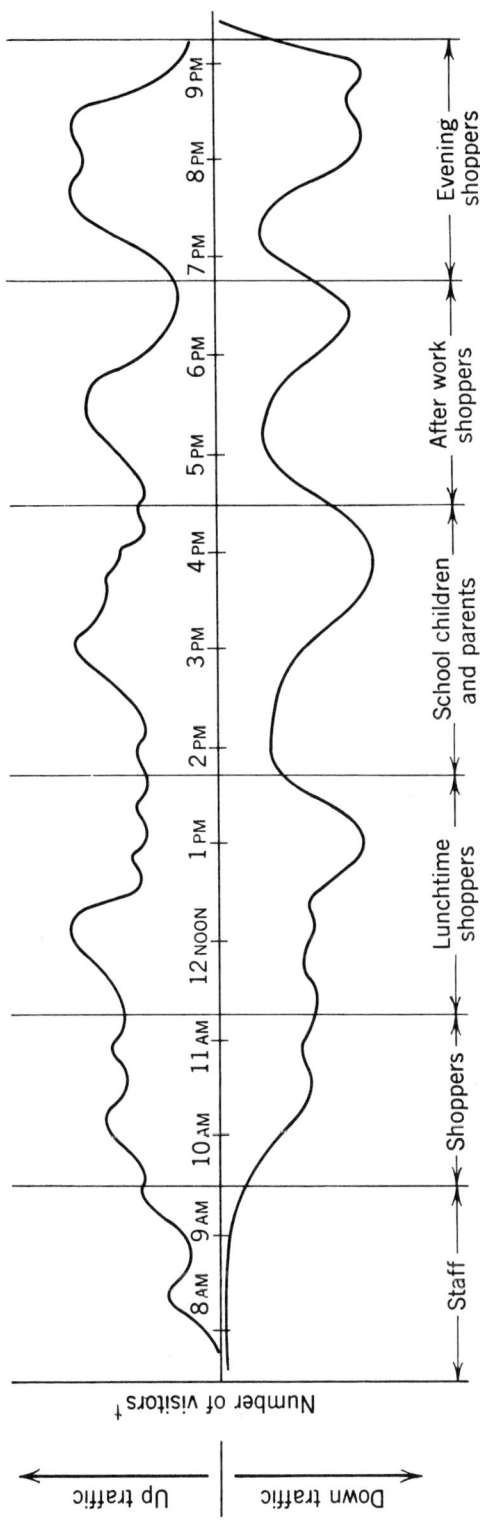

Figure 10.4. (*Continued*) (*d*) Department store or shopping mall. (Scale will depend on the total traffic, which will be divided approximately 90% escalators and 10% elevators.)

(d) Department store

Emergency requirements such as for fire safety, earthquake, riots, and power failures require additional operational considerations. These will be discussed fully in a later chapter.

Service Elevators

A separate service elevator or elevators is a necessity in any building of more than 250,000 ft² (2500 m²) of net usable area (300,000 ft² gross) or where the passenger elevators are in full use during most of the day. In smaller buildings, the need for freight service is light enough that one of the passenger elevators may be diverted to this purpose during off-peak periods. This is not feasible if much visitor traffic is expected, and a separate service elevator should be provided. One service elevator for each 300,000 to 500,000 ft² gross is suggested.

Transportation of myriad items used for tenant activities and building operation requires service elevators. These items may range from mail and office supplies to new furniture, soft drinks, coffee carts, lunch wagons, wall partitions during renovations, or masonry supplies. Increasing cost of service personnel makes it economical to schedule renovation of repairs during the day and, for tenant convenience, management wants to avoid passage of building material or maintenance workers through the passenger lobby.

In many buildings simple freight traffic can attain volumes amounting, in the form of mail, trash, office material, and supplies, to about 150 to 200 lb per employee per week. A 250,000-ft² net usable area building may have as many as 2000 employees and will require about 250 tons of material to be moved in and out each week. With a service elevator that has a 4000-lb capacity minimum, averaging trips every 10 min at half-capacity, handling 250 tons will require a full 40 hr.

Most organizations depend on mail and insist on its prompt delivery. Concurrent mail and passenger peaks will create a dilemma for management. An unsatisfactory solution is to spread the mail on the lobby floor to be picked up by tenants as they enter the building. A separate service elevator can expedite mail delivery and provide the additional handling capacity required for service needs during the rest of the day. Various mail distribution systems are available that utilize automatic unloading dumbwaiters or conveyors. These specialized systems increase the utilization of the service elevators especially if mail is a priority demand, as would occur in a single-purpose building. Automated mail-handling systems will be discussed in a later chapter.

The sizes recommended for service elevators vary with building use. The 4000-lb hospital-shaped platform, as discussed in Chapter 12 on institutional buildings, is the most satisfactory minimum. Planning the dimensions of the entrance should consider the largest-sized equipment expected to be in the building and the normal size of the building doors. If wider elevator entrances are required, two-speed sliding or two-speed center-opening doors should be provided. The use of vertical biparting doors is generally not recommended for commercial buildings and nonattendant operation.

A larger elevator with a capacity of 6500 lb is usually required in the larger buildings. The elevator should have a cab interior height of 12 to 14 ft (4300 mm) and doors 8 ft (2400 mm) or higher. A hoist beam at the top of the elevator cab allows rolls of carpet that are 12 to 15 ft (4000 mm) long to be moved and carried. The extra height is also an aid in building completion and renovation since it allows long pieces to be fabricated off-site and easily transported up into the building. The additional capacity and size allows the transportation of air-conditioning equipment, transformers, large office machines, or anything of a larger weight or size than can be carried in a passenger elevator.

In professional buildings, especially medical office buildings where people may arrive on stretchers, a service elevator is necessary to handle the stretcher or any unusual loads. For medical office buildings of any size, the 4000-lb hospital-shaped car is usually the most satisfactory service elevator. For a smaller medical building, a 2500-lb car with 42-in. side-sliding doors or a 3500-lb elevator with a 48-in. center-opening door may be used for mobile stretchers.

Easy access to the service elevator at the loading dock and at the upper floors is of prime importance in locating the service elevator. If one of the passenger elevators is to be used for part-time service, traffic flow between that elevator and the building service entrance must be considered. A rear entrance on the elevator may solve the access problem. If a rear entrance is to be used, it should be designed to be operative during controlled periods and with the use of an attendant. A separate landing call button that will operate an annunciator in the elevator is desirable.

Separate service elevators are of inestimable advantage during the building constructions periods. Once the building is topped out, the service elevator can be quickly completed sufficiently for temporary operation and allow early removal of an outside hoist. The cost advantage may pay for the investment in the elevator many times over.

Cafeterias and Restaurants

A club, restaurant, cafeteria, or any eating facility in a commercial building requires a separate service elevator. It is difficult to carry foodstuff, carts, garbage, or bottles on an elevator without damaging the sides, spilling, or leaving an odor. Protective pads can be used, but strict management control is required to ensure their use. A facility usually requires considerable time to transfer its supplies from a loading dock area, and such supplies often arrive at the height of the morning rush.

Eating facilities in a building also place extra demands on the passenger elevators. The size of the facility, its expected turnover, and its mode of operation must be considered in the initial elevatoring of the building. In general, any facility designed to serve about 300 people that is open to the public requires an additional passenger elevator, plus use of the building's service elevator, over the normal requirements of the building. An additional passenger elevator would be required for each multiple of 300. Because public restaurants are often located on the top floor and generally cater to people outside the building, the extra elevator(s) can be the shuttle type and travel only between the restaurant and the ground floor. Many successful top-floor restaurants use outside, glass-enclosed elevators to attract and accommodate their clientele. This type of equipment is discussed in a later chapter.

Cafeterias that are to be used by the building's tenants are preferably located in the basement or on the second floor. The main building elevators should not stop at the cafeteria floor, and the cafeteria should be served separately by escalators or by separate shuttle-type elevators. Cloakrooms on the cafeteria floor will reduce elevator use by employees who eat and leave the building. Reducing this traffic lessens interference with the next group of employees going to lunch. Separate kitchen-supply elevators should be located near a loading dock and are desirable for the basement or second-floor cafeteria.

Locating a cafeteria on an upper floor, such as the transfer floor between a high-rise and low-rise elevator group, leads to all elevators stopping, in both directions, at that floor during the noontime traffic period. Elevatoring calculations must consider this extra stop as well as the amount of traffic expected as a result of the cafeteria location. An example is given in Example 10.2D. Any upper-floor eating facility requires that a separate service elevator be provided in the building. A cafeteria should never be located at an intermediate floor within a group of elevators unless it is exclusively for the residents within that group and ample elevator service is available.

TYPICAL OFFICE BUILDING ELEVATORING

Before starting the planning of an office building of any size, the architect or builder should investigate existing buildings of a similar size and nature. The real estate boards can suggest comparable examples. The plot size and local zoning laws usually establish the per-floor area and height of the new building, as well as required use and other economic factors. Initial elevator requirements are usually based on very meager information, which typically includes the expected height and approximate floor area of the building and its intended use. With more specific information a better-elevatored building can result.

Sample calculations for elevatoring a diversified office building appear in Example 10.1, where the type and size of the building are indicated and its elevatoring is developed. Introduction of a garage facility in the basement changes an acceptable elevator solution to a marginal one. With the provision of a separate shuttle elevator between the lobby and garage floor, the four elevators are acceptable. In addition to enabling the four main elevators to serve the building, the use of a shuttle facilitates use of the garage. The shuttle elevator allows the garage to be operated independently of the main elevators and possibly used as a public facility on weekends and evenings without impairing the security of the building.

Solving elevatoring problems often takes the form of considerable compromise between the optimum size of the building and the number of elevators to be provided. Each elevator represents a cost factor in both initial equipment investment and rentable area sacrificed. With land selling as high as $1500/ft^2 in some areas of large cities, the greatest return on total investment is necessary. Commanding rents to yield an adequate return requires the best possible elevator service. Space is necessary, however, so for tall buildings the initial cost of elevators relative to all other aspects of a building is generally about 12% of total construction cost. The difference between the best and poorest elevators in regard to speed and capacity is only about 10 or 15% of their cost—less than 2% of the entire project—but the return in terms of satisfied tenants can be far greater.

Example 10.1. Diversified Office Building

A. Given: diversified offices, suburban location, investment-type building; 12,000 ft^2 gross per floor, 10 floors, 12-ft floor heights. Population: 12,000 × 0.80 = 9600 ft^2 net per floor 77 @ 125 ft^2 per person = people per floor.

Total population floors 2–10 = 693 people.

Assume: 10 passengers up per trip, elevators travel at 400 fpm, 48-in. center-opening doors, floors 1 to 10, 12 ft floor-to-floor, 9 × 12 = 108 ft, elevator rise.

Probable stops: 10 passengers, 9 stops = 6.2, no highest call return

Time to run up, per stop: $\dfrac{108}{6.2} = 17.4$ ft. rise per stop

$\qquad$ 17.4 ft at 400 fpm = 6.1 sec

Time to run down: $\dfrac{108}{1 + 1} = 54$ ft $\dfrac{(54 - 17.4) \times 60}{400} + 6.1 = 11.6$ sec

Elevator performance calculations:

Standing time

Lobby time $10 + 0.8$	$=$	10.8 sec
Transfer time up stops 6.2×3	$=$	18.6
Door time, up stops $(6.2 + 1) \times 5.3$	$=$	38.2
Transfer time, down stops 1×4	$=$	4.
Door time, down stops 1×5.3	$=$	5.3
Total standing time		76.9 sec
Inefficiency, 10%	$=$	7.7
	Total	84.6 sec

Running time

Run up 6.2×6.1	$=$	37.8
Run down 2×11.6	$=$	23.2
Total round-trip time		145.6 sec

$$HC = \frac{(10 + 1) \times 300}{145.6} = 22.7 \times 4 = 91 \text{ people}$$

Percent HC: $77 \times 9 = 693$, $91/693 = 13.1\%$

Interval: $145.6/4 = 36$ sec

Four 2500-lb elevators @ 400 fpm required as a minimum.

B. If a single basement garage for 200 cars is added: 200 automobiles at 1.5 people per auto = 300 people = about 50% of building population. Therefore, it is likely that elevators will travel to the basement every second trip.

Recalculate round-trip time: building B, floors 1 to 10

Additional time required for trip to basement garage

Stop at lobby—part of original round trip

Run to basement, 10 ft	$=$	5.1 sec
Transfer at B	$=$	4.0 sec
Add door time B and 1, 2×5.3	$=$	10.6 sec
Run up to 1, 10 ft	$=$	5.1
		24.8 sec
Assume every second trip to B, 24.8/2	$=$	12.4 sec

New round-trip time $145.6 + 12.4 = 158.$ sec

$$\text{New HC} = \frac{(10 + 1) \times 300}{158} \times 4 = 84 \text{ people}$$

New percent HC: $84/693 = 12.1\%$

New interval: $158/4 = 39.5$ sec

The interval, 39.5 sec, would be unacceptable.

Example 10.2 Example of Low-rise-High-rise Elevators in an Office Building

Given: single-purpose office building, downtown location, 20,000 ft² gross per floor, 20 floors, 12 ft floor to floor, 20-ft lobby height. Population factor average 125 ft² per person based on net usable area.

A. Population calculations—assumptions

Floors 2 to 11 $20,000 \times 0.75 = 15,000 \div 125 = 120$ people per floor
Floors 12 to 20 $20,000 \times 0.80 = 16,000 \div 125 = 128$ people per floor

B. Criteria: 5-min elevator capacity—15 to 20%, interval—30- to 35-sec

C. Elevator calculations

    1. Low rise: assume six-3500-lb elevators @500 fpm for floors 1 to 11, 48-in. center-opening doors.

Rise: 20 ft + (9 × 12) = 128 ft

Probable stops: 16 passengers, 10 stops = 8.2

Up floor-to-floor time $\dfrac{128}{8.2} = 15.6 = 4.7$ sec

Down floor-to-floor time $\dfrac{128}{1 + 1} = 64 = \dfrac{(64 - 15.6) \times 60}{500} + 4.7 = 10.5$ sec

Elevator performance calculations:

Standing time

Lobby time 16 up + 2 down = 14 + 1.6	=	15.6 sec
Transfer time, up stops 8.2 × 3	=	24.6
Door time, up stops (8.2 + 1) × 5.3	=	48.8
Transfer time, down stops 1 × 4	=	4.0
Door time, down stops 1 × 5.3	=	5.3
Total standing time		98.3 sec
Inefficiency, 10%	=	9.8
	Total	108.1 sec

Running time

Run up 8.2 × 4.7	=	38.5
Run down 2 × 10.5	=	21.0
Total round-trip time		167.6 sec

Population: 120 × 10 = 1200

$\text{HC} = \dfrac{(16 + 2) \times 300}{167.6} \times 6 = 193$ people

Percent HC: 193/1200 = 16.1%

Interval: 167.6/6 = 27.9 sec

    2. High rise: assume six-3500-lb elevators @ 700 fpm for floors 1, 12 to 20

Rise: 20 ft + 18 × 12 = 236 ft

Probable stops: 16 passengers, 9 stops = 7.6

Up floor-to-floor time: local run 12 to 20 = 8 × 12 = 96/7.6 = 12.6 = 4.5 sec

Down floor-to-floor time: $\dfrac{96}{1 + 1} = 48$ ft

$\dfrac{(48 - 12) \times 60}{700} + 4.5 = 7.6$ sec

Express run: $\dfrac{(236 - 96 - 12) \times 60}{700} + 4.5 = 15.5$ sec

Elevator performance calculations:

			Up Transit Time
Standing time			
Lobby time 16 up + 2 down = 14 + 1.6	=	15.6 sec	15.6
Transfer time, up stops 7.6 × 3	=	22.8	22.8
Door time, up stops (7.6 + 1) 5.3	=	45.6	45.6
Transfer time, down stops 1 × 4	=	4	—
Door time, down stops 1 × 5.3	=	5.3	—
Total standing time		93.3 sec	84.0 sec
Inefficiency, 10%	=	9.3	8.4
	Total	102.6 sec	92.4 sec

Totals from page 268 (102.6 sec)   *(92.4 sec)*

Running time

Run up express	=	15.5	15.5
Run up local 7.6 × 4.5	=	34.2	34.2
Run down local 2 × 7.6	=	15.2	—
Run down express	=	15.5	—
Total round-trip time		183.0 sec	142.1 sec

Population = 9 × 128 = 1152

$$HC = \frac{(16 + 2) \times 300}{183.0} = 30 \times 6 = 180 \text{ people}$$

Percent HC: 180/1152 = 15.6%

Interval: 180/6 = 30 sec

The foregoing calculations show that the six elevators low rise and six elevators high rise will serve the incoming traffic to the building. When the building is designed, a transfer floor should be provided at the eleventh floor on the high-rise group so people from the low rise can readily travel to the high rise and vice versa. This transfer floor does not affect the elevator calculations since the eleventh-floor population will continue to be served by the low rise and the eleventh-floor high-rise stop as a transfer floor will be served by allowing the elevators to stop only for up landing calls and down car calls. The eleventh-floor car button in the elevator should operate only in the down direction.

To elaborate on the building, a further assumption can be made that it is desired to locate a cafeteria on a middle floor, such as the eleventh floor, and have both the low-rise and high-rise elevators stop there. To determine whether this is feasible, a two-way traffic calculation must be made.

D. Two-way traffic
1. Assume: cafeteria serves 500 meals per seating, 3 seatings. Usual design for a cafeteria facility in a single-purpose building is to serve about 60% of the population. Total population 2352 − 1500 = 832 eat out
2. Assume cafeteria turnover ½ hr

$$\text{Cafeteria traffic} \frac{500 \text{ in} + 500 \text{ out}}{6(30 \text{ min})} = \frac{1000}{6} = 167 \text{ in and out per 5 min}$$

People not eating, interfloor ⅓(2352) = 784 × 10% = 78 per 5 min

People in and out of building

$$⅔ \times ⅓(832) = 185, \quad \frac{185 \text{ in} + 185 \text{ out}}{6} = 62$$

= 62 in and out per 5 min

Total lunchtime traffic (167 + 78 + 62) = 307 people on both elevator groups

$$\text{Prorate low rise} \frac{1200}{2352} \times 307 = 157 \text{ people, two-way demand on low rise}$$

$$\text{high rise} \frac{1152}{2352} \times 307 = 150 \text{ people, two-way demand on high rise}$$

3. Calculate two-way traffic on the low-rise elevator.
   Probable stops: assume 8 people up, 8 people down; 5.9 stops up, (.7 × 5.9) = 4 stops down

$$\text{Time to run up, per stop} \frac{128}{5.9} = 21.7\text{-ft rise per stop}$$

$$21.7 \text{ ft} = 5.4 \text{ sec}$$

$$\text{Time to run down} \frac{128}{4 + 1} = 25.6 \text{ ft} = 5.8 \text{ sec}$$

Elevator performance calculations:
Standing time

Lobby time 8 in + 8 out	=	16.0 sec
Transfer time, up stops 5.9 × 3	=	17.7
Door time, up stops (5.9 + 1) × 5.3	=	36.6
Transfer time, down stops 4 × 4	=	16.0
Door time, down stops 4 × 5.3	=	21.2
Total standing time		107.5 sec
Inefficiency, 10%	=	10.8
	Total	118.3 sec

Running time

Run up 5.9 × 5.4	=	31.9
Run down 5 × 5.8	=	29.0
Total round-trip time		179.2 sec

$$HC = \frac{(8 + 8) \times 300}{179.2} = 27 \times 6 = 162 \text{ vs. } 157 \text{ people, demand.}$$

Interval: $179.2/6 = 29.8$ sec

4. Calculate two-way traffic on the high-rise elevators.
Probable stops: assume 10 people up, 10 people down; 9 upper floors + cafeteria, up stops = 6.5, down stops (6.5 × 0.7) = 4.6

Time to run up, per stop $\dfrac{(96 + 12)}{6.5} = 16.6$-ft rise per stop

$$16.6 \text{ ft} = 4.7 \text{ sec}$$

Time to run down, $\dfrac{(96 + 12)}{4.6 + 1} = 19.3$ ft $= 5.2$ sec

Express run time $\dfrac{(236 - 108 - 19.3) \times 60}{700} + 5.2 = 14.6$

Elevator performance calculations:
Standing time

Lobby time 10 in + 10 out	=	18.0 sec
Transfer time, up stops 6.5 × 3	=	19.5
Door time, up stops (6.5 + 1) × 5.3	=	39.8
Transfer time, down stops 4.6 × 4	=	18.4
Door time, down stops 4.6 × 5.3	=	24.4
Total standing time		120.1 sec
Inefficiency, 10%	=	12.0
	Total	132.1 sec

Running time

Run up express	=	14.6
Run up 6.5 × 4.7	=	30.6
Run down express	=	14.6
Run down 5.6 × 5.2	=	29.1
Total round-trip time		221.0 sec

$$HC = \frac{(10 + 10) \times 300}{220} = 27 \times 6 = 162 \text{ vs. } 150 \text{ people, demand.}$$

Interval: $221.0/6 = 36.6$ sec

The foregoing elevator arrangement will support the personnel requirements for a cafeteria on the eleventh floor based on the assumptions made. An essential consideration is that sufficient service elevator capacity be provided to serve both the normal needs of the building and the food service requirements. A separate service elevator is a necessity for a building of this size and essential for a single-tenant operation. For a single tenant, an automated mail-handling system, independent of the service elevator, should be provided.

The final recommendation for the building would be six low-rise elevators, 4000 lb (1800 kg) at 500 fpm (2.5 mps), serving floors 1 to 11, six high-rise elevators, 4000 lb (1800 kg) at 700 fpm (3.5 mps), serving floors 1, 11 (transfer), 12 to 20, and one service elevator, 6500 lb (3000 kg) at 400 fpm (2 mps) serving all floors. If a basement is included, an additional shuttle passenger elevator, 3500 lb (1600 kg) at 125 fpm (0.6 mps), should be provided to serve from floor 1 to the basement.

RULES OF THUMB

Rules of thumb for evaluating the elevatoring of a planned office building quickly are helpful if requirements are relatively clear-cut and elevators need *not* serve more than one entrance, a garage, or a cafeteria, nor serve as part-time service elevators. In other words, the rules are valid only when the elevators serve direct traffic between the office floors and the lobby.

A rule of thumb is to provide at least one elevator for each 225 to 250 building occupants. This rule is related to the 35,000-ft^2 rule on the basis of population density of 125 ft^2 per person. Again, the rule fails for buildings of more than 20 floors or with floor areas of less than about 10,000 ft^2.

A better guide is given in Figure 10.5. By referring to the figure, the approximate elevatoring may be established for a diversified office building requiring the percentage of handling capacity and having the density of population shown. This chart is based on simple elevatoring without a garage floor, no double lobbies, no odd stops, and no other complexities. The size and speed of the elevators are to be established from the guides shown earlier in this chapter.

A detailed elevator study should be made where the building is near the upper extremes of the curves.

MAXIMUM ELEVATOR SIZE IN OFFICE BUILDINGS

Elevators of more than 4000-lb capacity should not be used in office buildings without special provisions for loading and operation. The 4000-lb car is about the maximum that people will fill to capacity when left on their own. Its nominal capacity is 19 people, which is quite a large group and will make the elevator appear crowded even though it is not. Other people will usually wait for the next car. In addition, if a 4000-lb (1800-kg) car serves more than 12 to 14 stops, the trip becomes too long, people are irritated, and building reputation suffers. The use of low- and high-rise elevators for buildings of 15 floors or more should be investigated.

Elevators larger than 4000 lb (1800 kg) can be effectively used in some buildings. The floor area per floor should be 25,000 ft^2 (2500 m^2) or greater, and each elevator group should be a minimum of six elevators to provide a suitable interval. Lobby space should

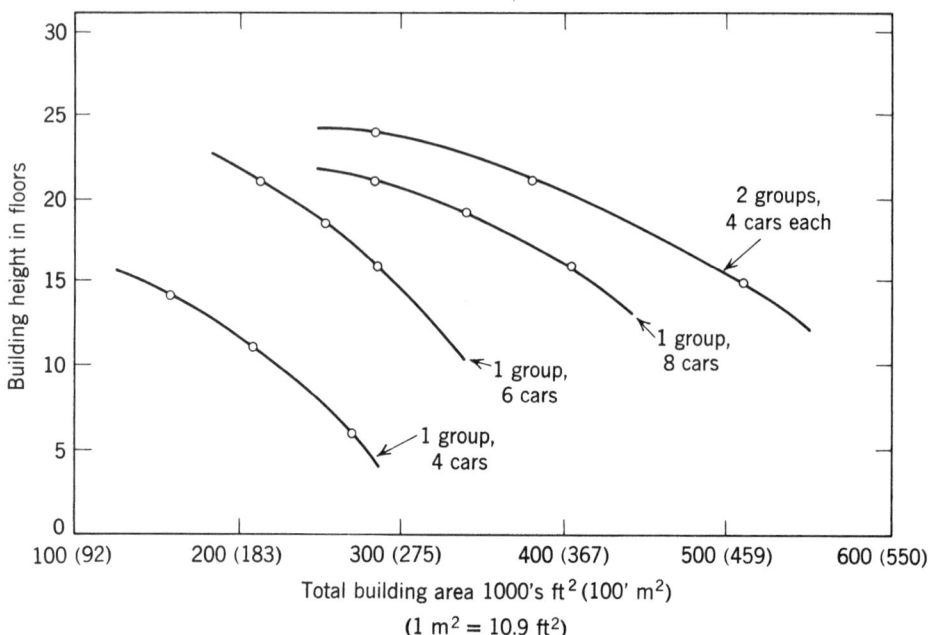

Figure 10.5. Office building elevatoring, diversified office building based on: 1, no basements, no upper-floor garages or eating facilities; usable office floors only. 2, 30-sec loading interval. 3, 200-sec maximum round-trip time. 4, 125 ft² (11.5 m²) per person. 5, 12 ft (3.5 m) floor-to-floor height. 6, 12% up traffic per 5 min (percentage of population). *Note:* One separate service elevator needed for each 300,000 ft² (27,500 m²) of office space.

be wide enough to accommodate the expected heavy loading and extra-heavy exiting traffic. Elevators should be operated so that all elevators are available for loading if they are at the main lobby, and loading time factors should be automatically adjusted to suit the varying loads expected.

Large insurance and banking operations centers have successfully used elevators of 5000- or 6000-lb capacities. Further discussion of larger elevators is given in a later chapter on double-deck elevators and sky lobbies.

Campus Office Buildings

Some organizations vie for an exclusive setting for their headquarters and supporting staff. A remote, country setting has great appeal, and thus a complex of buildings serving a single tenant may be built. Various two- and three-story (or higher) buildings may be built and connected by sheltered walkways. Parking is located on the periphery of these structures, and automobiles or car pools are the essential horizontal transportation. Alternately, the complex may be all under one roof, consisting of many wings with long corridors providing horizontal access for the occupants.

Such buildings, known as "campus offices," necessitate the same considerations for population and traffic as their downtown high-rise counterparts. The major difference is that various cores of vertical transportation must be considered, as well as the need to support the various functions located distant from each other. Centralized cafeterias and service facilities are usually provided, and the inclusion of automated mail and materials handling systems should be considered. Chapter 15 discusses this aspect and offers some guidelines. Among the management considerations should be the cost of full-time equivalent employees versus the cost of an automated system, also discussed in Chapter 15.

Elevatoring such buildings will involve careful study of expected employee movements: how they will enter and leave the building from the parking facility and how they will travel once within the building. Each movement will involve a percentage of the population. Planners must also consider the paths they will take on a pleasant day versus an inclement day. Will an employee park near an entrance, use the vertical transportation, and travel to his or her location on an upper floor corridor? This is a question that should be examined. In any case, the rules concerning the location and grouping of elevators apply. Each location of vertical transportation should provide ample handling capacity and should be planned so that horizontal walking is minimized. Centralized monitoring of elevator functioning is essential since the relative remote location of each unit may be such that observations are meager.

ELEVATORING STORES

As noted in the chapter on escalators, escalators are usually the preferred vertical transportation in stores. Modern merchandising depends on people visiting each floor to shop, and the exposure of patronage to the displays on each floor is by escalator. The escalator is a center of attraction, is constantly operating, and requires very little decision or patience on the part of the user.

Demands on the vertical transportation in a store vary with selling space (net selling area) on each floor, the density at which customers are expected to occupy that space, and the rate the per-floor customer population must be "turned over," that is, carried up and down to and from that floor.

Nominal densities in selling area vary from 10 ft^2 (1 m^2) per person on the "bargain" floors to 40 to 50 ft^2 (4 to 5 m^2) in exclusive departments. The normal time of turnover expected is an hour, about the time a person will spend if he or she is a serious shopper and seeking a particular item of any consequence. In some downtown specialty shops, where much patronage comes from office workers during lunchtime, a half-hour turnover may be more realistic.

Escalators should provide transportation for about 90% of the shoppers, with the other 10% expected to use elevators. The elevators should also be capable of serving about 10 to 15% of the staff in a 5-min period during working hours to provide for lunchtime or shift changes. If the store's offices are above the selling levels, the entire population of the office floors needs elevator service similar to a single-purpose office building. Store elevators should be located so that the view from the elevator encompasses prime selling area. When a store is large enough to require a group of four or more elevators, multiple groups of four elevators located in different quadrants of a large floor are suggested. The elevators should be in a line of four or less to avoid unnecessary holding of an elevator at a floor. With more than four elevators in a line, the distance between the first and last car tends to be too great and prospective passengers can miss the end cars if they were not waiting near them.

Department store passenger elevators should be large in size to accommodate shoppers with strollers and wheelchairs. If separate service or freight elevators are provided, the passenger elevators of 3500-lb (1600-kg) or 4000-lb (1800-kg) capacity with 48-in. (1200-mm) center-opening doors are generally ample and acceptable. If separate service elevators are not provided, the passenger elevators must be used as combination elevators and larger sizes must be provided.

Sufficient service elevators must be provided in a store for stock handling. A complete department store may require elevators of sufficient size to handle 12-ft (3600-mm) rolls of carpet, for which a 5000-lb (2250-kg) elevator with a 12-ft-high cab height can be provided. The more usual requirement is for racks of dresses or coats, which may be from 5 to 8 ft (1600 to 2400 mm) long. In general an 8000-lb (3600-kg) passenger-type elevator with a platform approximately 8×8 ft (2400×2400 mm) is recommended. This arrangement allows automatic operation and use by both passengers and freight. Providing two-speed center-opening doors 60 in. (1600 mm) wide will allow any large packages to pass with ease. Easy access from loading docks to the service elevator is a necessity.

Restaurants and cafeterias in department stores require the same considerations as in office buildings. If the escalators do not serve the cafeteria floor, ample elevator service must be provided.

Shopping Malls

Popular in suburban locations and in proximity to large cities are developments of complexes including large stores as well as specialty shops and extensive eating facilities. Some of these complexes have been extended to include amusement-park-type attractions and other recreational facilities. Transportation to and from these "shopping malls" is by private auto, and extensive parking facilities on their periphery is a basic requirement.

Upon entering the mall the visitor is attracted by the various facilities, and vertical transportation centers should be prominent and easily accessible. Larger stores in the complex usually have their own internal elevators and escalators, and here the rules for elevatoring department stores apply. Our concern is the pedestrian traffic to the various levels to encourage people to visit stores located on the upper levels as well as the food service facilities usually located there.

Centralized transportation cores consisting of both escalators and elevators should be provided. These should be located at approximately 300-ft (100-m) centers, depending on the locations of the entrances to the mall. It is essential that elevators be as attractive as the escalators and in the same location so as to discourage any use of escalators by people with strollers or carts and to avoid the inherent hazard such movement presents.

The traffic demands of such cores will be a function of the ability of the parking areas to create entering and exiting traffic. Extended parking areas without convenient horizontal transportation (buses or trams) will provide a more even flow than ones where groups of people are transported to and from the mall. Quantity will be a function of the expected turnover of the net selling area and the other facilities, as would be calculated for a department store. Convenience of vertical transportation is the prime consideration from a quality standpoint.

Service needs, such as the location of loading docks and "back of the house" access to the various stores, are a prime consideration. Below-level loading docks and elevators to serve the various levels, strategically located and connected by ample walkways, are expected to be part of the overall plan.

Because most shopping malls are expected to be of no more than three levels, hydraulic elevators are the equipment of choice. Observation-type passenger elevators with a view of extended pedestrian corridors provide a merchandising opportunity, as they allow patrons to see the stores that attract them.

PASSENGER ELEVATORS IN GARAGES

The number of passenger elevators required for a self-parking garage depends largely on the efficiency of the garage design. That efficiency in turn depends on how quickly the city streets can deliver or absorb autos from the garage.

If we assume that a well-designed garage has sufficient ramps to allow all its spaces to be filled (or emptied) within an hour, we will have some basis for elevator calculations. The nature of the garage is also important. Use by commuters who will park in the morning, leave their cars all day, and drive away in the evening presents a different problem than use by transients who will come and go during the course of a day and park while they shop or attend to business. The commuter garage has the simplest elevator traffic to serve, all out in the morning and all in when evening comes. The transient garage presents a two-way elevator traffic pattern, and shoppers with children usually comprise the greatest traffic volume.

A garage connected to an airport or downtown bus transportation terminal will have a combination of long-term and transient parking. In airport terminals, many people will park their cars and leave on early flights and return in the evening, while others will park to meet incoming flights and leave in an hour or so after they meet their party. Peak traffic is generally established by the transients, and the approach should be to determine their peak usage and establish the elevators accordingly.

Actual demand on elevators varies with the two types of traffic and is a function of the number of automobile passengers. The commuter may or may not have a partner and, on the average, each car has 1.1 to 1.4 occupants. The shopper usually is accompanied by other people for an average of 2 to 3 occupants. This varies with the type of parking facility, its location, and the area of the country. In airports, traffic counts have indicated about 1.5 occupants per auto.

If the garage can be filled, emptied, or turned over in an hour's time, the number of cars that can be parked times the average number of occupants divided by 12 (for 12 5-min periods) gives the expected demand for elevator service.

Example 10.3 gives two typical situations. Similar calculations would be used for basement or upper-floor garages in buildings where separate garage elevators are provided.

It is reasonable to expect some patrons to use stairs in a garage if they are conveniently located and well marked. With large garages connected with sports arenas or theaters the impact of the crowd exiting from an event is such that escalators may be warranted. For general convenience, in any large multilevel garage, escalators are recommended. If the garage covers a large area, multiple locations may be necessary. Multiple locations require additional security and reliability considerations.

For a garage up to about 50 ft of elevator rise, the hydraulic-type elevator operating at 150 fpm is usually cost-effective, especially when garage design warrants single elevators in a number of locations.

Example 10.3. Garages—Self-parking—Passenger Elevators

A. Commuter garage: 6 levels, 10-ft floor heights, 200 automobiles per parking level. Expected elevator demand: 5 (upper levels) × 200 = 1000 automobiles × 1.4 people per auto = 1400 people

$$\frac{1400}{12} = 117 \text{ people (all up or all down) in 5 min}$$

Assume: 16 passengers per trip, 150 fpm, 6 stops, floors G, 1 to 5, 10 ft floor to floor, 50-ft rise

Probable stops: 14 pass up, 5 stops = 5 probable stops

Time to run up, per stop $\dfrac{50}{5}$ = 10-ft rise per stop

$$10 \text{ ft} = 7.1 \text{ sec}$$

Time to run down, $\dfrac{(50 - 10) \times 60}{150} + 7.1 = 23.1 \text{ sec}$

Elevator performance calculations:

Standing time

Lobby time 16 in or out	=	14 sec
Transfer time, up stops 5 × 3	=	15
Door time, up stops 5.3 × (5 + 1)	=	31.8
Transfer time, down stops	=	—
Door time, down stops	=	—
Total standing time		60.8 sec
Inefficiency, 10%	=	6.1
	Total	66.9 sec

Running time

Run up 5 × 7.1	=	35.5
Run down 1 × 23.1	=	23.1
Total round-trip time		125.5 sec

$$HC = \frac{16 \times 300}{125.5} = 38 \times 3 = 114 \text{ vs. 117 people, demand}$$

Interval: 125.5/3 = 41.8 sec

The calculations as shown for all incoming traffic are essentially the same for outgoing traffic.

If three elevators are in the same location, handling capacity and interval is good. Three 3500-lb elevators should be provided.

Because of size of floor for 200 automobiles (60,000 ft² approximately), at least two locations are required.

Recalculate: demand = 117 + 20% (split-location inefficiency) = 140 or 70 people elevator demand per location

Assume: 12 passengers per trip, 200 fpm, 6 stops, floors G, 1 to 5, 10 ft floor to floor, 50-ft rise

Probable stops: 12 pass up, 5 stops = 4.6 stops

Time to run up, per stop $\dfrac{50}{4.6}$ = 10.9-ft rise per stop

$$10.9 \text{ ft} = 6.4 \text{ sec}$$

Time to run down, $\dfrac{(50 - 10.9) \times 60}{200} + 6.4 = 18.1 \text{ sec}$

Elevator performance calculations:

Standing time

Lobby time 12 pass in or out		=	11.0 sec
Transfer time, up stops 4.6 × 3		=	13.8
Door time, up stops (4.6 + 1) × 5.3		=	28.7
Transfer time, down stops		=	—
Door time, down stops		=	—
Total standing time			53.5 sec
Inefficiency, 10%		=	5.4
	Total		58.9 sec

Running time

Run up 4.6 × 6.4		=	29.4
Run down 1 × 18.1		=	18.1
Total round-trip time			106.4 sec

$$\text{HC} = \frac{12 \times 300}{106.4} = 34 \times 2 = 68 \text{ vs. 70 people, demand}$$

Interval: 106.4/2 = 53.2 sec

Two elevators each in two separate locations will provide sufficient handling capacity at an acceptable interval. Two 3000-lb elevators @ 200 fpm in each location should be provided.

B. Some garages may have complex traffic patterns, and an analysis of the various conditions should be made to determine the most critical. For example, assume a garage connected to a department store: 7 levels, 10-ft floor heights, bridge at fourth level, 75 cars per level; expected elevator demand, turnover entire garage in 1 hr; 5 (upper levels, exclude bridge floor) × 75 = 375 automobiles × 2 people per auto = 750 people.

750 × 2 (up and down per hr) ÷ 12 = 125 up and down per 5 min

Analyzing the possible traffic patterns that may predominate suggests three:

1. Most people enter and leave at the fourth level.
2. Most people enter and leave at the ground level.
3. Equal traffic in and out of both ground and fourth level.

For an example and to show a method, a detailed analysis will be made of the third condition, which appears to be the worst one.

In	Floor	Out	In	Floor	Out
	7	1a	1a	7	
	6	1a	1a	6	
	5	2a	2a	5	
4a	4	4a	4a	4	4a
2a	3	2a	2a	3	2a
2a	2	2a	2a	2	2a
4a	G			G	4a
	Up trip			Down trip	

a = groups of 2 people traveling to or leaving one automobile

1. Elevator on the up trip makes 5 intermediate stops and 2 lobby stops.
2. Elevator on the down trip makes 4 intermediate stops and 1 lobby stop at the fourth floor.

3. Assume each lobby stop is for 16 people in and 16 people out = 28 sec.
4. Assume 200-fpm elevator speed, 48-in. center-opening doors.

Time to run up, per stop $\frac{60}{6}$ = 10 ft rise per stop

10 ft = 6.1 sec

Time to run down, $\frac{60}{6}$ = 10 ft = 6.1 sec

Elevator performance calculations:
Standing time

Lobby time 3 × 28	=	84.0 sec
Transfer time, up stops, assume 4 × 5	=	20.0
Door time, up stops (6 + 1) 5.3	=	37.1
Transfer time, down stops 4 × 5	=	20.0
Door time, down stops 5 × 5.3	=	26.5
Total standing time		187.6 sec
Inefficiency, 20%	=	37.5
	Total	225.1 sec

Running time

Run up 6 × 6.1	=	36.6
Run down 6 × 6.1	=	36.6
Total round-trip time		298.3 sec

$$HC = \frac{(16 \text{ up } + \text{ 16 down) } 300}{298.3} = 32 \text{ people}, \frac{125}{32} = 4 \text{ elevators}$$

Interval: 298.3/4 = 74 sec, longer than 60-sec criteria

The interval of 74 sec for four elevators is long but may not be too long since the one possible traffic pattern investigated was the worst condition. Further investigation should be done in order to judge what the average interval will be between the best and worst. Additional consideration in the final judgment should be given to the fact that 100% usage of the garage will be a peak situation and the 1-hr turnover considered will be highly unlikely with the other traffic congestion that will probably be experienced.

It is recommended, and many local codes require, that the elevator lobby in underground garages be totally enclosed to avoid the possibility of gasoline fumes, which are heavier than air, accumulating in the elevator pit. It is also a good safety practice to isolate the elevator lobby from the traffic by a suitable enclosure on any garage floor. Moreover, it is good commercial practice to provide a warm lobby for waiting patrons and to avoid cold-weather elevator operating problems. In some areas, raising the entrance of the elevator above the parking floor level is sufficient to meet the code.

PROFESSIONAL BUILDINGS

Professional buildings are specialized office buildings generally devoted to medical and dental use. Their success depends on services offered and accessibility to a large population center. It is not unusual to find 200 or more doctors, dentists, optometrists, x-ray labs, and other services concentrated in one downtown building.

Traffic in such buildings consists of patients visiting these offices, and its volume is a function of the normal turnover of visitors. The average doctor may take care of about three to four visitors per hour, each visitor usually coming with a companion. Based on the number of professional offices, the elevator traffic will amount to about eight people in and eight people out per office per hour.

For example, if a building has 200 offices, the critical elevator traffic would be about 16 people times 200 offices divided by 12 5-min periods, or about 267 people in and out of the building in a 5-min period. Peak elevator traffic usually occurs between 2:00 and 4:00 in the afternoon (see Figure 10.4).

If the number of offices is not known at the time the building is being planned, an estimate on the basis of 300 ft² (30 m²) of net area per office can be made. A further estimate of two people per office and a 5-min elevator peak traffic of 20% of this population, two-way traffic, leads to an approximate solution of the elevatoring problem.

The suggested speed, capacity, and interval may be found in Tables 10.5, 10.6, and 10.7. In addition, consideration should be given to the possible need in some professional buildings for stretcher service on the elevator. A hospital-shaped service elevator or special provisions such as a wider passenger-shaped platform with wide doors should be considered. The 3500-lb (1600-kg) passenger cars with 48-in. (1200-mm) center-opening doors can accommodate a standard 76 × 22-in. (1930 × 600-mm) mobile stretcher, or any size wheelchair, as well as a number of passengers. For the smaller medical building, a 2500-lb (1200-kg) car, specially arranged with 42-in. (1100-mm) side-sliding doors can also carry mobile stretchers.

MERCHANDISE MARTS

The success of merchandise marts depends in part on the speed with which temporary exhibits of wares can be set up as well as the displays of tenants who are resident. A mart caters to people interested only in a restricted line of merchandise. Shows will run from one to three days, the usual buyer trying to visit each display in the building the first day and concentrating on a limited number on succeeding days.

A common pattern for buyers is to start at the top and work their way down, using the elevators, stairs, or escalators if available.

Escalators or elevators specified on approximately the same basis as those used in a diversified office building of the same area will provide approximately the vertical transportation needed for passengers—4000-lb (1800-kg) elevators are recommended. Ample, large freight elevators are necessary to accommodate the merchandise that must be moved for temporary or permanent display. The number and size of the freight elevators will depend on the requirements of the building and the time allowed to set up for the show. Combination elevators that can be used for merchandise movement and subsequent pedestrian movement may be considered but are usually not practical.

INDUSTRIAL BUILDINGS

Elevatoring of industrial buildings depends on their general function, the specific type of work to be performed, and the personnel practices of the occupying firm. There are no general rules, and the extent of each specific traffic problem must be determined and treated accordingly.

In larger plants escalators have proved advantageous in that they can accommodate large numbers of people in a short time and are ideal for shift changes. With a single-shift operation, they can be reversed to accommodate the traffic flow. Elevators have the advantage of being able to double as freight elevators during working hours and to serve the employees at incoming, luncheon, and quitting times. Employee demand for service is usually established by strict working rules, and because everyone usually starts or finishes at the same time, the peak traffic is exceedingly heavy.

SUMMARY

It is obviously impossible to cover all the vertical transportation situations that may arise in the elevatoring of commercial and industrial buildings. Particular attention must be paid to avoiding situations that will reduce the efficiency of the elevator plant, such as two entrances to the building, odd floors served, unnecessary special operations, too many priority services, and upper-floor cafeterias. The best elevator or escalator installation is usually the simplest and most direct, with one clear, unobstructed path for everyone.

11 Elevatoring Residential Buildings

Residential buildings are buildings where a number of people live either permanently or temporarily. They include hotels, motels, apartments, senior citizen housing, dormitories, and other residence halls. Elevator traffic in such buildings is generally not as intense as in commercial or institutional buildings, and a greater tolerance of waiting is found. Availability and capacity, as well as a prompt trip and furniture-moving capability, are the more important criteria.

POPULATION

Each occupant of a residential building is allotted a certain minimum space to live and sleep, as little as 100 ft^2 (10 m^2) in some low-cost housing projects and considerably more in luxury apartments. The average, when the layout and room utilization of a residential floor is unknown, is about 200 ft^2 (20 m^2) of net area per person. Design and use of a building will alter this average. For example, in hotels and motels the average occupancy per room is relevant. In apartments, because room count is often distorted by assigning half- and quarter-room values to such areas as foyers and closets, the number of bedrooms and the average occupancy per bedroom are the criteria. In dormitories, which may start as large single suites and change to two- or three-person bedrooms, the 200 ft^2 (20 m^2) per person average is best to use.

In most apartments and residences people have a more relaxed attitude toward vertical transportation than in commercial buildings and will tolerate a longer average wait for service and a longer trip time.

In hotels and motels, because people are paying for service, they expect more prompt and efficient vertical transportation. There is a paradox, however; during a convention or gathering when crowds are coming and going, the average patron is tolerant of delays and accepts them in the carnival atmosphere that prevails. Waiting for a minute or two may be acceptable, but beyond that irritation may become apparent.

CHARACTERISTIC TRAFFIC AND INTENSITY

Vertical traffic in residential buildings is predominantly two-way. In downtown apartment buildings and hotels substantial down peaks occur at the start of the working day, and up peaks as business people arrive home. In in-town and suburban apartments the peak traffic occurs in the early evening when people are leaving for evening activities and others are returning after shopping or other activities. In a hotel the late afternoon is often

The Vertical Transportation Handbook, Third Edition, Edited by George R. Strakosch
ISBN 0-471-126291-4 © 1998 John Wiley & Sons, Inc.

marked by a check-in peak, meetings breaking up, and people returning to rooms as others leave to seek refreshment, and in the morning, by breakfast and checkout.

In dormitories and residential halls the two-way peak may occur in the evening when residents are going to and returning from dinner. This peak is also influenced by whether the dining facilities are cafeterias or dining rooms. In senior citizen housing recreational periods may cause the greatest elevator activity.

The percentage of a building's population that the elevators must serve during the critical 5-min traffic periods varies with the facility. For a hotel that hosts frequent conventions, elevators must be able to serve 12 to 15% of the population during a 5-min peak period. In a luxury apartment where the number of children is expected to be low, the percentage of the population served during a 5-min peak period may be as low as 5%. Required capacity for other building types varies between those extremes. Information about recommended handling capacities, intervals, and population criteria appears in Table 11.1.

CALCULATING ELEVATOR REQUIREMENTS

Elevatoring a residential building can proceed as with any other type of building. The population is determined, and the number, speed, and capacity of elevators are assumed and verified for handling capacity and interval. Double entrances to the building, garage stops above or below the main entrance floor, and other services or facilities such as rooftop swimming pools, restaurants, and lounges all require additional elevator service. The extent, use, and capacity of such facilities must be determined and the impact on the elevator situation calculated.

Service requirements may be quite stringent in residential buildings. In hotels and motels there is always someone moving in or out, chambermaids at work, and, in the higher-class establishments, considerable room service for refreshments and food.

Luxury apartments hardly deserve that appellation without a separate service elevator. In a large apartment building the frequency of moves in and out may require use of a service elevator 4 to 6 hr daily just for moving.

A nominal 10% inefficiency factor in standing time was applied to office building elevatoring calculation when 48-in. (1200-mm) center-opening doors were used to reflect the expected promptness and attention of people in a business atmosphere. In residential buildings this inefficiency will be greater, depending on the nature of the tenancy. Suggested percentages are shown in the examples presented in this chapter and in Table 11.1, and are recommended for use in calculations.

Each major type of residential structure is reviewed here to show the impact of expected activity on the vertical transportation.

ELEVATOR EQUIPMENT AND LAYOUT

Rules of elevator location and grouping introduced in Chapter 2 apply to residential as well as other buildings. Elevators should be a center of attraction in the lobby and readily accessible on each floor. In hotels and dormitories, where many people come and go at the same time, ample elevator lobby space must be provided at the entrance and on other floors where people are expected to gather.

Service elevators should be located in separate alcoves, with ample lobby space at each floor to turn any carts to be carried as well as to accommodate waiting or stored ve-

TABLE 11.1. Residential Buildings

Type of Building	Population Criteria	Recommended 5-min Capacity (%)	Interval Range (sec)	Inefficiency Factor (%)
Hotel	1.5 to 1.9 people per room	12 to 15	40 to 60	10
Motel	1.5 to 1.9 people per room	10 to 12	40 to 60	10
Apartments				
Downtown	1.5 to 1.75 people per bedroom	5 to 7	50 to 70	15
Development	1.75 to 2 people per bedroom	6 to 7	50 to 90	20
Dormitories	200 ft² net per person	15	50 to 70	15
Residence halls	Same as dormitories			
Senior citizen housing	1.25 to 1.5 people per bedroom	6	50 to 90	25

Suggested Elevator Size

Type of Building	Passenger Elevators Size; Door Type and Size	Service Elevators Size; Door Type and Size
Hotel	3500 lb 48-in. center-opening	4000 lb 48-in. center-opening
Motel	2500 to 3000 lb 42-in. center-opening	3500 lb 48-in. center-opening
Apartments	2500 lb 36-in. single-slide	2500 lb 42-in. two-speed
Dormitories	3000 lb 42-in. center-opening	Use passenger elevators at off-peak times
Residence halls		
Senior citizen housing	2500 lb 42-in. two-speed	Suggest 4000 lb hospital type

Suggested Elevator Speed (fpm)

Building Height	Hotels-Motels	Apartments and Senior Citizen Housing	Dormitories and Residence Halls	
2 to 6 floors	150*	150*	150*	
6 to 12 floors	300	200	200	
12 to 20 floors	400 to 500	400	400	
20 to 25 floors	500	500	500	For buildings of this
25 to 30 floors	700	500	700	height, local and
30 to 40 floors	700 to 1000			express elevators
40 to 50 floors	1000 to 1200			should be considered.

*Use traction or hydraulic elevators.

hicles. This is especially important in hotels, where room-service tables and carts with their dirty dishes may be stored until picked up by service personnel.

Elevator hoistways should be isolated from sleeping rooms by lobbies, mechanical shaft space, or stairwells. Although elevators are relatively quiet, air noises of an elevator traveling through a shaftway are noticeable when other building noise is low. In addition, such mechanical noises as the opening of doors on a floor or the passing of the counter-weight within an inch or two of a wall are unavoidable. If a sleeping person's head is on the other side of a wall within 12 in. (300 mm) of a passing car or counterweight [assuming an 8-in. (200-mm) wall], the person will hear the noise and complain. If sleeping rooms must be placed next to the elevator space, ample sound insulation should be provided in the form of dead air space or a mineral wool blanket inside the finished wall, and sound isolation between the elevator rail and its bracket to the building structure should be considered.

Similarly, sleeping spaces should not be located next to elevator machine rooms. Electric motors starting and stopping and relays operating can be objectionable, especially in the middle of the night.

HOTELS AND MOTELS

From an operational point of view the distinction between motels and hotels is not always sharp. From the vertical transportation aspect a motel is defined as a lodging in which room service demands are minimal, and a hotel as one in which considerable service is expected. The major difference, then, is in the number of service elevators required.

A further distinction is necessary. If the establishment has considerable convention facilities, large dining or meeting rooms, or a ballroom that is not located on the ground floor, it is more a hotel than a motel.

Parking

Today, hotels and motels are expected to have ample parking facilities within or near the structure. Even if there are in-building facilities, they should be separated from guest rooms in the interest of security. This may require attendant parking with checking in and checking out at the entrance to the garage. With self-parking one of the best security arrangements is a separate shuttle elevator for the garage area. Not only can the desk clerk see who is coming and going, garage floor stops for the main elevators are eliminated, which improves their efficiency and may minimize the number required.

In spite of impaired security and the inefficiency created by garage stops, many hotel and motel operators insist that the main elevators serve both the garage and guest floors. If this is a building design criterion, a constant lobby stop plus additional stops both up and down in the garage area must be considered in elevator calculations.

Meeting Rooms

Other critical areas in elevatoring a hotel or motel are the meeting rooms or ballroom floors. When meetings are starting or breaking up, these floors will be constant elevator stops, so adequate allowance in elevator trip time must be made. The best arrangement is to locate the meeting floors where escalators can connect them to the street or lobby floor. (It is presumed that the lobby floor is at street level or connected with it by escalators.) Separate escalator service to the meeting room floor will permit public use of the facility with minimum interference with the hotel guests. When large meetings are breaking up, the elevators may be overwhelmed by many people wishing to get back to their rooms.

Kitchens

Kitchen facilities should be at the lower levels. A kitchen service elevator should connect the kitchen to the loading dock as well as to the restaurant level or ballroom level. Dumbwaiters can be used to connect the kitchen to the various food shops usually located at the main level. The kitchen service elevator should be a service type with wide, horizontally sliding doors that are operated automatically.

SERVICE ELEVATORS

Elevators for room service should be large enough to handle carts or portable tables. 4000-lb (1800-kg) cars with 48-in. (1200-mm)-wide center-opening doors are recommended. The number of service elevators depends on many factors, and experience has shown that a minimum of one service elevator for each 200 to 300 rooms in a hotel is necessary. This minimum number requires that schedules, deliveries, and movement of linen and other supplies to each floor be restricted to other than peak dining hours and that special functions be held to a minimum. If a number of special facilities such as rooftop restaurants or lounges must be serviced, or if there are a considerable number of hospitality rooms, additional service elevators are necessary.

A complete study of service requirements can be made by determining the average time per delivery and relating that to the number of deliveries in a given period of time and the average time required per elevator. Such factors as the number of room-service meals the kitchen can prepare in a given time, the number of service employees who must be transported, their shift changes, or local labor requirements, and frequency of special parties must be considered.

During the 1920s, the era of luxury hotel room service when many larger hotels were built, the rule of thumb for service elevators was one for each passenger elevator. Today, with swifter intercity transportation and a greater turnover in hotel guests plus considerably less room service, an approximate ratio of service elevators to passenger elevators should be 50 to 60%.

SAMPLE HOTEL ELEVATORING

Sample calculations for elevatoring a typical hotel or motel are shown in Example 11.1, using factors from Table 11.1. Note the impact of the garage, although minimum use was assumed. Also note that a restaurant on the top floor of a hotel or any other building requires considerable additional service. This impact cannot be minimized, because the restaurant will be used concurrently with other activities and by visitors as well as guests of the hotel. The traffic restaurant patrons create will often be opposed to other traffic in the hotel.

Example 11.1. Hotel-Motel

Given: 25-story + 3-basement hotel; first and second floors 20 ft, typical floors 10 ft; self-parking for 300 cars in basements; first floor, lobby and restaurants (kitchen in basement); second floor, meeting rooms and 1200-person ballroom; floors 3 to 25 guest rooms, 17 rooms per floor; located near college football stadium; will be used for conventions, may have 300-person restaurant on twenty-sixth floor.

Required: Number of passenger elevators and service elevators—recommended sizes

Calculations: 23 floors of rooms $\times$ 17 $\times$ 1.9 (convention occupancy) = 391 $\times$ 1.9 = 743.

743 $\times$ 12.5% = 93 people, 5 min, two-way peak demand

Assume: 3500-lb passenger elevators @ 500-fpm, 48-in. center-opening doors

Assume: 10 people up, 10 people down, average travel to twentieth floor (20% highest call and return)

Probable stops: 19 upper floors = 7.9 up $\times$ 0.7 = 5.5 down

$$\text{Time to run up, per stop: } \frac{(17 \times 10) + (2 \times 20)}{7.9} = \frac{210}{7.9}$$

$$= 26.6\text{-ft rise per stop}$$

$$26.6 \text{ ft} = 6.1 \text{ sec}$$

$$\text{Time to run down, per stop: } \frac{210}{5.5} = 38.2 \text{ ft} = 7.4 \text{ sec}$$

Elevator performance calculations:

Standing time

Lobby time 10 in + 10 out	=	18.0 sec
Transfer time, up stops 7.9 $\times$ 3	=	23.7
Door time, up stops (7.9 + 1) $\times$ 5.3	=	47.2
Transfer time, down stops 5.5 $\times$ 4	=	22.0
Door time, down stops 5.5 $\times$ 5.3	=	29.2
Total standing time		140.7 sec
Inefficiency, 10%	=	14.1
	Total	154.2 sec

Running time

Run up 6.1 $\times$ 7.9	=	48.2
Run down 5.5 $\times$ 7.4	=	40.7
Total round-trip time		243.1 sec

$$HC = \frac{(10 + 10) \times 300}{243.1} = 24.7 \text{ people}, \frac{93}{24.7} = 3.8 = 4 \text{ elevators}$$

Interval: 243.1/4=60.8 sec

The four elevators will provide the necessary handling capacity, but the interval is beyond the upper limit of 60 sec that is recommended as a maximum for good elevator service. Adding a garage stop will further increase the interval as follows:

Assume: 1 garage stop per trip

Travel to the garage below lobby floor 10 ft

10 ft	=	4.3 sec
Passenger transfer at garage	=	4.0 sec
Door operation at garage and at lobby (2 $\times$ 5.3)	=	10.6
Travel to and from garage (2 $\times$ 4.3)	=	8.6
Round-trip time without garage	=	243.1
New round-trip time		270.6 sec

New interval: 270.6/4 = 67.7 sec

The interval will exceed the criteria and become unacceptable. Therefore, separate elevator service to the garage must be considered.

Similarly, adding a large, 300-patron restaurant on the top floor will have a pronounced effect on the elevators. A new calculation must be developed since the original study included a substantial reduction in stops and travel for expected travel to the highest average floor call and return. With the upper-floor restaurant, frequent trips to the top landing will be experienced and is shown as follows:

Restaurant: 300 seats, assume turnover each 1.5 hr

$$\frac{300 \text{ in} + 300 \text{ out}}{90 \text{ min}} \times 5 \text{ min} \qquad\qquad = \qquad 33 \text{ people}$$

Hotel demand, per 5 min	93 people
Total demand	126 people

33 is about $\frac{1}{3}$ of 93, so each third elevator trip will be to the top. Therefore, highest call is reduced by one-third.

Travel 270 ft total, 210 ft used, $(60 \times \frac{1}{3}) = 20$, now 230 ft.

Upper floors 25 total, 19 used $(5 \times \frac{1}{2}) = 2$, now 21

Recalculate:

Assume: 10 people up, 10 people down

Probable stops: 21 upper floors $= 8.1$ up $\times 0.7 = 5.7$ down

Time to run up, per stop: $\dfrac{230}{8.1} = 28.4$-ft rise per stop

$$28.4 \text{ ft} = 6.2 \text{ sec}$$

Time to run down: $\dfrac{230}{5.7} = 40.4$ ft $= 7.6$ sec

Elevator performance calculations:

Standing time		
Lobby time 10 in + 10 out	=	18.0 sec
Transfer time, up stops 8.1×3	=	24.3
Door time, up stops $(8.1 + 1) \times 5.3$	=	48.2
Transfer time, down stops 5.7×4	=	22.8
Door time, down stops 5.7×5.3	=	30.2
Total standing time		143.5 sec
Inefficiency, 10%	=	14.4
	Total	157.9 sec
Running time		
Run up 8.1×6.2	=	50.2
Run down 5.7×7.6	=	43.3
Total round-trip time		251.4 sec

$$\text{HC} = \frac{(10 + 10) \times 300}{251.4} = 23.9 \text{ people}, \frac{126}{23.9} = 5.2 = 5 \text{ elevators}$$

Interval: $251.4/5 = 50.3$ sec

The addition of the top-floor restaurant requires a fifth elevator, which has the further benefit of improving the interval to 50 sec, which is acceptable. With the fifth elevator, a garage stop could also be served.

Additional considerations will include the necessary service to the ballroom and kitchen, and the final recommendations should include the following, assuming a top-floor restaurant:

Passenger elevators: Five 3500-lb (1600-kg) @ 500 fpm (2.5 mps) serving B, 1 to 26 with 48-in. (1200-mm) center-opening doors

Service elevators: Three 4000-lb (1800-kg) @ 400 fpm (2 mps) serving B, 1 to 26 with 48-in. (1200-mm) center-opening doors

Kitchen service: One 4000-lb (1800-kg) @ 125 fpm (0.6 mps) serving floors B, 1 and 2
Ballroom: Two 48-in. (1200-mm) wide escalators serving floors 1 to 2
For disabled persons: One 3500 lb (1600-kg) @ 150 fpm (0.75 mps) passenger elevator serving floors 1 to 2, adjacent to escalators
Ballroom freight (for displays): One 12,000 lb (5500 kg) @ 50 fpm (0.25 mps) freight elevator with 10-ft (3-m)-wide by 22-ft (6.7-m)-long platform (for automobile displays), 10-ft (3-m)-wide by 10-ft (3-m)-high vertical biparting doors stopping at the ballroom and a loading dock

The ballroom freight elevator recommendation is based on the fact that the 1200-person ballroom will accommodate many more people than the hotel and is expected to be used for conventions and exhibits that will attract many outsiders and may include large exhibits.

APARTMENTS

During the 1970s a minor revolution took place in the design of low-rise apartment houses—the inclusion of an elevator in apartments of three or more floors. This was made possible by the intensive development of the hydraulic elevator, which allows low-cost installation for a limited rise of up to 70 ft (21 m). Legislation for disabled persons has required developers to include elevators in any project. The need for intensive land development has increased the height of higher-rise downtown apartments to 30 or more floors, which encourages the use of local and express elevators in such buildings. In some areas, combination buildings, featuring stores on the lower three or four floors, office space above, and apartments above the offices, connected to the street with a sky lobby arrangement, are being developed. Sky lobbies will be discussed in a later chapter.

Many apartment buildings are now sold as condominiums or cooperatives representing the ownership of the building as compared with a landlord-tenant relationship. The clientele of the higher-priced condominiums will be the same as that of the high-rental apartments, and the lower-priced cooperatives will be similar to lower rental and development apartments. (This section will use the term "rental" for cooperatives and condominiums.)

The demands for vertical transportation in apartments follow the normal day of the building's residents. Outgoing traffic is heavy in the morning as people leave for work, with corresponding incoming traffic in the early evening as they return. These are the average peak traffic periods of apartments that house primarily business people.

In apartments and housing projects with a predominantly family occupancy, the needs of children and homemakers influence the traffic pattern. Traffic reaches a forenoon peak as shopping expeditions take place and a distinct afternoon peak as the children return from school and go out to play.

In both downtown and family-type apartment buildings the critical peak traffic that determines elevator capacity occurs in the late afternoon and early evening when people are returning for meals and others are going out for evening recreation. Studies have shown that this traffic is two-way in nature and amounts to about 5 to 7% of the building's population in a peak 5-min period.

Population

The population of an apartment building is based on the number of bedrooms provided in the building, counting so-called efficiency (one-room) apartments as one bedroom. Occupancy per bedroom varies with the type of apartment building and its rental range. The building housing business people will be almost as densely occupied as the family type. Occupancy of the former may average 1.75 people per bedroom because of the considerable apartment-sharing by working people. With low-rental apartments occupied by families with many children, an average of people per bedroom should be used. When the rents are high and rooms are spacious, an average occupancy estimate of 1.5 per bedroom is acceptable.

Elevator Capacity

The minimum recommended size of an elevator for any apartment building is the 2000-lb (900-kg) car with 68-in. wide by 51-in. deep (1750×1300-mm) inside dimensions, which is also the minimum size required to accommodate a wheelchair. Minimum doors are 36-in. (900-mm) single-slide side opening (Figure 11.1).

In some localities, building codes require that at least one of the elevators in a residential building be large enough to accommodate a mobile stretcher. A 2500-lb (1200-kg) elevator with a car having 82-in. wide by 51-in. deep (2100×1300-mm) inside dimensions with 42-in. (1100-mm) one- or two-speed side sliding doors will accommodate a 76 by 22-in. (1950×560-mm) mobile stretcher. The 2500-lb (1200-kg) elevator can also expedite moving operations, which may be quite frequent in some areas. If only one elevator is required for the building, it should be 2500 lb (1200 kg). If more than one is required, all should be the 2500 lb (1200-kg) elevator described above.

Elevators must be able to carry at least 5% of the population for the high-rent apartment buildings and 6 to 7% for the moderate- and low-rent apartments. The more economical the rent, the more children are likely and the higher the traffic peaks that will occur.

Service elevators are a necessity for a high-rent apartment building and essential for any apartment building with 500 or more apartments. Recent statistics show that one out of every 10 families moves once a year. Assuming that a move out and a move in will tie up one elevator most of a day, 500 families in a building will tie up an elevator one day per week just for moving—in addition to normal deliveries of furniture, rugs, groceries, cleaning, and various services as well as normal building maintenance. If building management imposes any restriction that all deliveries must be made on service elevators, one service elevator for every 300 units should be provided.

A service elevator in any apartment house may pay for itself in improved rental income. This may begin with the opening of the building. If it has only two elevators and the owner is trying to rent before the building is completed, the following situation frequently occurs. To eliminate an outside hoist, the building contractor will take an elevator from the elevator contractor to use temporarily, once it is usable and before it is finally completed, for use by construction workers and the rental agent to show apartments. If an apartment is rented, the tenant usually wants to move in as quickly as possible and the owner is more than willing to accept the rent. Because the completion of the building and moving in cannot take place during the same hours, the owner is faced with overtime moving at premium rates. This can persist for some time, because when the building contractor relinquishes the elevator he has been using, it will require considerable finishing to make it ready for tenant use. In addition, if both elevators are to operate as a group, their controls must be tied together by the elevator contractor, who must work on both elevators at night, at the owner's expense, to complete the installation. A third car, or a service

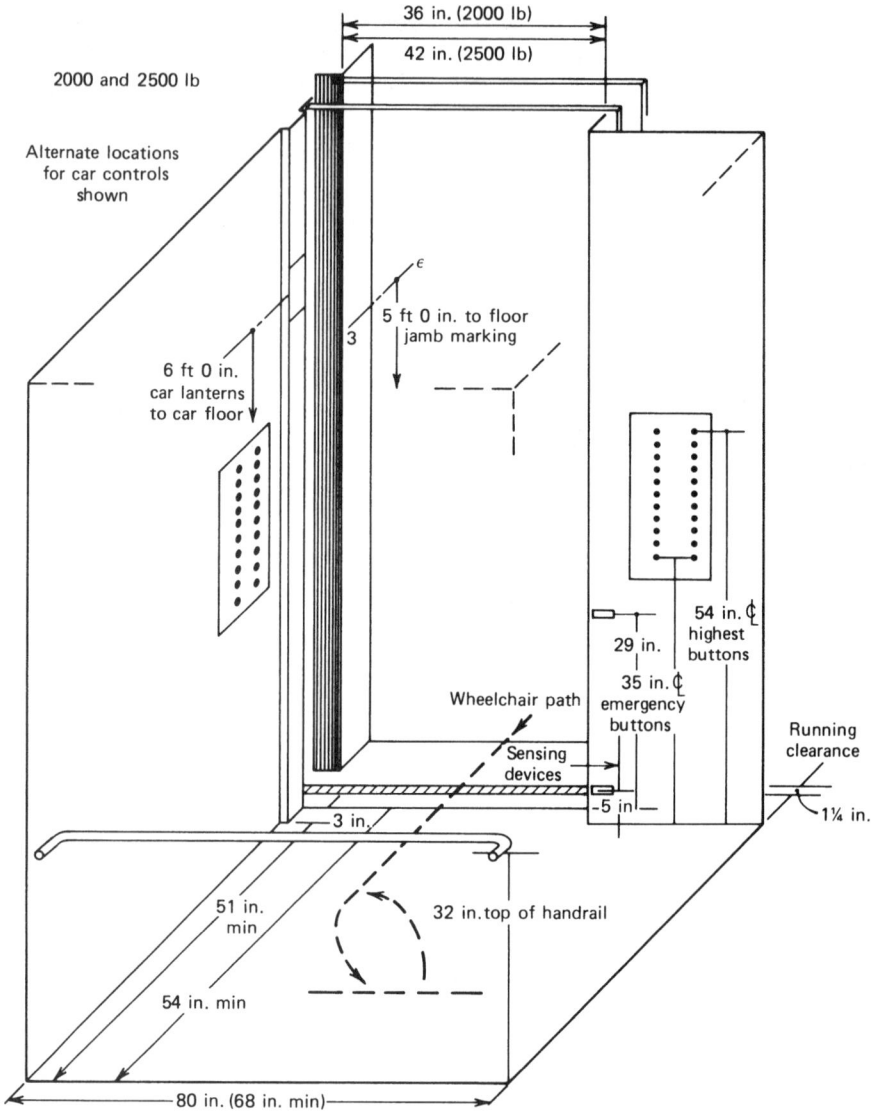

Figure 11.1. Handicapped elevator requirements (Reprinted with permission from the former *NEII Handicapped Guide*. Copyrighted © 1992 and 1994, National Elevator Industry, Inc., Fort Lee, New Jersey. This reprinted material is not the complete and official position of National Elevator Industry, Inc. on the referenced subject which is represented only by the standard in its entirety).

elevator, would avoid overtime costs for both moving and elevator installation, which have often amounted to a substantial part of the cost of the extra elevator! The tenants enjoy the convenience of a service elevator, and rental return per apartment is likely to be higher.

Tenant Garages: Public Use and Penthouse Floors

All floors in an apartment building should be served by all main elevators. Lower-floor garages in apartment houses exact the same elevator trip time penalty as in any other type of building. If an apartment has such garage floors, they should be served by all the elevators. If the tenants must wait for, say, only one of two cars to get to and from the garage, they are inconvenienced whenever they use such facilities. A separate shuttle elevator should be provided if security is expected to be a problem.

The same consideration of having all elevators serve any public-use floors or the use of a shuttle elevator will also apply to basement laundries, storage areas, recreation areas such as rooftop swimming pools, and terrace floors. It is most important that penthouse floors be served by all the elevators of a group; if not, the highest-rental floors in the building will have substandard elevator service. Alternatively, a shuttle elevator from the top common floor to the penthouse can be provided.

SAMPLE CALCULATIONS

The same calculation procedure employed for other buildings is used for apartment houses. Since the 36-in. (900-mm) single-slide door is frequently used, a standing-time inefficiency factor of 10% plus the suggested inefficiency of 10 to 15% for apartment houses should be used. Frequent elevator reversal below the top floor is expected and should be estimated at the floor that is the top of 75% of the expected population.

An apartment house elevator is seldom filled to capacity. Unlike the situation in office buildings where elevator passengers seldom have parcels, someone always has something such as a baby stroller, food parcels, luggage, or laundry in an apartment elevator.

The 2000-lb (900-kg) elevator has a nominal full capacity of 8 people and the 2500-lb (1200-kg) elevator a nominal full capacity of 10 people. Elevator calculations for apartment houses are based on two-way traffic. The normal probable stop values can be used, or probable stops estimated at one stop per person carried to develop a short method of calculation, which will be demonstrated.

Example 11.2. Apartments

A. Given: 20-story apartment building, moderate rental, 12 bedrooms per floor, 9-ft floor heights

Demand: 12×1.75 (people per bedroom) = 21 per floor $21 \times 19 = 399$ total $\times 6\%$ = 24 people, 5-min demand

1. Probable stops: assume 8 people per trip; 4 up, 4 down; high call floor $400 \times 75\% = 300 \div 21 = 14$ floors above, floors 1 to 15, 14 floors, 4 people = 3.6 stops up $\times 0.7 = 2.5$ down

Time to run up, per stop: $\dfrac{14 \times 9}{3.6} = 35$-ft rise per stop

Assume 300 fpm, 35 ft = 10.4 sec

Time to run down: $\dfrac{14 \times 9}{2.5} = 50$ ft, 50 ft = 13.4 sec

Elevator performance calculations:

Standing time

Lobby time 4 in + 4 out	=	8.0 sec
Transfer time, up stops 3.6 × 3	=	10.8
Door time, up stops (3.6 + 1) × 6.6	=	23.8
Transfer time, down stops 2.5 × 4	=	10.0
Door time, down stops 2.5 × 6.6	=	16.5
Total standing time		69.1 sec
Inefficiency, 20%	=	13.8
	Total	82.9 sec

Running time

Run up 3.6 × 10.4	=	37.4
Run down 2.5 × 13.4	=	33.5
Total round-trip time		153.8 sec

$$HC = \frac{8 \times 300}{153.8} = 16 \times 2 = 32 \text{ vs. 24 people, demand}$$

Interval: 153.8/2 = 76.9 sec

2. Short method of calculating apartment house elevators
   1. Assume 8 people per round trip = 8 stops
   2. Assign a per-stop value, for example: 36-in. (900-mm) single-slide doors

a. Door time	6.6 sec
b. Transfer time	3.0
c. Acceleration and deceleration	2.0
	11.6 sec
d. +20% inefficiency	2.3
e. Total (approx.)	14.0 sec per stop

3. Calculate. Per stop 8×14

Per stop 8×14	=	112 sec
Lobby time 8 people	=	8.0
Running time $\frac{(14 \times 9) \times 2 \times 60}{300}$	=	50.0
		170.0 sec

Round-trip time, total

$$HC = \frac{8 \times 300}{170} = 14 \times 2 = 28 \text{ people}$$

Interval: 170/2 = 85 sec—marginal

The results compare with the longer calculations. The interval at about 80 sec will be a source of tenant complaint since it will be a measure of waiting time for an elevator. Increasing the elevator speed would not provide any appreciable degree of improvement, and the only real solution is to add a third car. However, this is not economically feasible. The 3-sec minimum dwell (car stop transfer) time at each floor will seem long to the more energetic tenant as would be the minimum 4-sec landing call transfer time. These times are mandated by disabled requirements and must be considered. Our conclusion would be two 2000-lb (900-kg) elevators @ 300 fpm (1.5 mps), minimum, and two 2500-lb (1200-kg) elevators @ 350 fpm (1.65 mps) recommended.

B. Given: 30-story apartment building downtown, 18 bedrooms per floor, 9-ft floor heights
   Demand: 18 × 1.75 = 32 per floor, 29 × 32 = 928 total × 6% = 56 people, 5 min demand.

Probable stops: assume 10 people per trip, 5 up, 5 down; high call floor 928 × 75% = 696 ÷ 32 = 22, 22 floors, 5 people = 4 stops × 0.7 = 3 stops down, assume 500 fpm

Time to run up, per stop: $\dfrac{22 \times 9}{4}$ = 50-ft rise per stop   =   8.8 sec

Time to run down: $\dfrac{22 \times 9}{3}$ = 66 ft, 66 ft   =   10.7 sec

Elevator performance calculations:
Standing time

Lobby time 5 in + 5 out	=	10.0 sec
Transfer time, up stops 4 × 3	=	12.0
Door time, up stops (4 + 1) × 6.6	=	33.0
Transfer time, down stops 3 × 4	=	12.0
Door time, down stops 3 × 6.6	=	19.8
Total standing time		86.6 sec
Inefficiency, 20%	=	17.4
	Total	104.2 sec

Running time

Run up 4 × 8.8	=	35.2
Run down 3 × 10.7	=	32.1
Total round-trip time		171.5 sec

$HC = \dfrac{10 \times 300}{172} = 17$ people, $\dfrac{56}{17} = 3$ elevators required

Interval: 172/3 = 57 sec, good

C. Given: 30-story apartment building, low rental, 20 bedrooms per floor, 9-ft floor heights
Demand: 20 × 2 = 40 people per floor × 29 = 1160 × 7% = 81 people 5 min demand
Calculating the high call floor would give the twenty-third floor (22 above the lobby), the same as in Example 11.2B. Further calculation would show that a 10-passenger elevator would have the same round-trip time and handling capacity as in Example 11.2B. Therefore 81/17 = 4.8, 5 elevators would be required.
Further study as follows would indicate that five elevators is neither necessary nor cost-effective.
Probable stops: assume 16 people per trip, 8 up and 8 down
High call floor 1160 × 75% = 870 ÷ 40 = 22, 22 floors, 8 people, 6.8 stops up 0.7 = 4.8 down; assume 500 fpm, 48-in. center-opening doors

Time to run up, per stop: $\dfrac{22 \times 9}{6.8}$ = 29-ft rise per stop

29 ft   = 6.3 sec

Time to run down: $\dfrac{22 \times 9}{4.8}$ = 41 ft, 41 ft = 7.6 sec

Elevator performance calculations:
Standing time

Lobby time 8 in + 8 out	=	14.0 sec
Transfer time, up stops 6.8 × 3	=	20.4
Door time, up stops (6.8 + 1) × 5.3	=	41.3
Transfer time, down stops 4.8 × 4	=	19.2
Door time, down stops 4.8 × 5.3	=	25.4
Total standing time		120.3 sec
Inefficiency, 15%	=	18.0
	Total	138.3 sec

(Total from page 293, 138.3 sec)

Running time

Run up 6.8×6.3	=	42.8
Run down 4.8×7.6	=	36.5
Total round-trip time		217.6 sec

$$HC = \frac{16 \times 300}{218} = 22 \text{ people}, \frac{81}{22} = 3.7 = 4 \text{ elevators}$$

Interval: $218/4 = 55$ sec, good

Four larger 3500-lb elevators @ 500 fpm (1600 kg @ 2.5 mps) would be required to provide the necessary handling capacity and would provide an excellent interval of 55 sec. These larger elevators are, however, expensive, and a further study should be made for the most cost-effective solution as follows, using the short-form calculation.

Assume low-rise elevators serve floors 1 to 14.

$13 \times 18 \times 1.75 = 410 \times 75\%\ 308 \div 32 = 10$ floor high call

8 people per round trip

Stops 8×14	=	112
Lobby	=	8
Run $\dfrac{(10 \times 9) \times 2 \times 60}{200}$	=	54
Round-trip time	=	174

$$HC = \frac{8 \times 300}{174} = 14 \text{ people, use 2 elevators} = \frac{28}{410} = 6.8\%$$

Interval: $174/2 = 87$ sec

Assume high-rise elevators serve floors 1, 15 to 30.

$16 \times 18 \times 1.75 = 504 \times 75\% = 378 \div 32 = 12$, 26 floor highest call

8 people per round trip

Stops 8×14

Lobby	=	8
Run $\dfrac{(26 \times 9) \times 2 \times 60}{500}$	=	56
		176

$$HC = \frac{8 \times 300}{176} = 14 \text{ people, use 2 elevators} = \frac{28}{504} = 5.5\%$$

The four elevators, two 2500-lb @ 200 fpm (1200-kg @ 1.0 mps) low-rise serving floors 1 to 14 and two 2500-lb @ 500 fpm (1200-kg @ 1.5 mps) high-rise serving floors 1, 15 to 30 will provide an elevator solution to this building. A conservative approach would be to use the low-rise and high-rise arrangement plus a fifth, larger elevator to operate as a full-time service elevator serving all floors, and operate either with the low-rise or high-rise in the event one of the elevators is out of service for any reason.

Variations in apartment building sizes and heights are endless. In addition, certain minimum standards for elevators have been established by various lending and governmental agencies as well as those required by codes for disabled persons. The criteria for elevator service shown here generally comply with these regulations, but local requirements may differ and should be ascertained.

DORMITORIES AND RESIDENCE HALLS

Most of the aspects of apartment elevatoring already discussed also apply to residence halls and dormitories. The essential difference is that dormitory residents are on a more fixed schedule as far as working and mealtimes are concerned. They are usually required to attend classes at certain hours and often take meals in minimum time, which results in a more severe impact on vertical transportation.

Experience has shown that about 10% of a dormitory's population seek elevator service during a critical 5-min peak, which usually occurs before suppertime and varies with the type of dining facility. If a formal dining room is provided where students must be seated at a given hour, the peak will approximate 15% of the population, will be a down peak (assuming lower-floor dining), and will utilize the full capacity of the elevators. If there is a cafeteria, only a 10% peak is expected.

Elevators of 2500-lb (1200-kg) capacity with 48-in. (1200-mm) center-opening doors are recommended for dormitories, to provide ample size for students as well as for moving operations and servicing the building. Separate service elevators are seldom required, as their function can be performed by the passenger elevators during slow periods of the day.

Dormitory elevators should be as student- and vandal-proof as possible, as should many apartment elevators, especially in development-type housing. Car operating fixtures should be of substantial construction, extraneous switches should be omitted, and operating buttons should be made of solid metal. Landing buttons, too, should be of solid metal, for it is not unusual for students to use their feet to register calls. Such construction is designated as "vandal-proof."

The protective edge on the car doors should be of metal to foil attempts to carve it with a pocket knife. Indicators can be picked out with a sharp instrument and thus should be inaccessible. For anything that is fastened with a screw, the screws should have spanner or allen heads or be avoided entirely; young minds are challenged by what may be behind the cover.

Key-operated switches are a challenge and are best omitted. If a floor must be blocked off at certain times, it is best to provide a locked elevator lobby. A load-weighing switch that will sound a loud alarm if the car is filled to 125% of capacity is recommended.

Elevator requirements should be calculated on the same basis as for apartment houses. The elevators should be determined for the apartment house percentage of 6 to 7% two-way traffic and a 10 or 15% capacity at mealtimes, usually a down peak. Population should be based on approximately 200 ft^2 (20 m^2) of net area per student, as shown in Table 11.1.

Floors with special facilities such as lounges, recreational areas, laundries, and cafeterias require special consideration because of their expected impact on vertical traffic.

As discussed in Chapter 2, skip-stopping is seldom recommended in dormitories or apartments. There is almost no saving in number of elevators, and the saving in entrances can create moving and building service problems. With any emphasis on serving disabled people, skip-stop arrangements increase management and elevator assignment problems.

SENIOR CITIZEN HOUSING

Apartment dwellings serving the needs of elderly and semi-invalid people are receiving considerable emphasis. Each suite may have its own kitchen facilities as in a conventional apartment house, or community-type dining may be provided. The buildings are designed for people of advancing years who can take care of themselves or require only minor aid.

Because these people are totally dependent on elevators for vertical transportation, each building needs a multiplicity of units to ensure continuity of service. In addition, the elevators must be designed to wait for the user, whose movements and reaction time may be slow. Many passengers may have poor eyesight and use walking aids.

Because occupants are subject to illnesses and may require the use of stretchers for medical care, at least one elevator should be planned for this contingency.

Handling capacity needs will be minimum, no more than about 5 to 6% of the population in a 5-min period. Waiting will be no problem, provided the wait is comfortable; a convenient bench in the elevator lobby on each floor is suggested. Room occupancy is very light; many apartments will have only one person, which indicates an average of 1.25 to 1.5 people per bedroom, depending on the expected use of the facility.

The elevator cars should be wide and shallow and equipped with handrails. The 2500-lb (1200-kg) car, 82 in. wide by 51 in. deep (2100 by 1300 mm) inside is suggested, and 42-in. (1100-mm) two-speed doors should be used. Light-ray devices are required for door protection in addition to the safety edge to avoid closing before the entrance way is clear. An elevator of this size can easily handle wheelchairs and mobile stretchers. A separate service elevator, if provided, should be of hospital shape.

Additional time should be allowed at each elevator stop for the expected slow transfer of passengers, resulting in a standing time inefficiency of about 25%. A prominent signal at each floor to indicate whether the elevator is going up or down is required. Raised floor numbers on the doorjambs or sight guards are necessary in addition to the normal in-car position indicator. Most of these requirements are part of handicapped regulations.

Elevator calculations can proceed as for normal apartment houses, with the necessary extra time values. An elevator should never be considered for more than about six passengers per trip.

Housing for the elderly and nursing homes are similar, the essential difference being the increase in personal service and the operating staff requiring more elevator capacity. A nursing home should be considered as an institutional building and is discussed in Chapter 12.

ELEVATORS IN RESIDENTIAL BUILDINGS

Operation and Control

In any modern apartment house it has become essential that the elevator control system provide leveling. This is both compliance with good practice and legislation for disabled persons that makes leveling mandatory. All hydraulic elevators are equipped with leveling, and alternating current control elevators can be equipped with two-speed motors or VVVF Control to provide leveling. The hydraulic elevator is perfectly acceptable for rises of up to 50 ft (15 m) and speeds of 150 fpm (0.75 mps).

In buildings of six or more floors, elevators with generator field control or solid-state motor drives should be used to provide the higher speeds expected and necessary for that height. Even in a lower building, generator field control or solid-state motor drive for a traction elevator should be considered for a long-term economical elevator installation.

Many thousands of single-speed traction elevators were installed during the late 1940s and well into the 1950s. None of these had leveling capability, and many are still providing satisfactory service in the six-story apartment houses popular during that era. Recent developments in solid-state motor drives provide a means to upgrade those elevators, using most of the original equipment and adding the electronic devices and leveling provisions needed for the improvement (see VVVF Control discussed in Chapter 7). An upgrade can often be accomplished with a minimum of shut-down time and is relatively economical. It is a worthwhile modernization effort and can ensure the level stopping that has become a standard with modern equipment.

For apartments, dormitories, and housing for the elderly, normal operation during the day will include long periods of minimum elevator traffic. For this reason an on-call type of operation should be used. Additional features for heavier traffic periods should be provided if heavy elevator use is expected. For example, if three elevators are required to provide an adequate interval of operation in a tall building and the average passenger load per trip is expected to be no more than 4 or 6 people, long periods of intensive use are unlikely and a minimum operating system should suffice. On the other hand, if the elevators are expected to carry 10 or 12 passengers per trip and the expected interval will be near the maximum, special attention to incoming and outgoing peaks is recommended.

In hotels and motels whose occupancy must be 80% to provide a good investment return, intensive elevator traffic frequently occurs and more refinement in group operation is necessary, Programmable features such as incoming and outgoing traffic operations plus two-way traffic operations are required.

Special floors may require special operations. For example, if the ballroom floor is served by the main elevators and is on other than the main floor, it will be necessary to change the dispatching floor to the ballroom floor when meetings are breaking up even when escalators are provided to a main lobby.

Traffic flowcharts for the various types of buildings (Figure 11.2) are typical and will vary for particular buildings. Necessary group control can be determined in a manner similar to that described for commercial buildings.

For example, typical traffic in a hotel–motel follows an in-and-out pattern of movement as shown, depending on the activities of the guests during the day. If a convention is in progress, guests generally go to and from their rooms between meetings and business sessions and frequently after meals. In the late afternoon and evening there is considerable traffic between floors and to and from the lobby, consisting of guests visiting hospitality rooms or local friends and people leaving the building for evening entertainment and other activities.

If no convention is in progress and the motel or hotel is patronized by business people, substantial two-way peaks occur in the morning and outgoing peaks in the evening. In addition, there is a check-in peak in late afternoon, whereas checkout activity takes place as early as 7:00 A.M., with most facilities having an 11:00 A.M. or noon deadline.

Traffic curves shown for apartments and dormitories are also altered by the location and use of the facility. If the dormitory is convenient to the classrooms, it is frequently used by the students during the day. If it is remote, morning outgoing and late afternoon incoming traffic peaks may be similar to those in an in-town apartment occupied by business people. Each building has its unique traffic.

Traffic in housing for the elderly approximates that found in an apartment house but with greatly reduced intensity. This is partially offset by the relatively long time spent at each stop by the elevators designed to accommodate the elderly.

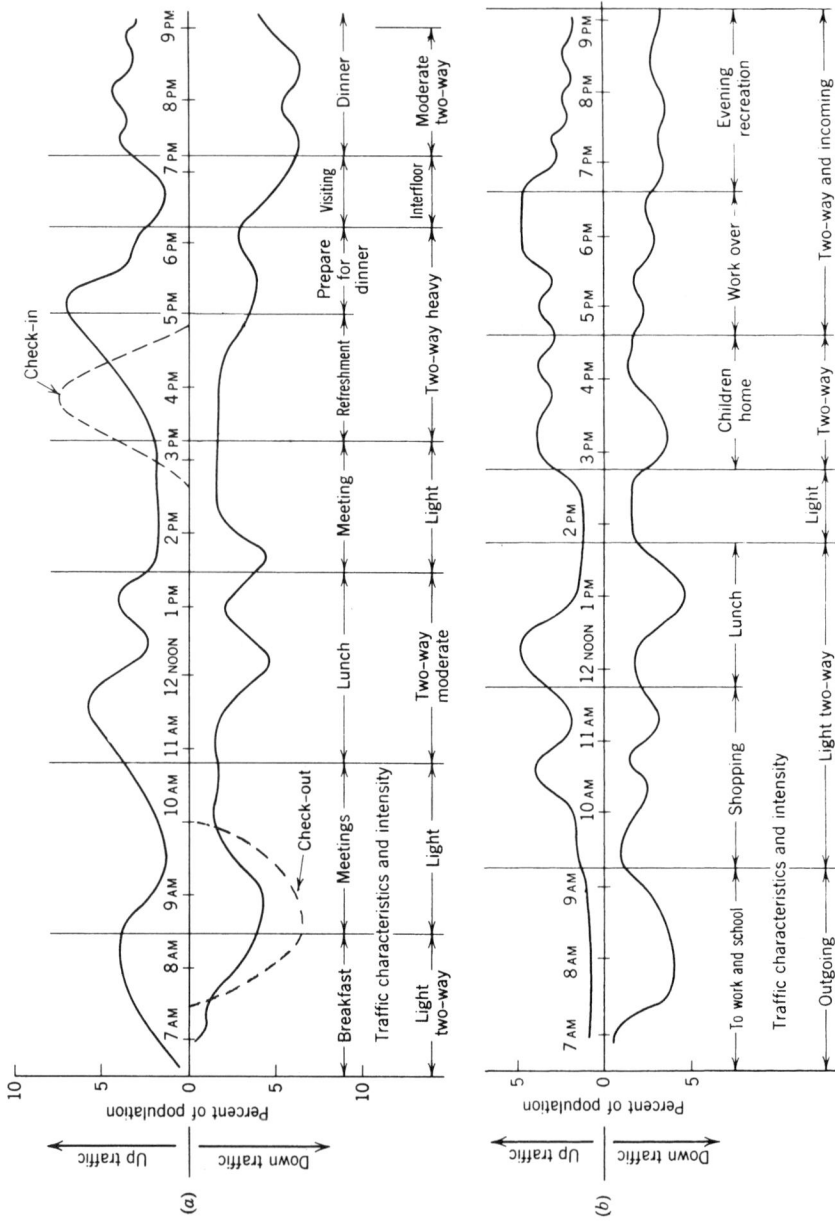

Figure 11.2. Traffic flowcharts—traffic in and out of elevators at a main lobby. (*a*) Motel—hotel convention type; (*b*) apartment building; (*c*) dormitory.

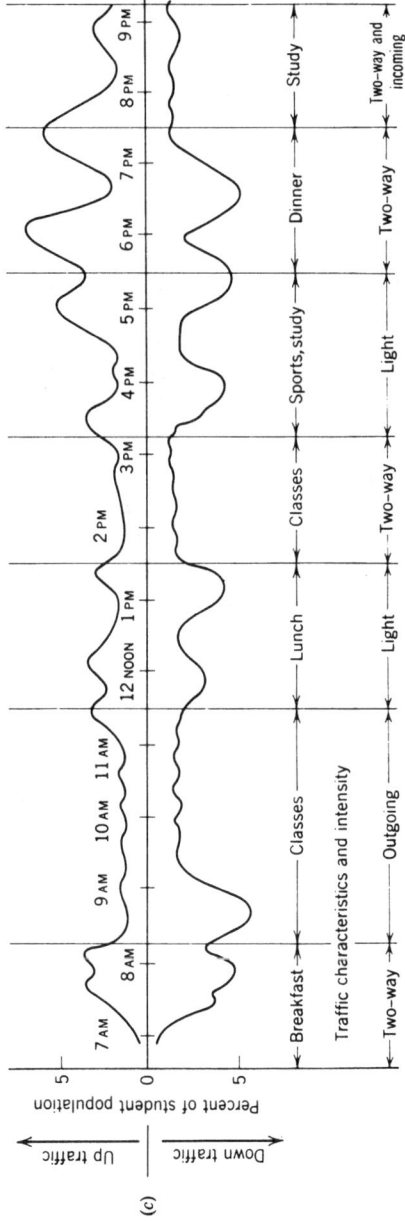

Figure 11.2. (Continued)

299

Features of Residential Elevators

Because elevators are used by many people not familiar with them, installations in motels and hotels should have the maximum features that make them easy to use. Such features include prominent landing buttons that illuminate when operated, readily seen directional indicators (landing lanterns) at each floor for each elevator, and well-lighted lobbies and elevator cars. Car operating panels should clearly present the floors the elevator serves, and both a position indicator in the car showing where the car is stopping and a floor numeral on the sight guard of the hoistway door are recommended. Requirements for disabled persons include use of raised designations adjacent to the car operating buttons, and use of a single-stroke gong in the up landing lantern and a double-stroke gong in the down landing lantern to audibly indicate the direction of the elevator.

In other residential buildings, because turnover is minimized, only the minimum informational features are necessary. These are a landing button that lights to acknowledge the registration of a call, position indicators in the car, and prominent directional lanterns in the car or at the landing to show the entering passenger which way the elevator is headed.

Service elevators should be large enough to carry their intended loads: carts, tables, or baggage. Additional considerations must be given to moving furniture in all residential buildings, including cars with ceilings 10 or more ft high as well as higher and wider hoistway and car doors.

If the service elevator is normally a passenger elevator, protective pads should be provided. An independent service switch to remove the elevator from group operation with other passenger elevators should be provided, and the elevator should operate with an attendant in response to car calls only. If frequent service use is expected, a separate landing button system and an annunciator in the car is an economical means to provide the special operation.

Other chapters in this book also include suggestions for special operations or features that may be necessary for a particular building. Many of them can be used in any building to solve a present or expected problem. Competent elevator engineers should be consulted to help solve operational or other problems. The following section discusses platform lifts for wheelchairs and stair climbers, which are necessary in some buildings to meet requirements for disabled persons and are often used in public buildings.

ELEVATORS IN PRIVATE RESIDENCES

One area of residential elevatoring that is growing in popularity with the introduction of economical equipment is the installation of an elevator in a private residence. Residential elevators have long been prohibitively expensive and were found only in the most palatial homes. Increased use and the popularity of multifloor houses have created a demand, and many elevator manufacturers offer specialized equipment.

The average home elevator is a small car of no more than 700-lb (320-kg) capacity and operates at a speed of 35 fpm (0.18 mps) maximum. It is designed with a collapsing car gate and swing-type hoistway doors and can be either supported by the building structure or installed in an independent tower. The car interior is designed to accommodate a standard-size wheelchair.

The installation and design varies with each manufacturer. The A17.1 Elevator Code, recognizing that the elevator will be infrequently used, has allowed certain easing of rules. One important rule is imposed, however; the elevator must contain a telephone capable of reaching an outside agency should the need ever arise. This is a life-safety rule for any elevator in any unattended building.

Over the past decade or so there has been increased application of various specialized lifting devices. Generically known as "handicapped lifts," these include platform lifts and stairlifts. Platform lifts are limited in size to accommodate a person in a wheelchair, and rises are generally up to about 5 or 6 ft (Figure 11.3). The American Society of Mechanical Engineers (ASME) A17.1 Elevator Code limits the maximum to 12 ft. In 1997, the provisions for wheelchair and platform lifts will be removed from the American Society of Mechanical Engineers (ASME) A17.1 code and developed into a separate code designated ASME A18. This is being done as part of the harmonization effort between the United States and Canada regarding elevator codes since Canada has a separate code for these devices (see Chapter 16). Until such a code is adopted by local enforcing authorities, these devices will continue to be covered by the published A17.1 code in effect in affected jurisdictions. Versions include platform lifts designed to operate along the incline of a stairway with a platform that will fold out of the way to permit normal use of the stairs. These are often found in public buildings and are often too cumbersome for a residence.

Stairlifts in residences are quite popular and may range from a portable device set onto a stairway that has a traveling seat on which the occupant rides, to permanently installed devices, some of which are designed to turn corners. Figure 11.4 shows a typical installation.

The need to provide a means of vertical transportation in any multistory building has prompted concern about usage of handicapped lifts. Although supposedly limited to the use of people in wheelchairs or those who are mobility impaired, they have been used for material transportation and by the general public. The elevator code mandates many safety features and controlled access; however, this later aspect is often ignored. Another concern is an installation that may penetrate a floor which, per building code, would mandate a fire-retardant, fully enclosed hoistway.

In an effort to provide an elevator with full protection for anyone who may use it and to meet all applicable building codes, the industry has developed what is called a limited use, limited application (LU/LA) elevator. Such an elevator is limited in size, speed, and travel, and reduction in many of the emergency-type and specialized maintenance-type

Figure 11.3. A platform-type lift installed adjacent to common stairs (Courtesy The National Wheel-O-Vator Co., Inc.).

Figure 11.4. An inclined lift set in an existing stairway with a fold-down seat for the user (Courtesy The National Wheel-O-Vator Co., Inc.).

features of a conventional elevator has been allowed. It is ideal for a two- or three-story residence or multiple dwelling, and its application should be considered. The LU/LA elevator has been accepted for us in public buildings. Various manufacturers are offering it as a standard.

SUMMARY

We can foresee a growing emphasis on the need for expanded residential building in both downtown and suburban areas. The difficulties of commuting and the aging of the population are two factors that can effect such change. Limited convenient suburban space for "ranch-type" housing will prompt more and more multiple-storied dwellings, and the changing 24-hour amenity of downtown areas will make living there more attractive. The architect and developer need to be aware of the importance of vertical transportation in any residential building.

This chapter has pointed out the many considerations needed in either constructing or renovating a building space. The major amenity of any building will be its elevators and the service they provide. The guidelines presented here provide the basis for the judgment of those aspects and major direction in application. Whether in an apartment house, hotel, dormitory, or any building where people will live and sleep, access to the upper floors must be the primary consideration.

12 Elevatoring Institutional Buildings

DEFINITION

Institutional buildings are defined as those in which people obtain a particular care or service. They are generally public and are designed to perform specific functions. Those functions influence vertical traffic, which consists of a combination of staff, visitor, and vehicular traffic, as in a hospital, or primarily visitor traffic, as in a museum or exhibition building.

The major types of institutional buildings are hospitals, both general and specialized, nursing homes, schools, courthouses, museums, sports arenas, exhibition halls, observation towers, and jails. Combinations are occasionally found, for example, a governmental office building and courthouse. In such a case vertical transportation must serve both the office and courthouse traffic.

Today's institutional buildings usually have extensive parking facilities that may be served. Parking integrated in the building is best served by shuttle elevators. Parking located adjacent to the building requires vertical transportation similar to that required by commercial parking discussed in a previous chapter. Residential areas, such as living quarters on upper floors in a hospital, may be integrated into the building, and the impact of these areas on vertical transportation must be determined.

Existing facilities, especially in hospitals, may often be expanded to serve growing needs. A basic rule in planning the expansion of any facility is to determine the adequacy of its existing vertical transportation. Too often the vertical transportation load of the entire complex is placed on the new facility, which overwhelms otherwise adequate service. One of the best ways of evaluating the impact of the new facility is to assume that the entire facility is being built today and, perhaps by expansion of existing equipment, to develop a total solution. A traffic study of the existing facility is required.

POPULATION

The amount and type of service to be rendered often determines the population of a facility. This is best described by examples of each type of institutional building and is summed up in Table 12.1.

Hospitals have more critical vertical transportation demands than most other types of institutional buildings. Vertical transportation in hospitals is related to the staffing of the institution, and the number of staff per bed has always been a good indication of demand. Statistics indicate that in 1954 the average was about 1.98 staff per bed, which increased to about 2.7 in 1973. Statistics in 1981 indicated about 3.7 per bed, which has since lev-

The Vertical Transportation Handbook, Third Edition, Edited by George R. Strakosch
ISBN 0-471-16291-4   © 1998 John Wiley & Sons, Inc.

TABLE 12.1. Institutional Buildings

Type of Building	Population Criteria	Recommended 5-min Capacity (%)	Interval Range (sec)	Expected[c] Inefficiency (%)
Hospitals	3 to 5 per bed (per facility type)	12	30 to 50	5
Long-term care facilities	1.5 to 2 per bed	8	40 to 70	10
School				
Classroom	15 ft^2 (1.5 m^2) per student	25 to 40	See text	0
Laboratory	100 ft^2 (10 m^2) per student	20	40 to 50	0
Library	Depends on seating	15	40 to 50	0
Courthouse	30 people per court- room plus spectators	15	40 to 50	5
Jail	Varies	See text	See text	
Museum, sport arenas, exhibition, observation tower	Varies	See text	See text	

	Recommended Elevator Sizes	
Type of Building	Passenger Elevators Size;[a] Doors, Type, and Size	Service Elevators Size;[a] Doors, Type, and Size
Hospitals	5000 lb, 54-in. center-opening	6500 lb, 60-in. center-opening
Long-term care facilities	3500 lb, 48-in. center-opening	4500 lb, 48-in. center-opening
School	6000 lb, 60-in. center-opening	6500 lb, 54-in. two-speed
Courthouse	4000 lb, 48-in. center-opening	6500 lb, 54-in. two-speed
Museum, sport arenas, exhibition, observation tower	6000 lb, 60-in. center-opening	8000 lb, 60-in two-speed center-opening
Jail	3500 lb, 48-in. two-speed	6500 lb, 54-in. two-speed

	Recommended Speeds (fpm)				
Building Height	Hospital	Nursing Home	Courthouse	Museum	Jail
2 to 6	200 to 400	200	400	400	200
6 to 12	400 to 500	400	500	500	400
12 to 20	700	500	700	700	500
20 to 25	800[b]	500 to 700[b]	800[b]	—	—
25 to 30	1000[b]	—	1000[b]	—	—

	Exterior Elevators	*Inside Elevators*
Observation towers (two-stop elevators)	100 ft (30 m)—200 to 400 fpm	500 fpm
	200 ft (60 m)—400 to 500 fpm	700 fpm
	300 ft (90 m)—500 to 700 fpm	1000 fpm
	400 ft (120 m)—500 to 700 fpm	1200 fpm (6 mps)
	500 ft (150 m)—700 to 800 fpm	1400 fpm (7 mps)
	600 ft (200 m)—800 to 1000 fpm	up to 2000 fpm (10 mps)

[a]Metric equivalent: 3000 lb (1400 kg), 3500 lb (1600 kg), 4000 lb (1800 kg), 6000 lb (2700 kg), 6500 lb (3000 kg), 8000 lb (3600 kg), 200 fpm (1 mps), 400 fpm (2 mps), 500 fpm (2.5 mps), 700 fpm (3.5 mps), 800 fpm (4 mps), 1000 fpm (5 mps).

[b]For buildings of this height, local-express elevators should be considered.

[c]Inefficiency percentage is in addition to door type and platform size inefficiency.

eled off at about that number. The changing nature of health care, the increased specialization of treatment and testing, and the growing emphasis on out-patient procedures have modified any increase, and the ratio of 3.7 continues to provide a good criterion of population. As with any statistical average, the needs and experience of particular institutions must be considered before a final judgment is made.

With the increase in out-patient functions, visitor demand in hospitals has decreased and ceases to be a major factor in establishing a basis for vertical transportation. The objective is to provide good service for the staff members, allowing them to make maximum use of their productive time.

In nursing homes and mental institutions the number of staff per bed is somewhat lower than in hospitals and the elevator requirements less severe. The number of staff per bed is a good population criterion unless other information is available.

Population ratios in schools vary with the type of facility. In primarily classroom buildings, population densities of 15 ft^2 (1.5 m^2) of net area per student are common. The total density in the building varies with the classroom utilization factor, which school authorities are striving to maximize and which may be as high as each seat being used about 80% of the time. At present, use factors of 50 to 75% are attained.

Laboratory buildings provide more space per student than do classrooms. The average is about 100 ft^2 (10 m^2). In advanced laboratories and graduate schools averages of 200 ft^2 (20 m^2) of net area per person are often allowed.

Population in library buildings varies with use. A general reading room may have densities as high as 15 ft^2 (1.5 m^2) per person or, in some specialized areas where many book racks or files of reference material are located, density may be as slight as one person per 100 ft^2 (10 m^2) or more.

Courtrooms are designed for a specific primary function—a jury trial. There will be spectators (varying in number with the importance of the trial), the judiciary staff (about 12 to 16 people), and the jury (12 to 16 people). By using a value of about 30 people per courtroom plus expected spectators, a reasonable population factor may be established. A probable use factor of 75 to 80% may be assumed with the prospect of improved judiciary scheduling.

Museums, exhibits, sports arenas, observation towers, and so on are designed for a stated spectator capacity. In addition, an anticipated "turnover" or entering and leaving of these spectators in a given time is considered in the overall design. Vertical transportation must be directly related to expected population and turnover.

Jails, especially holding facilities, are often located in high-rise buildings or on the top of governmental buildings. Population is easy to determine by the number of cells, but the major vertical transportation needs are related to the staff, number of visitors, type of visiting, security, service needs of the prisoners, and the type of prison. The best determination of vertical transportation need will be based on the prison staff provided plus allowances for visitors, transfer, trials, and so on.

ELEVATOR TRAFFIC IN INSTITUTIONAL BUILDINGS

Considerations

Elevator traffic characteristics vary with the type of institution. In hospitals emphasis is on interfloor traffic with considerable periods of two-way traffic. in a courthouse two-way traffic is the major form.

In classroom buildings traffic is primarily interfloor, especially during class change periods. Laboratories and libraries have substantial incoming and outgoing traffic as well as two-way traffic similar to that found in office buildings.

Traffic in arenas, theaters, and other entertainment facilities is entirely in at the start of a show and out at the end. In exhibits and observation towers flow is primarily two-way, and vertical transportation systems must be designed to turn over the expected number of visitors in the time allotted to view the exhibit or panorama. If a restaurant is located in the facility, the average time to be seated and served is an important consideration.

Because traffic in each type of institutional building varies with the facility, a few general rules can be stated, but each building type is individually evaluated. Table 12.1 in the previous section, on population, gives the expected intensity of the critical traffic, and the text gives examples of particular elevator problems.

As mentioned previously, parking or horizontal transportation facilities affect the vertical traffic in institutional buildings as in other types. Often it is necessary to locate such parking on various levels, which creates multiple entrance floors to the building. In this case it is best to provide shuttle transportation to the single lower terminal of the main elevators, or a necessary time penalty must be included in elevatoring calculations. Further consideration must be given to how quickly autos can fill or empty a parking facility. In sports arenas especially, little is gained by excellent elevator or escalator service if bottlenecks cause slow departure from the parking exits.

All elements of the circulation system must be planned as an entity. Each form of transportation—auto, escalator, elevator, bus, or subway—has a definite utilization factor that must be considered when each of these elements is integrated into the system. The vertical transportation system should be optimized in relation to the effective utilization of the facility.

Layout

Special traffic-handling requirements in institutional buildings, especially in suburban areas, are posed by visitors who may never have ridden an elevator or escalator before and may never do so again. These are the people who visit a hospital once, occasionally go to a tourist attraction, have never been in a courtroom, and seldom go to a large city. They are often confused by complex buildings, may be at a peak of emotion, and need the maximum in guidance.

Even more important than in an office building, the elevators in an institutional building must be a center of attraction. The directions and the signage associated with elevators and escalators must be clear and well defined. Doors should be wide and car interiors should be inviting. Ample signals should be provided, and the direction of elevator travel and elevator destinations plainly indicated. Desirable features include large floor markings at eye height on the jamb or sight guard edges of each landing door, large numerals to designate floors or function adjacent to each car button, and directional lanterns with arrows and lantern gongs that sound once for up and twice for down. Codes enacted in many areas for disabled persons require visual and audible features plus the provision that floor markings be tactile, that is, raised 0.030 in. (0.8 mm) so those who are visually impaired can feel them.

For the staff of institutional buildings less elaborate signaling and information devices are necessary, but vertical transportation must be located to minimize time consumed in horizontal movement. The transportation core should be central, located near the facilities it is expected to serve, and designed to save employee time.

When heavy public use is expected and elevators serve few floors, large cars of 5000- to 6000-lb (2300- to 2800-kg) capacity are often desirable. This procedure maximizes handling capacity because the average tourist or spectator does not expect high-frequency service or, in other words, accepts the quality of bus rather than taxi service.

Large elevators necessitate ample queuing space instead of alcove arrangements, and deep lobbies are required in front of elevators where people may congregate.

In hospitals at least some of the elevators will have to carry beds and stretchers, a common requirement for which a special hospital-shaped elevator has been designed. Because the hospital car is narrow and deep, it is, unfortunately, not the best shape for passenger traffic. In a large hospital the vehicular or bed traffic should be separated from the pedestrian traffic. The passenger elevators can also be designed to serve stretchers during a major emergency. In a smaller hospital both functions are usually performed by the same elevators.

Observation towers, long popular, are being built in more and more localities to provide the sensation of viewing the area from a point hundreds of feet above the ground. Vertical transportation is essential to the success of such a tower. One of the most dramatic uses of an elevator is its operation on the outside of the tower with a glass-enclosed car so that people may visually experience ascent and descent. Outside elevators require specialized design to avoid the hazards of wind and weather and to preserve architectural simplicity. These goals have been achieved in many such towers, notable among them the Seattle Space Needle, the Skylon Tower at Niagara Falls, the Hemisfair Tower in San Antonio, and the Canadian National Tower in Toronto (Figure 12.1).

Figure 12.1. Old and new towers depend on elevators to function: (*a*) Eiffel Tower, Paris, 1899; (*b*) Canadian National Tower, Toronto, 1975; (*c*) Seattle Space Needle.

TABLE 12.2. Notable High-Rise Structures

	Height
TV Tower, Delhi, 1988	235m
Eiffel Tower, Paris, 1889	300m
Empire State Building, New York, 1931	381m
Sears Towers, Chicago, 1974	443m
Menara TV Tower, Kuala Lumpur, 1996	420m
Petronas Towers, Kuala Lumpur, 1996	452m
TV Tower, Moscow, 1967	540m
CN Tower, Toronto, 1976	553m
*Proposed Millennium Towers and Skyscrapers**	
London Millennium Tower	376m
Shanghai Financial Centre	460m
New York Stock Exchange Tower	546m
Bombay TV Tower	560m
Tokyo Millennium Tower	840m

*These are only a few of the many high-rise structures that are believed to be in the planning stage in time for the year 2000.

Courtesy Concrete Society.

Observation elevators are not limited to towers. Many commercial buildings, such as the Empire State and the World Trade Center in New York City and the Sears Tower in Chicago, enjoy revenue from observation decks on their uppermost floors. In addition, as shown in Table 12.2, both towers and multipurpose buildings (featuring retail, commercial, residential, and transient functions) are being built or planned in various countries. The attraction of the highest structure in the area and the desire for people to ascend to that highest point provide the impetus for the continuing reach for the "highest." Some examples of the "highest" are shown in Figures 12.2a and 12.2b. Even during the publication of this edition, new proposals are being developed, as indicated in Table 12.2.

Suggestions for elevator layout are offered in the following pages. As with building designs, variations are infinite and competent elevator engineers should be consulted whenever a problem is perceived.

HOSPITALS

Vertical traffic in hospitals may be separated into two distinct parts: (1) pedestrian traffic, consisting of staff, doctors, technicians, volunteers, visitors, and ambulatory patients, and (2) vehicular traffic, made up of patients on stretchers or in wheelchairs, linen carts, dietary carts, supply carts, portable equipment, and so on. Separating these two types of traffic is by far the best approach to hospital elevatoring.

If pedestrians and stretcher patients use the same elevators, patients suffer delay and discomfort. People must squeeze past a stretcher patient to get in and out of the elevator, thus slowing the entire trip. If all the elevators in a hospital must be designed to carry patients on stretchers, they must be narrow and deep, a shape that reduces the efficiency of passenger transfer and slows the elevator trip.

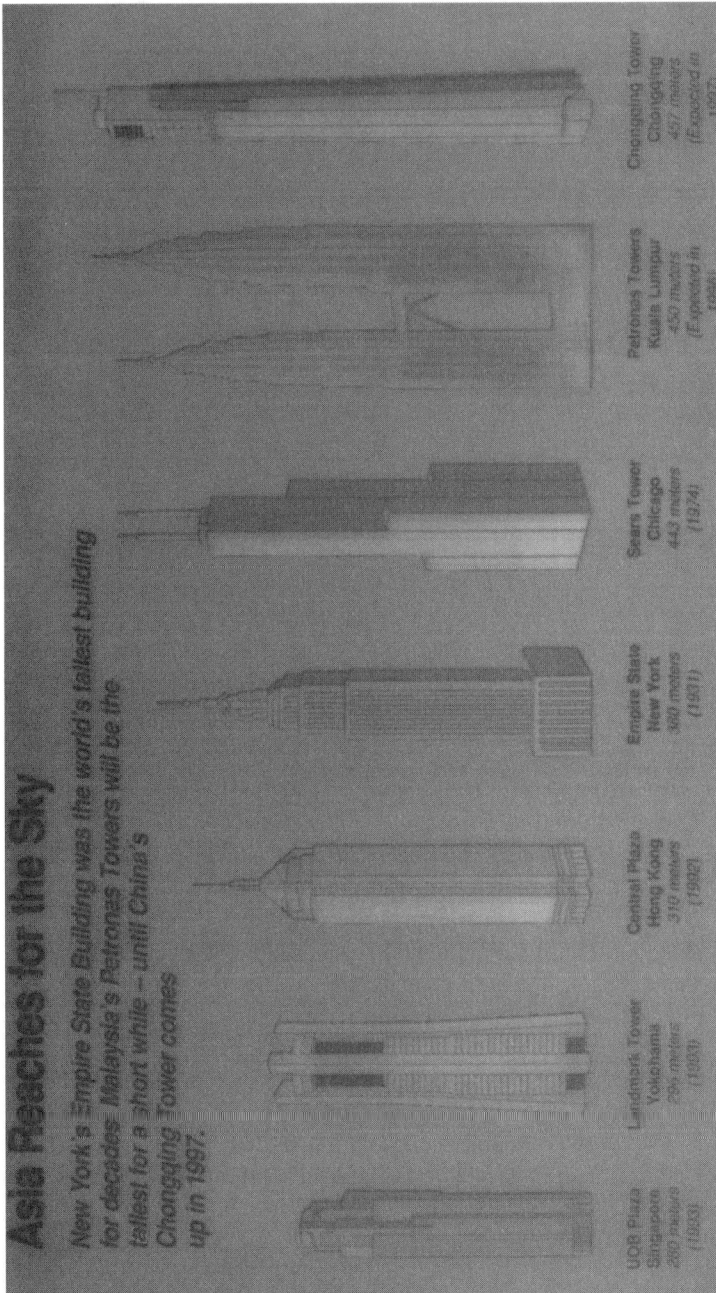

Asia Reaches for the Sky

New York's Empire State Building was the world's tallest building for decades. Malaysia's Petronas Towers will be the tallest for a short while – until China's Chongqing Tower comes up in 1997.

| UOB Plaza Singapore 280 meters (1993) | Landmark Tower Yokohama 295 meters (1993) | Central Plaza Hong Kong 310 meters (1992) | Empire State New York 380 meters (1931) | Sears Tower Chicago 443 meters (1974) | Petronas Towers Kuala Lumpur 450 meters (Expected in 1996) | Chongqing Tower Chongqing 457 meters (Expected in 1997) |

(a)

Figure 12.2. (*a*) Examples of the "highest" (Courtesy *Far Eastern Economic Review*).

Figure 12.2. *(Continued)* (*b*) The Petronus Towers, Kuala Lumpur (Courtesy *Elevator World Magazine*).

Although most of the elevator traffic in a hospital may be of a pedestrian nature, that is, staff and visitors going to and from patients, the most important traffic is the movement of patients to and from treatment or to evacuate them in the event of a building emergency. A compromise elevator design is required, which dictates a wide elevator to expedite passenger transfer but with sufficiently wide doors so that a stretcher may be moved and with a car depth to accommodate that stretcher. This size forms the basis of passenger elevator service in a hospital and leads to a unique "hospital pedestrian elevator," which will be described as opposed to the conventional, narrow and deep, "hospital elevator."

For these reasons the first approach to planning proper hospital elevatoring should begin with two separate groups of elevators: pedestrian-shaped cars for pedestrian traffic and hospital-shaped cars for vehicles. Because the first rule is that more than one elevator in a hospital must be capable of handling patients, the minimum plant for any multistory hospital is established as two hospital-shaped cars.

The second rule is that elevators be sufficient to provide an operating interval of 30 to 50 sec. This interval is attained by the design of the elevator plant, as is shown by Example 12.1.

A third consideration is to have a sufficient number of elevators so that at least one is accessible within a minimum time. If food service requires exclusive use of an elevator during a particular time, a separate food service elevator should be provided. This elevator can be used at off-times for linen delivery and other supplies that can be scheduled at convenient times. A major priority is that any elevator may be commandeered for use to transport a critical patient.

Automated material and food-handling systems are available and, because they can save many employee man-hours, these systems are frequently applied to hospital requirements. Such equipment does not materially decrease the need for stretcher- or vehicle-type elevators in hospitals, but allows controlled scheduling of necessary deliveries. These systems will be suggested in this chapter and discussed fully in the chapter on material handling.

Lifts with automated loading and unloading ability are recommended for use between surgical supply and operating suites and between various nursing floors and pharmaceutical supply. A simple, manual dumbwaiter connecting the kitchen and cafeteria or coffee shop is also useful. If the loading dock is at a different level from the kitchen, a kitchen service elevator should be provided, of the same type as described for hotel kitchen service.

Vehicular Traffic

Vehicular traffic in hospitals consists of patients on stretchers or in wheelchairs, dietary carts, equipment such as portable x-ray machines, pharmaceutical, linen, and exchange carts, library and sundry carts, plus maintenance carts. Surveys have shown that vehicular traffic can be as high as 4 carts up and 4 carts down for each 100 beds in a hospital in a 5-min period. Unless the cart delivery system is automated, each cart or vehicle will require an attendant, who, for elevator purposes, is considered as a partner to the vehicle and not part of the pedestrian or staff traffic. Vehicular traffic usually requires a longer transfer time on and off an elevator than pedestrian traffic and often causes elevator delays, especially when an elevator filled with pedestrians stops and there is no room for the vehicle.

The reverse is the more frequent occurrence. An elevator can be filled in bulk with an empty stretcher, which limits the space to pick up additional pedestrians; hence, a useless stop is created.

Elevator manufacturers should be challenged to develop a means to reliably recognize when an elevator is filled by area rather than by weight, as is the current practice. In this way many useless stops could be eliminated and hospital elevator efficiency improved.

Layout

Various elevator layouts are effective in separating vehicular and passenger traffic. In small hospitals, which may require only three or four elevators, the efficiency of the passenger-shaped elevator is sacrificed to obtain the necessary flexibility of operation. In any hospital requiring five or more elevators, the use of two separate types of elevators is generally feasible and should be investigated.

Hospitals are often planned for future expansion, and the layout should always include additional elevator hoistways or framing so that the elevator plant may be expanded. The ultimate expansion should be determined and the elevator plant projected to that end. If vertical expansion is contemplated, elevator machinery of the basement type can be installed initially. The overhead equipment can then be installed with minimum interference to the operating plant and the transfer made in minimum time. Alternatively, the elevator machinery can be installed in its permanent location atop a shaftway tower for the building's full ultimate height and adjusted when the upper floors are activated.

A sample layout in Figure 12.3 shows a suggested 6000-lb hospital-shaped elevator. The center-opening sliding doors provide the optimum door opening and both efficient vehicle and passenger transfer.

The 6000-lb car is the minimum that accommodates a motorized hospital bed with a patient in a body cast with traction devices and leaves room for people riding alongside the bed. A 6500-lb car should be considered if architecturally possible.

Passenger elevators can be of the conventional size and shape as for commercial buildings. For minimum efficiency in passenger handling, the 3500-lb elevator, shallow and wide, is recommended. As an additional feature, an elevator of this size accommodates a mobile stretcher (22 × 76 in.) in an emergency. A 5000-lb elevator with a platform 102 in. wide by 82 in. deep with 54-in.-wide center-opening doors is an ideal size for a larger hospital.

A compromise 3500-lb (1600-kg) size with a platform is shown in Figure 12.4. This size and arrangement of elevator allows a stretcher to be carried if necessary but provides a reasonable-sized elevator for the expected pedestrian traffic. It is ideal for the newer out-patient facilities being built or planned.

Groups of four or more cars should be installed in alcoves as recommended in Chapter 2. Service elevators require additional lobby space if elevators are placed facing each other. A minimum lobby of 14 ft (4.2 m) is recommended for sufficient room to swing two stretchers unloading simultaneously.

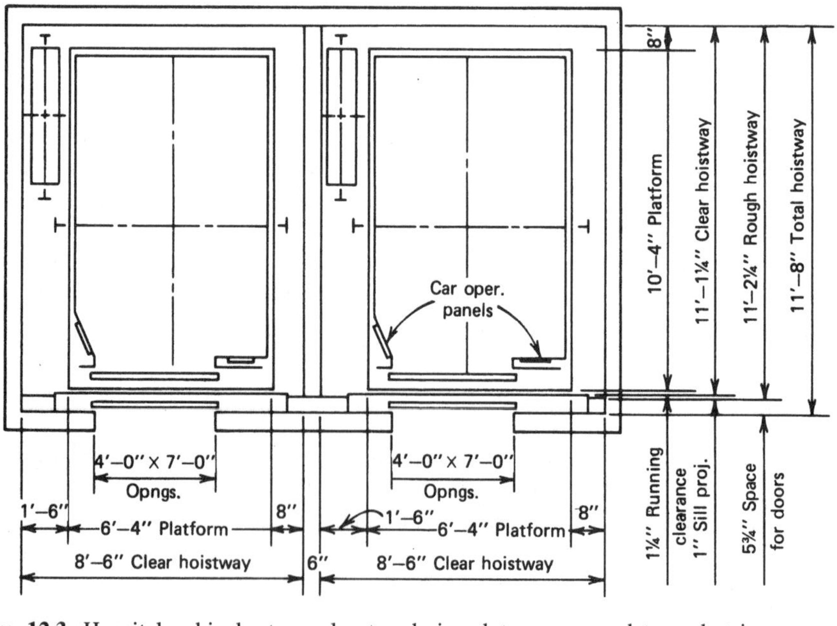

Figure 12.3. Hospital vehicular-type elevator designed to accommodate pedestrians: capacity, 6000 lb, speed, 400 fpm.

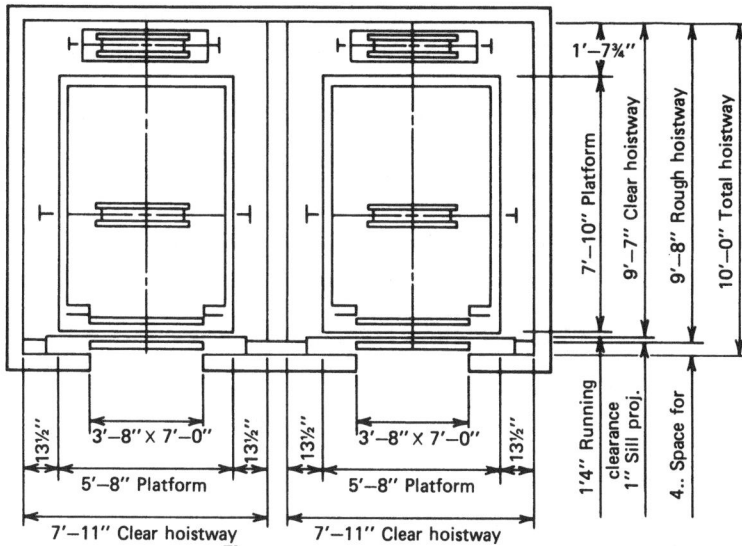

Figure 12.4. Hospital passenger-type elevator designed to accommodate a stretcher: capacity, 3500 lb, speed, 500 fpm.

In hospitals of more than 16 to 18 floors, local/express elevators may be suggested, especially for passenger-type elevators that would have an excessively long trip time if called on to serve more than 16 floors. Service elevators may be in a single group if sufficient capacity is attained with six or eight cars. If more service cars are needed, consideration should be given to separating them according to their functions, possibly by restricting all food service and supply to a single group and reserving another group for stretcher and staff traffic. Automated cart-handling supply systems become a definite requirement for hospitals of more than about 12 floors.

Outpatients

Facilities for outpatients are receiving increased emphasis. To reduce the time spent in hospitals and to serve community needs, separate outpatient facilities are usually established and should be served by separate vertical transportation.

If the facility is large, escalators may be considered to enhance the use of floors above the lobby for this function. A smaller facility may be served by an elevator or elevators with limited stops. The cars should be of ample size, 3500 lb (1600 kg), with 48-in. center-opening doors to facilitate the loading and unloading of wheelchairs and people on crutches. Any upper-floor outpatient facility requires service by an elevator in view of the limited mobility of most of the patients.

Calculating Elevator Requirements

As noted in Table 12.1, the expected passenger traffic demand will amount to 12% of the hospital population in a 5-min period. This population, for elevatoring purposes, should be based on a minimum of three people per bed and adjusted to reflect expected staffing in the facility under study whether it be a community hospital, special treatment, teaching, or central regional hospital.

Example 12.1A shows how the service demand is translated into elevator facilities. The characteristic hospital traffic is a combination of two-way and interfloor traffic. Calculations are based on a two-way traffic approach.

Critical traffic periods occur at a number of times during the day. One of the most important periods occurs at about 3:00 P.M. when hospital nursing staffs are changing shifts and a coincidental visitor peak may occur.

Critical vehicular traffic reaches a peak at about 8:30 to 9:00 A.M. when patients are transferred for operations and therapy, the cleanup from the morning meal is in progress, supplies are transferred, and maintenance activity is in full swing.

Example 12.1. Institutional Buildings—General Hospital

A. Given: 350 beds, 10 floors, 12-ft floor heights. Passenger traffic: 350×3 per bed = $1050 \times 12\%$ = 126 people per 5 min; vehicular traffic: 350 at 4 per 100 beds = 14 vehicles per 5 min.

   1. *Passenger traffic*

     Assume: 10 passengers up, 10 passengers down, 500-fpm elevators, 48-in. center-opening doors

     Probable stops: up, 10 passengers, 9 floors = 6.2 stops $\times$ 0.7 = 4.3 down

     Time to run up, per stop: $\dfrac{9 \times 12}{6.2}$ = 17.4-ft rise per stop

               17.4 ft = 5 sec

     Time to run down: $\dfrac{9 \times 12}{4.3}$ = 25 ft = 5.8 sec

     Elevator performance calculations:

Standing time		
Lobby time 10 in + 10 out	=	18.0 sec
Transfer time, up stops 6.2×3	=	18.6
Door time, up stops $(6.2 + 1) \times 5.3$	=	38.2
Transfer time, down stops 4.3×4	=	17.2
Door time, down stops 4.3×5.3	=	22.8
Total standing time		114.8 sec
Inefficiency, 15%	=	17.2
	Total	132.0 sec
Running time		
Run up 6.2×5	=	31.0
Run down 4.3×5.8	=	24.9
Total round-trip time		187.9 sec

     $\text{HC} = \dfrac{(10 + 10) \times 300}{188}$ = 32 people, $\dfrac{126}{32}$ = 3.9, 4 elevators

     Interval: 188/4 = 47 sec

   2. *Vehicular traffic:*

     Assume 400-fpm vehicular-service elevators.

Time to load vehicle	15 sec
Run one-half building height (average trip) $5 \times 12 = 60$ ft	11.5
Unload vehicle	15
Run	11.5
Door time, two stops 2×5.3	10.6
Average trip	63.6 sec

$$HC = \frac{300}{63.6} = 5 \text{ vehicles per elevator per 5 min}$$

$$\frac{14}{5} = 3 \text{ elevators required}$$

Minimum elevators:

Four 3500-lb passenger elevators @ 500 fpm, 48-in. center-opening doors.

Three 5000-lb service elevators @ 400 fpm, 48-in. center-opening doors.

Recommendations:

Four 5000-lb passenger elevators @ 500 fpm, 54-in. center-opening doors

Three 6500-lb service elevators @ 400 fpm, 60-in. center-opening doors.

B. Given: 200 beds, 6 floors, 12-ft floor heights. Passenger traffic: 200×3 per bed = $600 \times 12\% = 72$ people per 5 min; vehicle traffic: 200 at 4 per 100 beds = 8 vehicles per 5 min; combine passengers and vehicles

Assume: 6 passengers up, 6 passengers down, 1 vehicle per trip, 300-fpm elevators

Probable stops: up 6 passengers, 5 floors, 3.7

Time to run up, per stop: $\dfrac{5 \times 12}{3.7} = 16.2$-ft rise per stop

$$16.2 \text{ ft} = 6.6 \text{ sec}$$

Time to run down: $\dfrac{5 \times 12}{2.6} = 23 \text{ ft} = 8 \text{ sec}$

Elevator performance calculations:

Standing time

Lobby time 6 in + 6 out	=	11.0 sec
Transfer time, up stops (3.7×3) + vehicle 15	=	26.1
Door time, up stops 3.7×5.3	=	19.6
Transfer time, down stops (2.6×4) + 15	=	25.4
Door time, down stops 2.6×5.3	=	13.4
Total standing time		95.5 sec
Inefficiency, 15%	=	14.3
	Total	109.8 sec

Running time

Run up 3.7×6.6	=	24.4
Run down 2.6×8	=	20.8
Total round-trip time		155.0 sec

$$HC = \frac{(6 + 6) \times 300}{155} = 23 \text{ people}, \quad \frac{72}{23} = 3 \text{ elevators required for passengers and,}$$

$$HC = \frac{1 \times 300}{155} = 2, \quad \frac{8}{2} = 4 \text{ elevators will be required for vehicles and passengers}$$

together.

Interval: $155/3 = 52$ sec

The conservative recommendation will be to provide four elevators. The interval is at the upper limit of recommended elevator service and, with three elevators, there will be times when vehicles will have to be delayed so that passengers can be accommodated or vice versa. Additional investigation should be made to determine whether the nonpatient vehicle transfer can be accomplished with an automated cart lift system. These vehicles would include dietary needs, supplies, and linens. With only three elevators, the judgment of the elevator engineer must be shared with the hospital administration and the limitations recognized. One important consideration will be the necessity of operating the hospital if one of the three elevators is out of service for maintenance or repair.

The vehicular peak may amount to three to four vehicles per 100 beds in a 5-min period, depending on the scheduling of treatment and delivery of supplies. For utmost efficiency, vehicle traffic should be scheduled so that the fullest possible elevator utilization is attained.

Example 12.1A shows the most efficient way to serve vehicles, that is, separate vehicular and pedestrian travel. With a mixture of personnel and vehicles, considerable elevator inefficiency and delay will occur. Example 12.1B attempts to account for most of the expected delay by allowing 15 sec for one-way vehicle transfer. With exclusive use of the elevator, this transfer time can be substantially reduced.

Examples 12.1A and 12.1B are generalized studies of new hospitals. Many other vertical transportation considerations are necessary and will involve the overall and materials management in the proposed hospital. Materials management is of great concern since transporting necessary supplies and food from the stockroom and kitchens is labor-intensive; that is, an unsupervised person is often used to move carts from place to place. In a moderate-size hospital, it is not unusual to have 10 to 20 people involved in the materials movement activity, and with 24 hr a day, seven days a week operating requirements and with liberal employee benefits, a staff of 30 or more people is required to effectively accomplish this activity. Automated materials-handling systems properly applied and managed can effectively reduce the number of personnel. This aspect will be discussed fully in a later chapter.

Another aspect of hospital design that is becoming prominent is the construction of a new hospital adjacent to and integrated with an existing facility. The existing facility may remain or may be phased out when the new facility is complete. A typical program is to partially build the new facility and operate both the new and existing facility together and then complete the new facility once new patterns of operation are established. For example, the new facility may contain all the operating rooms, supply rooms, and laboratories, while the old facility retains the patient rooms and beds.

Considerable study will be required to ensure the continuing operation of the old facility, recognizing that some of the vertical transportation load is transferred to the new facility. In the design of the new facility, the eventual vertical transportation system must be planned even though it may not be finally installed until a future date.

INSTITUTIONS FOR LONG-TERM CARE

Institutions for long-term care includes hospitals for the chronically ill, mental institutions, postoperative care centers, nursing homes, homes for the aged, and other facilities in which patients or residents remain for an extended period and receive less intensive care than in a hospital. All such facilities can be considered as nursing homes for elevator requirements.

Elevatoring buildings of this type is approached much the same as for general hospitals. If the facility is large, with 400 to 500 beds, complete separation of vehicular and pedestrian traffic should be considered. If many patients are expected to be ambulatory, elevator traffic will be heavy especially if recreation and dining facilities are concentrated on the lower floors.

If definite staffing plans are not established for the facility, a population factor of 1.5 to 2 people per bed may be used, the higher figure related to the higher degree of patient care. For peak 5-min elevator capacity and a handling capacity of 8 to 10% of the population, two-way traffic should be provided. A longer interval range is permissible because of little urgency in day-to-day activities.

With many ambulatory patients, the hospital pedestrian elevators should be considered the main elevators. This will minimize the required hospital-type cars needed for the necessary transfer of supply carts and occasional stretcher movement. Elevator design should include all the features of elevators designed for disabled persons. This includes properly located elevator operating controls, markings for the visually impaired, audible and visible signaling, extra protection while doors are closing, and additional time at stops for passengers to transfer. In calculating elevator requirements, a standing time inefficiency of at least 20% should be used.

In mental institutions security should be provided by locked elevator lobbies. The practice of keying all elevator call buttons has led to various abuses, including forced locks, toothpicks and chewing gum in key ways, and forged keys, and is not recommended.

SECONDARY SCHOOLS, UNIVERSITIES, AND COLLEGES

As land for horizontal expansion becomes unobtainable and the inefficiency of a sprawling campus becomes apparent, more and more schools are expanding vertically. Possible height requires proper attention to vertical transportation. The University of Pittsburgh and the University of Moscow are both vertical schools in single buildings of about 30 floors, the latter with the living space, classes, and libraries all in one building.

Six- or seven-story buildings with classroom space for 2500 to 3000 students are perfectly feasible when served by escalators plus two elevators for necessary freight and disabled student needs. Science departments, with their long laboratory periods, are ideally located in high-rise buildings and require approximately the same elevator service as an office building. School libraries, by proper allocation of floor use, can be multistory buildings. General reading rooms can be located on the lower floors served by escalators, and specialized book collections can be located on upper floors with elevator service designed for the expected use, which is often minimal.

Classroom Buildings

In classroom buildings the demand for vertical transportation is most severe during the time allotted to changing classes. From 3 to 15 min may be allowed for this purpose, depending on the type of school. Secondary schools, usually in buildings of limited height, allow the least time between classes so as to discourage student mischief in the halls. Longer class change times are required for larger schools and extensive campuses.

During the peak 5-min of the class change period, as much as 50% of the student population may seek vertical transportation. As an average, 40% will be considered. Severe vertical transportation demands also occur when evening classes start in downtown colleges. These classes usually start about half an hour after businesses close, by which time the students must travel to the school and have something to eat. They will arrive about 5 to 10 min before classes start and create an incoming 5-min peak on the elevator system of 25 to 40% of the student population.

Student population is determined by multiplying the total number of classroom seats by a utilization factor. For day sessions in most colleges this factor ranges from about 50 to 80%, depending on classroom scheduling. For high schools and evening colleges, a use factor of 80 to 90% of the available seats is not uncommon. The use factor is an important determinant in elevatoring a classroom building and should be established by study of the expected or past practice.

Escalators are by far the best means of vertical transportation for a school with 1500 to 2000 students. They provide service to every floor, can be reversed to serve heavy incoming and outgoing peaks, and can be used for building heights up to about 11 floors. Typical applications appear in the chapter on escalators. In addition to the escalators, at least two elevators are required for disabled persons and for the movement of furniture and supplies. The elevators should be of substantial size (3500 lb minimum) and should be readily accessible. it may be necessary to provide an attendant during class change periods to limit use of the elevators to authorized personnel.

Many state laws and local building codes are requiring all public buildings that include schools to be either built with or retrofitted with means to serve the needs of disabled persons. A grade-level entrance and conventional elevator service to each floor is the necessary solution. For an existing facility this can usually be accomplished by adding a shaft adjacent to the building. In situations where this is not feasible, ramps, platform lifts, or stairway-type elevators may be applied. A discussion of these specialized lifts can be found in Chapter 11.

Laboratory Buildings

Science, computer, and other types of laboratory sessions are generally longer than normal classroom periods. A student with a project is required to complete it and usually may leave when finished. The net result is that the vertical transportation problem is one of an incoming peak with very little other concentrated traffic during the day.

If the science building is elevatored as an office building, based on sufficient elevator capacity to fill the building in a 30- to 40-min period, service should be adequate for all other traffic periods. Elevators should be designed for a 5-min incoming peak of 10 to 15% of population, based on expected occupancy ranging from 50 ft² (5 m²) to 200 ft² (20 m²) of net area per student, depending on the type of facility.

Laboratory sessions are generally well attended, so the utilization factor is high. In many laboratories, large and unwieldy apparatus is often moved in and out, requiring an extra-large passenger elevator. Rooftop observatories and greenhouses may also require elevator service, which can be provided by one of the main elevators with the necessary penthouse structure or by a two-stop underslung or hydraulic type of elevator of limited speed and capacity, operating between the top main elevator stop and the roof.

Library Buildings

A high-rise building is ideal for a library. The lower floors, connected by escalators, can provide ready access to the common reference works and general reading. The upper floors can house specialized references and rare book collections and may require minimum elevator service. The vary nature of a library, the long-term storage of information, permits this segregation of function and traffic.

The elevatoring of any library depends on the expected use of the various floors and will require extensive preliminary planning.

In an undergraduate library, considerable student turnover is expected in the general areas. Observations have shown that a full turnover of the available seats or spaces for students occurs about every hour. Therefore, 12 to 15% of the population of the library will seek vertical transportation in a 5-min period, two-way traffic.

Population is based on the number of reading spaces provided and the nature of the carrels, either locked or open. Population density also depends on the provisions for and use of seminar rooms, special collections, special libraries, either visual or audio, and other factors peculiar to modern libraries.

Elevator calculations should proceed as for the office-type of two-way traffic. Handling of book carts and delivery of books to the checkout desk may warrant consideration of an automated materials handling system or automated cart lift.

COURTHOUSES

Elevatoring a courthouse is directly related to the activities in the courtrooms, as well as the various functions connected with a courtroom trial. The court clerk's office, where necessary pretrial papers are filed, is a critical area with a constant flow of lawyers and messengers between it and the entrance to the building. Circulation is expedited by locating this facility on the main floor.

Another critical area is the reporting room for jurors. All the prospective jurors gather in this area and are moved in groups of 20 or so to the various courts for examination.

A usual requirement in courthouses is the positive separation of the judicial from the general public traffic, which calls for separate vertical transportation systems.

Transportation of prisoners in a criminal court requires security arrangements and may have to be restricted to a special elevator or elevators. Alternatively, this function may be combined with the service needs of the courts, which include normal maintenance and, possibly, the transportation of displays and records.

Finally, the general public and participants in civil suits as well as the courthouse staff require vertical transportation. If the courthouse also houses extensive office facilities such as those for the district attorney and investigative staffs, vertical transportation requirements in part resemble those of other office buildings.

These requirements suggest that elevatoring a courthouse begins with study of the general plan of the building and the extent of its various facilities. A major determinant is the number of courtrooms and the percentage likely to be in full use at a given time. Full usage is improbable, but a factor of 75% is reasonable.

For each court in session about 30 people will take part in a criminal trial or in a civil trial with a jury. In addition, spectators must be considered. In civil trials without juries only a few people are involved; hence, if jury trials are always considered as population criteria, conservative elevatoring will result.

Observation has shown that the starting and ending or recessing of trials overlap by about 30 min, so that about half the population enters the courtrooms while another half leaves. For example, if there are 20 courtrooms in a building, about 15 will be in use at one time. At 30 people per court, a population of 30 times 15 or 450 people plus spectators is possible. The demand for vertical transportation is 450 divided by 6, or 75 people every 5 min, two-way traffic. If elevators are filled to only half their capacity and operated at the acceptable intervals of 40 to 60 sec, capacity should be ample for spectators, whose numbers can be controlled.

Although judges may require separate transportation, their elevators can also be used to carry juries, either prospective or charged. At least two separate elevators in a private alcove may accomplish these tasks. Judges, because of their status and limited number, should receive priority service. Because jurors are usually escorted and each group may receive exclusive service, elevators large enough to transport the expected number of jurors in each group should be provided. This may be 16 or 20 people, so a 4000-lb (1800-kg) elevator or larger is recommended.

If the building includes office functions as well as courthouses and their related functions, additional elevator service may be needed. Elevators established for the courtroom traffic should be checked to determine whether they can also serve the incoming requirements of the office staff in a manner similar to that used for a single-purpose building. A further check for possible conflict during lunchtime should be made and the number of elevators and intervals adjusted accordingly.

JAILS

Short-term jails for holding people awaiting trial, indicted offenders, or short-term prisoners are possibly located above the courthouse. Transfer problems are minimized and, in a high-rise structure, all necessary facilities can be provided with maximum security. Consideration must be given to evacuation during a building emergency.

Short-term prisons located on the upper floors of courthouses or municipal buildings require a separate group of jail shuttle elevators for access to and from the street. Security is strengthened by separate local elevators from the jail sky lobby to the cell levels (Figure 12.5). Providing secured stairway escape routes to refuge areas in case of a building emergency is an important consideration, as in any high-rise structure.

Determining the elevators necessary for a prison depends on staff, operation, and number of transfers in and out during the course of a day. Transfers consist of people being moved to long-term prisons, prisoners reporting for trial, and new arrivals. Lawyers are allowed to visit prisoners in either the cell area or a visiting room. This entails additional transfer.

Recreation and sick call, if these facilities are located away from the cell floors, require vertical transportation. Meals, however, usually take place in the cell area.

Security is accomplished by locked elevator lobbies and by attendants on the elevators. Prisoners are usually escorted, with the escort calling the elevator, which should be equipped with a vision panel so the attendant can see what is taking place before opening the doors. A form of riot control, such as a foot-operated or elbow-operated switch to call for aid or send the elevator to a secure lobby, is advisable.

Closed-circuit television can be provided for in-car surveillance. Special controls to call the elevator to a protected floor, in the event of difficulty, should be provided.

The extent of vertical transportation requirements must be established and sufficient elevators furnished. Because of security requirements operation efficiency is poor, so ample extra transfer time should be allowed. Larger elevators, 3500-lb minimum, should be used to facilitate cart handling and the transportation of groups. The elevators should be used for all purposes, that is, by both passengers and vehicles, to minimize attendant staffing.

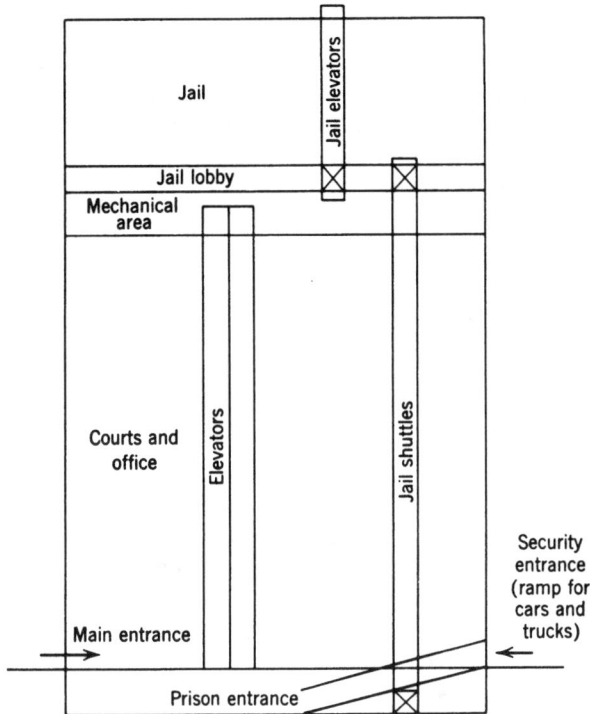

Figure 12.5. High-rise civic center complex.

MUSEUMS, EXHIBITS, SPORTS ARENAS, AND OBSERVATION TOWERS

The value of buildings such as museums, sports arenas, and observation towers depends, in part, on how quickly people can be moved into, through, or out of the facility. The vertical transportation demand in a sports arena is obvious: to put people in their seats in the shortest possible time for prompt presentation of the event. "The shortest possible time" varies with each facility, depending on a number of factors—parking, ticket selling, and other attractions, such as betting at a horse race. As a guide, transportation should be sufficient to serve an average crowd in 30 min both in and out.

In a museum or exhibition hall a more leisurely pace is expected. Aside from performances at particular times, an hour or two may be allowed for visitors to fill the building to capacity. Vertical transportation needs are for turnover rather than rapid filling or emptying as in a sports arena.

The key to the success of an observation tower is the vertical transportation and the area provided for viewing or dining. A visible indication of appeal is a line of people waiting at the ground level for elevator service up into the tower, and crowd control may require that people must come down before additional people are allowed up.

Observation areas placed atop taller buildings are often served by additional capacity in the high-rise group so that an elevator may be used for the visitors during peak hours without affecting tenants in the building. A separate entrance lobby is recommended for the observation-deck elevators.

Escalators

For low-rise, heavy-traffic applications, escalators are the recommended means of transportation. They can serve the most people, can be reversed for incoming or outgoing traffic, and require the least space in relation to their traffic-handling capacity. Local and express escalators have been furnished for some sports arenas to provide the necessary service to a succession of upper levels.

When escalators are applied, it is essential that adequate queuing area be provided at the entrance and exit to an escalator. An unobstructed area at least as wide as the escalator and 10 to 15 ft (3 to 5 m) should be provided at the entering and exiting areas. Escalators that feed into another rise of escalators or into an area that may be locked at times should be electrically interlocked so that the exiting area is open before the escalator can be operated.

Elevators

Two-stop elevators are the other preferred means of high-capacity vertical transportation. Because everyone boards at one level and exits at the other, they can serve structures of any height and minimize passenger confusion. Introducing a third stop exacts a substantial time penalty besides creating confusion for passengers.

If an observation tower facility has more than two main upper levels, serious consideration should be given to shuttle elevators between the various levels. All people would then go to one of the upper main levels by the major elevators and transfer to the shuttle elevator for the other upper levels.

Refreshment Service

Service needs, especially refreshments, are of prime importance in any facility. Refreshment facilities should be stocked between events, using service or freight elevators provided for that purpose. Those elevators should be located near loading docks or storage areas on the lower level and close to the refreshment stands on the upper levels to minimize truck transfer time. Elevators should be large and heavy enough to accommodate either industrial truck or handcart loading. Freight-type, vertical biparting doors can be used to gain the maximum width in the elevator interior.

Elevator Capacity

Passenger capacity rather than frequency of service is the prime objective in elevatoring facilities of this nature. Elevator speed should be proportional to the height of the structure, as shown in Table 12.1. Because elevators are the most critical and expensive equipment for observation towers, consideration must be given to employing the same elevators for both service and passenger use. Deliveries should be scheduled at off-peak visitor periods, and ample cab interior protection should be provided. Outside elevators are popular for observation towers and can be used as an attraction in other spectator-type facilities. A full discussion of outside elevators will be found in a later chapter.

GENERAL

All the general rules for equipment location and passenger information and guidance apply to the elevatoring of institutional buildings. The large number of visitors, many unfa-

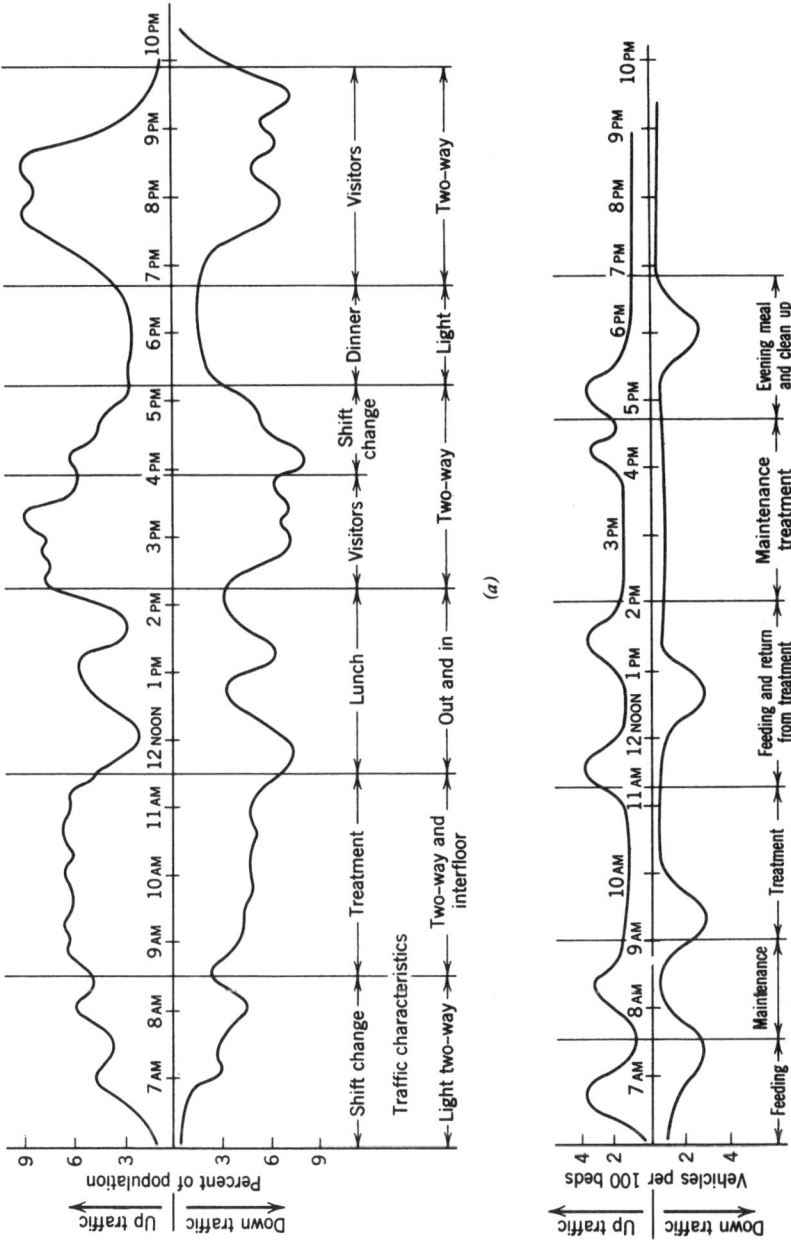

Figure 12.6. Hospital elevator traffic—in and out of the elevators at a main floor: (*a*) pedestrian traffic; (*b*) vehicular traffic.

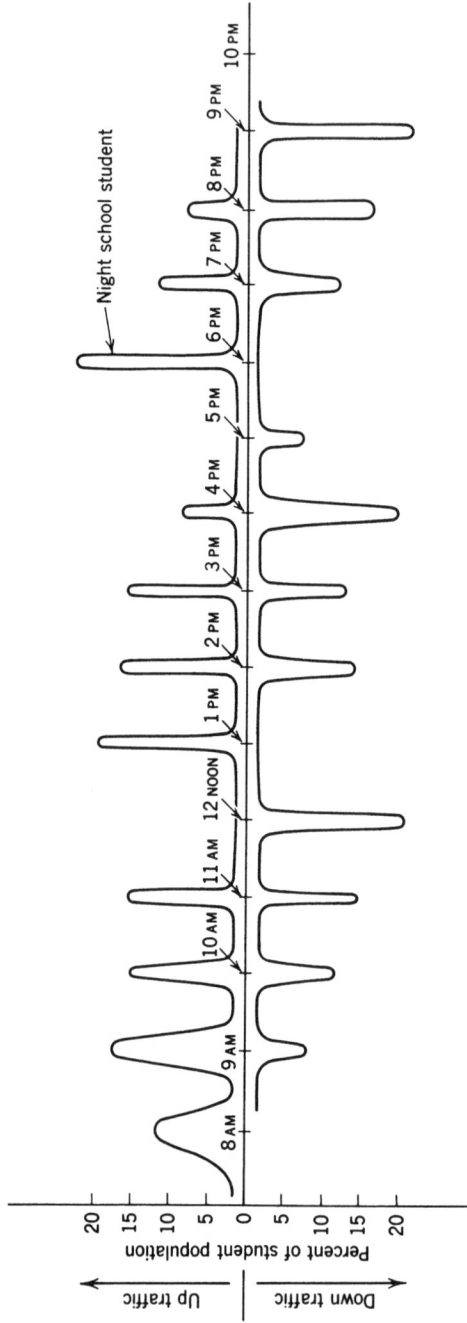

Figure 12.7. Elevator traffic—school, classroom building, day and night classes. Based on hourly class changes, 10-min class change time allowance. Traffic in and out of the elevators at the main floor.

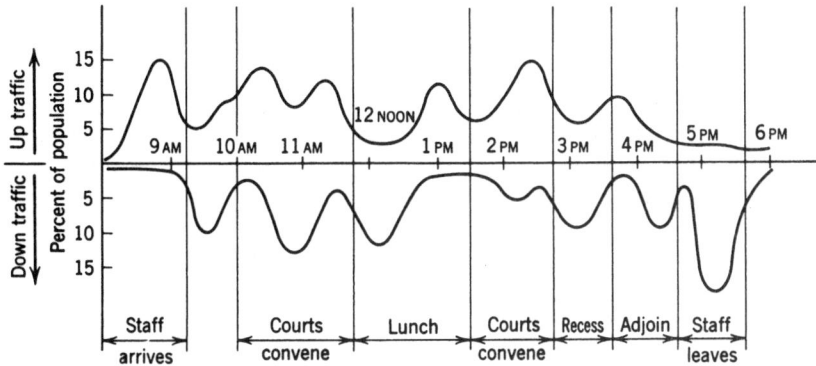

Figure 12.8. Combination courthouse and civic office building elevator traffic in and out at the main lobby.

miliar with elevators or escalators, requires that particular care be taken to avoid confusion and operating inefficiency and to provide ample passenger information.

Elevator operation must be simple and the signals explicit. An occasional visitor to a tall building is more inclined than the regular occupant to take any elevator that stops at a floor, no matter which way it is going. For this reason signs and directional arrows mut be prominent and unequivocal in indication. All buildings should have elevators equipped to serve disabled persons.

OPERATION

As described in Chapter 10, on commercial buildings, an operational study must be combined with equipment analysis to gain full utilization of the vertical transportation plant. It is essential to provide traffic-responsive operating features, and the programmable type of elevator control is most promising for the unique types of traffic expected in institutional buildings.

In Figures 12.6, 12.7, and 12.8 a series of traffic flowcharts is presented for principal types of institutional buildings. The charts also indicate some of the expected traffic conditions and the operations necessary to meet those conditions. People responsible for design as well as the operation of the building should see that these requirements are fulfilled.

Each chart shows the percentage of building population expected to be traveling in each direction at a particular time. This percentage will vary among the individual institutions and should be studied relative to the capabilities of their vertical transportation systems. If a system has substantial capability in relation to the expected traffic, minimum operating features for each traffic period are required. if traffic is heavy as compared with system capabilities, additional operating features are necessary.

Note, in Figure 12.6, the relation between expected activity in a hospital and its need for vertical transportation. The activities shown on the chart are typical for most hospitals, but the times they take place may be different. Observations of local practice should be made to construct a chart for the new or existing facility. Constructing a chart will help relate the building's activities to its vertical transportation system. This is clearly borne out in the school situation, Figure 12.7, where the class change periods create the critical traffic demands.

A relation between planning and operation is very important in institutional buildings. Customary practices are fewer than in office buildings, in that management can largely establish its rules as long as the desired function is performed. For example, if a sports arena is not opened until 15 min before a performance, vertical transportation needs are entirely different than in a facility that is open 1 hr before.

The successful elevatoring of institutional buildings is predicated on understanding relationships between all factors involved and early involvement of the people who will operate the facility. Any compromises must be well documented for the guidance of later management.

13   Service and Freight Elevators

AL SAXER

SERVICE AND FREIGHT ELEVATORS

There are only two types of elevators that are recognized by elevator codes. One is a passenger elevator that has well-defined requirements regarding the usable area of the car platform and the load that must be carried. The other type is a freight elevator which, depending upon its classification, can have various platform loadings versus area, depending on the intended use of the elevator. Freight elevators are prohibited by elevator codes from carrying any passengers other than those required to handle freight. They can carry passengers if the platform area related to load meets the requirements for passenger service.

A service elevator is a passenger elevator, usually designed with a rugged interior and intended to carry both passengers and freight. All the rules for passenger elevators must be met, and the equipment can be designed to meet certain freight-loading requirements, such as the use of industrial trucks or to serve one-piece loads. These will be discussed later in this chapter.

The classification "service elevator" is generally used to designate an elevator in an office building primarily intended to move material rather than people, a stretcher or vehicular elevator in a hospital, or the "back of the house" elevator in a hotel.

FREIGHT ELEVATORS

Size and Capacity

If a building requirement calls for the ability to lift large, light-weight loads such as furniture or clothing, a freight elevator may be an ideal application. The various loading classifications are explicit in the A17.1 Elevator Code and are amply illustrated in Figure 13.1.

As may be noted in Figure 13.1, most freight elevators have a minimum platform loading of 50/lb ft^2 (245 kg/m^2). Where the elevator is intended to carry motor vehicles, such as the elevators in a garage or a truck elevator serving a basement loading dock area from the street, a loading of 30 lb/ft^2 (145 kg/m^2) may be used. With a truck elevator it will be important to determine the loaded weight of the truck intended to be carried, which may require loadings as high as 150 lb/ft^2 (750 kg/m^2), as may be experienced with armored trucks in a bank or fully loaded trash trucks in an office building.

The Vertical Transportation Handbook, Third Edition, Edited by George R. Strakosch
ISBN 0-471-16291-4   © 1998 John Wiley & Sons, Inc.

][CLASS "A"

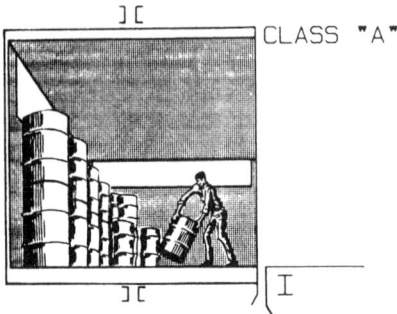

][I

GENERAL FREIGHT LOADING
WHERE NO ITEM (INCLUDING LOADED TRUCK)
WEIGHS MORE THAN 1/4 RATED CAPACITY

| RATING NOT LESS THAN |
| 50 LBS/SQ-FT - 244.10 kgs/SQ-M |

][CLASS "B"

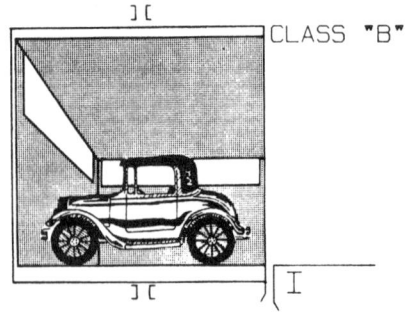

][I

MOTOR VEHICLE LOADING
(AUTOMOBILES, TRUCKS, BUSES)

| RATING NOT LESS THAN |
| 30 LBS/SQ-FT - 146.46 kgs/SQ-M |

][CLASS "C1"

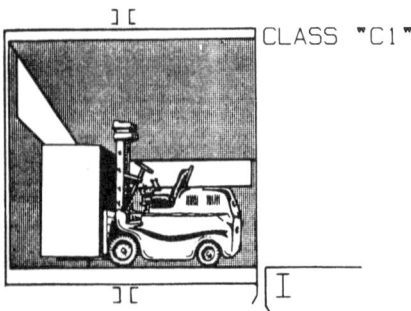

][I

INDUSTRIAL TRUCK LOADING
WHERE TRUCK IS CARRIED

| RATING NOT LESS THAN |
| 50 LBS/SQ-FT - 244.10 kgs/SQ-M |
| MUST HAVE AUTOMATIC LEVELING |

THIS LOADING APPLIES WHERE CONCENTRATED
LOAD INCLUDING TRUCK IS MORE THAN 1/4
RATED CAPACITY BUT CARRIED LOAD DOES NOT
EXCEED RATED CAPACITY.

][CLASS "C2"

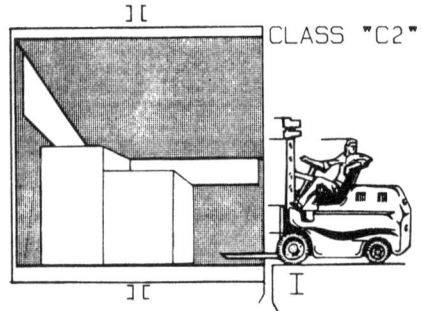

][I

INDUSTRIAL TRUCK LOADING
WHERE TRUCK IS NOT CARRIED, BUT IS
USED FOR LOADING AND UNLOADING

| RATING NOT LESS THAN |
| 50 LBS/SQ-FT - 244.10 kgs/SQ-M |
| MUST HAVE AUTOMATIC LEVELING |

THIS LOADING APPLIES WHERE CONCENTRATED
LOAD INCLUDING TRUCK IS MORE THAN 1/4
RATED CAPACITY BUT CARRIED LOAD DOES NOT
EXCEED RATED CAPACITY.
THIS LOADING ALSO APPLIES WHERE INCREMENT
LOADING IS USED, BUT MAXIMUM LOAD ON CAR
PLATFORM DURING LOADING OR UNLOADING DOES
NOT EXCEED 150% OF RATED LOAD.

Figure 13.1. Classes of freight loading on elevators (class "A," "B," and "C" loading); refer to ANSI-A17.1, rule 207.26. Note: Both freight and passenger elevators can be designed for these various loadings. (Reprinted with permission from *NEII Vertical Transportation Standards* 7th Edition, including 1994 Supplement. Copyrighted © 1992 and 1994, National Elevator Industry, Inc., Fort Lee, New Jersey. This reprinted material is not the complete and official position of National Elevator Industry, Inc. on the referenced subject which is represented only by the standard in its entirety.)

In contrast, passenger elevator platforms are rated from about 70 to 130 lb/ft^2 (105 kg to 200 kg/m^2), depending on total platform area. If a freight elevator is intended to be used for passenger service, it must be rated as a passenger elevator and special door operation provided, as will be described later. In European countries both passenger and freight elevators have the same area load requirements.

NOTE

THE PICTORIAL FREIGHT LOADING SHOWN FOR THE DIFFERENT CLASSES OF LOADING IS INTENDED FOR BOTH ELECTRIC AND HYDRAULIC ELEVATORS. SOME DIFFERENCES OCCUR IN RAIL REACTIONS.

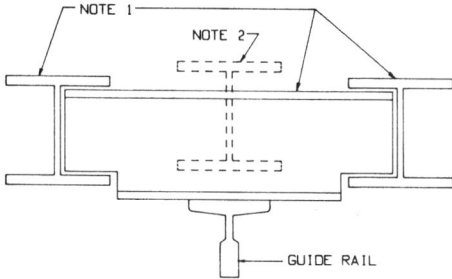

NOTE 1

NOTE 2

GUIDE RAIL

CLASS "C3"

CONCENTRATED LOADING

(NO TRUCK USED) BUT LOAD INCREMENTS ARE MORE THAN 1/4 RATED CAPACITY. CARRIED LOAD MUST NOT EXCEED RATED CAPACITY

RATING NOT LESS THAN 50 LBS/SQ-FT - 244.10 kgs/SQ-M

MUST HAVE AUTOMATIC LEVELING

NOTE 1   VERTICAL RAIL COLUMN SUPPORTS AND CROSS TIE MEMBERS ARE REQUIRED AND PROVIDED BY OTHER THAN THE ELEVATOR SUPPLIER WHEN RATED LOAD EXCEEDS 8000 LBS./3628.7 kgs. THE SIZE OF THE RAIL COLUMNS ARE DETERMINED BY OTHERS FROM RAIL REACTIONS FURNISHED BY THE ELEVATOR SUPPLIER.

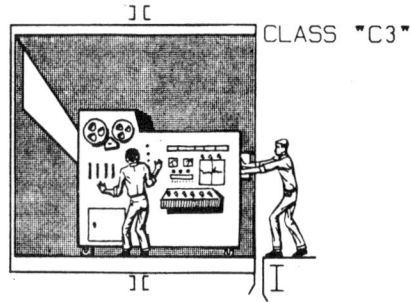

NOTE 2   ALTERNATE METHOD OF RAIL COLUMN SUPPORT. WHEN RATED LOAD IS 8000 LBS./3628.7 kgs. OR LESS. THE SIZE OF THE RAIL COLUMNS ARE DETERMINED BY OTHERS FROM RAIL REACTIONS FURNISHED BY THE ELEVATOR SUPPLIER.

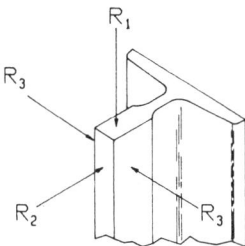

GUIDE RAIL REACTIONS (FOR A SINGLE RAIL)	ELECTRIC ELEVATOR		HYDRAULIC ELEVATOR	
	U.S. LBS.	S.I. kgs.	U.S. LBS.	S.I. kgs.
R_1				
R_2	FOR THESE REACTIONS,			
R_3	CONSULT ELEVATOR SUPPLIER			

Figure 13.1. (*Continued*) Reprinted with permission from *NEII Vertical Transportation Standards* 7th Edition, including 1994 Supplement. Copyrighted © 1992 and 1994, National Elevator Industry, Inc., Fort Lee, New Jersey. This reprinted material is not the complete and official position of National Elevator Industry, Inc. on the referenced subject which is represented only by the standard in its entirety.

Freight elevators are available in sizes of 2500 to 25,000 lb (1200 to 12,000 kg) or more (Tables 13.1 and 13.2). Some can carry loads of 100,000 lb and others can carry fully loaded trailer trucks. It is quite common to provide 60,000- or 70,000-lb (27,500- or 32,000-kg) capacity hydraulic elevators with platforms 12 ft by 40 to 45 ft (3 m by 12 to 13.6 m) long to take trailer trucks from street level to a basement loading dock in congested downtown areas.

TABLE 13.1. Loading by Hand or by Hand Truck (Class A Loading)

| Capacity, lb (kg) | Platform Size | | Maximum Speed (Standard Equipment) |
	Width, in. (mm)	Depth, in. (mm)	
2,500 (1,200)	64 (1,650)	84 (2,250)	No limit
3,000 (1,400)	76 (1,930)	96 (2,450)	No limit
3,500 (1,600)	76 (1,930)	96 (2,450)	No limit
4,000 (1,800)	76 (1,930)	96 (2,450)	No limit
5,000 (2,250)	100 (2,540)	120 (3,050)	Usually 700 fpm (3.5 mps)
6,000 (2,700)	100 (2,540)	120 (3,050)	Usually 700 fpm (3.5 mps)
8,000 (3,650)	100 (2,540)	144 (3,660)	500 fpm (2.5 mps)
10,000* (4,550)	124 (3,150)	144 (3,660)	300 fpm (1.5 mps)
12,000* (5,500)	124 (3,150)	168 (4,270)	300 fpm (1.5 mps)

*Elevators of this size should always be considered for industrial truck loading.

Because of the relatively long time required to load or unload a freight elevator, speed is generally of secondary importance. Elevator travel at high speed is not cost-effective if 4 or 5 min are necessary to transfer its load. Practical speeds are proportional to building height and required loading time, with quicker transfers warranting higher travel speed. Speed should be sufficient so that traveling will consume only about 25% of total round-trip time.

Hydraulic elevators are especially attractive for freight elevators. Lower speeds are the most cost-effective since motor horsepower is directly proportioned to speed. The load imposed on a hydraulic elevator is directly transferred to the pit through the plunger; hence, less structural cost is incurred by a hydraulic elevator versus an overhead electric elevator (Figure 13.2). Starting about 1980, hydraulic elevators with telescoping jacks were applied and, in later years, roped hydraulic elevators. Application of the roped approach has increased since the cost of drilling for rises in excess of about 30 to 40 ft (9 to 12 m) especially multiple plungers, can be offset even though the rope hydraulic, since it is suspended, requires a car safety.

The direct-plunger-type remains the favored application for the larger trailer truck elevators required to meet the growing demand for off-street truck loading, which is a necessary feature of any urban office building. Because the rise is usually 20 ft (6 m) or so, the

TABLE 13.2. Loading by Industrial Trucks (Class C[b] Loading)

| Capacity, lb (kg) | Platform Size | | Maximum Speed (Standard Equipment) fpm (mps) |
	Width, in. (mm)	Depth, in. (mm)	
10,000 (4,550)	100 (2,540)[a]	144 (3,660)	300 (1.5)
12,000 (5,500)	164 (3,150)[a]	168 (4,270)	300 (1.5)
16,000 (7,300)	124 (3,150)[a]	168 (4,270)	300 (1.5)
18,000 (8,200)	124 (3,150)[a]	192 (4,880)	300 (1.5)
20,000 (9,100)	148 (3,160)[a]	240 (6,100)	200 (1.0)

[a]Vertical biparting doors of this width less 4 in. (100 mm) can be accommodated.
[b]Class C1 or C2 depending on static load requirements.

Figure 13.2. Hydraulic freight elevator capable of handling trucks (Courtesy Dover Elevators).

drilling cost is not a major factor. We emphasize, as we have done previously, that underground protection of the cylinder is an important consideration with the use of any direct hydraulic elevator.

Loading and Unloading

The loading and unloading of a freight elevator usually takes the greatest percentage of freight elevator round-trip time, depending on the loading method used. Loading pallets by handtruck may require 15 to 20 sec to transfer each pallet load on or off the elevator. if the car is to be loaded with four pallets, more than a minute will be spent in loading or unloading. This time can be improved if power truck loading is employed. Time for door operation must be added to the loading time.

Freight elevators usually have vertical biparting doors, consisting of upper and lower panels that counterweight each other (Figure 13.3). With manual operation, a door of this

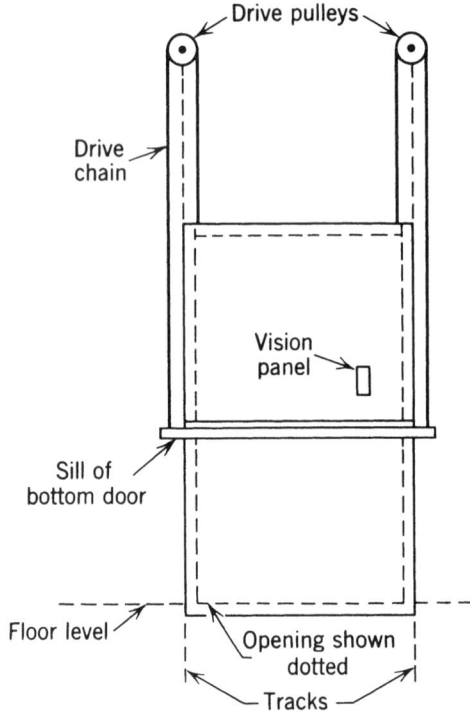

Figure 13.3. Vertical biparting doors.

type can be opened or closed in 6 to 8 sec plus about 6 sec to open or close the vertical lifting gate on the elevator. With power operation the door and gate can be simultaneously opened or closed at a speed of about 1 fps. The average 8-ft-high opening can thus be closed or opened in about 4 sec, as each panel has to travel only half the height.

Vertical biparting doors allow an opening almost the full width of the platform, and the structural steel frames of the opening are less susceptible to damage than the formed metal frames of passenger-type entrances. The door panels are completely hidden before loading can begin and thus are less likely to be struck by a moving truck, as can happen with horizontally sliding doors. Sills can be reinforced to withstand truck wheel impact (Figure 13.4).

The time required to transfer a single pallet load with an industrial truck obviously varies with the transfer distance. If we assume transfer directly off or on the elevator, each pallet can be moved in 10 sec and total loading or unloading of four pallets will require no more than about 40 sec. The typical cycle would be as follows:

Door opening time	4 sec
Unloading (or loading) time	40 sec
Door closing	4 sec
Time per stop	48 sec
inefficiency, 15%	7 sec
Total time per stop	55 sec

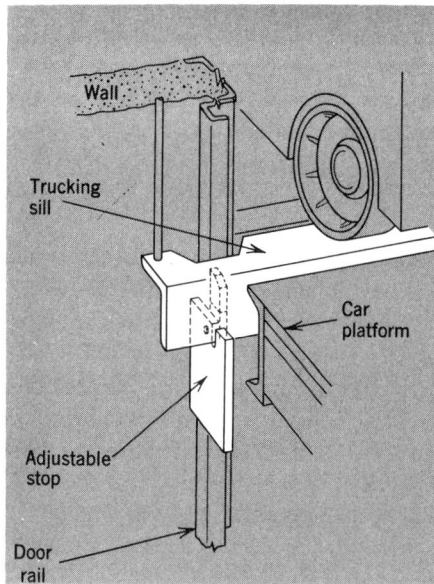

Figure 13.4. Truckable sill for vertical biparting door.

If loading at one stop and unloading at another takes place, running time is simply added to twice the time per stop indicated in the preceding list to determine the time required to handle four pallet loads. If a running time of 1 min is assumed (which will vary with speed and rise), the total round-trip time is 3 min, or 20 trips per hour or 80 pallets per hour. If the elevator is to be unloaded and then reloaded at the same stop, the additional loading time should be allowed and the longer round-trip time calculated.

Personnel on Freight Elevators

If people are to ride on any freight elevator, elevator codes require certain provisions to be made. If only personnel necessary for freight handling are allowed on the car and prominent signs to this effect are displayed, the elevator is considered strictly freight. If other employees use the elevator for transportation to upper floors, the A17.1 Elevator Code requires that the elevator platform area be rated the same or greater than a passenger platform area.

Biparting doors must have sequence operation so that the car gate closes before the hoistway door can close and the hoistway doors must open before the car gate opens. A protective safety edge must be added to the car gate so that the closing gate will stop and reopen if it encounters an obstruction in its descent. Additional protection can be a light ray across the entrance that will stop and reopen the gate when interrupted. When sequence operation is provided, an elevator with biparting doors may have an automatic closing operation. This means that, as with a passenger elevator, the doors may begin to close a time interval after they are opened so that the elevator can be called from floor to floor without an attendant. Without these provisions the doors must be closed by a constant pressure button after each stop. Because people tend to be careless about reclosing doors, a door-open bell should be provided that automatically rings when someone calls for an elevator at another floor and the elevator doors are not closed.

The door-close button is often hung on a pendant in front of the elevator so that the driver of an industrial truck does not have to dismount to close the door.

Structural Requirements

Freight elevators with class A loading are loaded and unloaded manually and require structure in the hoistway equal to that of a passenger elevator. If the elevator is to be rated for class B or C loading with vehicular loading, tractor and trailers, or industrial truck loading and unloading, extra rail support must be provided (Figure 13.5).

The entire impact of the truck stopping, running, and reversing is transmitted to the building structure through the elevator rails. The diagram in Figure 13.5 shows how these forces react. An example of an extra rail structure to absorb these forces is shown in Figure 13.6. The necessary structure for each installation must be calculated and engineered for the particular elevator size and capacity.

The elevator engineer normally calculates the forces, and the architect or structural engineer is responsible for providing the building supports. The following memorandum containing Figures 13.7 and 13.8 show a form for transmitting the necessary information.

All other parts of the elevator must be designed to withstand the extra loads of industrial truck loading. The platform, for example, requires extra-heavy flooring either of steel plate or industrial-type wooden blocks on end; the platform structure requires a multiplic-

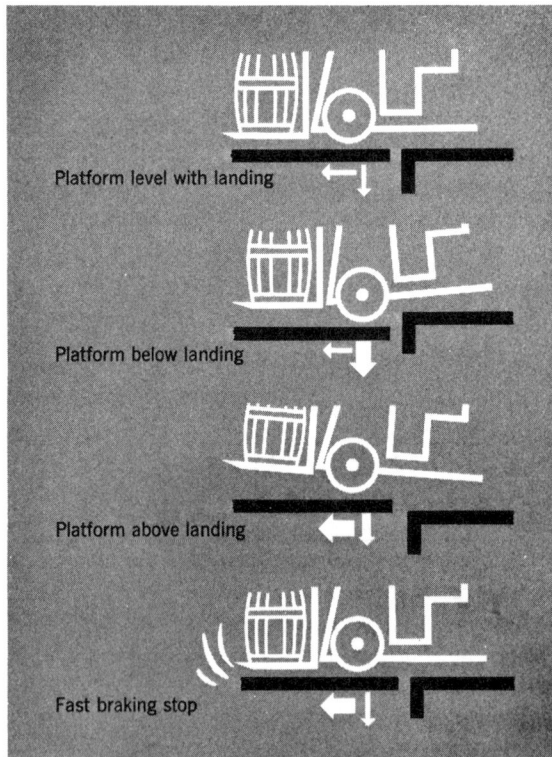

Figure 13.5. Reactions on the edge of a freight elevator platform during loading.

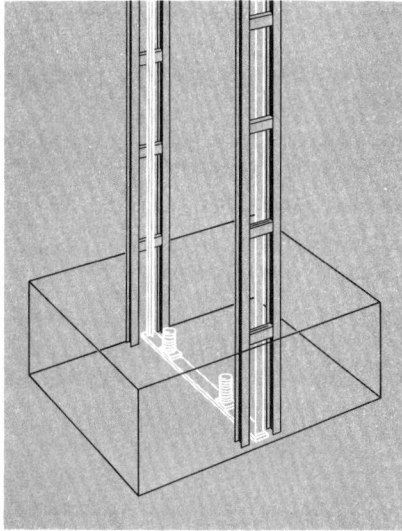

Figure 13.6. Rail column support.

ity of stringers to distribute the load and prevent distortion; and the car frame and hoisting machine or hydraulic pistons must be of heavy-duty design. With extremely large loads, multiple guide shoes and, possibly, two sets of rails may be required.

Memorandum

Manufacturing Company
Anywhere

Subject: Industrial Truck Loaded Elevator
Building Number
Anywhere

Freight elevators carrying industrial trucks, automobile trucks, or passenger automobiles exert forces of large magnitude on the building structure. In order to assure a safe and satisfactory installation, it is essential that the supports you provide for the guide rails will be of adequate strengths and stiffness. We are therefore furnishing you the rail forces and stiffness requirements for your elevator. A brief explanation of the effects of these forces is outlined below.

Forces at Loading.   When the loaded truck enters the elevator, it exerts heavy loads at the front edge of the platform. These loads produce a force couple—the top guide shoe exerts a force on the guide rail toward the loaded edge of the platform and the lower guide shoe exerts a force away from it. With entrances at the rear of the elevator, these forces will be reversed. These rail forces (designated R_2) produce bending moments in the rail, and its support, the double H columns, acts as a beam placed vertically.

Capacity	20,000 lb	Safety type	Instantaneous
Car size	12 ft 0 in. × 18 ft 0 in.	Safety load	42,000 lb
Service	Industrial truck	Vertical height between guide shoes	152 in.

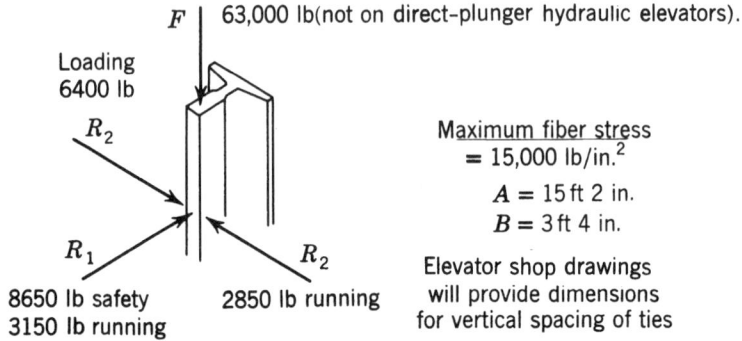

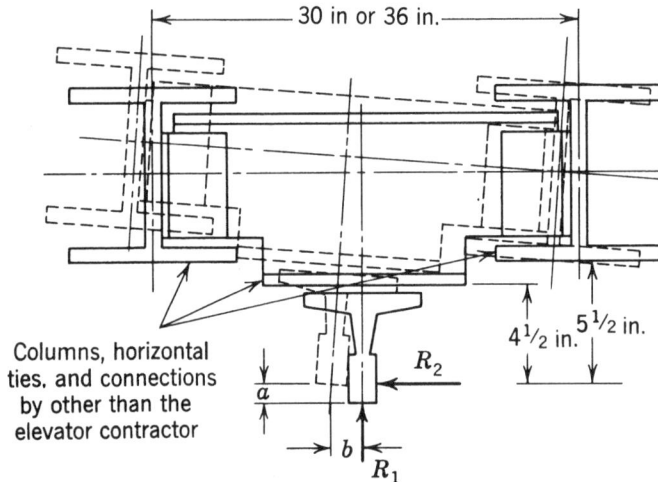

F | 63,000 lb(not on **direct-plunger** hydraulic elevators).

Loading
6400 lb

R_2

Maximum fiber stress
= 15,000 lb/in.^{2}

A = 15 ft 2 in.
B = 3 ft 4 in.

R_1 R_2

8650 lb safety   2850 lb running
3150 lb running

Elevator shop drawings
will provide dimensions
for vertical spacing of ties

30 in or 36 in.

$4\frac{1}{2}$ in. $5\frac{1}{2}$ in.

Columns, horizontal
ties, and connections
by other than the
elevator contractor

R_2

a

b

R_1

Section X–X
See table below for values

Condition	Net forces	Maximum allowable deflection
Loading a	R_1 = 3150 lb	$a = \frac{1}{8}$ in.
(as indicated)	R_2 = 6400 lb	$b = \frac{3}{16}$ in.
Running a	R_1 = 3150 lb	$a = \frac{1}{8}$ in.
(at S)	R_2 = 2850 lb	$b = \frac{3}{16}$ in.
Safety a *	R_1 = 8650 lb	$a = \frac{5}{16}$ in.
(at S)	F = 63000 lb b	–

a These conditions need not be considered as acting simultaneously.
b For design of pit supports, this force must be doubled for impact.

*These forces are not present with
direct–plunger hydraulic elevators

Figure 13.7. Freight elevator rail reactions—plan.

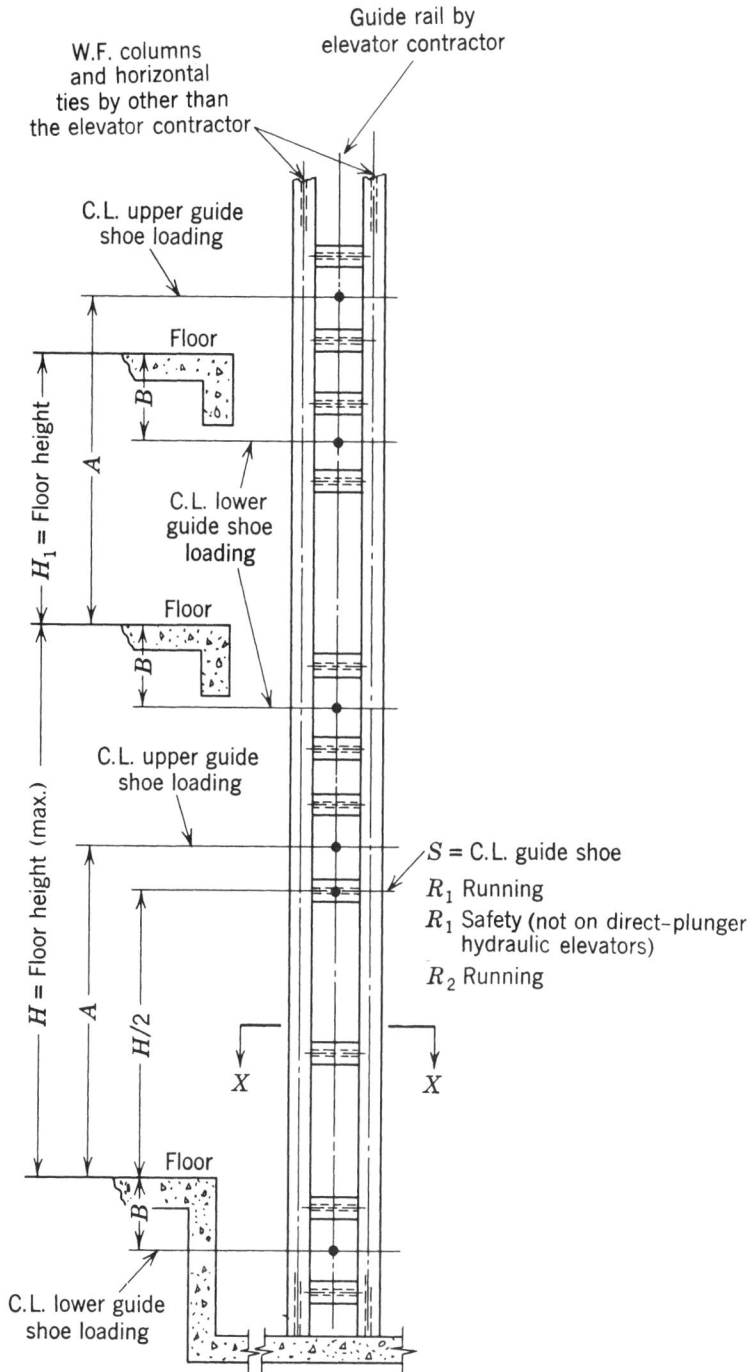

Figure 13.8. Freight elevator rail reactions—elevation.

The R_2 forces also produce turning moments in a horizontal plane, bending one column toward the hoistway wall and the other away from the hoistway wall.

In addition to force R_2, there is a force R_1 tending to spread the rails apart due to eccentric loading of the platform. This force produces direct bending moments in the columns (Figures 13.7 and 13.8).

Since these forces are at loading and are transmitted to the columns through the guide shoes and the rails, they occur at determinable points above and below the loading levels. Refer to Figure 13.7 for dimensions and table for values of forces.

Forces at Running. Forces R_1 and R_2 also occur during running of the car and have the same effects. Force R_1 can be the same magnitude as for loading or somewhat larger. Force R_2 will be somewhat smaller. However, since the running forces R_1 and R_2 occur during the entire travel of the elevator, it must be assumed for design purposes that they will be applied to the center of each column span and each support for the column (floor beam, spreader beam, etc.). Refer to the table in Figure 13.7 for magnitude of forces.

Forces at Safety Application. At safety application when the car is provided with a safety device, the stopping action of the safety exerts a large force F vertically on the rails in a downward direction. Such forces are transmitted through the rail to the pit floor. There is also a large force R_1 due to eccentric loading, which tends to spread the rails (similar to loading R_1 and running R_1). The loads resulting from safety application can occur anywhere in the entire car travel. See the table in Figure 13.7 for size of these forces. These forces do not apply in direct-plunger hydraulic elevators.

ELEVATOR DESIGN

Car interiors deserve particular attention. Steel or oak rubbing strips should be mounted along the sides, adequately supported to withstand blows from carelessly driven trucks, and designed to be renewable. If trucks are to be carried, the location of the landing button fixtures demand special consideration, such as the use of pendant mountings. Car operating panels should always be recessed into the side of the elevator.

Car lighting should be bright to make the interior of the elevator fully visible, which will tend to reduce the likelihood of trucks hitting the side or rear. A curb can be effectively used to limit the distance a truck can be driven into the elevator. An effective protection for the sides of the car is to mount one or two 12-in. (300-mm) channels on end on the lower sides. In that way, the forks of a forklift truck cannot damage the sides.

The car platform should be provided with durable or easily renewable flooring. Checkered steel plate or aluminum plates with abrasive surfaces have been used. These should be designed in sections of about 200 lb (100 kg) which can be removed and replaced. Many safety codes require the use of high-visibility markings (for example, yellow and black striping) on movable parts such as doors and gates. These markings act as a warning and the contrast in color may act as a caution and minimize impact.

Materials-handling elevators also have to be installed in "problem" environments ranging from an explosive atmosphere to an area in which normal cleaning with hot water and steam requires a completely waterproof installation. Corrosion control and other elements of maintenance of structural steel in the elevator installation must also be considered. All safety devices as well as ropes and electrical equipment must operate properly in these difficult environments, and special precautions taken in the specifications for elevator equipment to guard against premature wear and danger.

The contractor or structural engineer must provide the necessary structural framing at each level, which will become the frame and sill for the vertical biparting door. Because these doors as well as the lifting car gate may require extra space above the top landing or in the pit, the elevator overhead and pit must be structured to suit.

Structural requirements for freight elevators are considerably more critical than those for passenger elevators. They are numerous and variable and require early consultation with elevator engineers when large freight elevators are planned.

SERVICE ELEVATORS

Service elevators in commercial, residential, and institutional buildings are generally passenger elevators with special provisions to handle oversized loads and both hand- and motor-assisted trucks. They should have abuse-resistant interiors, must comply with code requirements for carrying passengers, and should be equipped with horizontal sliding doors that allow full automatic operation with maximum efficiency. Service elevators should be classified according to the types of loading expected: class A freight loading if hand- or motor-assisted trucks will be used, class C 1 or 2 if powered industrial trucks are used (the elevator should be large enough to accommodate such a truck), and class C3 if an occasional heavy single-piece load, such as a transformer in a commercial building, is to be carried. This classification ensures that the structure and the platform of the elevator will be designed to accommodate such loading.

General rules of application are that at least one service elevator should be provided for each 300,000 ft² gross (3000 m²) of office space, each 100 beds in a hospital, and each 200 rooms in a hotel or any apartment house that deserves the "luxury" classification. These are general rules and should be modified by the specific requirements of the building under study and the needs of the occupants. For example, a headquarters building of a large corporation of 300,000 ft² (3000 m²) may require two service elevators because of the expected activity, or a small single-purpose office building of 100,000 ft² (1000 m²) may have a definite need for a service elevator. An automated materials handling system may modify this requirement, as discussed in Chapter 15.

If enough information is available, a time and motion study can be made of the expected number of required trips on the service elevators. For example, if it is known that mail will be delivered twice daily, coffee service will take place to each floor, supply requisitions will be delivered on schedule, and a certain amount of maintenance movement and a predictable number of moves and renovations are possible, a time allowance can be assigned to each movement; these can then be totaled and compared against available service elevator time. A judgment must then be made as to the consequence of not having the service elevator available versus an alternate means of moving essential material.

Service elevators can be of any practical platform size or ceiling height provided that the platform area conforms to elevator code requirements. Elevators can have an interior height of up to 15 ft (4.5 m) if sufficient elevator hoistway overhead exists. Such a height can accommodate long carpet rolls (for an office building) and oversize furniture. To aid in loading and unloading, a hoist beam or lifting eye can be mounted in the ceiling so that a chain fall may be used to haul the load into the car (Figure 13.9). Elevators of this height have contributed to the early completion of buildings by allowing off-site prefabrication of large building components and office interiors. For the larger office building, an elevator of 6500-lb (3000-kg) capacity with 54-in. (1400-mm)-wide center-opening doors is recommended. A minimum recommended size service elevator is 4000 lb (1800 kg) with 48-in. (1200-mm) wide doors.

The size of a building service elevator should be established by determining the most frequently expected loading. This may be movable partitions for office renovations or large-size

Figure 13.9. Service elevator with high cab and hoist eye.

supply carts. The area of the car inside will establish the rated load-carrying capacity of the elevator. As a second consideration, if this size area capacity can be moderately increased to carry the heaviest load, such as a transformer or compressor, it should be done.

OPERATION OF SERVICE AND FREIGHT ELEVATORS

The grouping and operation of service and freight elevators depend on their role in the overall building industrial or storage system. Service or freight elevators are seldom used in large groups, the usual installations being individual units. Two major forms of automatic operation are available as well as a number of manual operations.

Automatic operations are either collective or single automatic. With collective operation, each call is remembered and the elevator answers all the up landing calls on the up trip, reverses at the highest call, and answers all the down calls on the down trip. The operation is useful if one loading is not expected to fill the elevator and additional loads can be taken on at other stops. If full loads are expected, a bypass feature is used, allowing an attendant to operate a special switch within the car to bypass landing calls.

Attendant operation is almost a necessity on a busy freight elevator in an industrial setting and can often be used to an advantage on a service elevator in a large commercial building or hospital. The attendant can move material on and off the elevator into an adjacent lobby and can act as a supervisor if a great deal of material delivery is being made, so as to avoid individual messengers tying up the elevator. Where a move in or move out is in progress in a building, the attendant can supervise and provide service for the movers, plus allow other material movements to take place when the movers are away from the elevator.

"With Attendant" operation transfers the control of a fully automatic elevator to controls located in the car operating panel. This is usually done by a keyed switch located in the car. The attendant will have full control to close the doors, select the direction of travel, and to start the car. Once the car is started, the car will automatically stop to answer landing calls in the direction chosen to travel, reverse at the highest registered call, and proceed to answer calls in the opposite direction if started by the attendant at each stop. The attendant may choose to reverse direction at any stop and proceed in that direction. A nonstop button is usually provided that allows the attendant to bypass any landing call for whatever reason, usually because the car is loaded.

The actual controls will vary depending upon the age and manufacture of the equipment but, in general, will include the following features.

If "with attendant" is to be used on a building service elevator, an annunciator is recommended to indicate the floors at which loads are waiting. In addition, an illuminated sign in the landing button or landing lantern fixture should inform prospective passengers that the car is "In Freight Service" and not available for general use. A "house phone" should be provided in the service elevator so that the attendant can be called if a special material movement is necessary.

If more than one service elevator is installed, the elevators should be next to each other and provided with a duplex collective or group operation. If security is a consideration, the loading dock attendant's office can be provided with elevator position indicators and means to recall and monitor the elevators, as well as a dedicated intercommunication system. All service elevators should be provided "with attendant" operation.

Single automatic operation allows the person who has control of the elevator exclusive use of the car for that trip. The landing call buttons incorporate illuminated signs to indicate whether the elevator is in use. When the light goes out, the elevator may be called to a particular floor and be used by the next person. This operation is preferred whenever the load is such that it would fill the elevator and additional stops would be unproductive. A variation of this operation is to have a central dispatching station through which a dispatcher controls the elevator use and destinations. The dispatcher is called by telephone or intercom to send a car to a given floor and to dispatch it to the unloading floor. This operation is often used in garages when a number of elevators in a group are required to serve the garage capacity. For any multiple automatic elevator group, a landing lantern or indicator light associated with each entrance should be provided to inform the truck operator as to which elevator will be available.

A semiautomatic operation called "double button" is often used with freight elevators as a cost savings approach. The operation is such that a car can be called to a floor by constant pressure on an up or down button at the landing. The elevator can also be operated from within the car by similar constant pressure on either an up or down car operating button. Leveling may be automatic or accomplished by means of "inching" buttons used to jog the car to floor level within a restricted zone at each floor with the doors open, and to operate the car only toward floor level, never away from it.

An optional operation for any freight elevator that opens on a street level is a system known as "tail-board inching," which can be used to stop the elevator platform above the actual floor level, at the same height as a truck backed up to the elevator. An extra-long guard is installed below the elevator platform to avoid the danger of a person or object falling into the hoistway.

Some factories and warehouses use a freight handling system where the elevator, similar to a dumbwaiter, is unattended and sent from floor to floor using controls mounted at each floor. The elevator code permits such operation as automated materials handling per part 14 of the ASME A17.1 code. The attendant loads the elevator at a floor, closes the doors, and sends the car to the desired floor using a button in a hall station. The elevator may be equipped with automatic doors or may indicate its arrival by a bell or light where it can be unloaded.

Such systems have been used in munitions factories and similar hazardous locations where it is not prudent to have personnel ride the elevator. Such operation is only permitted from the landings and there is no in-car operating device. In some applications the on-car safety device is omitted since its use could present a possible sparking hazard.

If an industrial plant or warehouse requires a group of elevators, many of the operations outlined for passenger service may be adapted for freight service. The final choice will depend on the door-operating system employed. Because it is difficult to ensure that people will close biparting doors after they use an elevator, the accepted system is to employ a door-open bell that will sound if someone is calling the car while the door is open. use of sequence operation with biparting doors and a safety edge on the car gate is one means of providing automatic operation.

Large openings with horizontal sliding doors will ensure prompt automatic operations of the elevators. If loads are light and bulky, such as racks of garments in a department store or clothing factory, the horizontal sliding entrances are desirable.

ELEVATORS IN INDUSTRIAL PLANTS

The typical piano factory constructed many years ago was a multistory building. Light parts of the piano were manufactured on the upper floors, all raw material being taken to the top. The foundry and finishing shops were on the lowest floors. The myriad small wooden pieces were assembled as they traveled down, becoming larger and larger sections until they became the completed piano on the shipping floor.

Like the piano factory of old, many modern industrial processes now follow a similar sequence. Because land costs were moderate when many manufacturers started, their plants were usually horizontally oriented. Transportation of raw materials and finished products was also economical. However, transportation and land costs have been rising, so that the vertical factory is becoming more and more economical and, with dependable vertical transportation, quite feasible.

By engineering vertical transportation on the basis of the production process, a completely integrated factory can have compact vertical design. Large elevators can carry raw materials to upper floors, and conveyors, dumbwaiters, or small elevators can bring the finished parts to the lower floors for assembly. Personnel can be moved swiftly to and from their jobs by elevators or escalators.

The role of gravity cannot be ignored in a vertical factory. In a cannery, for example, raw tinplate sheets are moved to upper floors for shaping into cans and rolled down to the food-processing floors for filling, sealing, labeling, and shipping.

The number, size (area), and capacity, as well as required speed, of manufacturing plant elevators can be determined by simple calculation based on indicated need. Each area requiring transportation must be studied and the expected volume estimated for an applicable time period. Elevators to handle this volume can be based on a number of factors, as discussed in the following paragraphs.

Time to Load

The time to load varies with the type of loading: industrial truck, hand truck, cart, or hand. Industrial truck loading is the fastest but also creates the greatest stress on the elevator equipment.

Door-close Time

Door-close time depends on the type of door, the height or width of the opening, and the use of power or manual operation. Vertical biparting power-operated doors can be operated at 1 fps (0.3 mps) per panel average speed, which allows an 8-ft (2400-mm) opening to be opened or closed in 4 sec. Horizontal sliding doors must comply with the kinetic energy limitations of the A17.1 Elevator Code, and door-open and door-close times are listed in Table 4.3.

Time to Start and Run the Elevator

Time to start and run an elevator in an industrial plant is the same as calculated for passenger elevators of the same given speed and is shown in Chart 4.2. Speed is not a critical factor for freight elevators, 50 fpm (0.25 mps) being acceptable for a two-story building and 100 fpm (0.5 mps) for four stories. Most freight elevators are of the low-rise hydraulic type because of the usual large loads they carry. The disadvantage of the hydraulic elevator is the relatively large electrical demand imposed on the power supply during starting in the up direction. If frequent use of the elevator is expected (many trips per hour), the more cost-effective elevator may be the counterweighted electric-type elevator.

Unloading Time

Unloading time depends on the factors considered in loading. The combination of front and rear entrances will expedite loading and unloading of the elevator, especially if industrial-type small tractors and trailers are used to pull loads around the plant.

Typical industrial elevator situations are analyzed in Examples 13.1 and 13.2. An infinite number of variations are possible to satisfy the requirements of various plants, but using equipment in line with manufacturers' standard equipment is recommended as being the most cost-effective.

Example 13.1. Elevators in Industrial Plants—Pallet Loads

Required to move: pallets 4×4 ft $\times 6$ ft high, each weighing 2500 lb, approximately 200 per day. Loading by industrial truck, truck weight 8000 lb, 80% of weight on front wheels. Distribution: 50% load to fourth level from dock, 36-ft rise; 50% load to second level from dock, 12-ft rise
Determine: elevator size and speed
Size: assume 4 pallets per trip; minimum platform size 8×8 ft interior, 10,000 lb; recommended standard size 8 ft 4 in. $\times$ 12 ft 0 in.

Rating:

1 pallet and truck	2500	+ 8000 = 10,500 lb
2 pallets and truck	5000	+ 8000 = 13,900 lb
3 pallets and 0.8 truck	7500	+ 6400 = 13,900 lb
4 pallets and 0.8 truck	10,000	+ 6400 = 16,400 lb

16,400-lb static load required. 12,000-lb car with static loading (50% over capacity). Number of elevators required (assume 100 fpm)

To fourth level:

Load at dock 4 pallets at 15 sec per pallet	60 sec
Door close	4 sec
Run to fourth level, 36 ft @ 100 fpm = 23.8 sec	24 sec
Door open	4 sec
Unload	60 sec
Door close	4 sec
Return	24 sec
Door open	4 sec
Total time, 4 pallets	184 sec, say 3 min

Total time, 100 pallets, 25 trips $\times$ 3 min = 75 min

To second level:

Same as above except running time changes:

12 ft @ 100 fpm = 9.4 sec, 24 − 9. = 15 $\times$ 2 = less 30 sec

Total time, 4 pallets 184 − 30 = 154 sec = say 2.5 min

Total time, 100 pallets 25 $\times$ 2.5 = 62.5 min

Total time, 200 pallets 75 + 62.5 = 2 hr 17.5 min

One elevator, 12,000 lb @ 100 fpm with static loading, will provide ample service.

Example 13.2. Elevators in Industrial Plants—Carts

Required: Determine the number of industrial carts that can be moved an average 6 floors (72-ft rise) by one elevator per hour. Maximum load each cart, 6000 lb, tractor load 8000 lb. Front and rear openings on elevator.

Assume: tractor can pull and deposit 2 carts on elevator; alternate tractor and 1 cart each trip (sufficient tractors)

A. 2 carts per trip:

Tractor and 1 cart 6000 + 8000 = 14,000 lb

½ cart, cart and ½ tractor 3000 + 6000 = 4000 = 13,000 lb

2 carts 2 $\times$ 6000 lb = 12,000 lb

Time to load 2 carts	15 sec
Uncouple	15 sec
Drive off, close doors	10 sec
Run 72 ft (assume 100 fpm)	45 sec
Couple tractor and drive off	25 sec
Close doors	15 sec
Run back to lower level, doors open	45 sec
Total per 2 carts:	170 sec

Number of carts per hour $\dfrac{3600 \times 2}{170}$ = 42 carts

B. 1 cart per trip and tractor:

Drive on	15 sec
Close doors	4 sec
Run	45 sec
Open doors	4 sec
Unload tractor and cart	15 sec
Second tractor drives on while first is unloading	
Close doors	4 sec
Run	45 sec
Open doors	4 sec
	136 sec per cart

$$\text{Number of carts per hour } \frac{3600 \times 1}{136} = 26$$

Result: 2 carts per elevator best way. Elevator should be rated at 14,000 lb class C1 loading for occasional cart and tractor trip. Static loading not required. Platform size depends on size of cart and tractor.

ELEVATORS IN WAREHOUSES

The nature of the warehouse operation determines the type of freight elevator required. If it is used for long-term, low-turnover storage with bulky but light loads, a large, slow-speed elevator with manual operation and manual doors may be sufficient.

If the warehouse is a fast-turnover facility making extensive use of industrial forklift trucks, its elevators must be so designed. Their number and size will be a direct function of the expected turnover.

As indicated earlier in this chapter, a time cycle of elevator loading and operation must be established to determine the average time per load. This is projected for a number of loads that must be transported and the elevator requirements determined accordingly.

A number of automated elevator systems are applicable to materials handling and may be considered for a particular warehouse problem. The systems may be divided into two categories: those in which the load is moved by means of a horizontal conveyor to the elevator and those in which the elevator moves, in essence, to the load.

Conveyor systems to feed a stationary elevator are generally supplied by conveyor specialists (Figure 13.10). Initiating and limit switches on the conveyor control elevator response. The elevator arrives, and conveyor equipment on the elevator platform transfers the load. A limit switch signifies the completion of transfer. Disposition of the load is controlled either manually or by program, with the elevator moving the load to another floor and automatically unloading. A programmable controller may be employed to keep track of and direct multiple loads to various floors and to return waiting loads at upper floors to a main floor or other destination.

One type of automatic transfer includes the use of an elevator with a tilting platform. Cylindrical loads, like rolls of newsprint, are rolled onto the elevator at the loading floor by either a pusher or a floor tilter. When the elevator arrives at its destination, the elevator platform is tilted and the load rolls off.

An elevator in a tower moving on tracks has been applied to a number of automated material-handling systems. The elevator contains a platform with a pallet transfer device, and by a combination of vertical elevator motion and horizontal tower movement, the

Figure 13.10. Automatic pallet loading on an elevator by means of pallet conveyors.

platform is indexed to various stalls and the transfer device either deposits or extracts the load from the stall.

Such systems have been utilized in warehouses handling pallet loads, and one such unit is operating in a frozen-food storage warehouse at 0°F (−15°C). A programmable computer system is employed to direct the tower and elevator to various stalls for loading and unloading and to keep inventory current. Inputs to the computer are used to assemble truck loads consisting of different pallet loads at an unloading station as orders for the materials are received. A similar system has been applied in an air freight terminal to receive and stage loads so as to minimize the turnaround time of large cargo-carrying aircraft.

ELEVATORS IN GARAGES AND FOR OFF-STREET LOADING

The history of garage elevators can be traced to the early 1900s when both horses and carriages were carried on elevators to upper floors for stabling or storage. Early horse elevators had a gate between the horses and the elevator operator and channel troughs in the platform so it could be washed down. Modern garage elevators do not need the trough, but protection against the vehicle's being driven into the extreme end of the elevator is advisable.

A garage elevator should be designed for operation by an attendant who parks or retrieves automobiles at an upper or lower floor. Landing buttons should be pendant mounted and the operating panels in the elevator located so that the attendant can operate them by leaning out of an automobile window.

Time can be saved by a garage elevator with both front and rear openings. The attendant can drive in the front and out the back, thus saving 10 to 20 sec per elevator trip.

A typical sequence of operation of a garage elevator with a single entrance, drive in, back out, is as follows:

Door open	4 sec
Drive in	10 sec
Door close	4 sec
Run to upper stop depends on rise and speed of elevator	
Door open	4 sec
Back out	20 sec
Door close	4 sec
Total standing time	46 sec to park one automobile

The cycle is then repeated by adding the running time to return for the next automobile.

With the drive-through type of elevator, back-out time is replaced by drive-off time, which can be as little as 10 sec.

The number of elevators required for a particular garage is a function of the number of floors, the number of automobiles to be turned over (parked and unparked in a given time), the number of attendants available, storage space at the entrance floor, and the speed with which customers can pay their bills and drive out to the street.

All these factors are important and are reflected in the economics of the elevator installations. Little is gained by an elevator plant that can deliver a car a minute if the local streets are so congested that it takes two minutes to leave the garage.

If the garage depends on elevators for its operation, a minimum of two is recommended to maintain continuity of service. A garage elevator usually has a 7000-lb capacity to handle large limousines. The dimensions of the 7000-lb elevator are a platform 10 ft (3050 mm) wide by 24 ft (7300 mm) long. If an attendant is to operate the elevator without leaving the automobile, the width should be 8 ft (2500 mm); 21 ft (6400 mm) is the minimum length for a garage elevator. If larger cars or trucks are expected to be handled, capacity and dimensions should be increased.

An important outgrowth from garage-type elevators is the use of off-street loading facilities in office buildings to minimize the area needed for street-side loading docks. Because street space is valuable, trucks can be taken by an elevator to a basement area, turned around by turntable if necessary, and unloaded at a basement loading dock. The size of the elevator is exceedingly important, and most of the truck elevators are hydraulic because of the loads involved, their location, and the short travel distance involved.

Truck-type elevators can range in size from a platform 10 ft (3050 mm) to 12 ft (3700 mm) wide by 24 ft (7300 mm) long, which will handle small delivery trucks, to a platform 12 ft (3700 mm) to 14 ft (4300 mm) wide by about 40 ft (12,200 mm) long to handle a large trash truck. If tractors and trailers are to be served, lengths up to 60 ft (18,300 mm) are necessary. Capacities range from 20,000 lb (9100 kg) for the smaller trucks to 70,000 lb (32,000 kg) for a fully loaded trash truck with wet trash. An important consideration in the design of a larger building is the way it intends to handle its trash. With extensive food service facilities, wet garbage can present a problem and the considerations may include the use of compactors and associated containers, which must be loaded on and off special trucks.

If the building is dependent on a basement-level off-street loading dock facility, two elevators are necessary. Traffic control must be provided so that a truck leaving the loading dock area will not be in the way of a truck entering either at the street level or at the elevator exit level below.

Consideration should also be given to emergency operation in the event of a power failure. The building's standby generator can be designed to have sufficient capacity to run one of the truck elevators at a time, or a low-horsepower auxiliary pump can be designed into the elevator system so that one of the hydraulic-type truck elevators can be raised to street level at a much lower speed than normal operation to get a truck out of the building.

OTHER FREIGHT AND SERVICE ELEVATOR CONSIDERATIONS

In addition to the total load and methods of loading freight and service elevators, the environment in which loading takes place must be considered. Unless special hazards in a plant dictate otherwise, durability is a prime requirement. If there are hazards such as chemical atmospheres, abrasive or explosive dusts, or moisture, the elevator equipment is governed by the same considerations as any other electrical and mechanical installations in the hazardous area. This is more fully discussed in Chapter 14 on special installations.

To gain the greatest efficiency from an elevator plant, sufficient space for access to the elevator and for unloading must be allowed. This is one advantage of an elevator with both front and rear openings, provided one side is used for unloading and the opposite side for loading. A minimum consideration must be sufficient room to maneuver either hand or industrial lift trucks. With a loading dock, sufficient space to hold material to be moved must be provided and related to the capability of the elevator.

Elevators in industrial plants are production equipment and must be treated as such. They are subject to downtime for maintenance and repair just as a lathe or a press would be. The fact that elevators are unusually vital, not only to the productive function but to the movement of people, must be recognized. If the entire operation of a plant depended on one machine, a standby would probably be considered, and the same logic should be applied to plant elevators. The alternative would be to have available spares for any parts that might fail to permit replacement in minimum time.

A systematic consideration of all the requirements that the use of a service or freight elevator must fulfill should be made before the actual equipment is specified. The choice of equipment and subsequent structural and building requirements will follow once the quantity and quality of the loads are known. The next consideration is a failure analysis—what would be the consequence or alternative if the elevator were not available.

As with any vertical transportation analysis, sufficient documentation should be provided so that the people responsible for building operation are aware of the basis of original design.

As may be seen in this chapter, a variety of designs are available to the architect for making a proper application of vertical transportation to a new facility. In Chapter 15 automated means of materials handling will be fully discussed.

UNUSUAL INSTALLATIONS

The following chapter deals with nonconventional elevators and their installation, but there are also a number of unusual applications and modifications of design specifically for freight elevators. If there is a need to solve a problem, the ingenuity of the plant engineer and the elevator engineer can often provide the answer to the particular application.

Examples of unusual installations abound. One notable instance is the assembly of jet engines, wherein the engine frame is vertically mounted on a platform that is lowered as various components are mounted and piped or wired to related devices. The completed unit is then raised and rolled to the next station for final fitting. The equipment used is hardly an elevator, but it employs an application of freight elevator technology, as would any vertical positioning device where heavy loads are involved.

Industrial work is often done in extremely hazardous locations, and any elevator equipment must be designed to operated and be maintained safely in that environment. Obsolete as it may sound, wood guide rails may be needed as well as safety devices that will operate on wood, reminiscent of the late 1800s. Bronze guides on steel rails, beryllium copper safety jaws, stainless steel or fiberglass rails, plastic guide shoe gibs, and other nonsparking combinations have been applied to the moving parts of an elevator car. Structures can be made of composite material where normal steel would be subject to extreme corrosion.

Electrical equipment has included controllers that are oil immersed; placed in explosion-proof or rated boxes; pressurized with nitrogen; and even remotely located to avoid exposure. Simply recognizing the extent of the hazard may be sufficient. Approaches such as pressurizing a machine room, mounting electrical equipment above a floor level or pit where heavy gases may accumulate, or providing protection only in those areas may be sufficient.

Vintage Equipment

As both heavy and light manufacturing facilities and warehouses have been closed as a result of the changing economics in certain areas, many of these buildings are being recycled into residential buildings, retail outlets, and a variety of other uses. Notable examples are the former textile mills in New England. The elevator equipment in these buildings, often dating back to the nineteenth century, continues to be used for purposes other than its intended use.

Such equipment includes open platforms, wooden lifting gates, handrope operations, and a number of ancient belt-driven machines. Although efforts are made to outlaw such reuse, the architect and engineer should also be aware of the inadequacy of such elevators for modern use. A platform area may be rated very lightly for the loading that may occur, safety devices may be completely inadequate, and hoistway protection nonexistent. Elevator codes, as well as building codes, are seldom retroactive, and the solution may be a complete replacement of this antiquated equipment. Designation of a converted older building as a historic site may inhibit dramatic improvement, but a great deal can be done to preserve the original nature of the installation and provide the modern safety features and operation expected from the equipment. A particular requirement is the leveling mandated by legislation such as the Americans with Disabilities Act if the building becomes a public facility.

An effort to establish a national elevator code was made in 1986 with the issuance of the American Society of Mechanical Engineers' (ASME) A17.3, the Safety Code for Existing Elevators and Escalators. This standard is available for adoption by various jurisdictions and provides a minimum level of safety for existing installations. Certain operations, such as handrope control, are prohibited and belt machines are outlawed. Electrically released and mechanically applied brakes are required. Car enclosures and adequate hoistway protection are required, as well as interlocks on landing doors and car door contacts. Emergency operation in case of fire and top of car operating devices to allow inspection are also mandated.

Maintaining the vintage look of an installation while meeting the updated requirements of the code is a challenge for both the architect and the elevator engineer. Fire-resistant glass can be used to back up a grill hoistway or car interior retaining their original appearance. Car doors can be designed and installed to complement the vintage style. Fixtures can be designed to retain their antique look. The old car switch or handrope operation can remain but becomes noneffective and decorative only. Machinery, because it is not in the public eye, can be totally replaced. The challenge is to make the area visible to the public appear old but add the necessary structural, mechanical, and electrical aspects that are part of a modern elevator. Many buildings have done this with complete success. What may be lacking is the legislative encouragement that all existing elevators comply with this farsighted code effort, as well as some ingenious compromises that elevator engineers may suggest. An elevator can still look like an old-fashion freight elevator while operating as safely as a modern passenger elevator.

Some historic buildings desire to show the vintage machinery in its original operation and condition to operate the elevators. The elevator engineer faces the challenge of meeting current safety codes and standards with minimum additions and modifications to the existing equipment. The minimum would include the addition of interlocks to ensure that doors are closed and locked before the car is allowed to move, reducing clearance between the car and landing entrances, ensuring safety against over travel, plus a myriad of other conditions possibly not recognized when the original equipment was installed. Code authorities as well as the curator of the site will have to be involved to participate in the necessary variances to the current codes when conditions will not permit exact compliance.

Freight Elevator Use and Misuse

When a building is constructed and the equipment installed, there may be design criteria for the lighter loading allowed for a class A or B freight elevator. As time goes on, this may be forgotten, and the changing nature of the manufacturing operation may foster the use of power truck loading. The building design may allow the increased floor loading, but the elevator, while capable of handling the loads, is being strained. Even if the initial design allowed the heavier class C type loading, age and continued heavy use may result in structural deterioration.

Inaccurate leveling allows the truck wheels to impact the platform edge, and braking stops strain the car structure. Indications of weakness will include loosening of guide rail fastenings, possible cracking of the car frame structure, or simple deformations. Doors and sills are particularly vulnerable, and continued need to keep the elevator in service may lead to repeated troubles. This may often encourage the totally unsafe practice of defeating the electrical door protective circuits.

Periodic review and complete appraisal of the elevator system for wear and tear are necessary. Continued abuse can lead to disaster, but it is often possible to make corrections if early warnings are heeded. Structure can be reinforced, doors can be upgraded from manual to power operation, and appraisal of the capability of the hoisting system and of the building structure may reveal that heavier elevator construction, such as metal floor plates, can be added. Management can also preserve equipment by limiting loading with such simple steps as making more frequent trips with lighter loads. An effective solution may be to designate an individual as the "freight handler" and have on-floor personnel leave loads in front of the elevator for supervised loading, travel, and unloading.

Each installation is different, each presents its own condition, and solutions must be achieved by an elevator-by-elevator approach.

Modernization Potential

Some of the possible modernization aspects were alluded to in the previous section. A competent elevator engineer who can look at all the relevant factors in use of the equipment, as well as the needs of the building's occupants, can often provide an economical solution once the appraisal is complete.

Automatic control substituted for an attendant may be appropriate for a particular operation, especially where a cart handler travels with the load. Many freight elevators designed for heavy manufacturing facilities have a single automatic operation whereby each trip is exclusive. A change to light manufacture may warrant a collective-type operation in which more than one call is served per trip. The reverse is also true if the assemblies are large and additional stops would be unproductive.

Power-operated doors and gates of the biparting type allow use of the full width of an elevator. These may be automated so that calling a car to another floor from the landing will not be frustrated by someone's leaving a door open. Sequence operation must be added; that is, the car gate must be closed before the landing door is closed and the landing door must open before the car gate. In addition, a warning bell and a reopening device on the car gate are necessary. Light-ray devices to provide additional entrance protection are also available.

Although it is possible to add power operation to an existing manual biparting door, a number of earlier designs used a rigid astragal which was, usually, an angle iron that was fastened to the upper door panel. When the doors were in the closed position this rigid astragal overlapped the lower panel. The intention was to block the possibility of a floor fire passing through the gap between the closed door panels. Other designs used a hooked center latch to prevent the doors from falling open if, for any reason, the suspension chains failed. Both of these designs created a hazard and could cause serious injury if a person's hand were to be caught between the closing doors (see Figure 13.11). If such a condition exists it should be replaced with a nonshearing astragal consisting of a fireproof flexible material as shown in Figure 13.12. Vision panels should be protected with grills, and it is essential to ensure that landing doors have an approved interlock.

In buildings in the older sections of cities and in a number of factories with their own power plants, the power to the elevator should be checked to ensure that the original characteristics have been maintained. The older dc machines may now run off converted ac power, and any increased voltage will affect speed. In addition, a dc machine regenerates while traveling with an overbalanced load, and the means to absorb this regenerated power must be provided.

Fixtures can be upgraded to be of the vandal-proof type and can be relocated if the elevator is intended to meet requirements of the Americans with Disabilities Act. A communication system between the car and a central station is necessary with any automatic elevator. Visual indicators, remote monitoring, and other amenities to make the lowly freight look like a first-class installation are further options, and a colorful paint job and improved lighting can often create an attractive conveyance.

Maintenance

No plant manager would think of operating a production machine continuously without periodic maintenance and shutdown time for critical replacements. Too often plant elevators, which are major production tools, do not enjoy this benefit, and panic sets in when their use is denied.

Periodic maintenance, with the necessary lubrication and replacement of worn parts, is required. Whether this is done during a shift change, late at night, or on weekends, it is

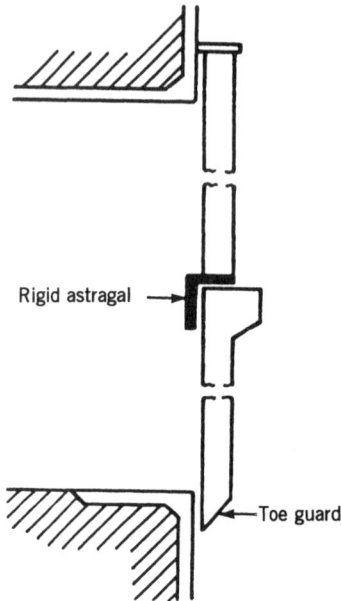

Figure 13.11. Vertical biparting freight entrance with rigid astragal (Courtesy A17 Handbook).

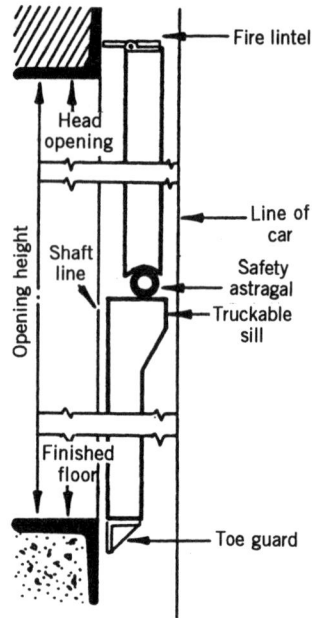

Figure 13.12. Vertical biparting doors equipped with safety (nonshearing) astragal (Courtesy A17 Handbook).

the only way to ensure that an elevator will run when needed. Major repairs and replacements such as re-roping, gear overhaul, and motor reconditioning have to be scheduled to coincide with plant shutdown periods and must be planned to minimize out-of-service time. If the elevator is critical, a spare motor, set of ropes, replacement guides, and other items that require advance orders, should be on hand.

It is important to have a trained elevator mechanic on staff or to contract with a local elevator contractor for continuing maintenance. The latter is the favored approach, since the elevator contractor can have access to skilled personnel if an emergency occurs. As with any industrial plant or complex, the approach will be unique to the facility and strategic planning is a prerequisite.

ABOUT THE AUTHOR

AL SAXER is an independent elevator consultant who has more than 50 years of experience in the elevator industry and has held a variety of field and management positions with the Otis Elevator Company. His positions have included those of mechanic, adjuster, field education supervisor, regional service supervisor, manager of field engineering, and director of modernization operations. Prior to his retirement from Otis, he was director of field services for the entire North American Operations. Mr. Saxer has also served on various ASME A17 Code committees as chairman of the existing elevator committee, A17.3, and was instrumental in overseeing the issuance of that document. He continues in that position and is currently chairman of the A17.1 Part XII working committee responsible for the rules for alterations and upgrading elevators, chairman of the Mine Elevator Committee, and chairman of the National Elevator Industries Performance Standards Committee. Mr. Saxer also is an instructor in the ASME seminar series for existing elevators. Few people can match the extent and depth of his experience in the field.

14 Nonconventional Elevators, Special Applications, and Environmental Considerations

GEORGE W. GIBSON

NONCONVENTIONAL ELEVATOR APPLICATIONS

The previous chapters described the bulk of the conventional elevator applications wherein the elevators were within a building and the building was used for a single function, that is, commercial, residential, or institutional use.

A growing trend toward multipurpose buildings is apparent. A hotel or apartment can be built on top of an office building and the same office building built on top of a store. A school can be located on the lower floors of an office building or an apartment house. By multiple use of the building the efficiency of a 24-hr-a-day operation can be gained from a single capital investment. This chapter discusses the vertical transportation aspects of such multipurpose buildings.

Other nonconventional elevator applications include the trend toward observation-type elevators in either the atrium of a building or on the outside of a building, either exposed to the weather or in a glass-enclosed hoistway. Such elevators not only provide the necessary vertical transportation but serve to enhance the uniqueness of the building they serve.

In addition to vertical elevators, inclined elevators are being favored for a number of applications both commercial and scenic. This type of elevator is described later in this chapter.

Many nonelevator-related vertical transportation systems are being increasingly used. Such applications include stage lifts and platform lifts. The latter is often needed to develop multipurpose room areas, such as in raising the end of a gymnasium to create a stage or providing a platform for a TV camera.

Special industrial applications are frequently found, which include the use of rack and pinion elevators alongside smokestacks for service personnel or inside the caissons of offshore oil production platforms to serve maintenance needs.

Depending on the type of building occupancy and the use to which the elevators will be subjected, there are many other technical issues that must be addressed by the elevator designer, including environmental issues, such as wind, flooding, and earthquake, as well as those peculiar to high-rise, high-speed elevator systems whose construction may require special design considerations to address building compression, building sway, and car ride. These issues are discussed in this chapter.

The Vertical Transportation Handbook, Third Edition, Edited by George R. Strakosch
ISBN 0-471-16291-4   © 1998 John Wiley & Sons, Inc.

Elevators are applied in all types of buildings and structures. Elevators are found anyplace where there is a need for personnel and material to travel any distance vertically or at an incline. Applications include TV transmission towers, which may be 1000 ft (300 m) high, smoke stacks to maintain emission control systems, mines thousands of feet deep, oil production platforms resting on the bottom of the sea, and nuclear plants where the elevator may be submerged under cooling water to be used for periodic plant overhaul, to cite just a few. Each represents a distinct set of environmental considerations. Even the environment in a full commercial building may present some unique concerns, including building sway, stack effect, life safety, or earthquake considerations.

Not only does the environment affect the design and installation of an elevator, but also the way the elevator can be used. In a downtown building it may be desirable to use an elevator for the handling of freight and mail or to move a single large piece of building equipment from the roof to the ground. In any structure, operation during an emergency may be totally different than during normal use. For example, an emergency in an oil refinery may require access to valves located at an upper level and commandeering an elevator may be the most effective approach. If hazards are present, such as explosive gases and dust, or if the elevator is exposed to wind and weather, certain measures in addition to necessary special operation must be taken. In this chapter many of these aspects are discussed and guidelines established to suggest trouble-free elevator application and operation.

MULTIPURPOSE BUILDINGS

With the increasing emphasis on land usage and energy conservation, the concept of the multipurpose building has become popular in recent years. A building may well combine lower-floor commercial space and residential space above. Hotels or apartments on top of office buildings and/or stores may permit 24-hr use of the structure.

The secret to successful multiuse of a building is separation of its multiple functions by separate vertical transportation systems and lobbies. People entering the residential area, for example, need not interfere with people using other areas of the building.

Elevatoring of such multipurpose buildings should be based on separation of the several functions and elimination of interfering traffic. If the expected elevator traffic patterns for the functions do not coincide, elevatoring may be established for the major function and sufficient service made available for the minor function. If the two functions are expected to coincide, as in combination office and apartment buildings, separate groups of elevators should be provided (Figure 14.1).

Figure 14.1 shows two alternatives for a residential section. The apartment elevators can start from the ground floor and operate express to the first apartment floor with local stops thereafter. Alternatively, shuttle elevators can be provided, stopping at the ground floor and operating nonstop to the sky lobby. Apartment amenities such as a package room, a concierge, health clubs, or swimming pool can be conveniently located at the sky lobby.

The service elevator functions for the office building and the apartment should be provided on two separate elevators. The nature of apartment service needs which consist of maids, deliveries of clothing, furniture and food, catering services, and movers are usually in conflict with office needs such as supplies, mail, and renovations. In addition, apartment tenants expect a greater degree of 24-hour security, whereas office tenants expect greater nighttime security. This can be effected by locking out the separate entrances to the elevators except for emergency purposes.

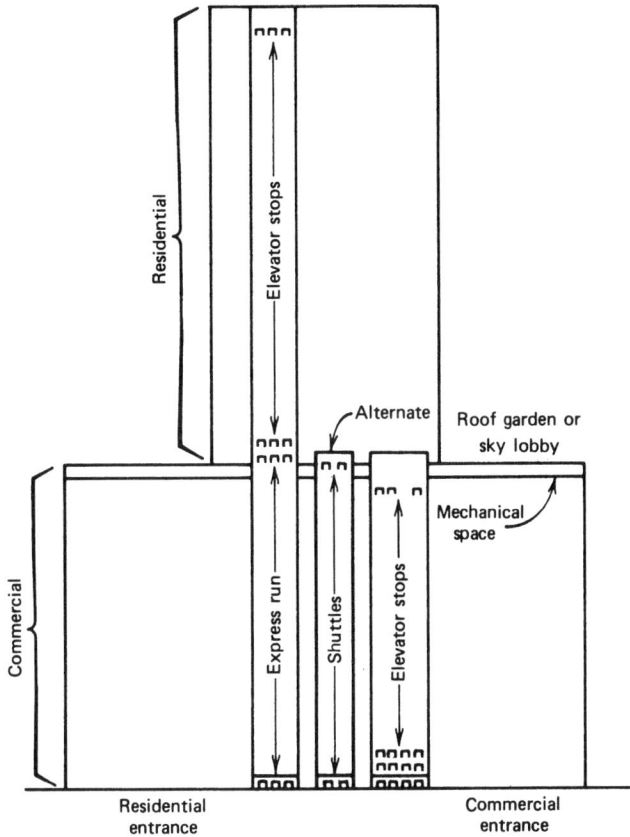

Figure 14.1. Multipurpose building—office and residential.

The sky lobby concept has been a successful approach to the office–apartment combination. A separate lobby is located on the lowest apartment floor and connected to the street by shuttle elevators. Apartment tenants ride these elevators to the sky lobby and change to the local elevator, which takes them to their floors. The sky lobby is enhanced by swimming pools, shops, or a restaurant. This is the elevatoring arrangement of the John Hancock Center in Chicago, where 44 floors of apartments are located above a 40-story office, store, and garage building (Figure 14.2).

Water Tower Place, also in Chicago, combines stores on the lower levels surrounding an atrium, which includes three observation-type elevators in glass hoistways, office space above the stores with separate elevator service, and a luxury hotel with a sky lobby above the office space. A third similar-type building is planned in the same area.

A hotel with extensive convention facilities, such as ballrooms and meeting rooms often used by others than the hotel guests, can be considered as a multipurpose building. Separate vertical transportation for outside guests increases the value of the meeting facilities and minimizes their interference with hotel guests. Guests, even when they are attending the meeting room functions, appreciate the reduced congestion on the main passenger elevators.

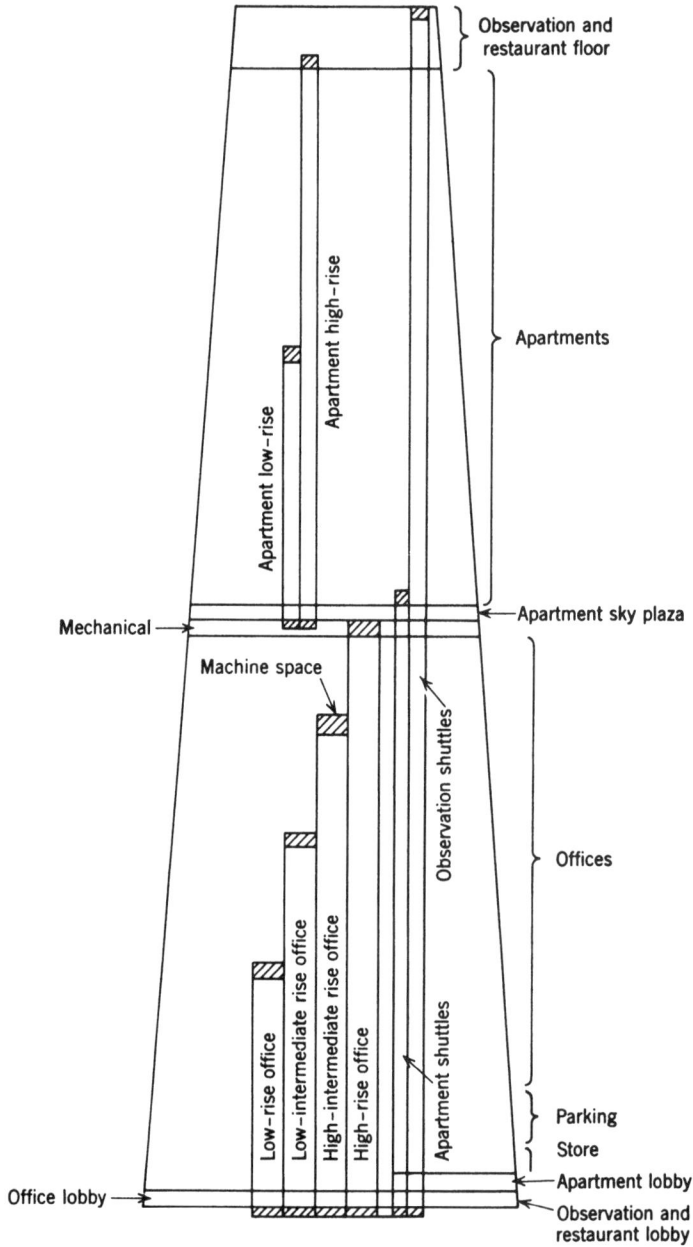

Figure 14.2. John Hancock Center, Chicago.

In estimating the vertical transportation requirements of the separate functions in a building, the expected maximum usage of each facility and the possible time of use of each must be considered. If periods of maximum use do not overlap, it may be possible to combine functions, and the vertical transportation is calculated for the maximum use. In addition to the quantity of service required, the nature of the traffic must be determined. Its direction is either in unison or opposed, and proper elevator provisions must be made.

An important example is the inclusion of restaurant facilities on the top of an office building. The normal lunchtime traffic of the tenants is usually in direct opposition to that of the restaurant patrons.

SKY LOBBY ELEVATORING

Sky lobbies have been used in many large office buildings, in addition to multipurpose buildings. The space starting at the sky lobby is elevatored as a conventional building, as if the sky lobby were the first floor. The elevators from the ground to the sky lobby are calculated as two-stop elevators for a single sky lobby. This has been done in a 17-story building as well as 100-story buildings.

It is possible to have the same shuttle elevators serve more than one sky lobby. A 28-story building has been built with two sky lobbies and a single group of shuttle elevators serving the ground and the two lobbies.

In calculating shuttle elevators the total handling capacity of the elevators must match the handling capacity of the local elevators and the required handling capacity for the population of the sky lobby floor. Elevator calculations should be based on the nominal full load requirements, and then the elevators increased in size to reflect what is required if one of the shuttle elevators is out of service. The high-rise elevators of the section below can be arranged to serve the sky lobby floor to provide an alternate route to the street in the event of an emergency, or alternatively, suitable stairways can be provided to connect the sky lobby with the lower floor.

The high-rise elevators of the lower section that are arranged to serve the sky lobby should not be arranged to serve the sky lobby during peak traffic periods. If they do so, people will find it more convenient to ride to the sky lobby and take the lower-section elevators down to their floors, creating two-way traffic on the elevators during an up-peak period.

Service elevator needs for a building with sky lobbies can be served either by conventional all-stop service elevators from the ground or by creating "sky-docks" with shuttle service elevators from the street and conventional service elevators starting at the sky lobby level.

Each of the World Trade Center towers in New York City consists of approximately three 30-story "buildings" on top of each other. The lobby of each of these buildings is connected to the street by single-deck shuttle elevators. A person wishing to go, say, to the upper third of the building, will ride a shuttle elevator nonstop to the 77th-floor sky lobby and there change to a local or express elevator to the desired floor. Without this sky lobby arrangement the necessary shafts for conventional local and express elevators serving all floors would consume almost the total area of the lower floors (Figure 14.3).

The Sears Tower in Chicago is an example of sky lobby elevatoring with double-deck shuttle-elevators. People entering the building at the ground floor and wishing to travel to floor 34 or floor 67 will enter the upper deck of the shuttle elevators at the ground floor, while those wishing to go to floor 33 or floor 66 will ride an escalator to a lower shuttle elevator lobby corresponding to the lower deck of the shuttle elevators. At the two-level sky lobby, the terminal landing of the local elevators to the desired floor corresponds to the existing floor of the shuttle. If people wish to interchange, escalators are provided to connect the two levels of each sky lobby.

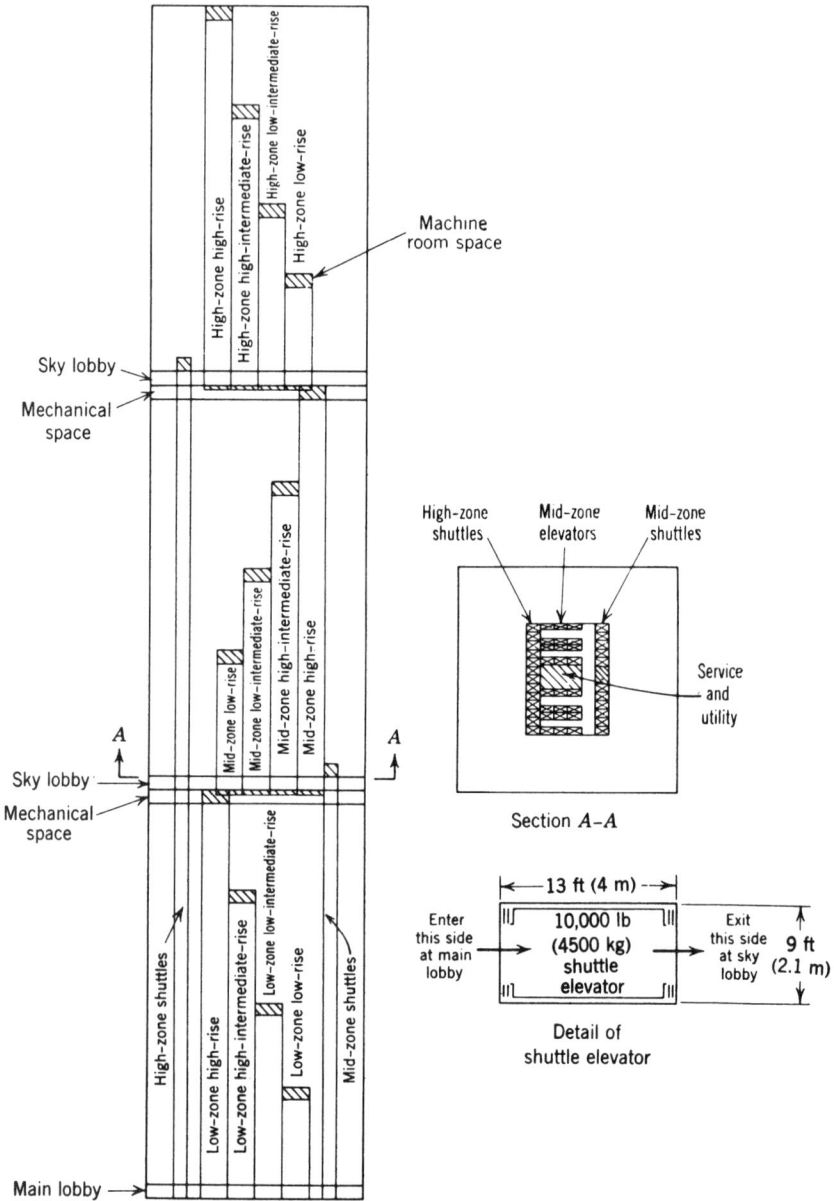

Figure 14.3. World Trade Center, New York.

DOUBLE-DECK ELEVATORS

Operation

Another approach to reducing the space required by elevators in taller buildings is the use of multideck or compartment elevators. Here the upper and lower decks of each elevator are loaded simultaneously (during the incoming rush, for example), with passengers destined for the odd-numbered floors entering the bottom deck and those for the even-numbered floors entering the upper deck. There are four typical dual-loading lobby variations used with double-deck systems: split-level lobby (Figure 14.4*a*), main lobby at the lower level (Figure 14.4*b*), main lobby at the upper level (Figure 14.4*c*), and entry/exit lobby serving each level (Figure 14.4*d*). When the elevator stops, passengers are discharged from both decks simultaneously (Figure 14.5).

This restricted operation is maintained for passengers entering at the lobby floor at all times. If the car stops for a landing call during its up trip, the operation of both decks is made unrestricted so that a person entering either deck at an upper floor is able to exit at any floor served by the elevator.

After operation becomes unrestricted, all the car buttons in the car on both decks are operative and the elevator is arranged so that the trailing deck, that is, the lower deck on an up trip and the upper deck on a down trip, is the one that responds to a landing call. In this way advantage may be taken of coincident stops, that is, stops wherein both a landing and a car call are served at the same time. The operation on the down trip is unrestricted, and people entering a down-traveling double-deck elevator do not know whether they are entering the upper or lower deck until they arrive at the lobby.

Brief History

Although the popularity of double-deck systems seemed to take hold in 1970 with the Otis installation at the Time-Life Building in Chicago, the roots of these innovative systems can be traced to the eight Otis double-deck elevators installed at the Cities Service Building, known as Sixty Wall Tower, in New York City in the early 1930s. At a height of 950 feet and with 67 floors, the building was the tallest in Manhattan's financial district and the third-tallest building in the world. The building was peculiarly adapted to double-deck operation, since the building site occupied a slight hill rising from Pearl to William Streets. The upper lobby was entered from Pine and Cedar Streets; the lower lobby from Pearl Street.

Designated as Cars 17 through 24, the eight double-deck elevators had a capacity of 2500/2500 lb at 1000 fpm, were arranged in two 4-car groups oppositely facing on both sides of the main elevator corridor, and comprised the high-rise elevator bank serving the street lobbies and the 28th to 60th floors. Only one car, No. 21, served the 28th to 63rd floors. Six cars, Nos. 18–20 and 22–24, had a rise of 750 ft 0 in.; car 17 had a rise of 719 ft. 0 in.; car 21 traveled 746 ft 6 in.

The double-deck operation was set up to receive passengers at both street levels, the lower-deck doors opening on the Pearl Street level and the upper-deck doors opening on the main lobby floor. From each street level four of the decks stopped at even floors and four at odd floors. Of the four cars in one group, whose upper decks stopped at even floors, the lower cabs stopped at odd floors. In the opposite group of four cars, the upper cabs stopped at odd floors, and the lower cabs stopped at even floors. Therefore, a passenger entering at either level could board a car to any floor desired.

The double-deck elevators had full signal control operation. The cars were run on double-deck operation during the morning starting hours, noon lunch, and evening rush hours. During double-deck operation, each cab had its own attendant. The elevator could start only when the operating handles of the car switches of both compartments were in the full "start" position and all car and hoistway doors were closed. The upper compartment was used for single-cab operation. Elevator starters manned both lower lobby levels. The eight cars were dispatched by a complicated mechanical starting system.

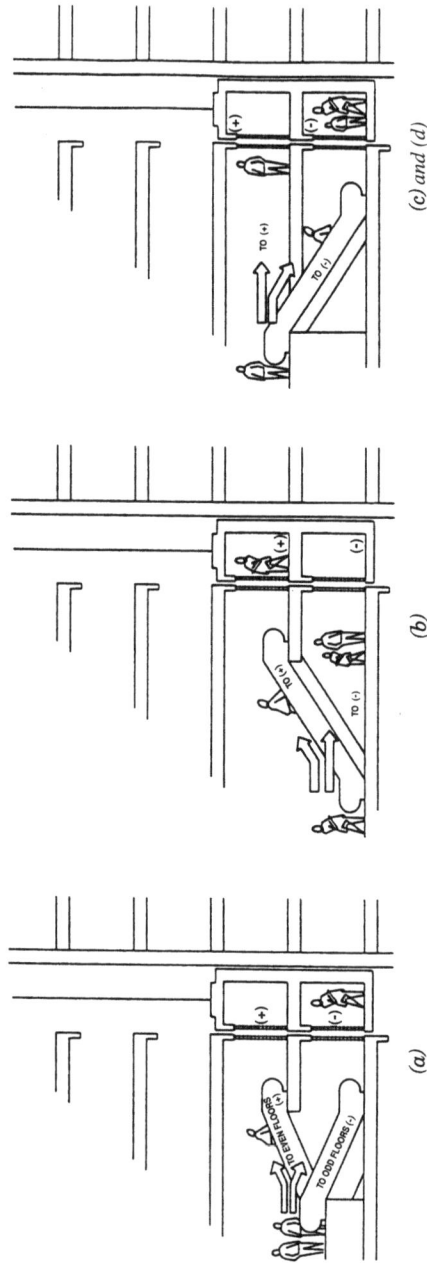

Figure 14.4. Typical double deck dual-loading lobby arrangements: (*a*) split-level lobby; (*b*) main lobby at the lower level; (*c*) main lobby at the upper level; (*d*) dual entry/exit lobby serving each level. (Note: + represents even numbered floors, − represents odd numbered floors.)

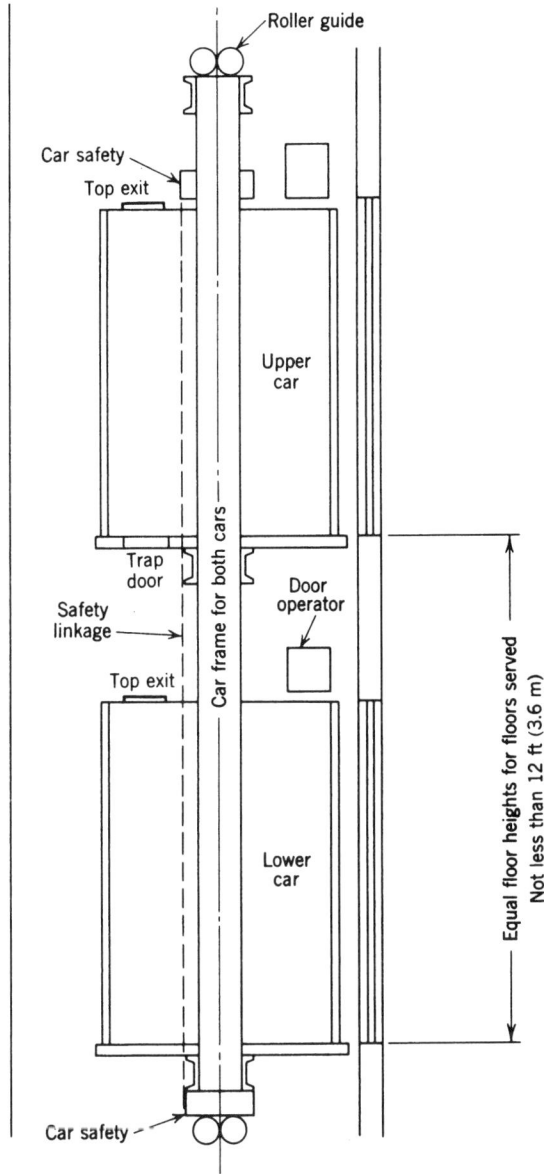

Figure 14.5. Double-decked elevator car.

A speaking tube was provided between the two upper and lower decks to enable the operators to communicate with each other. Trapdoors between the two decks were provided to enable the transfer of passengers between compartments in case of an emergency. Each double-deck car-frame measured approximately 25 ft in overall height and was provided with Otis Type 13A duplexed flexible guide clamp safeties.

In the article "Double-Deck Elevators Give New Joy to Wall Street," written by E. J. Smith, operating manager of the Sixty Wall Tower building, published by *Real Estate*

Magazine, January 1934, Mr. Smith states: "Probably the greatest single benefit obtained from double-deck operation is the increased rentable area available from the saving in elevator shaft space involved." As the operating manager, Mr. Smith gave his assessment: "It can be said without reservation that during the short period in which double-deck operation has been tried in Sixty Wall Tower, it has been successful."

Notwithstanding this success, the elevatoring plan was short-lived. The Great Depression continued to grip the economy, and the resulting lowered building occupancy could not justify the double-deck operation. The operating costs of running the double-decks, each with double employees, were reduced by running the elevators with only one active cab, thus postponing the era of double-deck systems almost four decades. The unused lower deck in each car was counterbalanced with sandbags, as ballast. The lower car doors were rendered inoperative. In the early 1970s the elevators were modernized as conventional single-deck operation. The unused decks were removed.

There are many advantages to using double-deck elevator systems, among which are the following:

1. More passenger-handling space per hoistway than with single-deck results in less core area for the elevators, thereby increasing the rentable space in a building.
2. Fewer double-deck elevator systems are required as compared with the single-deck alternative.
3. Up-peak operation is optimized because passengers can reach their destination floors faster than with single-deck operation owing to the number of typical stops being halved.

Elevator maintenance costs per building are less with double-deck systems as compared with single-deck. Double-deck elevator installations can be found in operation all over the world (Figures 14.6 and 14.7).

Double-Deck Elevatoring

The round-trip time of a double-deck elevator is calculated in a manner similar to a single-deck elevator. The probable stops will usually be equal to the number of stops the elevator can make. Double-deck elevators are rated by the load in each deck, for example, 3500 lb/ 3500 lb (1600 kg/1600 kg) is indicative that each deck is the size of a 3500-lb conventional elevator. Each deck will have a nominal capacity of 16 people. Quite often, one of the decks may be stopped for a passenger while no one transfers from the other deck. If the elevator stops for a landing call, the person entering may only wish to travel one floor, causing an additional stop. If it is assumed that the probable stops are equal to the minimum possible stops, these contingencies are recognized. Transfer time should be calculated as if an equal number of people leave each deck at a single stop. Door time is the same as for a single-deck elevator. Since both doors must be closed and locked before the elevator can move, some delay in coordination may take place. For this reason, an extra 10% standing time inefficiency should be used in calculating the standing time of a double-deck elevator system. Interfloor traffic can be quite detrimental to double-deck operation, especially during peak incoming traffic.

YEAR COMPLETED	BUILDING AND LOCATION	NUMBER OF STORIES	GROUPS OF ELEVATORS DOUBLE-DECK (DD)	YEAR COMPLETED	BUILDING AND LOCATION	NUMBER OF STORIES	GROUPS OF ELEVATORS DOUBLE-DECK (DD)
MITSUBISHI				**OTIS** Continued			
1972	OBAYASHI BUILDING OSAKA, JAPAN (skip-stop operation)	32	4-1350/1350 kg at 150 mpm 4-1350/1350 kg at 240 mpm	1985	SOUTHWEST BELL TELEPHONE COMPANY ST. LOUIS, MO	48	3 GROUPS WITH 18 DD ELEVS. 6-1600/1600 kg at 210 mpm 6-1600/1600 kg at 300 mpm 6-1350/1350 kg at 360 mpm
1983	PEMEX (NATIONAL OIL COMPANY) MEXICO CITY, MEXICO (skip-stop operation)	51	6-1350/1350 kg at 150 mpm 6-1350/1350 kg at 240 mpm 6-1350/1350 kg at 300 mpm	1986	TREASURY BUILDING SINGAPORE	50	3 GROUPS WITH 17 DD ELEVS.* 9-1350/1350 kg at 210 mpm 4-1150/1150 kg at 360 mpm 4-1150/1150 kg at 360 mpm * upper deck is "odd" service
1994	RIVERSIDE SUMIDA TOKYO, JAPAN	16	5-1600/1600 kg at 180 mpm	1989	SHEARSON LEHMAN PLAZA NEW YORK, NY	39	3 GROUPS WITH 20 DD ELEVS. 7-1600/1600 kg at 150 mpm 7-1600/1600 kg at 210 mpm 6-1600/1600 kg at 300 mpm
OTIS				1990	TORRE PICASSO MADRID, SPAIN	43	EITHER DD PASS OR UPPER-DECK SERVICE, LOWER-DECK PASS 3-1600/1600 kg at 150 mpm, 240 or 360 mpm
1970	TIME-LIFE BUILDING CHICAGO, IL	30	2 GROUPS WITH 12 DD ELEVS.* 6-1800/1800 kg at 180 mpm 6-1500/1500 kg at 240 mpm * presently running as single-deck units due to low population density	1992	THE CONCOURSE SINGAPORE	42	2 GROUPS WITH 8 DD ELEVS. 4-1150/1150 kg at 210 mpm 4-1150/1150 kg at 360 mpm
1972	COMMERCE COURT TOWER (C.I.B.C.) TORONTO, CANADA	57	1 GROUP WITH 5 DD ELEVS. (LOW-RISE GROUP ONLY) 5-1500/1500 kg at 240 mpm CONVENTIONAL ELEVATORS SERVE ALL FLOORS ABOVE THE LOW-RISE GROUP.	1992	GRAND 50 TOWER KAOHSIUNG, TAIWAN	50	2 GROUPS WITH 12 DD ELEVS. 6-1350/1350 kg at 240 mpm 6-1350/1350 kg at 360 mpm
1973	AMOCO (STANDARD OIL) OF INDIANA CHICAGO, IL	80	5 GROUPS WITH 40 DD ELEVS. 8-1600/1600 kg at 240 mpm 8-1600/1600 kg at 300 mpm 8-1600/1600 kg at 420 mpm 8-1600/1600 kg at 420 mpm 8-1600/1600 kg at 480 mpm	1992	TORE VILLA OLIMPICA BARCELONA, SPAIN	40	2 GROUPS WITH 8 DD ELEVS. 4-1250/1250 kg at 240 mpm 4-1250/1250 kg at 360 mpm
1974	JOHN HANCOCK BUILDING BOSTON, MA	60	5 GROUPS WITH 30 DD ELEVS. 6-1800/1800 kg at 210 mpm 6-1800/1800 kg at 240 mpm 6-1800/1800 kg at 360 mpm 6-1800/1800 kg at 420 mpm 6-1800/1800 kg at 450 mpm	1994	TORRE PUERTA EUROPA MADRID, SPAIN	27	2 GROUPS WITH 4 DD ELEVS. 2-1600/1600 kg at 150 mpm 2-1600/1600 kg at 240 mpm
1976	CITICORP BUILDING NEW YORK, NY * 46 occupied floors	56 *	3 GROUPS WITH 30 DD ELEVS.* 8-1350/1350 kg at 210 mpm 6-1350/1350 kg at 360 mpm 6-1350/1350 kg at 420 mpm * upper level is "odd" service	1994	D.B.S. BANK SINGAPORE	32	2 GROUPS WITH 8 DD ELEVS. 4-1150/1150 kg at 210 mpm 4-1150/1150 kg at 360 mpm
1976	NATIONWIDE INSURANCE COLUMBUS, OH	40	2 GROUPS WITH 12 DD ELEVS. 6-1600/1600 kg at 210 mpm 6-1600/1600 kg at 300 mpm	1995	REPUBLIC BUILDING SINGAPORE	60	7-1600/1600 kg at 300 mpm 8-1600/1600 kg at 360 mpm
1977	FIRST CANADIAN PLACE TORONTO, CANADA	70	4 GROUPS WITH 29 DD ELEVS. 8-1600/1600 kg at 240 mpm 7-1600/1600 kg at 300 mpm 7-1600/1600 kg at 300 mpm 7-1350/1350 kg at 300 mpm	1995	SHANGHAI TV TOWER SHANGHAI, CHINA	12	1-1750/1750 kg at 240 mpm
1977	C.E. HOWE (TREASURY BUILDING 5) OTTAWA, CANADA	14	2 GROUPS WITH 10 DD ELEVS. (GLASS CABS IN ATRIUM) 5-1600/1600 kg at 150 mpm 5-1600/1600 kg at 150 mpm NOTE: 2 GROUPS SERVE THE SAME FLOORS ON EITHER END OF THE BUILDING.	1996	PETRONAS TOWERS KUALA LUMPUR, MALAYSIA	88	TWIN TOWERS – UNITS/TOWER 6 GROUPS WITH 29 ELEVATORS 6-1600/1600 kg at 210 mpm 6-1600/1600 kg at 240 mpm 6-1600/1600 kg at 300 mpm 3-1600/1600 kg at 420 mpm 3-1600/1600 kg at 420 mpm 5-2100/2100 kg at 480 mpm
1982	NORTHWEST BELL TELEPHONE OMAHA, NE	17	6-1800/1800 kg at 180 mpm	**SCHINDLER**			
1982	PHILIP MORRIS BUILDING NEW YORK, NY	26	7-1800/1800 kg at 210 mpm	1991	SCOTIA PLAZA TORONTO, CANADA	72	6-1350/1350 kg at 210 mpm 6-1350/1350 kg at 300 mpm 5-1350/1350 kg at 360 mpm 5-1350/1350 kg at 425 mpm
1984	REPUBLIC PLAZA DENVER, CO	54	3 GROUPS WITH 22 DD ELEVS.* 7-1600/1600 kg at 150 mpm 7-1600/1600 kg at 240 mpm 8-1600/1600 kg at 300 mpm * upper deck is "odd" service	**ELEVATORS PTY. LTD. (KONE)**			
				1994	PACIFIC TOWER (T&G BUILDING) HYDE PARK CENTER SYDNEY, AUSTRALIA	30	6-1350/1350 kg at 210 mpm* 5-1350/1350 kg at 300 mpm * presently being modernized to Otis 411M
1985	NATIONSBANK PLAZA DALLAS, TX	73	6-1350/1350 kg at 150 mpm 6-1350/1350 kg at 240 mpm 6-1350/1350 kg at 300 mpm 5-1350/1350 kg at 360 mpm	**EXPRESS LIFT**			
				<>	NATIONAL WESTMINSTER TOWER LONDON, ENGLAND	<>	4 - <> 1-SERVICE/DIRECTOR EXPRESS LIFT
				HITACHI			
1985	INTERFIRST BANK (NATIONS BUILDING) DALLAS, TX	70	4 GROUPS WITH 23 DD ELEVS. 6-1350/1350 kg at 150 mpm 6-1350/1350 kg at 240 mpm 6-1350/1350 kg at 300 mpm 5-1350/1350 kg at 360 mpm	1973	TOYAMA KOWA BUILDING TOKYO, JAPAN (skip-stop operation)	18	4-1600/1600 kg at 180 mpm
				MODERNIZATIONS/REPLACEMENTS			
				1992	SUN HUNG KAI – OTIS HONG KONG (skip-stop operation)	60	4-1350/1350 kg at 360 mpm (HIGH-RISE ONLY)
				1995	U.O.B. PLAZA 2 – OTIS SINGAPORE	38	4-1150/1150 kg at 210 mpm 3-1150/1150 kg at 360 mpm

Figure 14.6. Buildings with double-deck elevators completed or under construction; pure double-decks (local zone service)—metric units—as of January 12, 1995.

YEAR COMPLETED	BUILDING AND LOCATION	NUMBER OF STORIES	GROUPS OF ELEVATORS
OTIS			
1992	U.O.B. PLAZA 1 SINGAPORE	66	(6) 1800/1800 kg at 300 mpm TOP/DOWN SKY LOBBY AT 37/38
SCHINDLER (WESTINGHOUSE)			
1974	SEARS TOWER CHICAGO, ILLINOIS	110	(8) 2250/2250 kg at 420 mpm SKY LOBBIES at 33/34 (6) 2250/2250 kg at 480 mpm SKY LOBBIES at 66/67
1984	FIRST INTERSTATE BANK (ALLIED TOWER) HOUSTON, TEXAS	72	(5) 1600/1600 kg at 300 mpm TOP/DOWN SKY LOBBY AT 34/35 (6) 1600/1600 kg at 300 mpm TOP/DOWN SKY LOBBIES AT 53/59
ELEVATORS PTY. LTD. (KONE)			
1981	CENTER POINT TOWER SYDNEY, AUSTRALIA	65	(3) 1150/1150 kg at 420 mpm

Figure 14.7. Buildings with sky lobby double-deck shuttles—metric units.

Example 14.1. Double-Deck Elevators

Given: an office building, 20,000 ft^2 (2000 m^2) net usable area per floor on floors 1 to 20, 12-ft (3.6-m) floor heights. Assume single-purpose occupancy, 150 ft^2 (15 m^2) per person average. How many double-deck elevators are required?

Population: $20,000 \times 20 = 400,000 \div 150 = 2666$ people

Required handling capacity: up-peak with 10% down, $2666 \times 15\% = 400$ people

Assume: 3500/3500 lb double-deck elevators, @ 500 fpm, 48-in. center-opening doors

Probable stops: 20 upper floors, 10 probable stops

Time to run up, per stop: $\dfrac{12 \times 20}{10} = 24$-ft rise per stop

$$24 \text{ ft} = 5.8 \text{ sec}$$

Time to run down: $\dfrac{240}{2} = 120$ ft $= 17.2$ sec

Elevator performance calculations:

Standing				Transit Time
Lobby time 16 + 16	=	14	sec	14
Transfer time, up stops 10×3	=	30		30
Door time, up stops 10×5.3	=	58.3		58.3
Transfer time, down stops 2×4	=	8		—
Door time, up stops $(10 + 1) \times 5.3$	=	10.6		—
Total standing time	=	120.9	sec	102.3
Inefficiency 15%	=	18.1		15.3
Total		139.0	sec	117.6

Total from page 364 (139.0 sec)   (117.6 sec)

Running time

Run up 10×5.8	$=$ 58.0	58.0
Run down 2×17.2	$=$ 34.3	—
Total round-trip time	231.3 sec	175.6 sec

$$HC = \frac{(16 + 16 + 2 + 2) \times 300}{231} = 47; \quad \frac{400}{47} = 8.5, \; 8 \; \text{elevators, minimum, required}$$

Interval: $231/8 = 27.5$ sec

A suggested elevator scheme would be eight 4000/4000 lb (1800/1800 kg) double-deck elevators @ 500 fpm (2.5 mps) serving lower lobby, upper lobby, and floors 1 to 20. The excess transit time over the desired 150-sec maximum would preclude locating executive offices at the top unless arrangements are made for special service or additional executive elevators. The total building requires escalators and a shuttle elevator to connect the two lobbies and sufficient service elevator capacity. Before a final recommendation is made, conventional and sky lobby elevators should be calculated and an analysis made of the difference in cost and space required for the conventional, sky lobby, and double-deck elevator arrangements. The final decision should be a result of a resolution of all the members of the building team.

Double-Deck Application

When double-deck elevators are considered for a building, a number of building restraints must be effected. All floor heights must be equal within a tolerance that can be adjusted by the setting of the landing door sills.

Lobbies for double-deck elevators require special considerations. They can be designed with either the upper or lower deck at ground level and must have both escalator and shuttle elevator service between the lobbies. Both lobbies should be equally attractive, and clearly visible signs should be provided to guide visitors to the proper elevator entrance level for the destination floors sought.

Because the vertical distance between guide shoes, that is, the elevator wheelbase, is almost double the comparable distance on a conventional single-deck car, the structural stiffness of the car frame uprights becomes critical, resulting in a more robust design that requires more hoistway space. The most efficient design of the car safeties and guide rail system results with the deployment of duplex safeties, that is, one set of safeties mounted to the car frame crosshead members and one set of safeties mounted to the lower plank members. Because the vertical distance between the car frame crosshead safety above the upper car and the plank safety below the lower deck will always be approximately 18 feet or greater, the column-loading capacity of the guide rails at safety application is double that which would be permitted if only one set of safeties were employed. This structural load allowance is specified in the A17 Code Section 200. For a high-rise building, the added cost of duplexing the safeties will be more than offset by the savings in the lowered cost of smaller guide rails than would be required with single safeties.

The space requirements for double-deck elevators must consider the increased structure and larger elevator machinery that is required to handle the heavier loads. The space in the hoistway allowed for rails and car frame structure is increased to 12 in. (300 mm) rather than the 8 in. (200 mm) allowed for a conventional elevator. Pit and overhead spaces are slightly more than conventional elevators for the speed specified.

Operating systems are the same as for conventional elevators, with additional features to accommodate the double-deck operation. When an elevator stops at a floor it can be arranged so that the doors on both decks open even though a passenger may transfer at only one deck. The people on the other deck, if they are not aware that it is a double-deck elevator, will think a false stop has occurred. Some buildings prefer to open only the required doors and light a sign in the other deck with a legend "Other deck loading."

In Chapter 18, the economic differences between double-deck and sky lobby elevatoring are discussed. In general, a full comparison between three schemes, conventional, sky lobby, and double-deck elevatoring, must be made for any major project before a final elevatoring recommendation is made.

OBSERVATION AND OUTSIDE ELEVATORS

An elevator traveling up the outside or in the atrium of a tall building or hotel is a dramatic sight. For the rider, the thrill of seeing the scenery in motion is unsurpassed; for the observer on the ground or in the lobby, the smoothness and majesty of the moving mass is incomparable. From the engineer's point of view, observation and outside elevators pose many challenging problems. Notable applications of observation and outside elevators have been made at the St. Francis Hotel in San Francisco, the observation towers mentioned in Chapter 12, hotels in Atlanta, Chicago, Nashville, San Francisco, and Cambridge, and a host of other installations throughout the world (Figure 14.8).

Observation elevators in a building atrium are less critical in design than an outside observation elevator exposed to the weather. The important consideration in design is making the parts of the elevator other than the cab unobtrusive. Counterweights are usually hidden at the side, the door mechanism and the backs of the hoistway doors are painted a dark color, and the top and bottom of the car are shrouded to hide operating mechanisms. The elevator should be arranged with a minimum, preferably one, traveling cable and the compensating chain or rope eliminated by the use of higher-horsepower hoisting motors or traveling cable compensation to the counterweight to match the fixed traveling cable to the car.

Figure 14.8. Outside elevators, St. Francis Hotel, San Francisco.

Elevators installed outdoors are subject to all varieties of wind and weather. Installations serving the general public may be found as far north as Niagara Falls as well as in less severe climates. Industrial installations are found in almost any climate, including Greenland above the Arctic Circle. Wind effect is a serious consideration and, in northern climates, icing may present a particular hazard.

The elevator hoist ropes, governor rope, and traveling cables are especially affected by wind. To overcome some of the affect, the hoist rope tension is increased by using smaller or fewer ropes, the governor rope is either provided with guides or located in a trough, and compensating ropes are eliminated by increasing the horsepower of the hoisting motor and electrically compensating for the changing load as the elevator travels from top to bottom (Figure 14.9).

From the design aspect the outside elevator must be weather- and windproof. Windproofing is helped by eliminating all possible ropes and troughing traveling cables and governor ropes. The compensating ropes are eliminated and extra motor horsepower provided for the uncompensated load. The top and bottom of the elevator cab are enclosed and streamlined, and if the elevator car doors do not face the inside, they are positively locked while the car is away from the landings.

All hoistway switches and electrical installations are completely moisture-proofed to withstand driving rain. For extreme winter weather where elevator operation may be essential, ice scrapers should be provided on the car frame and counterweight frames, usually as part of the guide shoe assemblies, to limit the buildup of ice on the rails, and operating systems designed to periodically and automatically run the elevator up and down may be considered. Rail heaters have been employed, but the cost of electrical wattage per foot of rail in extreme temperatures may prove uneconomical versus periodic operation of the elevators.

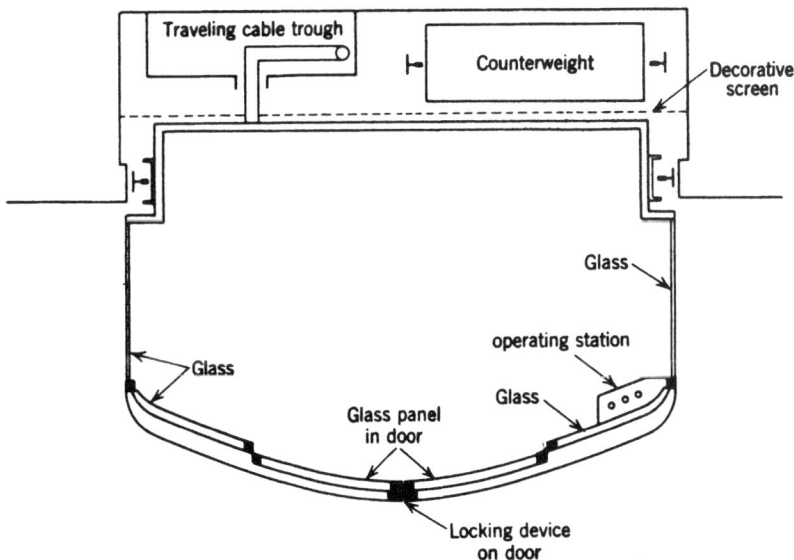

Figure 14.9. Outside observation-type elevator, which is exposed to the weather. Note the traveling cable troughing.

Wind problems are always present, and attempts have been made to overcome them by the use of wind gates, bridge-type rope guards, that automatically open when the elevator passes and then close to restrain the hoisting ropes from swaying and entangling the car or hoistway structure. An underslung elevator designed so that the ropes are semiprotected at the sides of the hoistway has been successfully used in one application.

An elevator cab exposed to the weather may require heating in the winter and air-conditioning in the summer. A heat pump mounted on the car is often used, and a receiver for air-conditioning condensate water is provided so that it can be emptied as the car travels to the bottom to avoid dripping on pedestrians from above. For winter operation, heaters are used in the door operator motor and on the car and hoistway sills.

Observation elevators are usually of the wall-climber design, in which the car frame and guide rails are set as close to the back of the car as possible. The final effect provides maximum viewing area for the passengers (Figure 14.10).

Glass used in outside elevators must be laminated and, preferably, tempered safety glass so that the hazard of broken glass falling or the elevator enclosure being opened is eliminated. Because the glass window wall in the elevator is critical to safe elevator operation, it is always recommended that spare glass of the correct size and shape be specified and packed for storage as part of the initial elevator contract so that prompt replacement

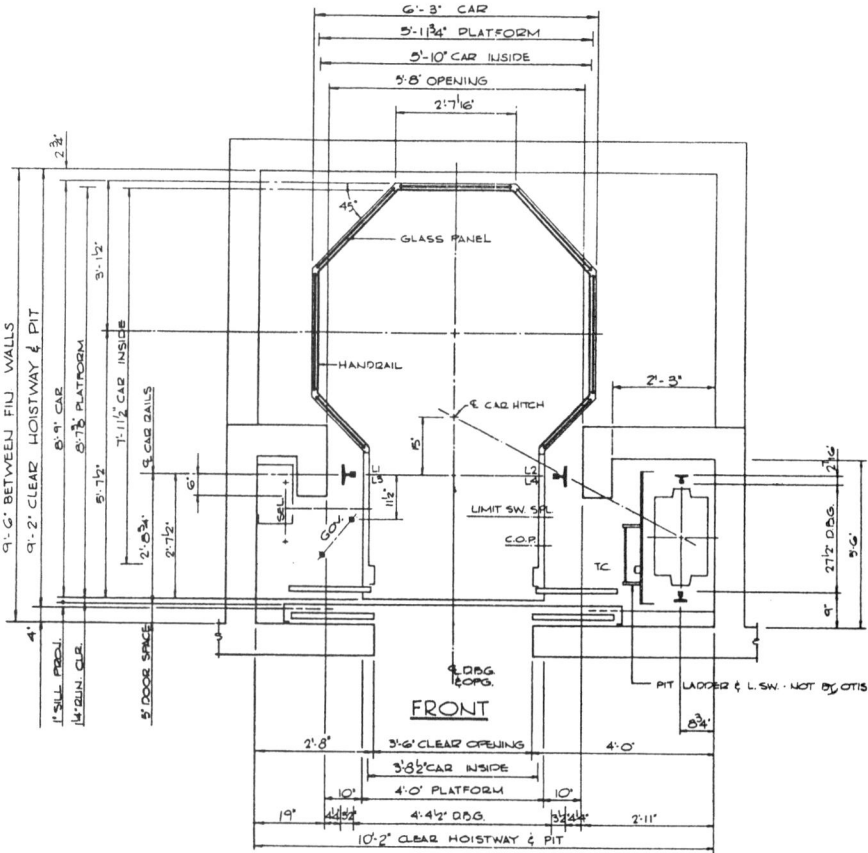

Figure 14.10. Sample layout of an observation-type elevator.

can be made. Recent code changes require the elevator to be shut down if any glass is missing.

Some observation-type elevators are designed to operate within a glass enclosed hoistway. Any glass that is used must be of the laminated safety type and should be tempered so that, even if broken, openings do not appear that may endanger passengers. Consideration must be given to cleaning the outside glass and both the inside and outside of the elevator enclosure.

Variations of observation elevators can be found in many applications (Figure 14.11). Some are found in shopping malls to serve upper and lower levels and to provide an attraction for patrons. Others are found at racetracks to serve the visitors to the upper clubhouse. Observation-type elevators can be found serving the floors facing an atrium and continuing into a glass-enclosed hoistway to serve a rooftop observation lounge and restaurant. One notable installation is in the Vertical Assembly Building at Cape Kennedy, where employees and guests can observe the assembly activity from the elevator, which rises more than 500 ft to the roof of the building.

When an outside elevator opens into an enclosed space, severe windage will occur in the gap between the elevator entrance and the car. In a number of applications this has been overcome by the use of a diaphragm around the perimeter of the door and weatherstripping on the hoistway door. In one installation at a rocket launching facility, the elevator opened into an airtight clean room. There, both the car and hoistway doors were installed in pockets and an inflatable diaphragm was used to seal the entrance before the car and hoistway doors were opened. Such diaphragms impose a time penalty to inflate and require stringent maintenance.

Figure 14.11. Observation-type elevator, Atlantic City (Courtesy Tyler Elevator Products).

Hoistway equipment must be weathertight and the machine room protected from the infiltration of rainwater being carried by the hoist ropes into the machine room area. Gaskets surrounding the ropes have been provided and equipped with wipers to clear off water from galvanized hoist ropes. Machine rooms can be designed to operate under a positive air pressure to minimize moisture. This latter practice is also valuable in locations where a hazardous or dusty atmosphere is present, such as at oil refineries or cement mills.

The nature of the atmosphere must also be considered when specifying elevators for an outdoor location. Temperature, expected winds and their direction, and extreme weather conditions should be part of the specification for any outdoor elevator. If the elevator is to be located near a body of salt water, corrosion-resistant materials must be specified. If an industrial application is being specified, statements as to any hazards or unusual operating conditions should be made for the guidance of the elevator design engineer.

INCLINED ELEVATORS

Inclined elevators were an outgrowth of the funicular railways that were quite prominent in the early 1900s, many of which have survived. An inclined railway was built on railway tracks with the car moved by a driving machine and ropes. The car traveling up was counterbalanced by a car traveling down, with a simple passing system at the middle of travel. The outside wheels had flanges that guided the car on the outside track. At the turnout the up car moved to one side and the down car to the opposite side as the inside wheels without flanges slid over on the rails and the outside wheels guided the car. The trip continued on the single track. The turnout had the formal name of "a Brown Turnout," named after the inventor (Figure 14.12).

Inclined railways had safety devices designed to dig into the rail ties or wooden strips along the rails if the tension on the ropes was released for any reason or if the attendant released an emergency brake. Surviving examples of inclined railways can be found in Chattanooga, Tennessee; Hong Kong; Bergen, Norway; and many European and South American countries.

An inclined elevator is a descendant of the inclined railway. It is presently designed to be automatic in operation, and the counterbalancing car and turnout are gone, replaced by a counterweight with its own set of rails. The car doors and landing doors are power operated and equipped with interlocks, and the attendant is also gone (Figure 14.13).

Elevators that travel in other than a vertical path are necessary for a wide variety of applications. Supplementing escalator service, especially for rises of 50 ft or more (15.2 m or more), to serve the people unable to use the escalator is a necessary application for inclined elevators. Many such inclined elevators have been provided in the Stockholm, Sweden, subway system for that purpose.

The development of a hillside for residential property indicates a need for inclined elevators. The alternative is to cut the side of the hill for roadways, which wastes space and creates the possibility of earth slides.

A resort located on a bluff above a beach is an application for inclined elevators which has occurred in many places in the world.

An inclined elevator can be considered for an application depending on the angle of incline if certain limitations are recognized. Up to about 10° from the vertical, conventional elevator equipment may be adapted to the incline; with more than 20°, considerations must include doors at the side, guide rollers for ropes to keep them from rubbing the

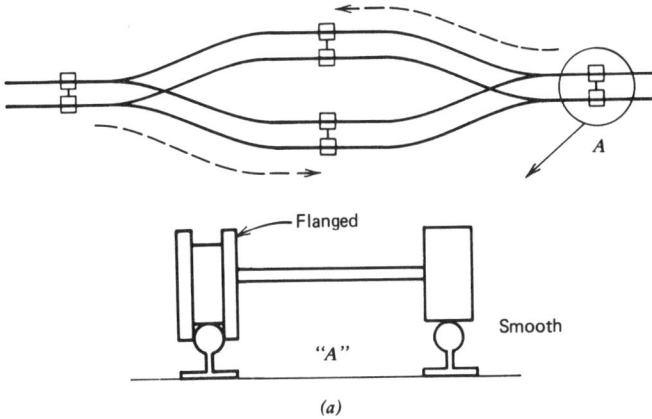

Figure 14.12. (*a*) Inclined railway "Brown Turnout;" (*b*) inclined railway, Pittsburgh, Pa.

back of the hoistway, and guide rollers reinforced for the load of the car and counterweight leaning in one direction (Figure 14.14). Traveling cables have to be provided with a guiding system to keep them from dragging on the structure.

The motion control of an inclined elevator must be designed for limited acceleration and deceleration, which will provide gradual starts and stops, and the stopping distance lengthened upon application of brakes during emergency stops depending on the horizontal accelerating forces expected to be exerted on standing people.

Unlike vertically moving elevators where the resultant acceleration or deceleration is directed vertically, inclined elevators are subject to a resultant acceleration or deceleration

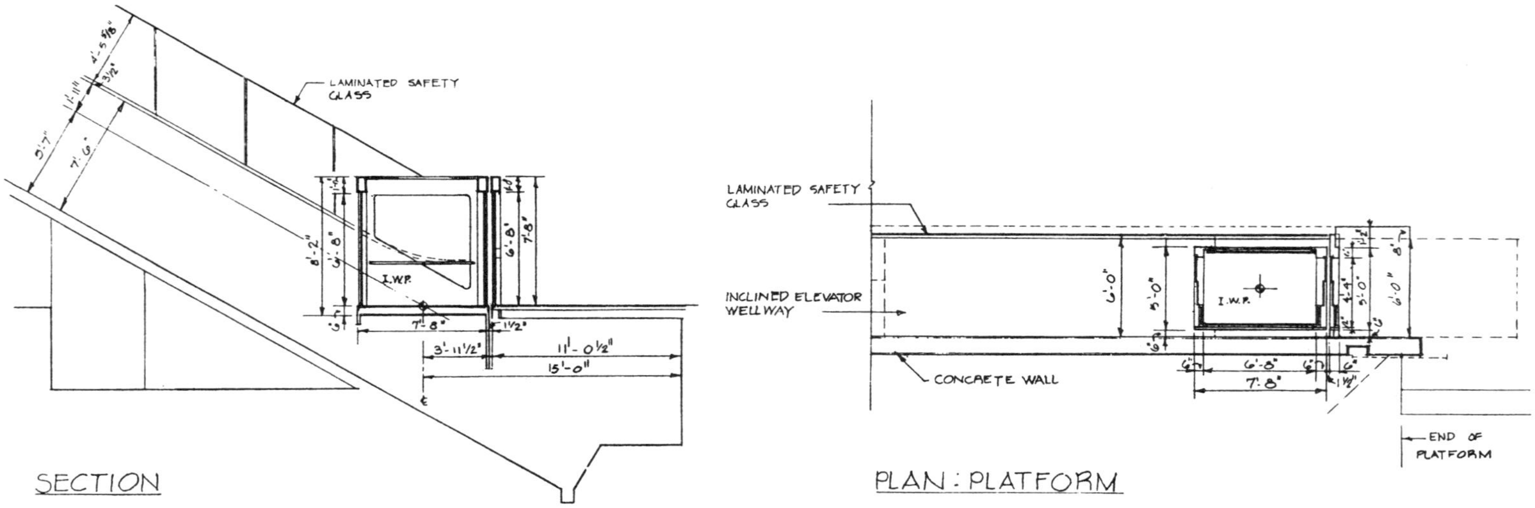

PLAN : PLATFORM

SECTION

Figure 14.13. Inclined elevator to be installed parallel to an escalator to serve mobility-limited persons in the same travel path. *Note:* The entrances are in line with the elevator travel as opposed to side entrances, as found on most inclined elevators.

(a)

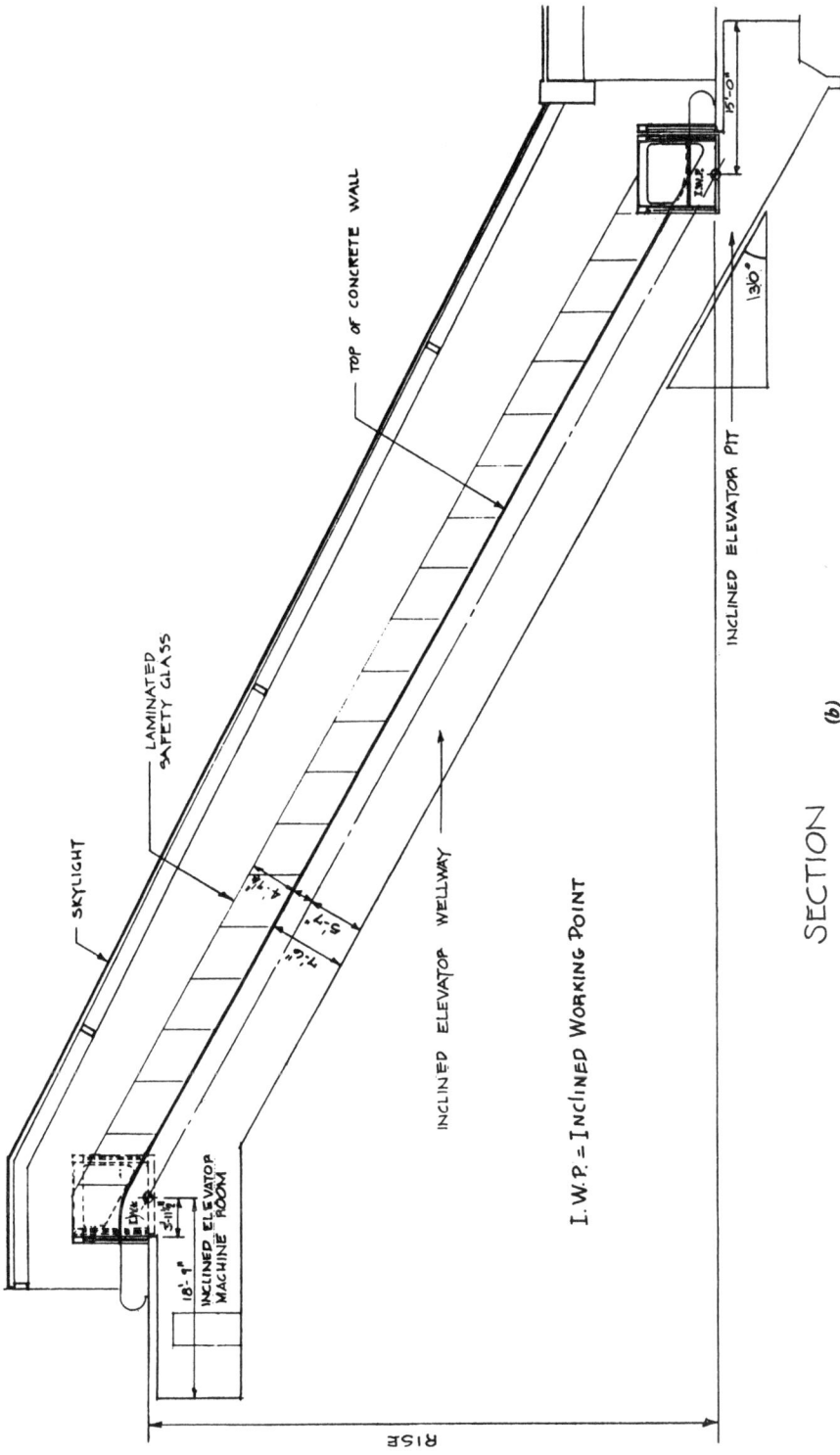

SECTION

(b)

Figure 14.13. (Continued)

I.W.P. = INCLINED WORKING POINT

373

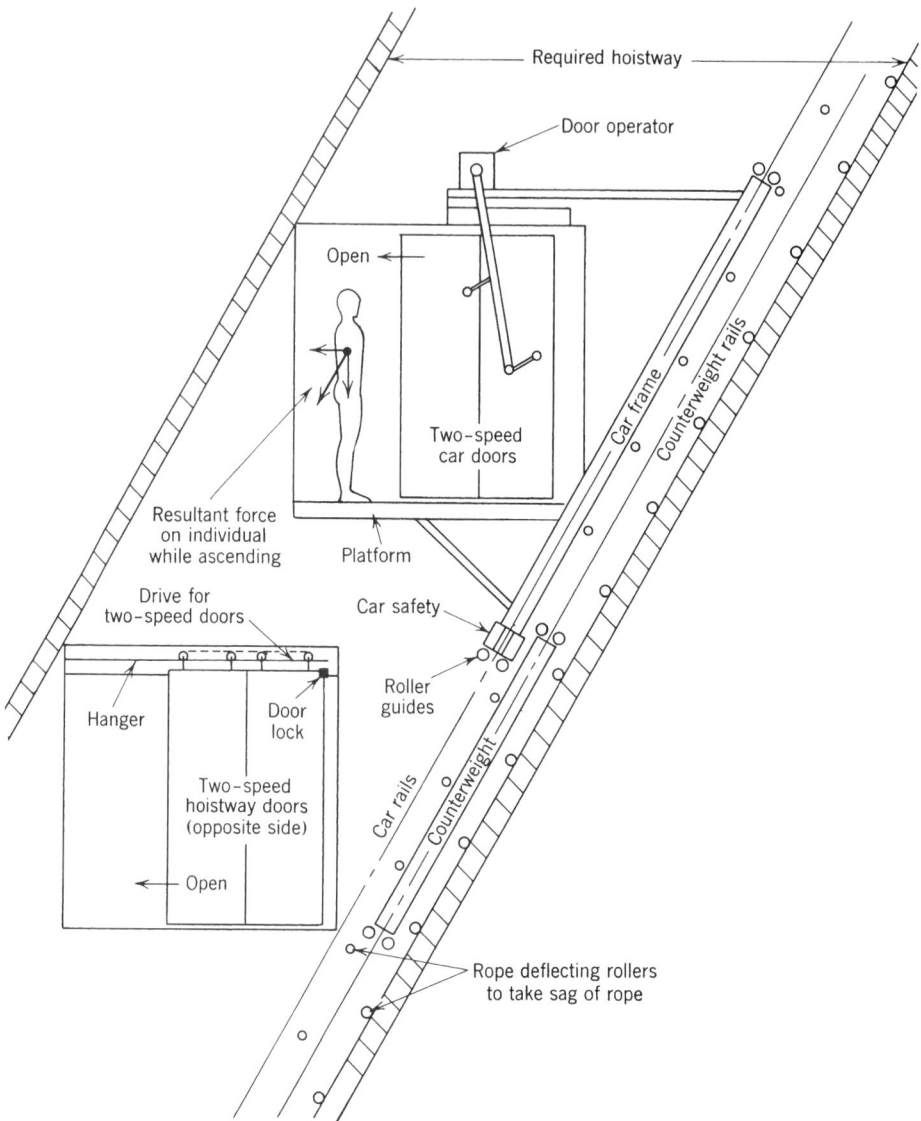

Figure 14.14. Forces and design considerations of an inclined elevator.

acting in the direction of motion, which is resolved into two components, one acting vertically, the other acting horizontally (Figure 14.14). It is the horizontal component that can destabilize passengers. At the time the A17 Code rules were written to cover inclined elevators, the assessment of retardation was done within the context of low-speed elevators (25 to 50 fpm) and their passengers would be subjected only to instantaneous horizontal retardations whose magnitude could reach 0.5g for a few milliseconds. High-speed inclined elevators did not exist and were not anticipated.

In the event that an inclined elevator is subject to an emergency electrical stop, the design objective is to impress a retardation on the elevator to stop it quickly, but at a retardation level that can be withstood by passengers without destabilizing them.

The development of the high-speed inclined elevators for the Luxor Pyramid Hotel in Las Vegas, Nevada, in the early 1990s necessitated a reassessment of the A17 Code rules and underscored the need to consider the body of contemporary research data relating to the stability of passengers involved in horizontal transport.

The A17 Code has now been revised to specify that no peak horizontal component of deceleration, that is, retardation, exceeding 0.1g (3.2 ft/sec²) shall have a time duration exceeding 125 milliseconds (⅛ sec) for an emergency electrical stop. The A17 Code also adopted rules limiting the maximum horizontal retardation at emergency mechanical stopping; that is, safety application or buffer engagement, with rated load in the car, measured over the total retardation time shall not exceed 0.25g (8.04 ft/sec²).

The term "inclined elevator" assumes travel in a single inclined direction in a single plane. Once a compound angle is introduced, such as the corner of a pyramid, multiple complications occur and an extremely difficult equipment layout is necessary. Such was the situation encountered in the Washington Masonic Monument in Alexandria, Virginia (Figure 14.15).

Inclined elevators can also transverse multiple angles in the same plane. The elevators on the legs of the Eiffel Tower in Paris travel up the approximate parabolic path of the legs of the tower. An installation that travels part of the way at an angle and then proceeds vertically is located in the tower of the Parliament Building in Ottawa, Canada. The compound angle was necessary to avoid a memorial in the base of the tower and to fit the elevator within the confines of the tower (Figure 14.16). Many commercial applications of inclined elevators may be found in Europe and residential applications in California (Figure 14.17).

Peace Tower

The inclined elevator developed and installed by the Otis Elevator Company in the early 1980s for the Parliament Hill Peace Tower in Ottawa, Canada, posed several design challenges. The space allocated to the elevator within the existing building structure required that the elevator start its travel vertically at the bottom floor and immediately ascend

Figure 14.15. Inclined elevator at the Washington Masonic Monument in Alexandria, Virginia. *Note:* This elevator travels at an inclined angle in two planes similar to the edge of a pyramid.

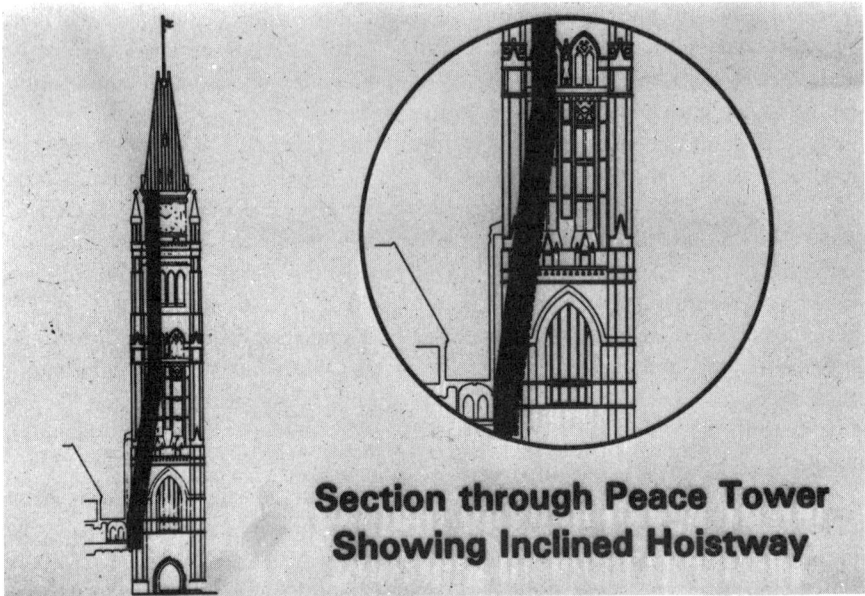

Figure 14.16. Combination inclined and vertical elevator. Peace Tower, Ottawa, Canada (Courtesy Otis Elevator Company).

Figure 14.17. Inclined elevators: (*a*) Commercial inclined elevator for eight passengers, running on steel track (Courtesy Telefuni AG); (*b*) commercial inclined elevator (Courtesy Telefuni AG); (*c*) inclined elevator, Inn at Otter Crest, Otter Rock, Oregon (Courtesy Dwan Elevator Co.); (*d*) inclined elevator system (European design) with steel cable guideway (Courtesy Telefuni AG).

Figure 14.17. *(Continued)*

through a transition curve until its travel path was inclined 10° from the vertical, where-upon it had to travel 98 ft (30 m), then travel through a second transition curve, so that the travel path became vertical again for approximately 60 ft (18 m). The passenger elevator has a 2000-lb (900 kg) capacity at a rated speed of 300 fpm (65 mps).

The elevator is roped 1:1, and the counterweight is roped 2:1 so that its travel is only half that of the car and is kept vertical in the upper part of the hoistway. A second coun-terweight, reeved 2:1, is used to tension the traveling cables, also deployed in the upper vertical hoistway (Figure 14.18).

Two sets of guide rails were used to guide the car in its travel (Figure 14.19). The car employed a gimbal-mounted double car frame, which in principle resembled a swinging birdcage, thus keeping the elevator platform horizontal at all times (Figure 14.20). Dur-ing the elevator's travel up the incline, its position shifts horizontally an overall distance of 12 ft 2¼ in. (3.7 m).

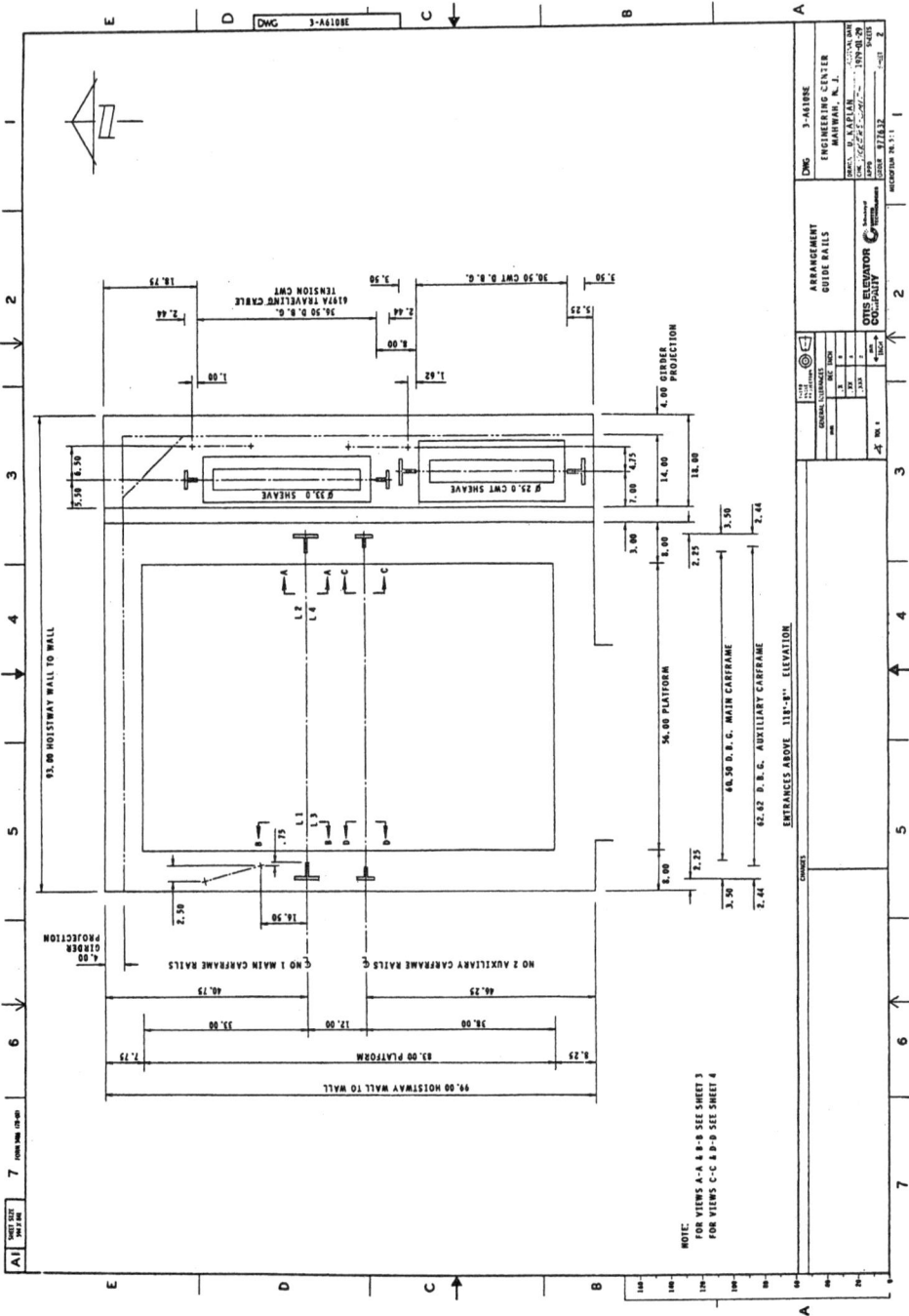

Figure 14.18. Peace Tower. Layout plan showing guide rails.

378

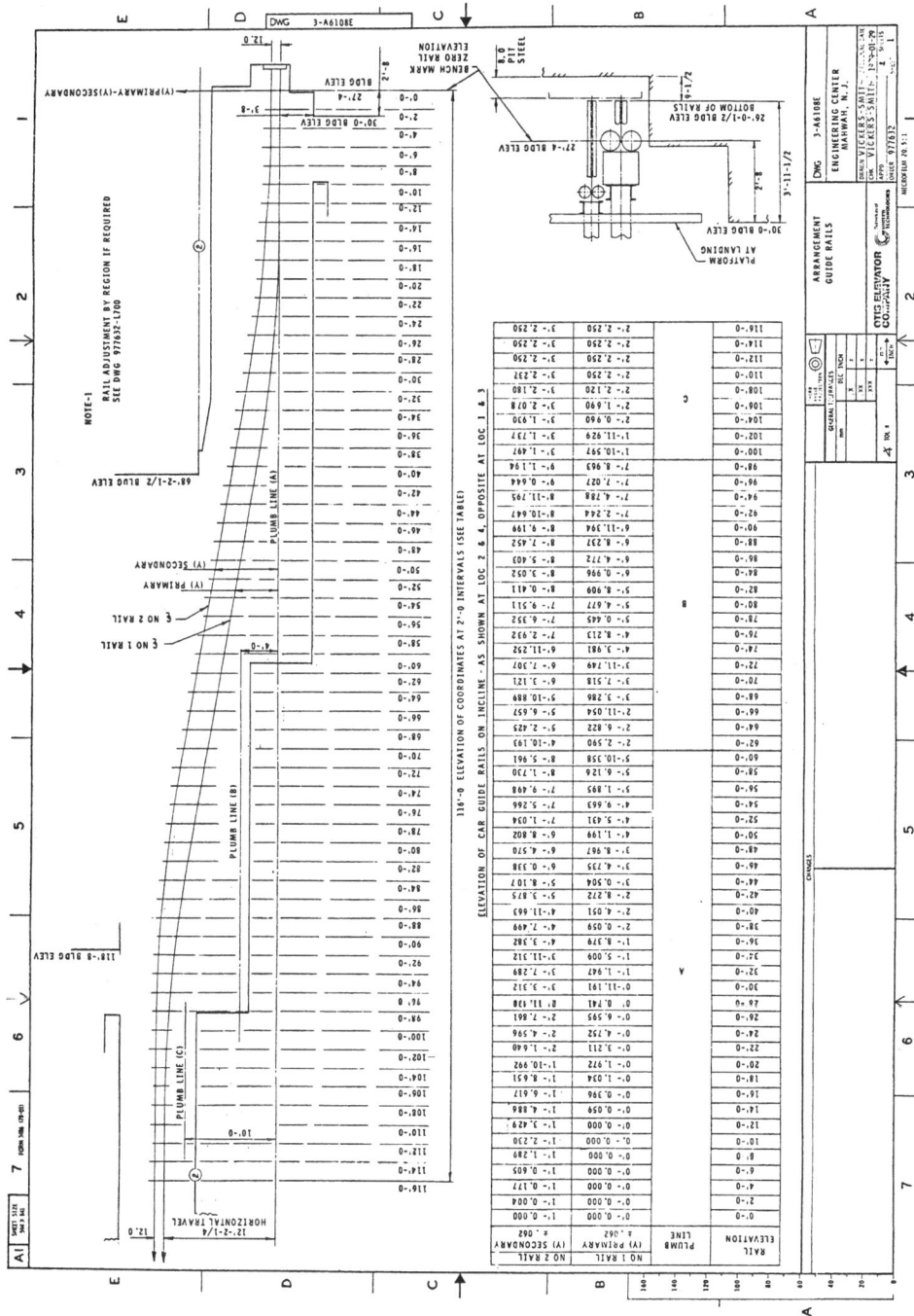

Figure 14.19. Peace Tower. Elevation showing guide rails.

379

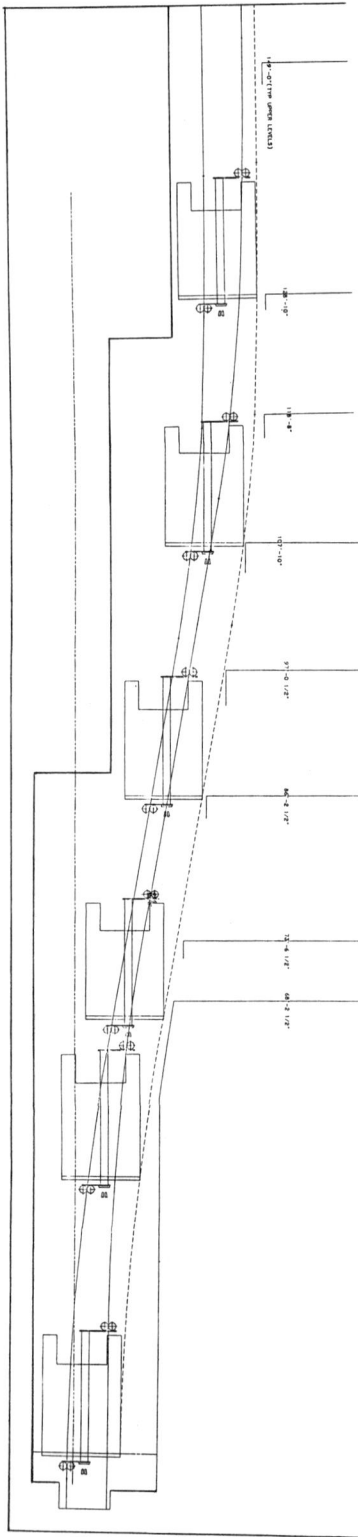

Figure 14.20. Peace Tower. Elevation showing car positions and guide rails.

The guide rail run was developed using a computer-aided design program that enabled the designers to see the exact motion of the elevator throughout its complete travel, as well as to assess the running clearances with all the suspended ropes whose catenaries had to be considered.

Because one of the most important design features involved the unimpeded swivel rotation of the auxiliary car frame at its connection with the main car frame, a switch was incorporated to detect malfunction of the auxiliary car frame pivot joint.

The slope of the wedges and guide blocks on the flexible guide clamp safeties were increased from their normal 5° to 7° to increase the space between the wedges, necessitated by the curvature of the rails in the transition curves. An additional safeguard included four switches, one at each lift rod, which were positioned to detect excessive premature vertical motion of the wedges and shut down the driving machine if this motion occurred.

Luxor Pyramid Hotel

Sixteen high-speed inclined elevators were designed and installed by the Otis Elevator Company in the pyramid-shaped Luxor Hotel (Figure 14.21) in Las Vegas, Nevada, in 1994. Each inclined edge of the pyramid contains a 4 car elevator core in which the elevators travel on a path inclined 39° from the horizontal (Figure 14.22). At this shallow angle, the elevators travel 1.23 ft (375 mm) horizontally for every 1.0 ft (300 mm) traveled vertically. The 3500-lb capacity cars run at a rated speed of 700 fpm (3.5 mps) measured along the incline, serving the 30-story structure (Figure 14.23).

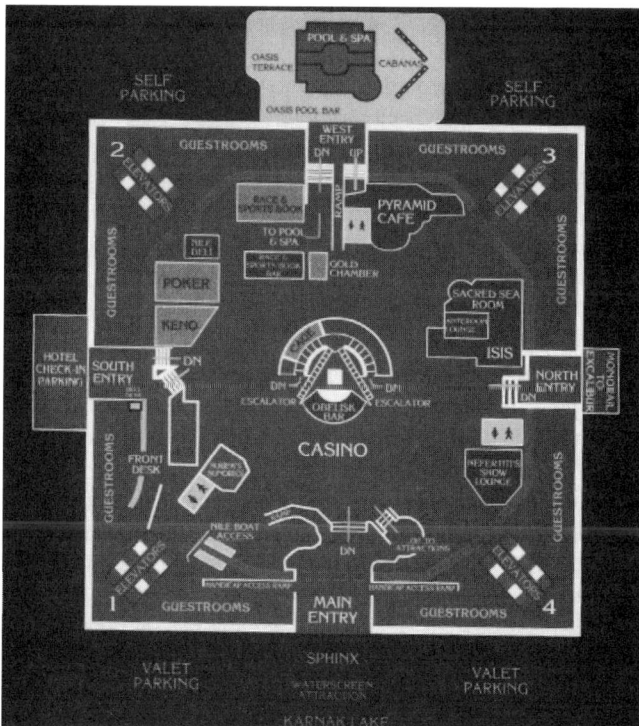

Figure 14.21. Luxor Hotel. Lobby plan view (Courtesy *Elevator World Magazine,* November, 1994, p. 66).

Figure 14.22. Luxor Hotel. Typical corner containing entry to elevator lobby (Courtesy *Elevator World Magazine,* November, 1994, p. 66).

A normal acceleration and deceleration is limited to 1.5 ft/sec^2 to minimize the horizontal component and provide the degree of passenger comfort and performance required by the hotel traffic. Recognizing that the opening of an electrical protective device, such as a hoistway door interlock, governor overspeed switch, or other device, requires the removal of electrical power to the driving machine motor and brake in accordance with the A17 Code, the horizontal component of the retardation was limited so that no peak value exceeding 0.1g (3.2 ft/sec^2) would have a time duration exceeding 125 milli-seconds ($\frac{1}{8}$ sec). These values were selected based on the concurrent A17 Code Committee research.

Conventional components were used in conjunction with innovative modifications. A single 18-lb (8 kg) guide rail was deployed along the bottom surface of the hoistway to accommodate two sets of 10-in. (250 mm)-diameter roller guides and a standard flexible guide clamp safety for the car. The safety is activated by a car-mounted governor (Figure 14.24) belt-driven by a roller that runs along the guide rail. The governor is accessible from the top of the car. For the primary guidance, the massive car frame that supports the elevator rides on train-style wheel trucks on two railroad tracks. The car weighs 12,000 lb (5400 kg).

Each car has a double-wrap traction gearless machine in a vertical orientation. Three hoist ropes run parallel to, and in the same plane as, the lower hoistway surface and are routed through an innovative reeving arrangement to deflect the hoist ropes over a multiple deflector sheave tree arrangement (Figure 14.25).

The weight of the hoist ropes hanging down the incline causes them to sag about 6 ft (1.8 m) in a natural catenary when the car is at the bottom of the hoistway. This sag was considered in setting up the clearances within the hoistway. The counterweight runs along its lengthwise edge, similarly supported by railroad-type wheel trucks running on a railroad track. Stability is provided by mounting roller guides on the opposite side of the counterweight frame and running them along a guide rail section.

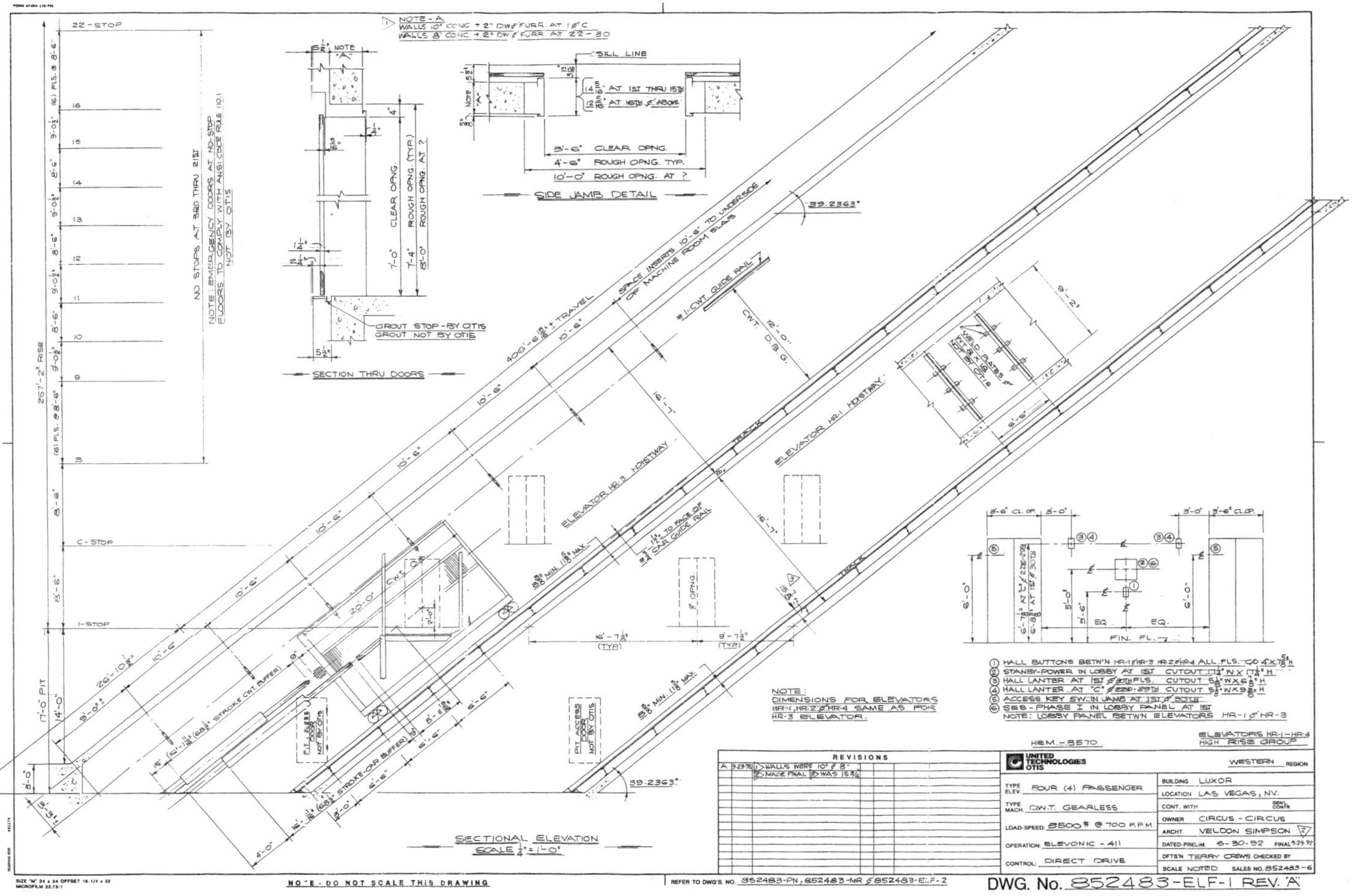

Figure 14.23. Luxor Hotel (Courtesy *Elevator World Magazine*, November 1994 pp. 70, 71). (*a*) Elevation: section through pit.

383

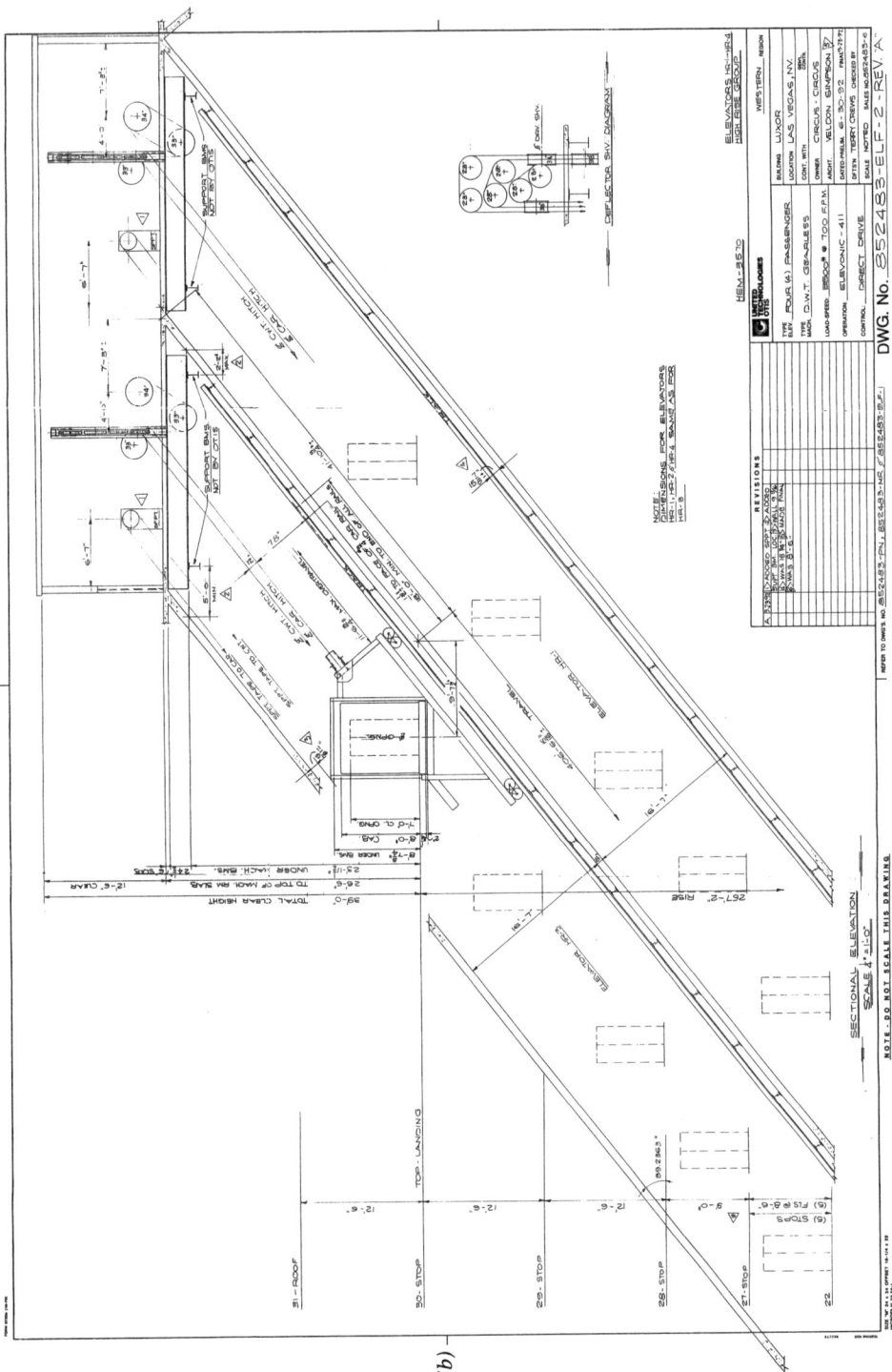

Figure 14.23. (*Continued*) (*b*) Elevation: section through overhead.

Figure 14.24. Luxor Hotel. (*a*) Car-mounted overspeed governor; (*b*) car flexible guide clamp safety (Courtesy *Elevator World Magazine*, November 1994, p. 70).

Figure 14.25. Luxor Hotel. Deflector sheave tree in machine room (Courtesy *Elevator World Magazine*, November 1994, p. 72).

Traveling cables are not used, but electrical power for the car safety circuits, lighting, and ventilation is brought down the hoistway on a set of four power rails. Brushes mounted on the car ride on the rails. Signaling and communication is accomplished by microwave transceivers mounted at the top of the hoistway and on top of the car.

Modified drive arms and blocks engage door vanes skewed at the 39° incline. Clutch rollers rotate through 180° to engage door vanes and drive doors to the open position. A retiring cam is used to unlock the hoistway door interlocks.

SHIPBOARD ELEVATORS

When elevators are installed aboard ships, many special factors must be considered. There are three internationally recognized safety standards used for the design of shipboard elevators, namely, the Internation Standards Organization (ISO) Lifts on Ships, the International Convention for Safety of Life at Sea (SOLAS), and the American Society of Mechanical Engineers (ASME) A17.1 Safety Code for Elevators and Escalators. The equipment must be moisture-proof and designed to resist corrosion, as these conditions are to be expected on a ship. Traveling cables must operate in troughs to avoid tangling when the ship rolls, pitches, or yaws. The elevator control equipment as well as the elevator car must be able to operate during these movements in moderate seas. The cab flooring must be slip-resistant to provide a secure foothold. All the equipment must be able to withstand the extra forces imposed by the motion of the ship.

Counterweighted elevators are usually used aboard ships for passenger cars. The counterweight must be equipped with a safety for the obvious reason that if it ever broke loose it might go right through the bottom of the ship. Layout conditions on shipboard are more exacting than on land. Space is at a considerable premium, and fitting a reasonable-size elevator in a hoistway trunk is quite a challenge. Often these elevators are roped 2:1 using underslung car sheaves to save overhead space. As with any special installation, competent manufacturers' representatives or consulting engineers should be called on for aid.

Cargo can be handled by elevator and, as in a warehouse on land, by industrial power truck (Figure 14.26). Because the elevator may be loaded by a truck weighing 8000 to 10,000 lb, the elevator must be rated for its duty load plus the load imposed by the truck. This may mean elevators capable of handling 15,000 lb or more. If a counterweighted elevator were used, the deadweight of the counterweight would decrease the loading capacity of the ship. For this reason cargo elevators on ships are usually drum machines, rated to lift the entire deadweight of the platform plus the capacity load, as well as impact loads of the truck during loading and unloading operations. Rack and pinion drives are especially suited to shipboard applications.

In recent years cruise ships have become prominent, and many are assuming the size and interior shape of landside hotels. A number feature atrium-type interior space with all the amenities of a grand hotel. Atrium-type observation elevators are being installed in addition to the more conventional passenger and service elevators. These observation elevators present all the challenges of their landside counterparts as well as their need to be able to withstand the various conditions encountered during a rough sea. Such conditions include the roll of the ship, pitch as the ship encounters waves, and yaw, the combination of pitch and roll. These factors must also be considered for the numerous service-type and other elevators present on any ship. A reasonable comparison would be a design that can withstand conditions comparable to a minor earthquake! Such earthquake conditions are discussed later in this chapter.

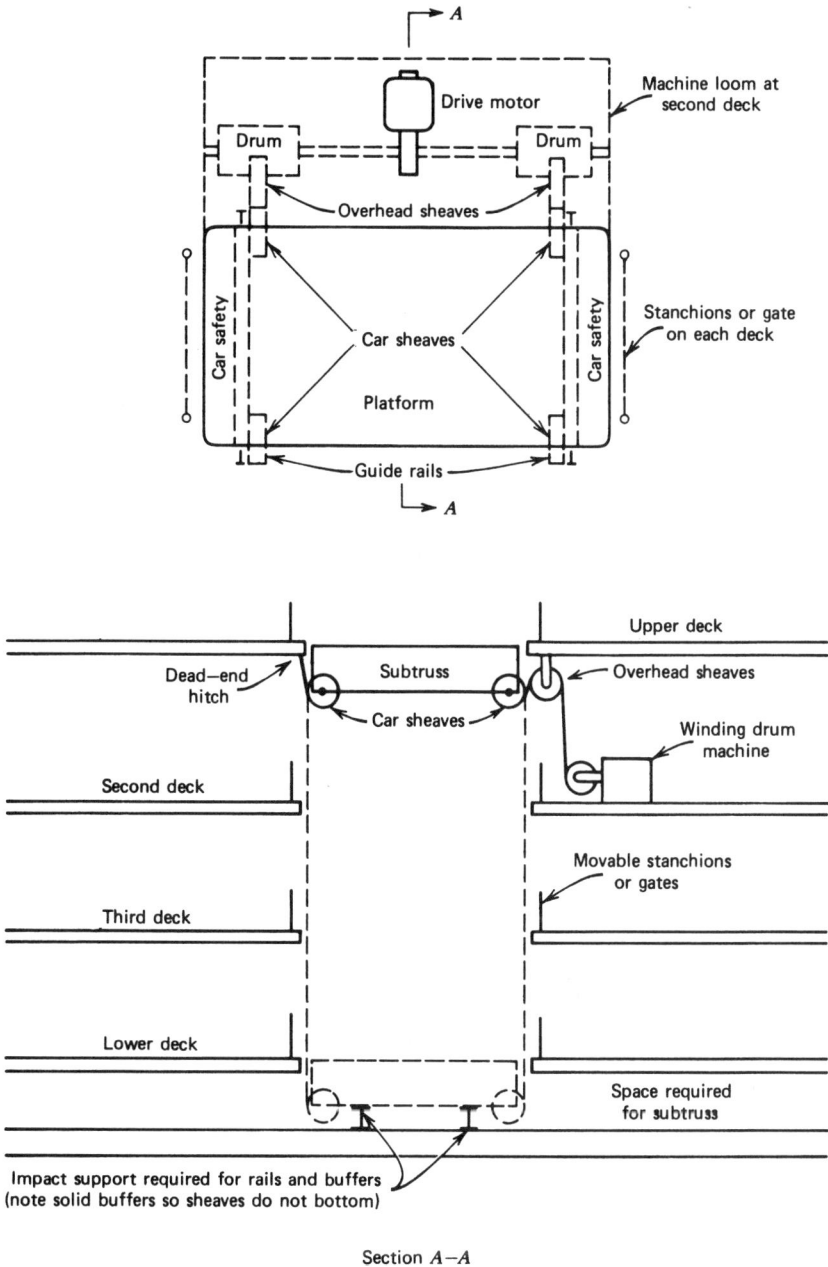

Figure 14.26. Shipboard cargo elevator. Subtruss and underslung design allow platform to be flush with the top deck. This design is ideal to lift helicopters to the landing pad.

ELEVATORS IN MINES

There is increasing recognition by mine operators of the value of providing high-speed passenger-type elevator service for miners, their tools, equipment, and mine supplies. Time spent traveling is nonproductive, and the well-being of the individuals contributes to their productivity. More stringent safety requirements recognize that the old mine "cage" hoist could be considerably improved (Figure 14.27).

Mine elevators typically have two stops. The machine room and hoistways are generally not of fire-resistive construction, nor are fire-rated hoistway doors required. Because of the high-moisture environment, moisture-proof wiring, limit switches, traveling cables, and fixtures in the hoistway must be used. Risers should have expansion joints.

The corrosive atmosphere of a mine necessitates corrosion protection for car and counterweight frames and safety parts. The guiding surface of the guide rails must be routinely cleaned to prevent rust buildup. A continuing problem with ice buildup on rails and brackets requires ongoing checks of the hoistway. It is common practice to inspect the hoistways at every shift change.

Hoist ropes must be of galvanized construction because the wet and corrosive environment causes accelerated deterioration of standard wire rope. Compensating ropes and tied-down compensation should be avoided, because of ice or water conditions, in favor of chain or unguided rope compensation. Pit oil buffers require frequent monitoring for the presence of water and operability owing to the frequent and sudden flooding of pits that cause buffers to be submerged.

Elevators in mines require additional operational and environmental considerations. Because temperature variation may be extreme, heaters are often required in door sills to prevent icing and the car equipment must be both temperature and moisture resistant. In a deep mine, considerable air pressure differential may be present, creating a stack effect in the elevator shaft. These air currents can cause traveling cable and ropes to sway, and precautions must be taken to prevent snagging. Very often the elevator shaft will also double as the ventilation shaft, and very high air velocities are intentionally employed. The design of the equipment must be suitable for low temperatures, because the elevators are usually installed in the intake air shaft.

Figure 14.27. Mine cars loaded on a double-decked mine elevator; upper deck serves personnel and lower deck serves mine cars.

Standby electrical power and means to move the elevator in case of unexpected shutdown or during an emergency are important considerations. In many mines, a single elevator installation in a single shaft requires an adjacent emergency ladder or stairs.

INDUSTRIAL-TYPE ELEVATORS

Elevators are often required to transport personnel on a periodic basis to the height (or depth) of such installations as smokestacks, cable shafts at the bottom of a dam, the operating cab of a high gantry crane, radio antennae, feed mills, grain elevators, the legs of a deep-sea oil production or drilling platforms, and in many other extraordinary locations.

A variety of small special-purpose personnel elevators designed for these special applications are available. This specialized field includes elevators of up to 1000 lb (454 kg) capacity with a net inside platform area not exceeding 13.0 ft^2 (3.9 m^2) and a rated speed not exceeding 150 fpm (0.75 mps) to carry maintenance personnel to service equipment, as well as for other applications. These elevators are of an overhead-traction-machine-type with counterweight installed with simple rails and a "telephone booth"-type cab integrated with the car guides and safety devices (Figure 14.28).

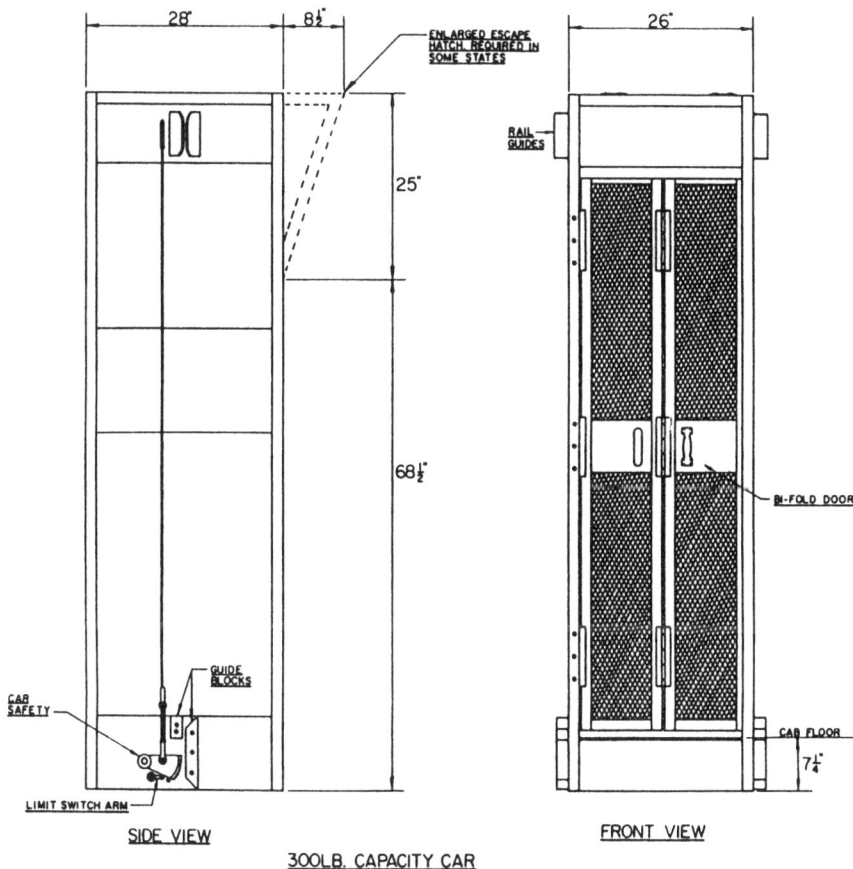

Figure 14.28. Special-purpose personnel elevator (Courtesy Sidney Manufacturing Company).

An alternative drive for the special-purpose personnel elevator utilizes the rack and pinion. A steel tower with guide rails and a continuous gear rack is erected adjacent to a smokestack, a crane, in a TV tower, or in a caisson, and the elevator is driven by a pinion gear and machine on the platform. Such elevators have unlimited travel capabilities. They can be adapted to travel vertically and then be inclined at an angle to follow the contour of the structure.

RACK AND PINION ELEVATORS

Rack and pinion elevators are elevators in which the driving means is carried on the car platform and provides motive power by having a motor drive a pinion gear operating on a stationary rack connected to either a self-supporting or supported tower. Guide shoes on the platform act on rails on the tower to maintain the pinion in alignment with the rack. A safety device that consists of a governor-operated brake and separate pinion operating on the same rack is provided. If the car overspeeds, the brake applies and the safety pinion locks the car on the rack.

Power to drive the machine is carried to the car by means of a traveling cable or trough trolley pickup. The operating device and controller are mounted on the car. In some applications the electric driving motor is replaced by a gasoline-powered motor—an outboard motor if you will—operating through a clutch and reversing gear. If the elevator runs out of gas or if the electric power fails, manipulation of the motor brake can be used to lower the car at a controlled speed.

The lack of ropes on the elevator and the possible use of a gasoline-powered motor make a rack and pinion elevator ideal for very tall TV towers, utility shafts in mines or underground power stations, maintenance elevators for smoke stacks and powerhouses, or any other location where periodic use of an elevator is necessary.

Since the rack and pinion elevator can be tilted to follow the contours of a structure, it is ideal where inclines or a combination of inclined and straight travel is required. Figure 14.29 shows an example of such an arrangement.

The rack and pinion elevator is designed to operate from a self-supporting structure, and its application as a temporary elevator on a construction site is quite extensive. By adding a counterweight and a multiplicity of drive machines, loads as high as 6000 lb (2800 kg) at a speed of 300 fpm (1.5 mps) have been obtained. For normal personnel lift work the standard load and speed are usually 650 lb (300 kg) at 125 fpm (0.62 mps).

The application of rack and pinion elevators has been expanded into more conventional uses. They are offered as passenger-type elevators for limited rises with all the attributes of any other elevator, such as power doors and automatic operation. One advantage is that they are self supporting, all the elevator loads being transferred to the floor of the pit with minimum horizontal or vertical load on the building itself.

BUILDING COMPRESSION

Any new building as it is being built, and when completed, filled with people and furniture, slowly compresses (Figure 14.30). The difference between initial and fully compressed height may be considerable, and it usually takes a year of heating and cooling to accomplish final compression. Because the elevator guide rails are erected before the

Figure 14.29. Rack and pinion elevator showing a combination of vertical and inclined travel.

building is heated, it is necessary to compensate for this compression in both the stacking of the rails and the use of rail clips that allow the compressive movement. In addition, provisions must be made in the elevator pit so that the excess rail length may be trimmed and the rails remain essentially bottomed to absorb safety impact if it occurs. This is accomplished by providing adjustable jack bolts under the rails (Figure 14.31).

Rail joints are usually erected with a slight gap to compensate for initial compression, and rail clips must be designed to allow rail slide as compression proceeds. Failure to de-

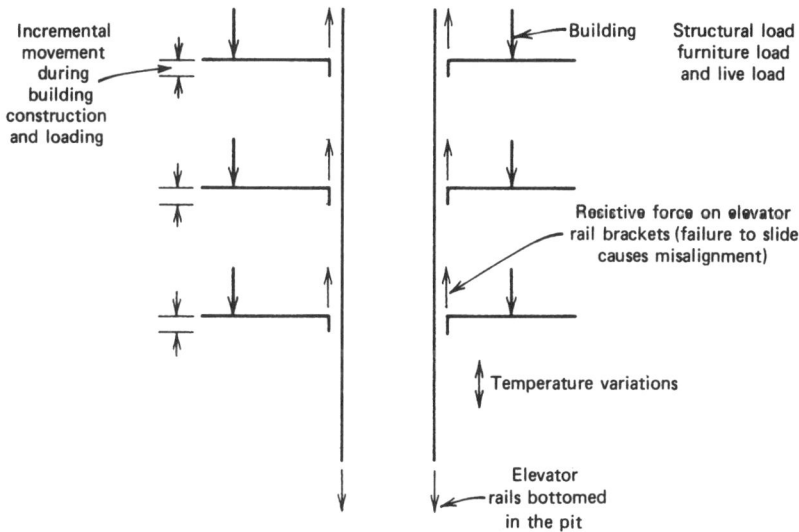

Figure 14.30. Building compression. Elevator rails erected during initial construction must adjust to the building compression during construction and furniture loading and the initial heating and cooling cycles.

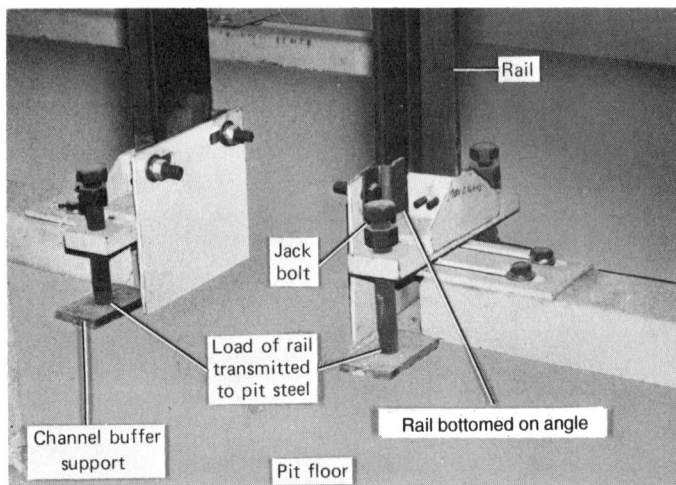

Figure 14.31. Jack bolts used under elevator rails to adjust for building compression so that rails may be maintained bottomed in the pit to transmit the force of a safety application.

sign for this action will cause the rails to distort and lose alignment, resulting in an elevator ride that produces excessive and noticeable horizontal accelerations. These accelerations are measured to judge the quality of the elevator installation, by instrumentation that produces a chart as shown in Figure 14.32.

CAR RIDE

The issue of car riding quality becomes important with high-rise, high-speed elevators because accelerations increase with the square of the speed for a given disturbance amplitude. The two components of motion that affect passengers are vertical and horizontal. Nonuniform vertical acceleration and deceleration profiles are manifested as a jerk disturbance and may be attributable to poor velocity dictation profiles and/or motor controller problems. Another vertical disturbance is due to car bounce when the car stops owing the elasticity of the hoist ropes. The resulting oscillation frequency depends on rope elasticity and car position in the hoistway.

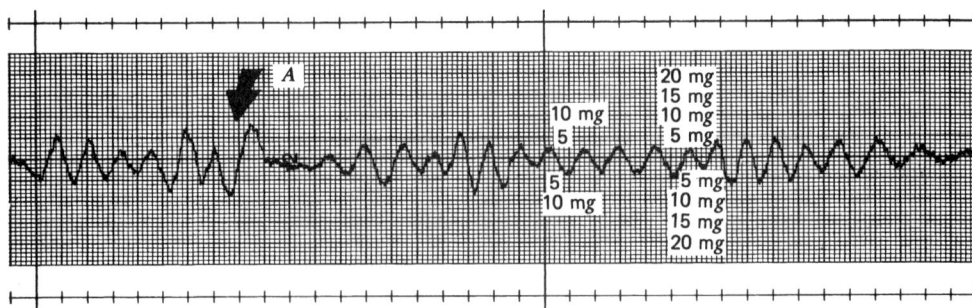

Figure 14.32. Rail ride characteristics as measured by a recording accelerometer. *A* indicates poor rail alignment with peak-to-peak horizontal acceleration exceeding 15 milli-*g*.

Horizontal accelerations and decelerations constitute the disturbance usually more discernible to the passenger. This issue will be discussed in detail

Car ride quality is measured in units of acceleration, usually in terms of milli-gs. Expressed quantitatively, 1.0 g is equal to 32.17 ft/sec^2 or, in metric units, 9.806 mps^2. 1.0 milli-g is one-thousandth of 1.0 g and is equal to 0.001 g, or 0.032 ft/sec^2. Car riding quality is usually expressed by elevator manufacturers and elevator consultants in milli-gs.

The horizontal acceleration that can be felt is about 10 milli-g at 500 fpm (2.5 mps) to 25 milli-g at 1000 fpm (5.0 mps). Measurements should be made both postwise (i.e., side to side between rails) and from front to back as the elevator travels up and down the hoistway at contract speed. The instrument should be mounted on the platform of the elevator at the approximate center. Readings that show excessive horizontal accelerations indicate that corrective rail alignment is required, and the length of the recording chart in relation to the chart speed and rise of the elevator will show the approximate location of the fault. Interpretation of the same chart can indicate other faults, such as poor rail joints and worn or flattened roller guides.

The principal factors that influence car ride are those that produce horizontal excitations, that is, accelerations and retardations, that act on the car, as follows:

1. Curvature and twist of the rails are caused by misalignment at installation or owing to building compression. Long, gentle curvatures to the rails occurring over several spans do not represent as big a problem as distortions occurring in localized spans; however, it should be recognized that, in any case, when the deflection of the guide rail exceeds the available float embodied in the roller guide, a horizontal excitation will be imparted to the car.

2. Steps at the rail joints resulting from the buildup of manufacturing tolerances on adjacent rail sections at the tongue and groove joint cause horizontal excitation of the car.

3. Unbalanced cars produce guide shoe forces at all positions in the hoistway. Setting up good car balance is extremely important. However, even when the car plus a certain portion of the rated load is assumed uniformly distributed over the platform so that its center of gravity coincides with the guide rails, guide shoe pressures will be induced when the car is at either the top or bottom of the hoistway owing to the unbalance caused by traveling cables and rope compensation. These effects induce guide shoe forces that deflect the rails in a front-to-back direction. As the car moves away from midspan, the deflection decreases over a time interval. Therefore, even with a balanced car and perfect rail job, deflections will occur and change in respect to time, as a function of car speed, thus inducing side acceleration of the car in a plane perpendicular to that of the car rails. The horizontal disturbance on the car is transmitted to the passengers through their feet.

4. Guide shoe forces produce deflections of guide rails resulting from any cause whose magnitudes increase and decrease as a function of car position which, in turn, is a function of car speed. These forces occur typically as a result of an eccentric live load in the car, which is the most frequent load condition that occurs, because of the random manner in which passengers distribute themselves.

5. Transverse rope vibrations are induced in the hoist ropes and/or compensating ropes owing to the hitch points moving sideways thus causing a transverse excitation to the end of the ropes owing to the horizontal deflection of the rails, or by

hoistway wind disturbances resulting from stack effects or building sway rope resulting from rope oscillations acting on short rope lengths as the car approaches the upper terminal. The lateral sway of the ropes may induce car motions.

6. Aerodynamic buffeting caused by the displacement of air around a single car produces air pressure forces against the side of the car. In the case of multiple cars, the turbulence produced by adjacent cars passing each other in adjacent common hoistways causes side loads on the cars which, in turn, sets up horizontal accelerations. Further, unevenness in air pressure produced by projections extending into the hoistway, which cause dimensional variations and irregular configurations in the hoistway, produces horizontal accelerations on the car.

7. Building structural response characteristics change from the new state to that of a fully occupied one. Building movements caused by compression, temperature distortion, etc., are transmitted to the rails. It would be unrealistic to hold the elevator manufacturer accountable for building movements, beyond those that state-of-the-art rail clips and car guidance/suspension systems can accommodate.

8. Out-of-round rollers in the guiding assemblies can cause vibrations to be felt in the car.

Some or all of the preceding factors will always be present to some degree in every elevator system and, in general, with the same running clearances, same roller guide float, and so forth, the horizontal accelerations and retardations that influence car riding quality vary as the square of the car speed. In terms of passenger sensitivity to motion, the longer the stroke permitted in the roller guidance system, the less the horizontal acceleration. However, the limiting values of these strokes are usually dictated by running clearances between the door clutch/vane and the hoistway door interlock, switches and cams, and safeties. Advances in car frame and guiding system designs have been deployed in the marketplace with excellent results.

The excitations or disturbances imparted to the car manifest themselves as a sequence of acceleration pulses. The amplitudes of these pulses indicate the severity of the car shaking, whose intensity is coupled with the frequency of occurrence, rate change of acceleration, and so on.

Research in the area of human response to motion suggests that acceptable peak values of acceleration alone are not sufficient to describe car ride unless coupled with additional parameters, namely, duration, or time interval over which the acceleration occurs; onset, which is the time rate at which an acceleration builds up, commonly known as jerk; frequency (Hz) and total excursions, that is, magnitude of the displacements.

A major difficulty in modeling and evaluating human body response to vibrations arises from the nature of the human body itself. There is no "standard" human body, owing to variations in size, age, sex, weight, and other factors. Therefore, any conclusion as to sensitivity levels should consider ranges of values that encompass these parameters. Although the establishment of car ride standards involves determining levels of comfort based on subjective opinion, an inexact approach at best, there has been extensive work done in the field of biomechanics to develop data on human sensitivity levels. In the absence of a rational basis for car ride standards, the acceptance of a range of values defining acceptable car ride should be pursued as opposed to setting a precise value, such as 0.5 ft/sec^2 (15.5 milli-g), which implies an exactness that does not exist.

Clearly, the objective of any car ride program should be the elimination of any stimulus that will cause negative perceptions of well-being. The best car ride should produce a neutral feeling on the part of the passenger. After all, when was the last time you encountered a passenger exiting an elevator proclaiming, "Whew! What a great car ride!"

Although there is no single fix that can ensure a top-quality car ride, experience in the U.S. and Japanese marketplace has shown that attention to several key issues, taken collectively, will result in good-quality car ride. These include providing a long wheelbase between upper and lower roller guides, extending the lower roller guides as far as possible below the platform, a well-balanced car, good rail alignment, good surface on the car and counterweight guide rails, relatively smooth rail joints (i.e., without discernible steps), stiff rails, joints and bracket system, elimination of projections into the hoistway as far as possible, adding airflow deflectors to the top and bottom of the car, advanced roller guide systems, utilizing double rollers of smaller diameter to increase the frequency of the shoes, adding rope follower guides (i.e., traveling rope carriages to effectively reduce free rope lengths, thus altering vibration nodes), realignment of guide rails after the building has been fully occupied and the natural building compression has occurred, and routine readjustment of the roller guides during the ongoing maintenance program.

Study initiatives to standardize the methodology of measuring the quality of car ride are under way within different countries. Because of the broad interest expressed by many countries, the International Standards Organization Technical Committee on Elevators and Escalators (ISO Technical Committee 178) is expected to develop and publish an ISO standard within the next few years.

WIND EFFECTS AND BUILDING SWAY

The modern trend in building construction is ever-increasing height, and the design of lightweight buildings has been made possible through the use of sophisticated analytical methods via computer. Accordingly, more efficient use is made of structural materials. The resultant tall and flexible buildings may have large wind-induced motions that not only are perceptible within human thresholds but also impact on the elevator systems, principally on the hoist ropes, compensating ropes, traveling cables, governor ropes, and selector tape. Those most affected are the lightly loaded compensating ropes and traveling cables. Still, a well-designed elevator system is expected to operate under these conditions without hazard to the passengers.

Damage to elevator equipment as a consequence of building sway is a function of a number of factors, the most significant of which include the natural frequency or period of a building, elevator travel, and various cable loadings.

The specific danger occurs when cable and/or ropes, because of their individual lengths and tensions, tend to sway at a frequency at or near the building frequency. This resonance, or near resonance, condition can, in a very short time, induce large cable and/or rope displacements, resulting in a variety of damaging conditions to the elevator equipment.

Once large building sway-induced rope or cable motions in the hoistway are induced, the cables will swing out into doors, interlocks, and hoistway switches, snag on rail brackets, and twist around one another, thus becoming entangled. In the case of compensating ropes, this tangled set of ropes cannot travel over the compensating rope sheave in the pit without causing serious rope or sheave frame damage.

If an elevator is in the upper part of a building when the onset of building sway occurs and excites the compensating ropes so that they sway, once the car starts to descend at full speed, centrifugal forces, which are proportional to the square of the speed, are induced, thus causing more violent swaying/whipping of the decreasing rope length. (Figure 14.33). The increased excursions of the compensating ropes may cause them to strike components or brackets in the hoistway.

Building deflections vary considerably as a function of construction. The Empire State Building, in New York City, deflects about 6.5 in. (165 mm) from center in an 80-mph (35

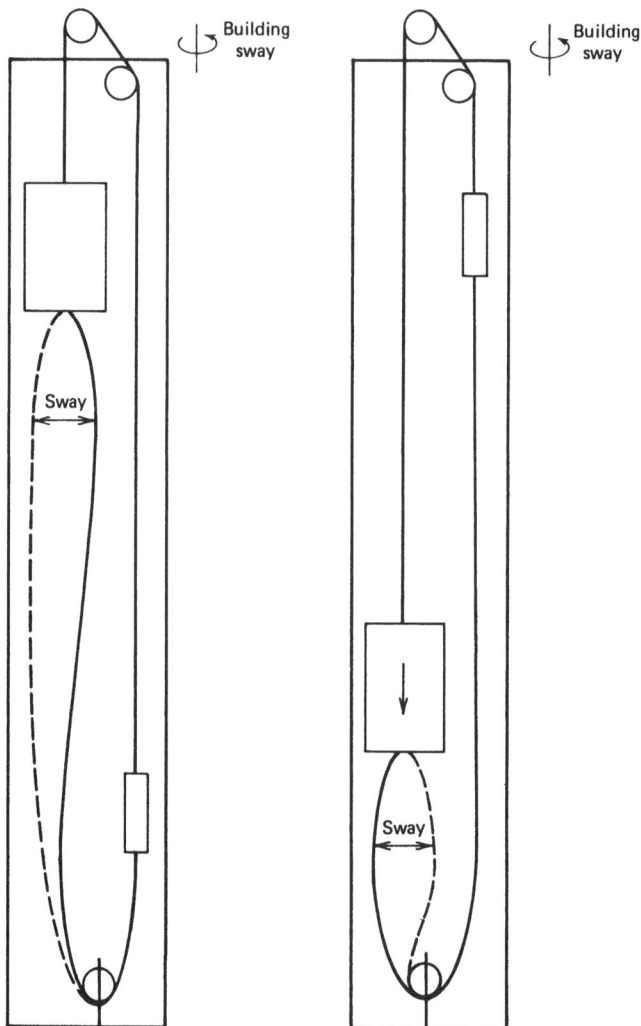

Figure 14.33. Rope sway in a high-rise building.

mps) wind. On August 9, 1976, instrumentation in the World Trade Center (New York City) recorded 18.5-in. (470 mm) building motion from center at the 110th floor during Hurricane Belle at comparable wind bursts.

Because newer buildings utilize less concrete, plaster, brick, masonry, and like materials for load-bearing structures, as were typically found in older buildings, the new buildings possess less damping, so that once building motions are induced by wind, they persist through many cycles.

Elevators, being confined within guide rails, sway with the building in a horizontal plane but, in addition, have a vertical velocity as well. This creates a waveform in those flexible items that hang from the elevator frame. The forcing frequency impressed on the elevator system is the natural frequency of the building itself. The horizontal motion of the elevator forces the various flexible media (i.e., hoist ropes, compensating ropes, traveling cables, governor ropes, and selector tape) to vibrate.

If the various flexible members, such as compensating ropes and traveling cables, have a natural frequency that matches the forced frequency imposed on the elevator system, resonance will take place with resultant high magnitudes of amplitude. Because these flexible members have very little damping, they tend to accumulate large amounts of energy if allowed to swing through many cycles. This will take place even if the elevator is not moving vertically. In fact, the longer the elevator remains at a given location, the more energy will develop. Repositioning the elevator will change the natural frequency of the various cables to noncritical values in a majority of cases.

Several items contribute to the amplitude of vibration of the various flexible ropes and cables: principally, the maximum elastic deflections of the building caused by wind gusts, the number of successive gusts occurring in a moderate time period, the natural frequency of the building in relation to the forcing frequency, and the damping present.

Mathematical analysis has shown that the factors affecting rope or cable motion are all related to the mass of the rope, its length, and its tension. Accordingly, there are a number of solutions available to address the building sway problem. These include the detuning techniques of changing the rope tension, which alters the natural frequency of the rope; and modifying the effective length of the rope, which changes its natural frequency. The latter measure, changing rope lengths, is easily accomplished by parking the elevators at certain floors, thereby altering the free length of rope or cable hanging below the car. In addition, reducing the car speed at the onset of detected building sway will reduce the severity of swaying ropes and cables, either manually by the physical activation of a switch, or automatically via the control system.

In general, the building sway problem seems to become critical as buildings approach the 40-floor mark. In planning the elevator equipment for such large buildings, it is important that building oscillation-related data be obtained from the building designers so that appropriate elevator design considerations can be included. This data includes the periods of building oscillation (sec) for the two principal axes (i.e., directions of the building), as well as twist, percentage of critical damping, the expected peak amplitude of building deflection along both building axes, and wind speed. This data should be based on full occupancy in the building. The building axes in respect to the North-South direction should be included.

In specific cases of superhigh-rise buildings where the aforementioned detuning techniques are insufficient, the elevator designer may have to consider the use of follower rope guides.

A follower rope guide, also referred to as a traveling carriage, is a device used to prevent compensation ropes and traveling cables from large-amplitude swing during high-wind weather. Roped 2:1, the follower rope guide travels at half the speed of the elevator car. Its purpose is to divide the free length of the hanging ropes, cables, and so on, effectively in half, irrespective of the position of the car, thus changing the natural frequency of the ropes or cables. It is hoisted upward by the elevator car and lowered by gravity of its own weight (Figure 14.34).

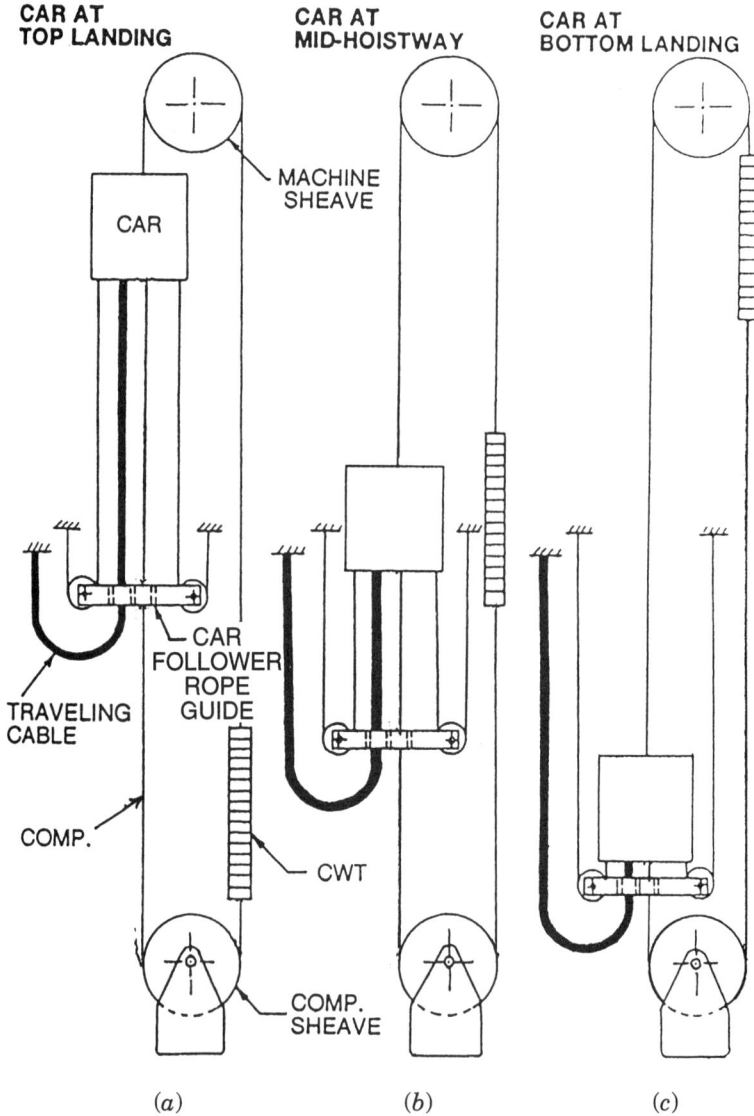

Figure 14.34. Schematic of elevator system arranged with follower rope guide. (*a*) Car at top landing; (*b*) car at mid-hoistway; (*c*) car at bottom landing.

When the car is near the bottom of the hoistway, at which point the follower is very close to the car plank, restraint to lateral motion of the compensating ropes and traveling cables is no longer effective. However, the part of the compensating ropes between the counterweight and the rope compensation sheave is close to the hoistway wall. Any lateral motion causing sway of this run of ropes would result in the ropes hitting the hoistway wall, causing them to detune.

An important factor in the design of the follower rope guide is to have sufficient weight to enable it to track properly as it rides on the car guide rails. Heavier follower weight smooths its downward motion, especially when rail alignment is not good. Accordingly, heavier structural members are purposely selected for the construction and, as a result, the strength of every component is usually many times greater than actually required for any loads. If the traveling carriage is designed symmetrically about its own geometric center, it will be subjected to no external forces except its own weight. With a symmetrical follower, no appreciable forces can be expected to be impressed on the roller guides.

Another important design consideration is to have the proper relationship between the vertical and the horizontal distances between the roller guides of the follower. The ratio of the vertical wheelbase to the horizontal distance between guide rails (DBG) must be at least 1.0, preferably 1.25, in favor of the vertical distance. This is a basic technical requirement to make sure that the carriage rides up and down freely without hanging up, breaking ropes, jumping rails, and so on. Follower rope guides have been successfully installed at the World Trade Center in New York City (Figure 14.35) and at the Sears Tower in Chicago (Figure 14.36).

The structural frame of the follower is suspended by a system of ropes, with one end of each rope hitched to the car frame plank and the other near the middle of the hoistway. An important design consideration relates to the proper tensioning of the traveling carriage guiding ropes. On one hand, it is undesirable to develop slack rope. Therefore, a slack cable switch is important. On the other hand, it is also undesirable to overtighten the ropes. Therefore, an additional switch is necessary. An effective arrangement is obtained by using a bidirectional cam and switch to monitor both conditions.

STACK EFFECT

An elevator hoistway can be a free path for air currents throughout a building unless some special construction considerations are made. For example, assume that an elevator is standing with its doors open and a window or door is open to the atmosphere in the elevator machine room at top. A door open to the street at the lobby allows air to flow from the street to the roof through the lobby space around the elevator entrance, through the necessary holes in the elevator machine room floor for the ropes, and out of the machine room (Figure 14.37). The effect may be a sufficient airflow so that the elevator doors cannot be closed by normal power means.

This stack effect is also a severe hazard if a fire occurs on a lower floor. Smoke or gases will be accelerated up the hoistway and, unless sufficient venting is provided at the top, will engulf upper floors.

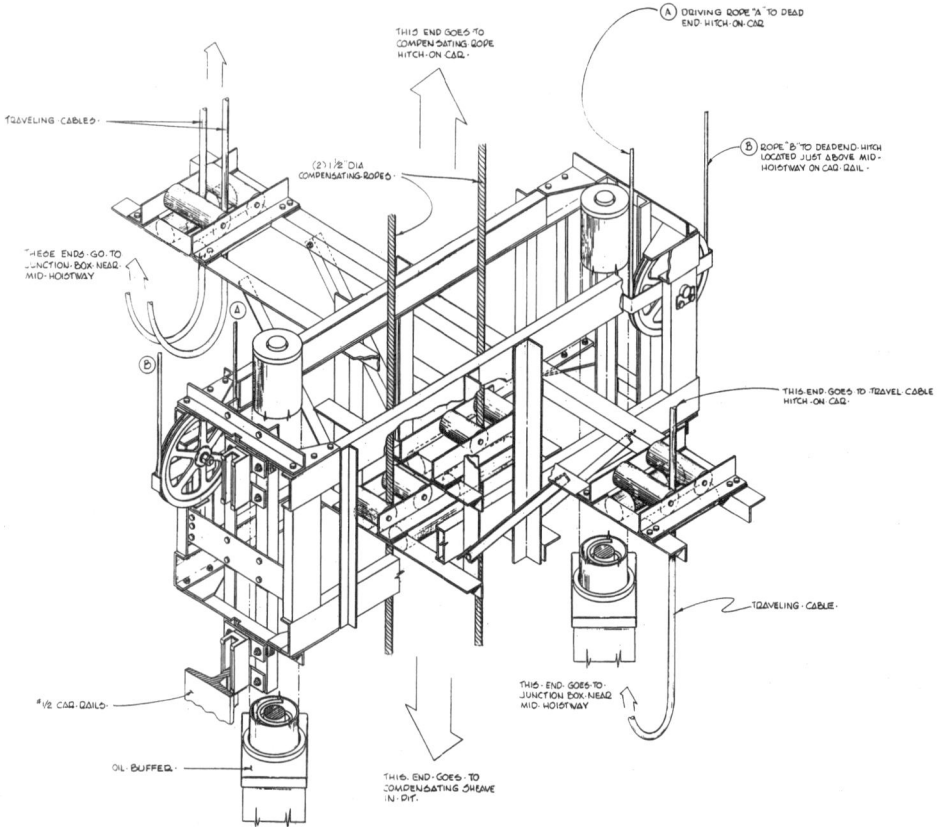

Figure 14.35. Follower rope guide assembly, World Trade Center, New York City.

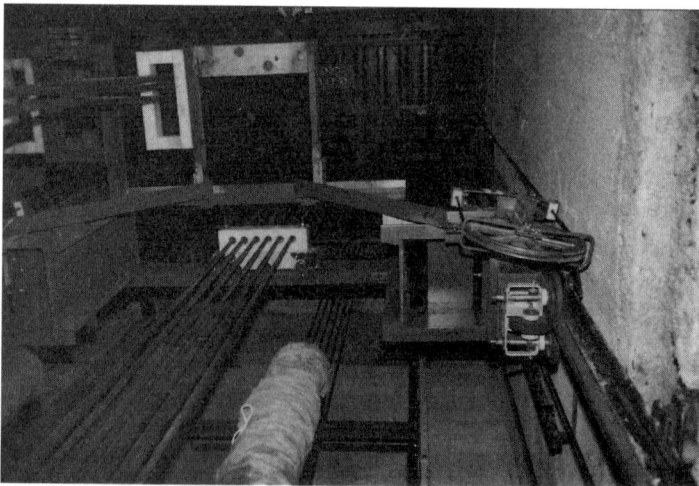

Figure 14.36. Follower rope guide assembly, Sears Tower, Chicago (Courtesy *Elevator World Magazine,* May 1995, p. 87).

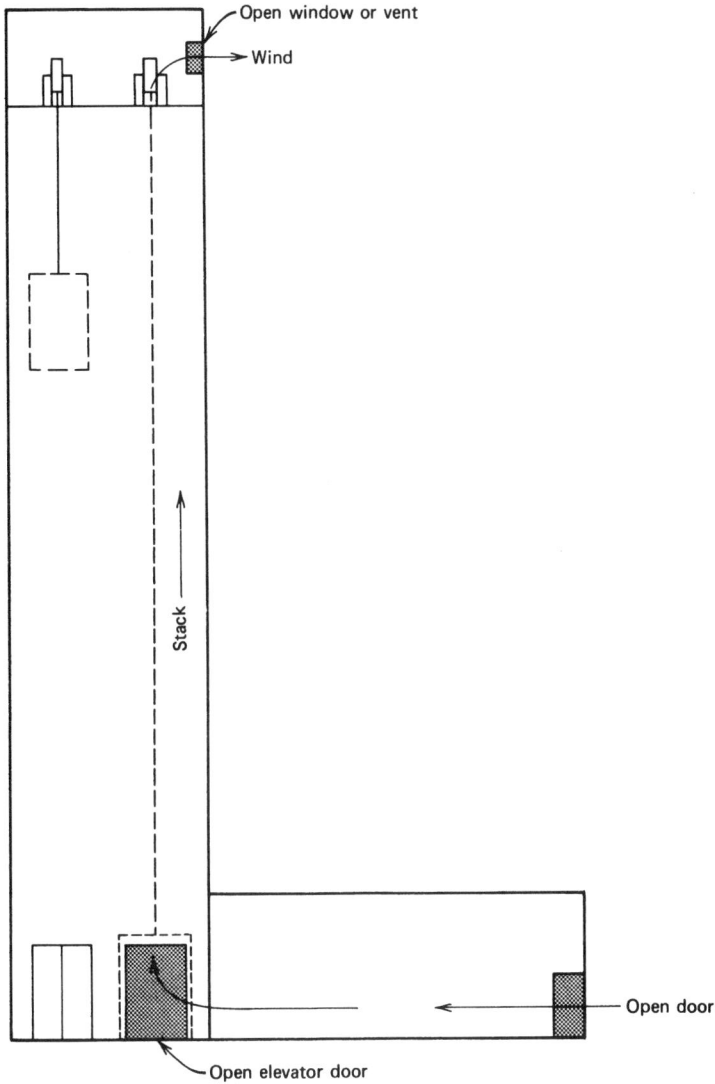

Figure 14.37. Stack effect.

The stack effect is particularly troublesome in winter weather when buildings are heated and service elevators may serve open loading dock areas. The use of double doors or revolving doors on the lobby floor may help to minimize this effect. Special closing devices to assist closing of lower-floor elevator doors, roller guide shoes in place of gibs in the door track, or special door operations to provide additional closing force may be employed to overcome the problem. The building design, venting, use of indirect air handling in the elevator machine room, and proper balancing of the building's air-handling systems must all be considered to reduce or eliminate the stack effect.

EARTHQUAKE DESIGN

In a number of large cities in the world seismic disturbances may be expected. The hazardous areas have been mapped, and both probability and severity ratings are available for most areas. Any building in a seismic-risk area should be designed to withstand earthquake shock, and elevators and their hoistways require certain additional considerations over normal installation practices.

One potential danger is the possibility of a counterweight becoming disengaged from its rails and swinging into the hoistway. If an elevator is running, there is the potential hazard of the car colliding with the free-swinging counterweight. A number of preventative means are available. One is to reinforce the counterweight rails by means of box brackets so the counterweight is restrained from swinging out (see Figure 14.38). Another is an electrical detector, a taut wire from top to bottom that will be electrically intercepted and cause an electrical circuit to stop the elevator if the counterweight is displaced (Figure 14.38).

In addition to the counterweight protective measures, the elevator hoisting machine and other machine room equipment is tied down with fastenings sufficient to withstand the expected shock. Rope guards are provided to prevent the ropes from jumping from the sheaves, and the car-to-counterweight compensating rope system is tied down with an arrangement to prevent the car and counterweight from bouncing upward during an earthquake shock. The elevator car is also equipped with retainer plates to maintain it within the rails. Antisnag guards are required in the hoistway to prevent swinging ropes and traveling cables from hanging up.

The rails and rail structure must be reinforced to withstand horizontal shocks in addition to the normal forces of loading and safety application that the rails are initially designed to withstand.

Seismic engineers have determined that there are distinct earthquake ground waves that precede a severe earthquake shock. Detecting equipment is available to sense those waves and provide a slightly early warning. This information is used to cause all elevators in the building to stop at the next available floor and remain with their doors opened. An elevator traveling in an express portion of the hoistway is stopped and designed to proceed at a slow speed toward the next landing in a direction away from the counterweight. For example, if the elevator system has an express run from floors 1 to 9 and serves floors 10 to 20, and an early warning occurs as the elevator has just passed the eleventh floor in the down direction, it could possibly encounter an up-traveling counterweight. It would be stopped and slowly travel up to the eleventh or twelfth floor before finally stopping. In this way the counterweight would travel down, away from the elevator, and the hazard of a collision would be avoided.

Once earthquake shock waves have ceased, the elevator hoistways should be inspected for displaced equipment. Emergency personnel have the option of opening the top of car escape hatch from inside the car by key to make a visual observation before the car is moved. In the same manner, they can move the elevator upward at slow speed and observe the hoistway to determine any damage before restoring the elevator to normal high-speed operation. As a first step, a reset switch on the elevator controller is provided to restore normal operation once the early warning seismic device is reset after damage has been appraised or repairs are accomplished.

The A17.1 Code has developed elevator safety requirements for seismic risk zone 2 or greater. The state of California has additional rules, as do many other seismic-risk areas. These rules must be considered as a minimum in designing elevators for those areas.

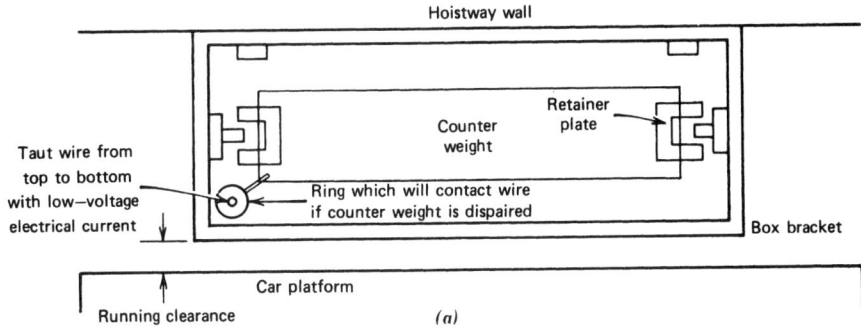

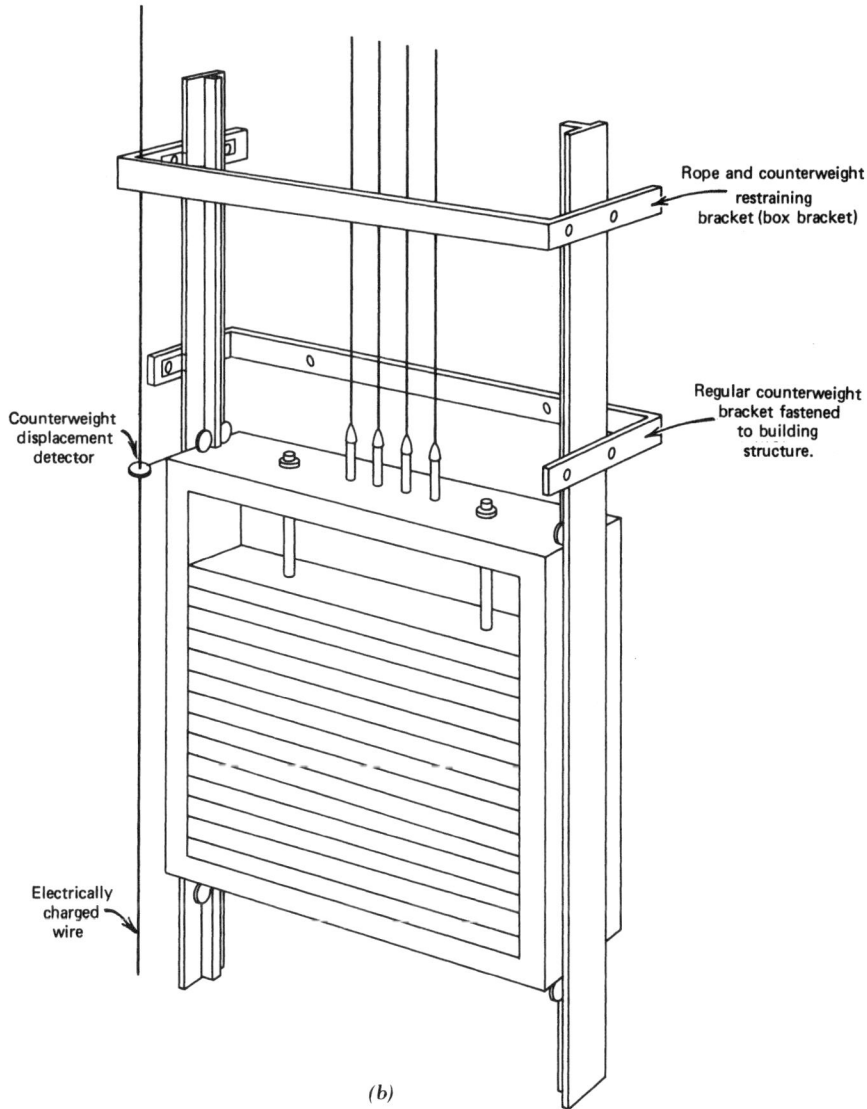

Figure 14.38. Counterweight box bracket and counterweight displacement detector; (*a*) plan, (*b*) elevation.

NOISE

The noise of an elevator causes, operating with the normal ambient noise in a building, is usually nonobjectionable. Elevator noise may be noticed and objected to under quiet conditions, such as in a residential building or hospital during the quiet periods. The counterweight or elevator passing sleeping rooms adjacent to the elevator hoistway will cause disturbing noises, and any vibration that may be present can be transmitted through the walls and structure. This can be avoided by providing sound and vibration isolators on the elevator brackets where they are fastened to the building structure (Figure 14.39). The best approach is to architecturally place such rooms away from the hoistways.

Elevators operating during the night create a certain amount of noise when they stop at a floor and doors open and close. Operating systems that park elevators at upper floors should be designed so such parking is done without door operation. Door operations should happen only in response to a car or landing call.

Structure-borne elevator noises and vibrations may be transmitted to places remote from the hoistway through the building structure. The elevator machine should be isolated from its support and the secondary or deflecting sheave attached to the machine rather than supporting structure. Electrical equipment such as isolation transformers used with solid-state motor drives should to be isolated from the building structure to break up sound paths and guard against the transmission of objectionable structure-borne frequencies.

As a high-speed elevator travels through the hoistway, various noises are generated. A "puff" may occur when the elevator passes each floor as the air currents eddy at landing sills, and various pockets and structural members in the hoistway can create additional "puffs." These can be overcome by aerodynamic design and structural considerations.

Wind noises around hoistway doors where stack effect is present will create whistling noises. These can best be overcome by minimizing the stack effect or as a last resort,

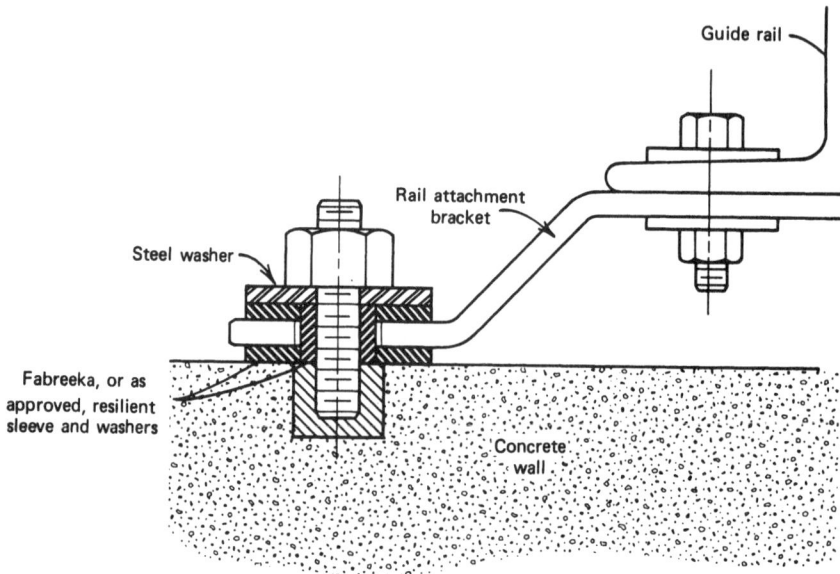

Figure 14.39. Structure-borne noise isolation arrangement of elevator rail brackets—necessary if sleeping rooms or sensitive equipment is located adjacent to an elevator hoistway (Courtesy Cerami Acoustical Associates).

weatherstripping the hoistway doors, which usually requires additional maintenance considerations.

Each noise problem must be individually approached. Sensible initial design can avoid the expense of later correction.

ELEVATORS IN PROBLEM LOCATIONS

Safe vertical transportation must often be provided in areas where hazards exist, detrimental to either the equipment or its operation. Such locations may be found in storage facilities for flour or other dusty, often abrasive materials, or in petroleum refineries or chemical processing plants handling corrosive or explosive substances. Wet locations include elevators in mines, those exposed to the weather, as in the observation towers previously mentioned or located at waterside, or, perhaps, used to handle wet ashes in a power plant. Elevators operating in the vicinity of rockets fueled with liquid hydrogen and liquid oxygen must endure special environmental hazards (Figure 14.40).

Figure 14.40. Columbia launch, Cape Kennedy. Elevators in the launch service tower (*left*) are often damaged and must be rebuilt after each launch.

Whatever the hazard or location, conventional elevator equipment would be subject to deterioration from the elements that present potential dangers to personnel.

Classes of hazard are recognized by the National Electrical Code, which sets forth rules for the treatment of electrical equipment in such locations. Common sense and engineering considerations must be applied to the other parts of the elevator exposed to hazards. For example, if all the electrical equipment is required to the watertight, all the structural parts of the elevator including the ropes, machine, and rails should also be protected against deterioration from moisture. Similarly, if the elevator equipment must operate in a corrosive atmosphere, as in a fertilizer or chemical plant, all metal parts as well as the electrical equipment should be treated to withstand corrosion. These considerations may often lead to installing wooden rather than steel guide rails and providing a safety mechanism that operates on wood. A damaged section of wood rail can be replaced, as in the early days of "safety" elevators. Safety devices with nonsparking metals such as beryllium or aluminum have been developed and successfully applied to steel rails.

Specifying elevator equipment for installations requires that the type and extent of the hazard be established. As an example, *Class I,* Group B, would indicate a possible explosive hazard of hydrogen gas as outlined in the National Electrical Code ANSI/NFPA 70. Equipment manufacturers should be consulted to determine the necessary precautions, and elevator specifications should be prepared accordingly.

FLOODING

In some areas flooding may be a hazard that will require special design and elevator operating considerations. This is prevalent in cities along rivers that are subject to flooding or in coastal cities during extreme high tides or hurricanes.

The elevators in such cases will normally serve a ground floor, but when a water level rises as indicated by a float switch in the pit, the elevators are prevented from traveling to the ground floor and the uppermost floor so that neither the car or the counterweight can enter the flooded pit area.

Secondary limit switches are provided, and all pit equipment such as buffers and electrical switches are designed to withstand being submerged. Operating speed is reduced because compensating ropes or chains and traveling cables may pick up water and soak other parts of the hoistway.

Such operation should be limited to emergency use since part of the elevator protective system is reduced. If for any reason the elevator were to travel to the pit area, the buffer might be submerged and rendered ineffective.

TRAILER TRUCK ELEVATOR

The loading imposed on a freight elevator system designed to transport a loaded straight or trailer truck is extremely special. The length of such an elevator can easily run 40 ft and, in some cases, can be as high as 60 ft. The structures for the guide rails, platform, bracing, and car frame can be expected to be quite large. Such elevators are especially suited to be hydraulic, but if the travel become significantly larger than conventional hydraulic elevator rises, traction elevators roped 2:1 and 3:1 have been designed and installed.

When the rated loading for a heavy truck elevator exceeds 30,000 lb and the front-to-back length of the elevator exceeds approximately 25 ft, the elevator design engineer must resort to methods of optimizing the car's structural design by placing supports at intermediate points, not normally supported. This is accomplished by utilizing a bridge-type truss bracing (Figure 14.41) in place of the conventional single long braces normally running from the edge of the platform diagonally up to a point close to the car frame crosshead.

The most efficient means of transferring loads from the loading position at the edge of the platform to the guide rails is to have the centerline of the long side braces intersect the guide shoes at their midpoint where they contact the rails. The problem of transferring extremely high loads to the guide shoes is that the magnitude of the loads may be much greater than the guide rail section can safely withstand. The 30-lb (13.6 kg) guide rail is the largest commercially available rail.

When the side loads being transmitted to the guide rail are too great, dual guide rails have been deployed in conjunction with equalizer guide shoes (Figure 14.42) at each of the four corners of the car frame. Otis Elevator Company developed this innovative design in the late 1950s. The linkages act to evenly distribute the guide shoe force in the front-to-back direction between the two adjacent rails, thus allowing the use of the 30-lb (15.6 kg) rails without having to fabricate a single special rail.

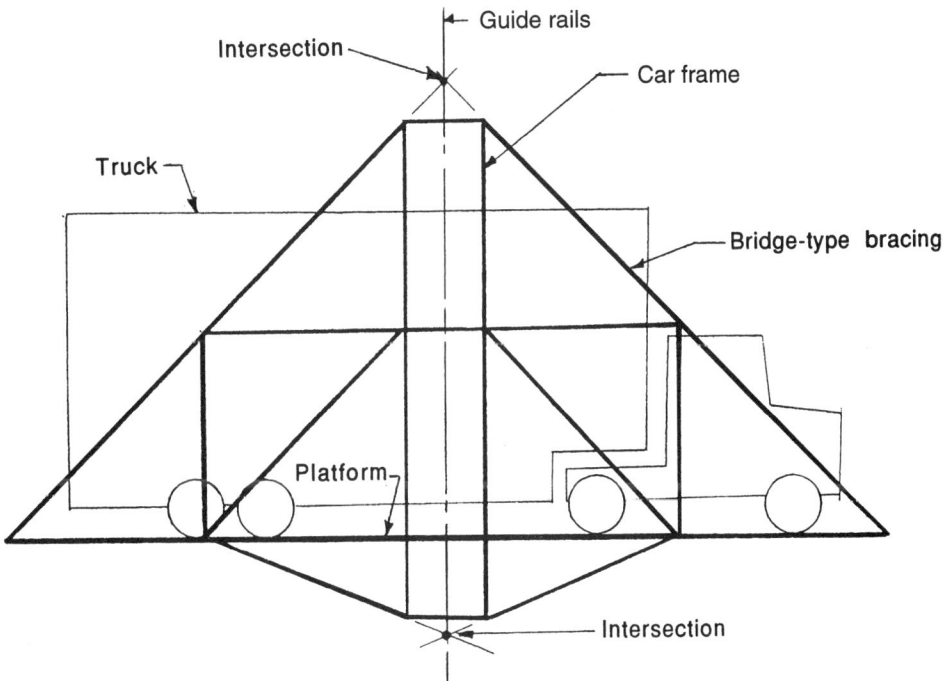

Figure 14.41. Trailer truck elevator. Elevation.

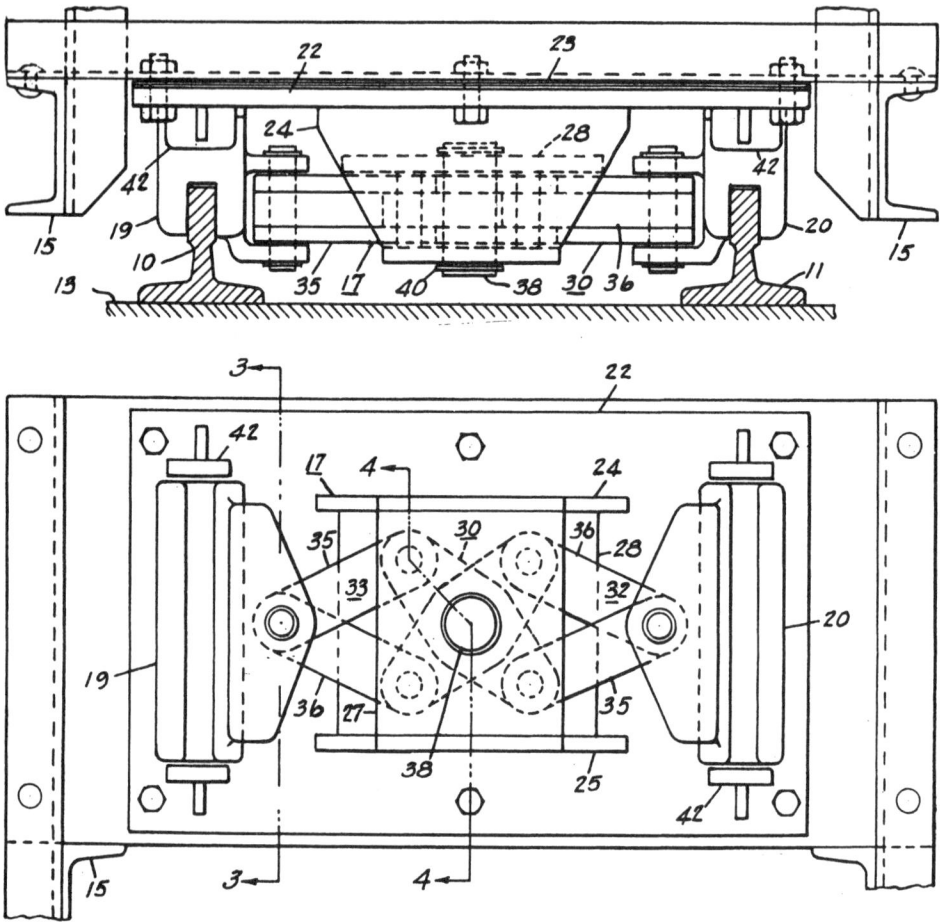

Figure 14.42. Equalizer guide shoes with dual guide rails.

PARKING GARAGE ELEVATOR WITH HORIZONTAL CARRIAGE

One of the more interesting garage elevator applications was installed in a medical center in Spokane, Washington. The parking system consisted of an elongated traction elevator, 10,000 lb (4500 kg) at 350 fpm (1.7 mps), rising vertically in a stationary hoistway, between two banks of parking stalls. With each floor serving 12 parking stalls, the garage capacity was 180 cars spread over 15 floors (Figure 14.43).

As the elevator was moving vertically, the horizontally moving carriage mounted on top of the elevator platform would be moved toward its destination stall by an attendant stationed at a console in the center of the parking carriage.

The elevator measured approximately 50 ft (15 m) wide in the postwise direction and 21 ft (6.4 m) in the front-to-back direction. The platform was designed as a three-dimensional space frame employing a system of 6-ft-deep (1.9 m) bridge-type trusses for the platform perimeter and 5-ft (1.5 m) deep trusses within the interior. The Otis type 30 FGC duplex safeties were used with a specially built car frame arranged with split "rock"

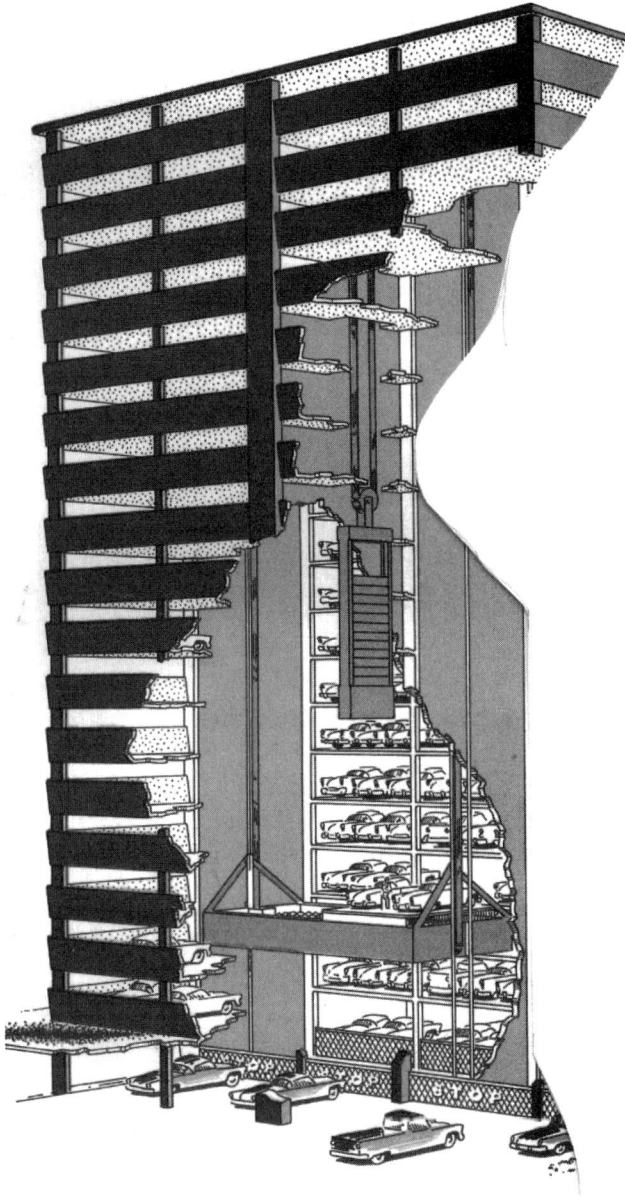

Figure 14.43. Cut-away view of parking garage.

shafts to ensure that the safeties at both sides of the car would be deployed simultaneously to prevent wracking of the immense structure.

The roping system employed four crosshead truss sheaves roped 2:1, together with a system of equalizer ropes and sheaves running in the postwise direction of the car at both the front and rear of the platform. The equalizer rope system is based on the principle used to guide long straightedges on drafting boards. The horizontal edges always remain parallel to the long direction of the drafting board. This principle was utilized in this elevator to ensure its horizontal stability. This equalizer system worked so successfully that

Figure 14.44. Operation of carriage and dolly: (*a*) operator positioning and aligning dolly; (*b*) dolly retracting automobile; (*c*) parking automobile at typical level.

it became the prototype for the bidirectional (i. e., front-to-back as well as side-to-side) equalizer rope systems used later on stage lifts (Figure 14.45b).

Figure 14.44(*a*) shows the console operator positioning the dolly horizontally to line up with the automobile. The dolly is then extended under the car, where it raised by two roller arms and retracted onto the carriage. Figure 14.44(*b*) shows the auto being retracted onto the elevator. Figure 14.44(*c*) shows the auto being deposited in a stall. Note the structural trusses.

Other Parking Garage Elevator Approaches

Automated parking of automobiles in high-rise structures has intrigued developers and elevator engineers for decades. Original attempts were made in the 1929 with the "Ruth" system in the Pure Oil garage installed in Chicago (See *Elevator World*'s publication "In Search of the Past" referenced in the appendix).

Later applications, prominent in the 1960s, included the Speedpark system and the Bowser system, both employing elevators installed in horizontally moving towers, which both vertically and horizontally "indexed" automobiles into stalls. The Speedpark system employed automatic depositing and retrieval, whereas the Bowser system (still in operation in various cities) relied on an attendant to drive the auto on and off the elevator. The first edition of *Vertical Transportation* (John Wiley & Sons, 1967) details these systems. Their popularity was short-lived, and most such systems were either abandoned or demolished.

Surviving is a "ferris wheel"-type system which, as the name implies, utilizes that principle whereby individual carriages holding single automobiles travel vertically on a continuous chain up and around and down. This form of installation can be found in many South American and Asiatic cities and, we can expect, will start to show up in North America. As with "the tallest building in the world" the development of automated parking systems is perennial and we can expect a new proposal or design at any time.

NONELEVATOR VERTICAL TRANSPORTATION

Safe lifting, which means utilizing other than a fully enclosed elevator in a fully enclosed elevator hoistway, is often desired for many applications. Such installations can include stage lifts in theaters and auditoriums, platform lifts in industrial applications, raising and lowering the bottom of swimming pools for various activities, lifting aircraft on an airplane carrier, plus lifts for extremely special applications.

Conventional elevator lifting approaches have been used, including elevator-type traction machines with counterweights; drum-type machines, both with and without counterweight; hydraulic plungers, both direct and indirect drive using chains or ropes, screw jacks, and hydraulic or electric rack and pinions.

Detailed descriptions are beyond the scope of this book, but a few examples are presented.

Stage Lifts

Stage lifts are generally platforms ranging from 10 to 20 ft (3 to 6 m) wide by 30 to 60 ft (9 to 12 m) long with a vertical rise ranging from a few feet to two or three floors. Multiple direct-hydraulic plungers are the preferred means of lifting, and two, four, or more such plungers may be employed depending on depth and width. When more than one plunger is provided, displacement of each plunger must be equalized, and this is done either by rope or rack and pinion equalization.

Figure 44.45*a* and *b* shows the rope equalization system between four plungers lifting a 10 × 36 ft (3 × 11 m) platform. Note the "four-way" considerations. The same equalization can be accomplished by rack and pinions. Two racks would be located next to each plunger, and pinions and shafts would connect the racks in the same plane.

If two such platforms are to be located next to each other as shown in Figure 14.45*c* and *d*, the platforms can be pinned together by toggles if rope equalization is used, or the locking can be accomplished by clutching the pinion shafts on each lift together if rack and pinion equalization is used. In one such installation, three adjacent platforms are clutched together and caused to move as a single unit. The configuration of the platforms could be flat or steplike, depending on the location of the platforms when they are clutched together.

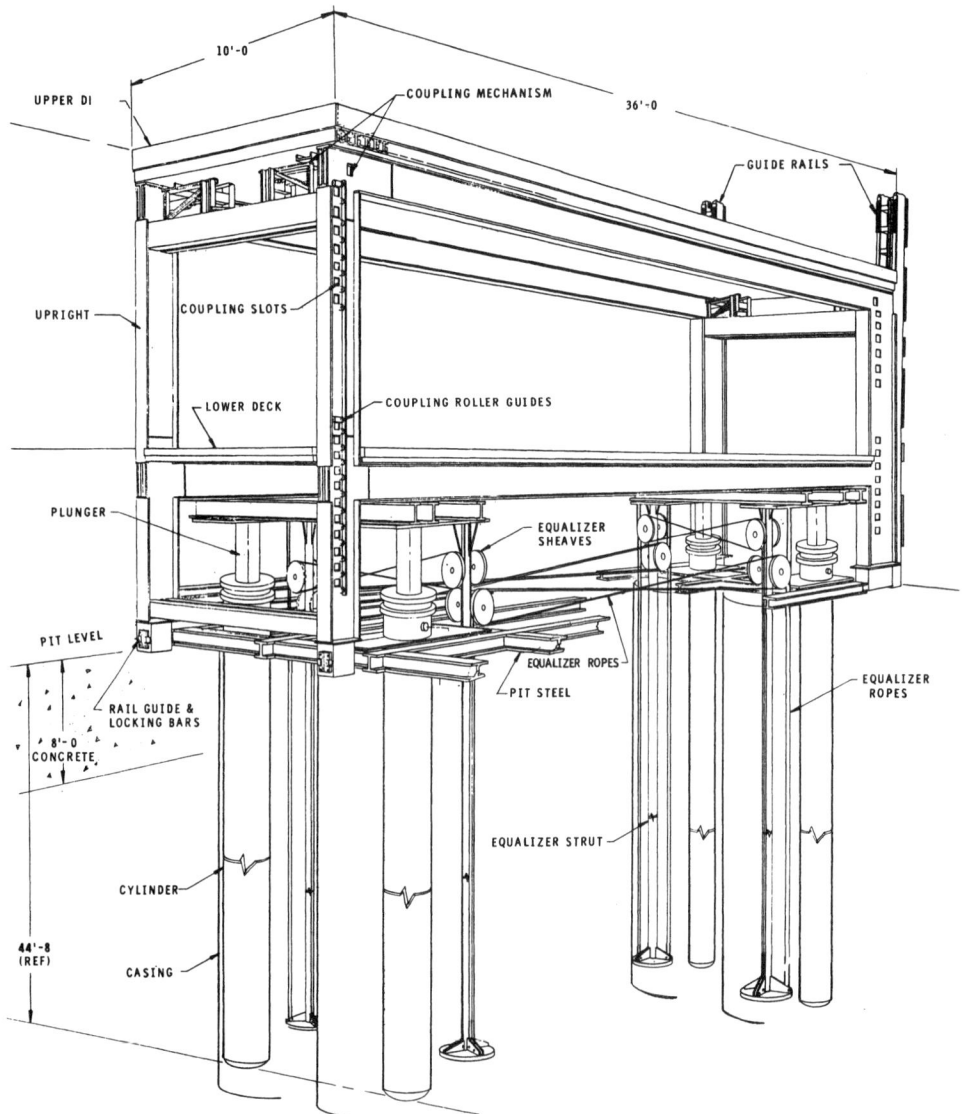

(a)

Figure 14.45. Stage lift. (Courtesy Otis Elevator Co.). (*a*) A single section of a three-section stage lift showing the main components.

Stage lifts can be quite elaborate. In Vienna, a stage lift is mounted on a three-story-deep turntable, and the entire stage can be revolved as the two-section lift raises and lowers. The effect is spectacular in that four different scenes can be shown in a matter of minutes by the combinations of lifting, lowering, and revolving the stage.

Moving Floors

Gymnasiums can be transformed to auditoriums with stages by lifting one end of the room and creating a stage. Screw jacks are recommended because they are self-locking, can be simply equalized, and require very little horsepower inasmuch as screw jack speed is very low.

Aircraft Carrier Elevators

Aircraft carrier elevators employ roped by hydraulics to lift the heavy load of an aircraft, which can be as great as 300,000 lb (140,000 kg), at a high speed of 300 fpm (1.5 mps). Figure 14.46 shows an arrangement of such a lift.

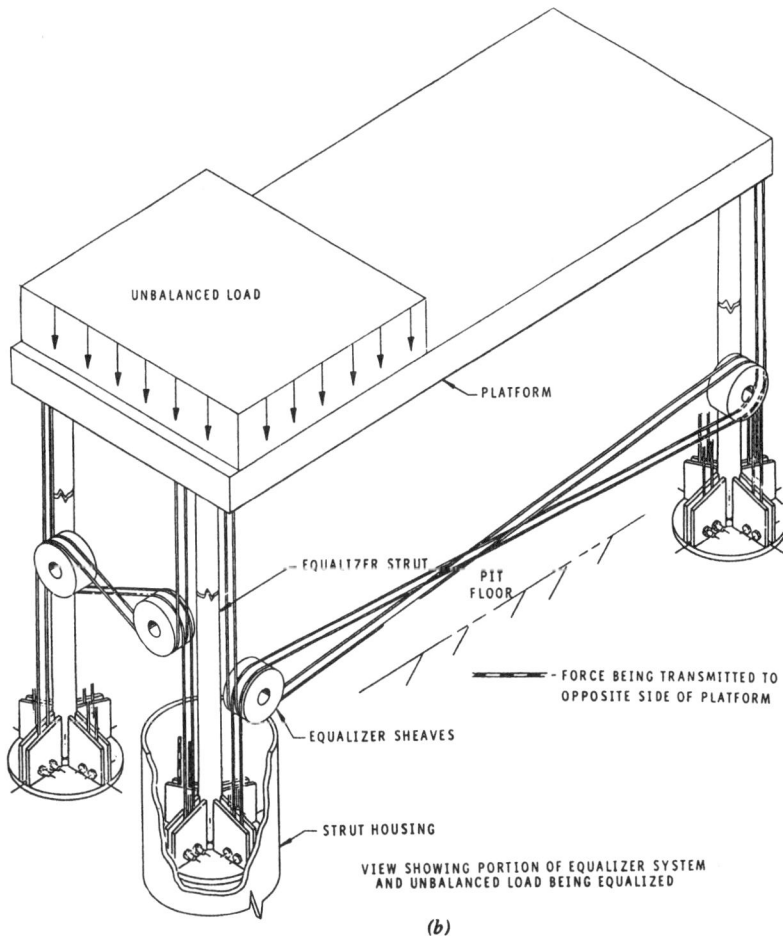

(b)

Figure 14.45. *(Continued)* (b) Details of the rope equalization system.

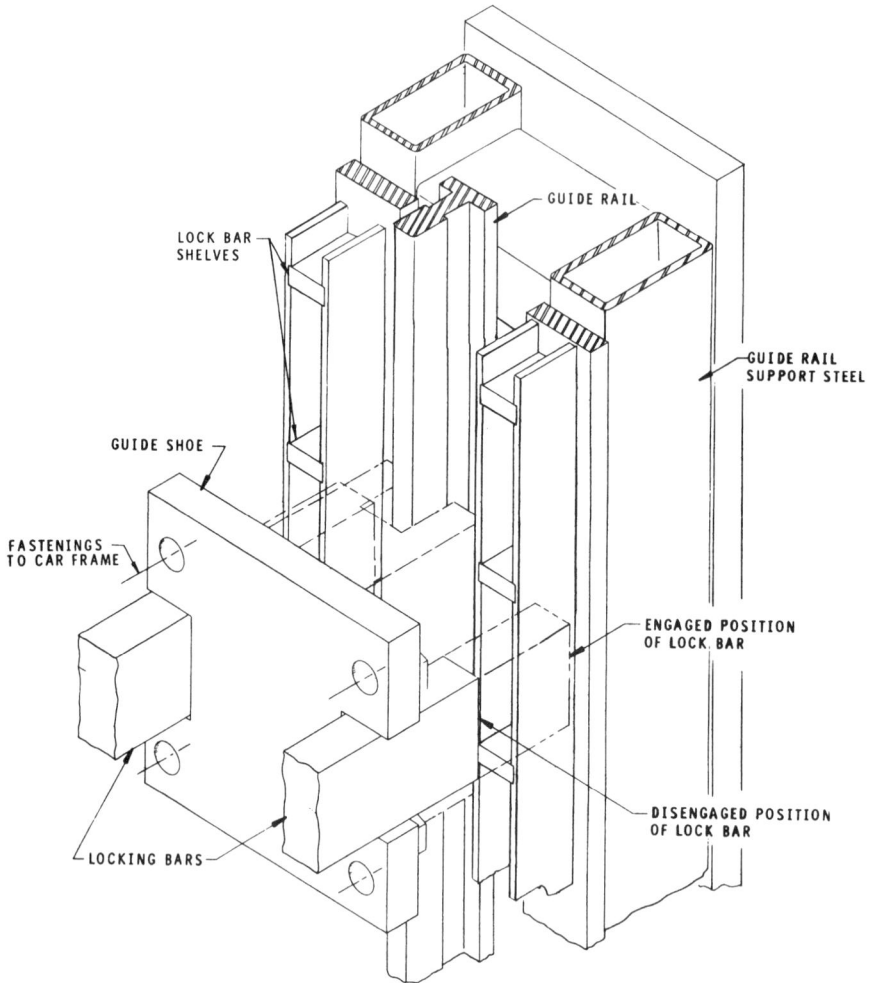

LOCK BAR
SHELVES

GUIDE RAIL

GUIDE RAIL
SUPPORT STEEL

GUIDE SHOE

FASTENINGS
TO CAR FRAME

ENGAGED POSITION
OF LOCK BAR

DISENGAGED POSITION
OF LOCK BAR

LOCKING BARS

(c)

Figure 14.45. *(Continued)* (*c*) Locking bar arrangement to lock platform so heavy loads can be transferred.

Platform Lifts

Platform lifts are often needed to provide a means to unload a truck or to position heavy equipment from one level to another that is a short distance within a floor height of 4 to 6 ft (1.2 to 1.8 m). Such lifts are usually hydraulic, equalized as described for stage lifts, with full skirting to minimize shear hazards. They are frequently found in industrial plants and often in the loading dock space of commercial buildings.

Other Lifts

The equipment available to provide vertical transportation can generally be adapted to any lifting need. Familiarity will suggest means of employing such equipment, and the

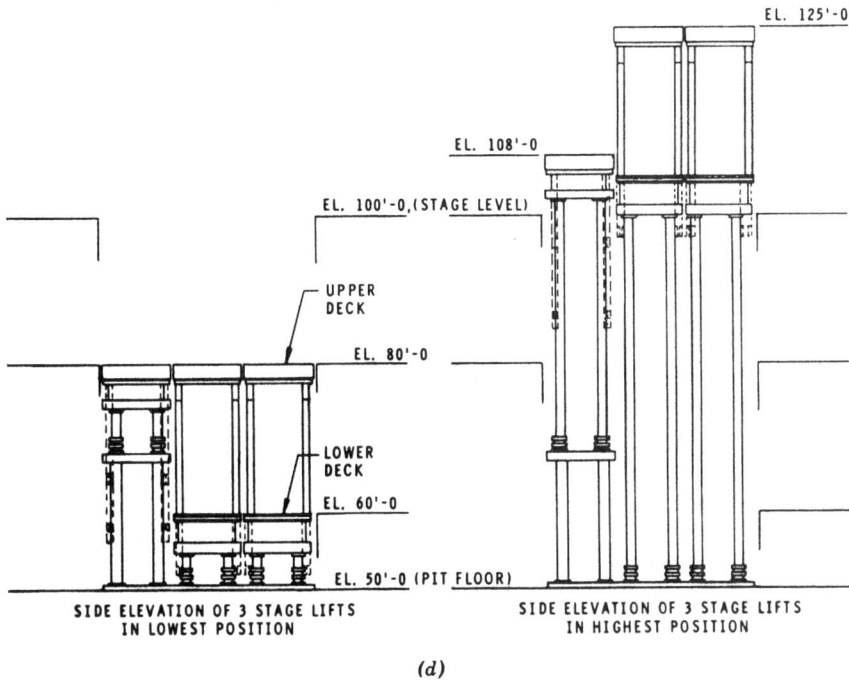

Figure 14.45. *(Continued)* (*d*) Possible configurations of single- and double-decked stage lifts.

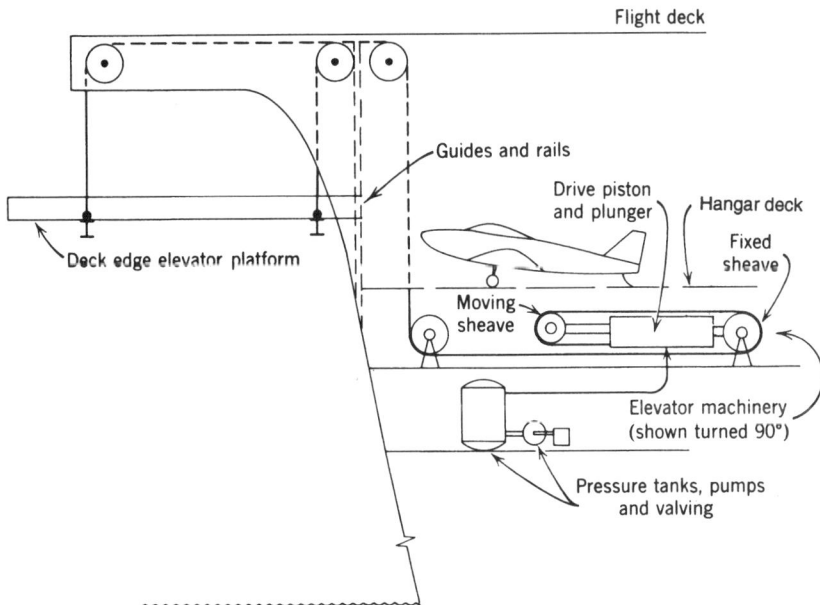

Figure 14.46. Aircraft carrier deck—edge elevator. High-pressure hydraulic fluid moves piston to actuate the roped hydraulic arrangement.

knowledge of elevator consultants and elevator engineers can be called upon to add to the development of the solution to fulfill the need. The foregoing is but a brief description of some of the applications.

SUMMARY

There are more than 700,000 elevators installed in North America, most of which are conventional-type installations. Equally important are the nonconventional types described in this chapter. These are the elevators that get the most publicity because their application is either unique or spectacular, as well as those that are never mentioned but are needed to service the extensive infrastructures that make the cities run—the power plants, refineries, factories, and underground facilities. This chapter has described many of these applications as well as the conditions under which elevators may have to operate.

An elevator is installed in an environment created by an architect along with the numerous engineering disciplines that are needed to create a building or facility. Elevators are usually taken for granted and expected to operate under many adverse conditions while remaining unobtrusive. This chapter has discussed means to allow such operation and the considerations necessary to make an elevator the essential building equipment it is with the least impact on the building.

Also discussed here are environmental factors, a number of unique installations and their requirements, and systems that allow the tallest building to be built. Our objective is to create awareness of special elevatoring considerations and the reasons for them, for the guidance of anyone contemplating the need for any such vertical transportation.

ABOUT THE AUTHOR

GEORGE W. GIBSON has an extensive technical and managerial background in the elevator industry. He is a graduate civil engineer and has undertaken extensive postgraduate studies in advanced theories of mechanics and structures and mathematics. In a career spanning 37 years with Otis Elevator Company, he has held several design engineering, engineering management, and corporate management positions, including manager of Mechanical Research & Development, manager of Product Engineering, manager of Mechanical Engineering, director of Engineering Administration, and director of Codes and Product Safety. Following his retirement from Otis in early 1993, he formed George W. Gibson & Associates, Inc., an elevator consulting firm specializing in elevator technology, codes and standards, product safety, and technical support of litigation. He is a member of the A17 Elevator Safety Code Main Committee, chairman of the binational A17/B44 Mechanical Design Committee, chairman of the A17 International Standards Committee, chairman of the A17/B44 Earthquake Safety Committee, a member of several other technical committees and past chairman of the NEII Long Range Planning Committee on Codes & Standards. He is the head of the U.S. delegation to the International Standards Organization Technical Committee 178 on Elevators and Escalators, chairman of the Advisory Board to the National Association of Elevator Safety Authorities (NAESA International) and advisor to the NAESA Board of Certification.

Mr. Gibson is an instructor in the (ASME) Professional Development Programs on the A17 Safety Code for Elevators & Escalators and vertical transportation technology, and at the NAESA International Inspectors School. He is the chairman of the Technical Advisory Committee to the Elevator Escalator Safety Foundation. He is a member of the International Association of Elevator Engineers (IAEE) and the National Association of Vertical Transportation Professionals (NAVTP). He has authored numerous technical articles on various aspects of elevator technology, safety codes, and standards in elevator industry trade publications both in the United States and abroad. He has served for several years as chairman of the *Elevator World* Technical Communications Council and Editorial Resource Group, and is a director of the Board of Directors of Elevator World. Among his professional achievements, he holds patents for an elevator vehicle design and a door locking device, was cited by the James F. Lincoln Welding Foundation with an award for a unitized elevator car frame and safety design, and was the recipient of a special award from the Board of Directors of United Technologies Corporation in recognition of outstanding contribution to the corporation. In 1997, Mr. Gibson was awarded the Safety Codes and Standards Medal by the American Society of Mechanical Engineers for his contributions to the A17 Elevator and Escalator Committee, and for his international safety codes and standards work.

15 Automated Material Handling Systems

WAYNE A. GILCHRIST

DEVELOPING TECHNOLOGY

The first dumbwaiters have unknown origins, but it is known that Thomas Jefferson had one while he lived in Monticello in the early 1800s. Dumbwaiters evolved from a hand-powered device to an electric-powered lift in the early 1920s (Figure 15.1). They were often applied in hospitals and hotels to move food from a central kitchen to an "on-floor" pantry with a reasonable assurance that an attendant would be present to unload the food. As its name implied, the dumbwaiter was not only silent but needed to be waited on.

About 1960 an innovation was made that led to second and later generations that can hardly be called "dumb." This was a means to automatically eject a tote box when the car reached the desired floor. That, coupled with the addition of powered doors that automatically opened and then closed after the load was ejected, has led to a complete line of automated material-handling systems. The lowly dumbwaiter, both hand powered and electrically powered, is still in demand and readily available for less stringent applications.

Later developments include automatic loading and unloading of wheeled carts as well as tote boxes, so that systems of both dumbwaiter size (maximum 9-ft (2700 mm) platform area, 48-in. (1200 nm) high, and 500 lb (230 kg)) and elevator size are available. A separate section of the A17.1 Elevator Code has been developed for such systems, entitled Part XIV, Material Lifts and Dumbwaiters with Automatic Transfer Devices. Systems have been developed whereby overhead track conveyers or in-floor conveyers provide the horizontal transportation and the material lift, the vertical transportation. Under continuing development are systems wherein a self-propelled robot vehicle follows a path along the floor and automatically calls a material lift, waits, boards, and automatically exits at a destination floor, then travels to its programmed destination.

A number of systems are used to transport materials—such as paper, cash, small items, blood, medication, and so on—in a variety of buildings. These systems include pneumatic tubes, automated tote box systems, and selective vertical conveyers, as well as dumbwaiter- and elevator-related systems. In addition, conveying systems adopted from industrial applications are being installed in buildings. These include both overhead and in-floor track conveyers, automated, self-propelled vehicles, and pallet lifts.

The Vertical Transportation Handbook, Third Edition, Edited by George R. Strakosch
ISBN 0-471-16291-4   © 1998 John Wiley & Sons, Inc.

Figure 15.1. Cutaway of a dumbwaiter showing major components (Courtesy MATOT).

AUTOMATED MATERIAL HANDLING IN PUBLIC BUILDINGS

The scope of this chapter is limited to the types of automated material-handling systems that are found in public buildings and require minimum training for the users, the employees who work in the building. In industrial plants and warehouses, elaborate automated material-handling systems are found, and skilled people are expected to be involved. The systems used in public buildings generally employ elevator technology and are regulated by local building and elevator codes. In industrial applications, conveyor and crane technology is used and regulated by industrial safety codes, which generally recognize the high degree of skill of the systems' operators. Industrial material handling, both manual and automated, is a field in itself and is a continuous growth technology. It is expected that many of the developments in that field will be applied to the public building area as the trend toward a more automated commercial or institutional building continues.

PNEUMATIC TUBE SYSTEMS

History

Transport systems have been a part of the world's industry since invention of the wheel; however, the pneumatic tube system is a relative newcomer to the scene, and as the technology develops, their role in automation increases. The first pneumatic tube system originated about 1875 in a Lowell, Massachusetts, department store. It evolved from a hollowed-out croquet ball mechanically transported up and down in a trough and was used to move cash from the sales counter to a central cash station located in the mezzanine. Though the primary application remained the transport of cash and paper, technology advanced to individual send and receive tubes, with containers called "carriers" propelled through the tubes by vacuum pressure.

Pneumatic tube systems are found in two forms. The first is a point-to-point system, whereby a carrier is inserted in one end, comes out the other, and can be returned via the same tube to the sending end. Such systems are generally used when the recipient must take some action. The more elaborate systems are networks whereby the carrier is taken to a processing station and its path changed to direct it to the recipient. Its destination is coded on the carrier itself, in earlier systems by movable rings and in later systems through a microprocessor used to input signals at the sending station.

The first type of system was introduced in the United States in the early 1950s by Airmatic/ITT. The system had a series of rings in the carriers, which deflected into a station location by matching the spacing of the dialed carrier destination with identical spacing in the tubing network. The early systems produced high-speed travel but lacked control of the carrier. The carriers (Figure 15.2) were propelled through a maze of high-impact deflectors, belt conveyors, and storage units and finally arrived at the destination through a high-impact valve used for deceleration. The carriers were designed to be unidirectional, weighted on one end, and padded to better crash through the high-impact deflectors and valves. The cargo was subjected to considerable vibration and shock.

Pneumatic tube system companies continued on into the 1970s with variations on the relay-controlled systems while attempting to retrofit the old carrier-directed systems. The attempts were many and varied and caused confusion in the industry. Failure to analyze the user's needs and the lack of user training, service response, and availability of

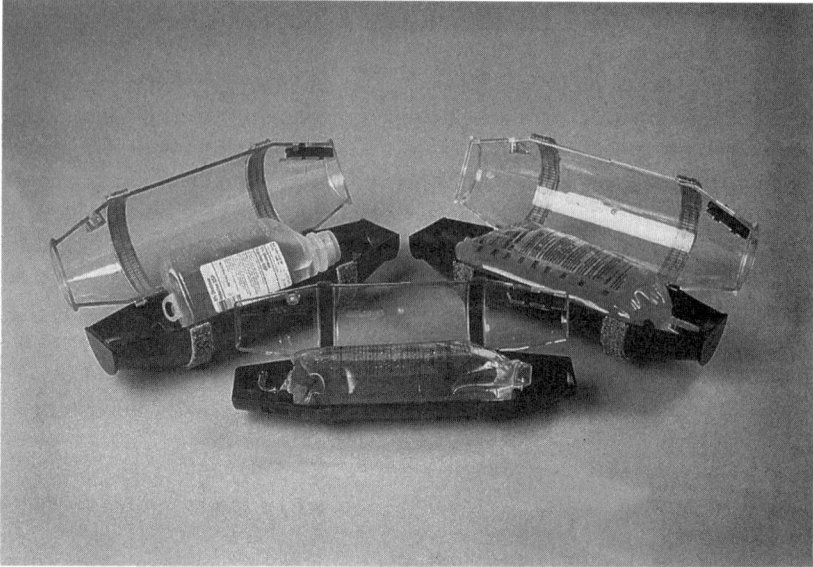

Figure 15.2. Pneumatic tube carriers (Courtesy Pevco Systems International, Inc.).

replacement parts resulted in a loss of credibility in the industry and a loss of confidence in the pneumatic tube system itself.

The first computer-controlled, single-tube vacuum-pressure system was introduced in the late 1970s. A prestigious hospital sterilizer manufacturer entered the market with a state-of-the-art product, which further intensified competition but, more important, by its entry, renewed confidence in pneumatic tube systems. However, the manufacturer's inexperience in the industry placed undue financial strain on the company, and in a few years it was forced to withdraw the product. Many systems that were under contract or close to completion were abandoned or completed by other newly formed companies.

The negative impact on the credibility and feasibility of pneumatic tube systems in hospitals was at its peak. The Veterans Administration, hospitals, and industry practically declared a moratorium on purchasing pneumatic tube systems from the late 1970s to the early 1980s.

Despite the negative influences in the marketplace, technology continued to advance through the 1980s. In 1987 the first IBM-compatible computerized system was introduced at the National Institutes of Health, Bethesda, Maryland, and the net result was more widespread use of the technologically improved transport systems.

From the 1980s to the present there have been two major manufacturers of computerized systems. The reliability of their computerized systems continues the revival of confidence in automated transport. These manufacturers began listening to the users and recognizing the special needs of the customer.

System Design

There are presently two basic pneumatic tube system designs, the point-to-point design and the branching-zone-type design.

Point-to-point systems connect two locations by a single tube. The ends of the tube, where the carriers are inserted, are called stations. There are two types of stations, the sliding sleeve and the vertical design. The bidirectional travel of the carrier through the tube is accomplished by applying pressure or vacuum from a power module that can be remotely located. Carriers, measuring from 3 in. (76 mm) to 7 in. (180 mm) in diameter and about 12 in. (305 mm) long, can be propelled up to 2,500 ft (760 m) at speeds of 25 ft (7.6 m) per sec. The station operation is controlled electronically by a printed circuit board, which simplifies its use and provides certain safety features such as electronic door locks, system-in-use signals, and carrier arrival signaling. After the carrier is inserted into the opening, it is automatically lifted off or sent with the push of a button. Remote signaling devices for carrier arrival are used at stations located outside the work area. Most are UL listed.

Point-to-point systems are simple to use and easy to maintain. If the carrier is inserted into the correct station, there is no opportunity for it to get lost or arrive at the wrong station because the tube is a direct link with the carrier's destination. These simple systems, generally speaking, provide the shortest transport times. It is up to the users to distribute the empty carriers to ensure most efficient use of the system. These direct systems usually require the least initial capital investment; however, transportation surveys can be used to determine whether multiple point-to-point systems or a branching-multiple-zone, multiple-station network would be more efficient and cost-effective.

Branching multizone systems connect multiple locations within a single building or a multibuilding complex. They are connected to each other and to one or more central locations. This complex type of system has undergone important technological advancements in the last decade, particularly in controlling and monitoring the carrier travel, which have greatly improved the safety of operation and reliability of the system. Even though they are computer controlled, the user does not have to be computer literate to use them. The user interacts with the system through the use of a simple touch pad. The carriers are the same as those used in the point-to-point systems. The stations, however, may have only one port as in the point-to-point system, or they may have multiple ports, depending on the location's volume of activity (Figure 15.3). When the carrier is placed in the station dispatcher, the operator keys in the destination, usually on a touch pad; the carrier is then entered in a queue and sent off to its destination. The carriers are propelled by vacuum-pressure along the tube and through devices that divert their direction to the proper zone and destination. Along the way they pass into interzone storage tubes until their place in the queue is cleared to go. In some systems, if the destination zone is not busy, the carrier passes straight through the storage unit without stopping. This, of course, shortens the travel time and makes a system faster. Each completed trip by a carrier is called a transaction. All of this network activity—the internal devices that change the direction of the carriers, the blowers, and the station operation—is controlled and monitored by a central computer usually located in the engineering or maintenance department.

Statistics

A pneumatic tube system's log can be a very useful tool. The log contains travel times, wait times, sources, destination station identification, user identification, and time of day for each transaction. From this vast amount of data various reports, charts, and graphs can be generated to assess system performance and improve its utilization. Transport turnaround time segments are easily identifiable for each specimen sent through the system.

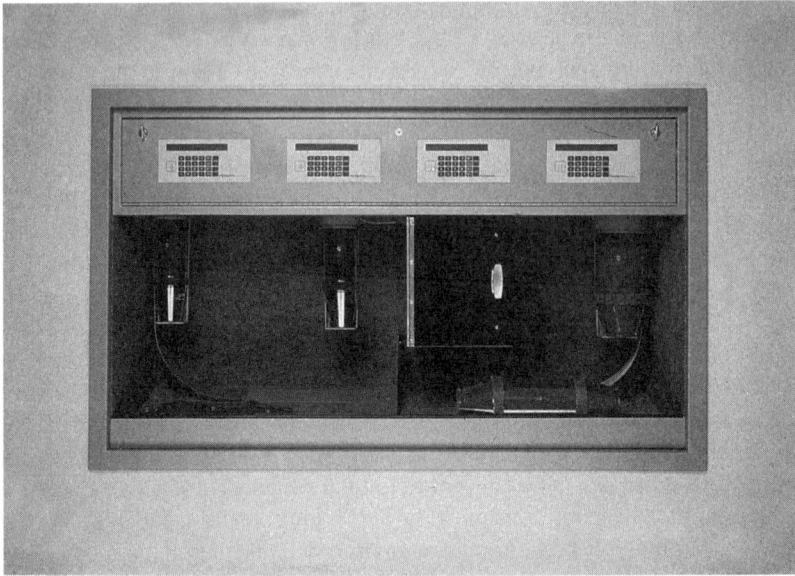

Figure 15.3. Pneumatic tube station front (Courtesy Pevco Systems International, Inc.).

This information can pinpoint the delays in requisition processing and improve laboratory efficiency. The information can also be useful in determining whether departments or individuals need more training or repeatedly make the same errors. The problem of locating lost carriers can be as simple as requesting a computer search by any number of parameters, such as source station identification and time of day. If the computer controlling the system is LAN (Local Area Network) capable, finding an errant carrier is even easier because the system's log can be accessed from the computer.

SELECTIVE VERTICAL CONVEYOR (SVC)

One aspect of the conveyor industry can be found in many office and hospital buildings. This is the selective vertical conveyor, which is designed to automatically load and unload tote boxes at various floors throughout a building. It can be interfaced with horizontal belt conveyors, gravity slides, both straight and spiral, and branch lifts so that the tote boxes can be delivered from and returned to a remote processing room by the system.

These systems have been used with considerable success in high-rise single-purpose office buildings for the delivery of mail and supplies from a central location. The operator places the filled box on a loading station for the conveyor (Figure 15.4), indicates the destination floor on a keypad, and walks away. The next empty carriage on the conveyor will activate a loading mechanism (Figure 15.5) and pick up the box on its upward trip. The box will be guided up and over to the down side, where the computer-controlled tracking system "tracks" its destination and activates an unloading arm. The box will be deposited at the unload station of the destination floor. At various floors, people wishing to send boxes to other destinations or to return them to the central station will use the same load-

Figure 15.4. Typical selective vertical conveyor station. Tote box to be dispatched is placed in the open on the right side, the destination indicated on the face place and the tote box is automatically accepted. Tote boxes are received on the stations run out on the left side (Courtesy Mathews Conveyor).

Figure 15.5. Selective vertical station loading mechanism (Courtesy Mathews Conveyor).

ing process at their floors. At a central station, if it is remote from the conveyor, the box will be discharged onto a powered belt or gravity conveyor at a high level and carried to the final destination, where it may accumulate with other boxes in a spiral gravity conveyor (Figure 15.6). The box is then removed, unloaded, and reused for another delivery.

A typical tote box size is 17 in. (430 mm) wide × 20½ in. (520 mm) long × 12 in. (305 mm) deep.

The approximate maximum capacity of a box is 60 lb (27 kg).

The maximum speed of a vertical selective conveyor is 80 fpm (0.4 mps). The tote boxes are picked up and delivered while the chain is in motion, and at full efficiency the conveyor can pick up and deliver (throughput) 8, 10, or 12 boxes per minute. The delivery cycle will vary with the length of the trip and can be estimated by the time required from the sending station, where the box is loaded on the up-traveling chain to the top of the structure and down to the unloading station where it is unloaded in the down direction.

Example 15.1. Tote Box Delivery Time

You have a 20-story building with a mail room at the basement level. How long will the delivery of a tote box to the 1st floor-station require? Assume 12-ft (3.7 m) floor heights. Assume sufficient empty carriers so that access to the system occurs within 8 sec after box is placed at the send station.

1. Accept box: 8 sec
2. Travel up: 250 ft (76.2 m)/68 ft (20.7 m) × 60 sec =   221
3. Travel down: 238 ft (72.5 m)/68 ft (20.7 m) × 60 sec =   210
                                                                             _____
                                                                             439 sec

439 sec/60 sec = *7.32 min Total Delivery Time*

Figure 15.6. Vertical conveyor spiral accumulator for receiving tote boxes into a high-volume area, such as a mailroom facility.

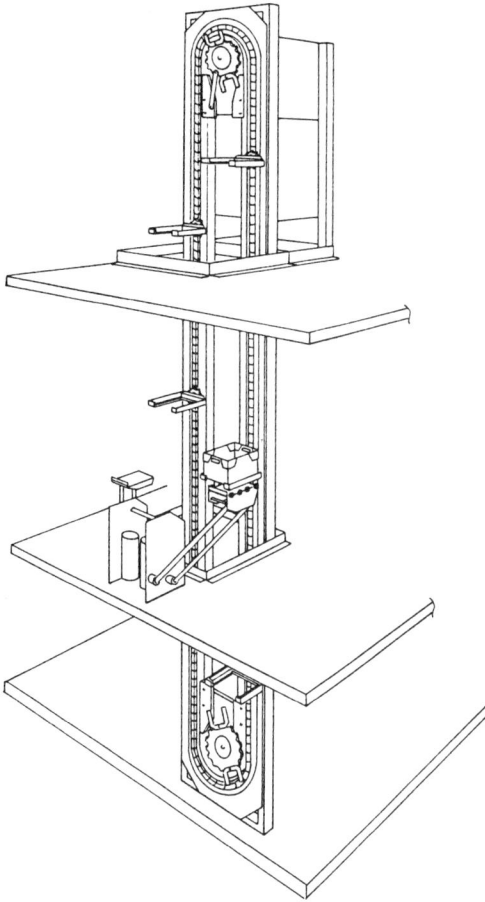

Figure 15.7. Cutaway of a selective vertical conveyor system (Courtesy Mathews Conveyor).

Owing to the one-way nature of the chain conveyor (Figure 15.7) any tote box that is loaded will require a full trip to the top of the building and down to the unloading floor, even though the unloading floor may only be one floor away.

Selective vertical conveyor systems presently require that the entry and exit access doors to the conveyor shaft remain open. The openings are generally equipped with a fire door, which should be provided with a magnetic latch that will release if deactivated by a signal from a smoke detector system, by water flow if sprinklers are activated, or if the power fails. To reduce or avoid a stack effect in a conveyor system shaft with open doors in a tall building, the conveyor shaft and entry and exit stations should be located in a closed room. An arrival lantern can be located over the door to the room to indicate that a box has been delivered and is ready to be picked up.

Future generations of selective vertical conveyors may have automatically opening and closing doors as technology progresses. This will certainly enhance their acceptance in view of the serious consideration being given to fire and smoke control in buildings.

An advantage of a vertical selective conveyor is its ability to interface with horizontal transportation. Conveyor systems in remote buildings can be interconnected by horizontal belt conveyors so that a tote box placed on one floor in one building can be automatically delivered to a selected floor in an adjacent building. The horizontal belt can be located in a ceiling area, or a remote mail room can be connected to the vertical shaftway by horizontal belt conveyors. Discharge to the mail room can be accomplished by gravity roller conveyors or spiral conveyors to accumulate filled tote boxes until they can be emptied. An on-floor branch lift can raise a filled tote box to ceiling height, where it can be discharged to a horizontal belt for further travel to the selective vertical conveyor serving all the floors in the building.

ELECTRIC "CAR ON TRACK" CONVEYOR SYSTEMS

Electric track conveyor transportation is a convenient method to dispatch and receive mail, small supplies, parcels, or any material of limited weight and volume from one place to another, regardless of horizontal or vertical travel. It is, in essence, a briefcase or suitcase that is transported by mechanical means by conveyor, as described in the previous section, or by means of a self-propelled carrier, as will be described in this section.

The tote box described for a selective vertical conveyor has been modified and securely attached to a self-powered car designed to travel along a dedicated aluminum track system, which is configured as intricately as an Erector set or a model railroad and seamlessly connects various departments both horizontally and vertically (Figure 15.8). Items to be transported are placed in the car's container, the destination code is entered into the station's "operator console," and the car is dispatched by depressing the "enter" key on the keypad. The car will travel horizontally via a motor-driven gear/friction traction wheel, and as it approaches a vertical bend section, a toothed gear rack section dovetailed within the track's profile engages the metal drive wheel and the carrier proceeds vertically. The spiroid-type gearing locks the motor when stopped, eliminating the requirement for a motor brake and drifting when the car sits idle in the vertical mode of travel. As the car traverses the track network, it will enter a decision point called a switch, stop, and be transferred to the right or left to a parallel track that will take the carrier to its final destination as previously entered by the dispatching operator. Along the path of travel, the car transmits destination information to the computer, which causes the proper travel and switching commands to be set up.

At the receiving station, which may be a Re-Entry, bidirectional, first in-last out (FILO) station (Figure 15.9), or a Thru, first in-first out (FIFO) station (Figure 15.10), the carrier must be manually unloaded and the next transaction's destination coded. If the receiving station is of the Re-Entry type, when dispatched the carrier will automatically reverse and travel back to the switch, which will place the carrier back on the mainline track, and it will proceed in its normal direction via the shortest possible route to the selected destination.

Electric track conveyor systems have been sold to well over 500 facilities servicing health care organizations delivering pharmaceuticals, files, and specimens; to agencies in commercial and governmental office buildings, delivering mail and supplies; and for use in industrial applications, such as the delivery of semiconductor IC wafer discs. The system's modular building-block design allows easy changes in routing and configuration modification as the operation or facility grows. Proper proactive design should include sufficient routing to project areas of expansion so as to allow ease of future add-on capabilities. System expansion may be added without unnecessary downtime for the existing portions, and connections can be accomplished over a weekend or during off-hours. In

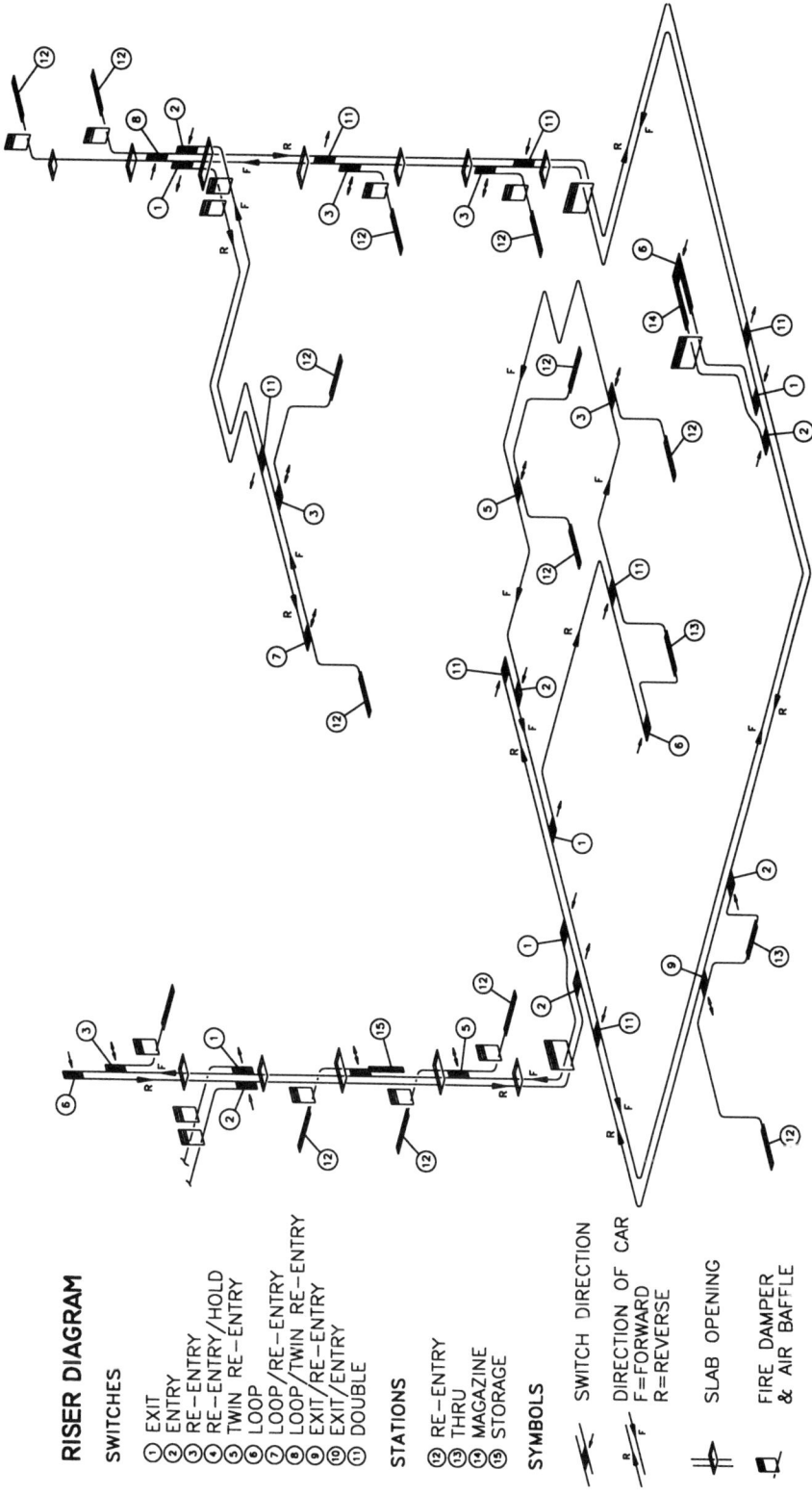

RISER DIAGRAM

SWITCHES

① EXIT
② ENTRY
③ RE-ENTRY
④ RE-ENTRY/HOLD
⑤ TWIN RE-ENTRY
⑥ LOOP
⑦ LOOP/RE-ENTRY
⑧ LOOP/TWIN RE-ENTRY
⑨ EXIT/RE-ENTRY
⑩ EXIT/ENTRY
⑪ DOUBLE

STATIONS

⑫ RE-ENTRY
⑬ THRU
⑭ MAGAZINE
⑮ STORAGE

SYMBOLS

SWITCH DIRECTION

DIRECTION OF CAR
F=FORWARD
R=REVERSE

SLAB OPENING

FIRE DAMPER
& AIR BAFFLE

Figure 15.8. Composite layout of hypothetical TELECAR system showing the variety of components (Courtesy Teledynamics, L.L.C.).

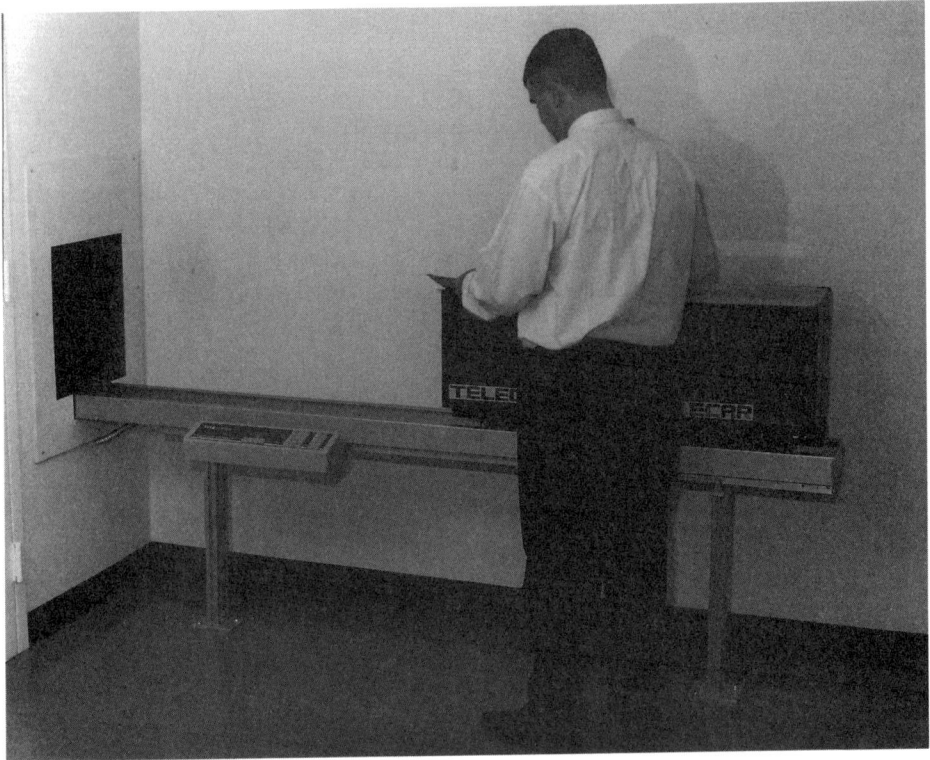

Figure 15.9. Typical (first in, last out (FILO)) four-car capacity TELECAR ReEntry station (Courtesy Teledynamics, L.L.C.).

designing a new or renovated facility, a system should be considered to contain the operating costs and provide an efficient mechanical means of delivery with a favorable return on investment as compared with the ongoing costs of manual delivery.

Currently, there are various types of containers available, carrying volumes of 720 (0.0123 m^3) to 5200 in.^{3} (0.085 m^3) while maintaining payloads of approximately 100 lb. (45 kg). Standard payloads for health care and commercial office marketplaces is 30 (13.6 kg) lb with a 20% safety factor yielding 36 lb. (16 kg). A partial listing of container sizes and capacities is given in Figure 15.11. Certain longer containers must use a special vertical "lift" (Figure 15.12) to provide the vertical path of transportation when the total length exceeds the radius allowed to pass through a bend section connecting the horizontal to the vertical section. This design also allows the transportation of fragile items, which must always be oriented horizontally while being delivered between floors with payloads exceeding the standard container weights.

A typical cross section of track and associated space requirements are indicated in Figure 15.13 for the most widely used 2400 model carrier. Various other clearances are required for horizontal curves and vertical bends. The majority of systems designed into facilities today run above the ceiling, making early system decisions and various trade coordination necessary to avoid structural beams, duct work, piping, and lighting fixtures. An isometric riser diagram of a typical system is shown in Figure 15.14.

Figure 15.10. Typical first in, first out (FIFO)) TELECAR Thru stations used in a large corporation's main mailroom environment (Courtesy Teledynamics, L.L.C.).

Model	Width, in. (mm)	Length, in. (mm)	Depth, in. (mm)	Capacity, lb. (kg)
2000	6.65 (169)	19.56 (497)	15.19 (386)	36 (16)
2400	13.00 (330)	18.75 (476)	10.25 (260)	36 (16)
2000	13.00 (330)	18.75 (476)	11.75 (298)	36 (16)
XR-1200	15.00 (381)	19.00 (483)	4.00 (102)	36 (16)
P-1600	12.00 (305)	12.00 (305)	11.67 (296)	36 (16)
3600	15.25 (387)	32.00 (813)	9.20 (234)	100 (45)
5200	15.25 (387)	32.00 (813)	13.31 (338)	100 (45)
Litho Plate	12.50 (318)	27.25 (692)	14.00 (356)	100 (45)
Semi-Conductor	11.63 (295)	24.88 (632)	14.18 (360)	100 (45)

Figure 15.11. Partial listing of various container sizes (inside box dimensions) and payload capacities for an electric track conveyor system (Courtesy Teledynamics, L.L.C.).

Figure 15.12. Teledynamics vertical lift for longer containers or heavier payloads (Courtesy Teledynamics, L.L.C.).

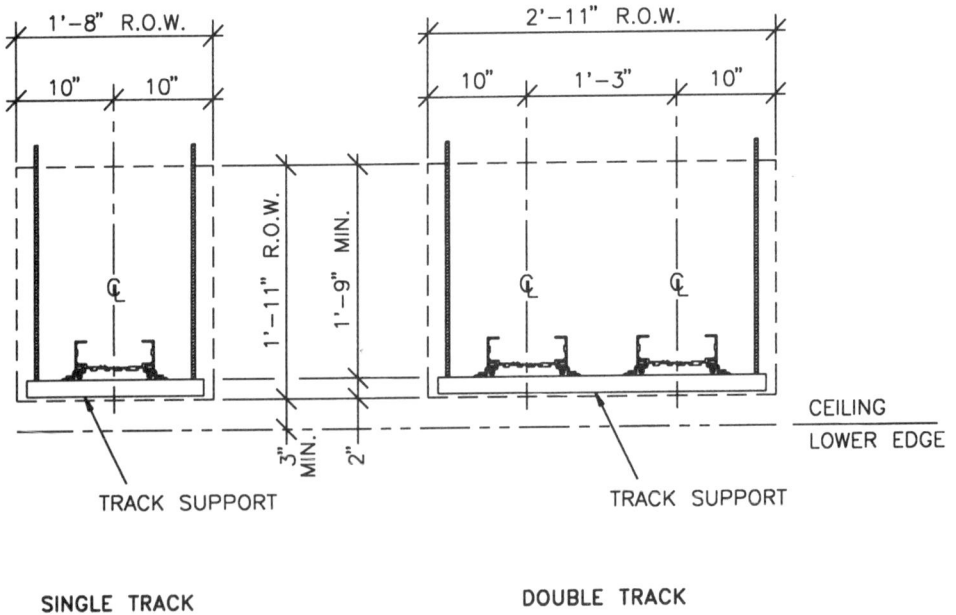

SINGLE TRACK

DOUBLE TRACK

Figure 15.13. Cross section of TELECAR track and required "Right of Way" (R.O.W.) clearances (Courtesy of Teledynamics, L.L.C.).

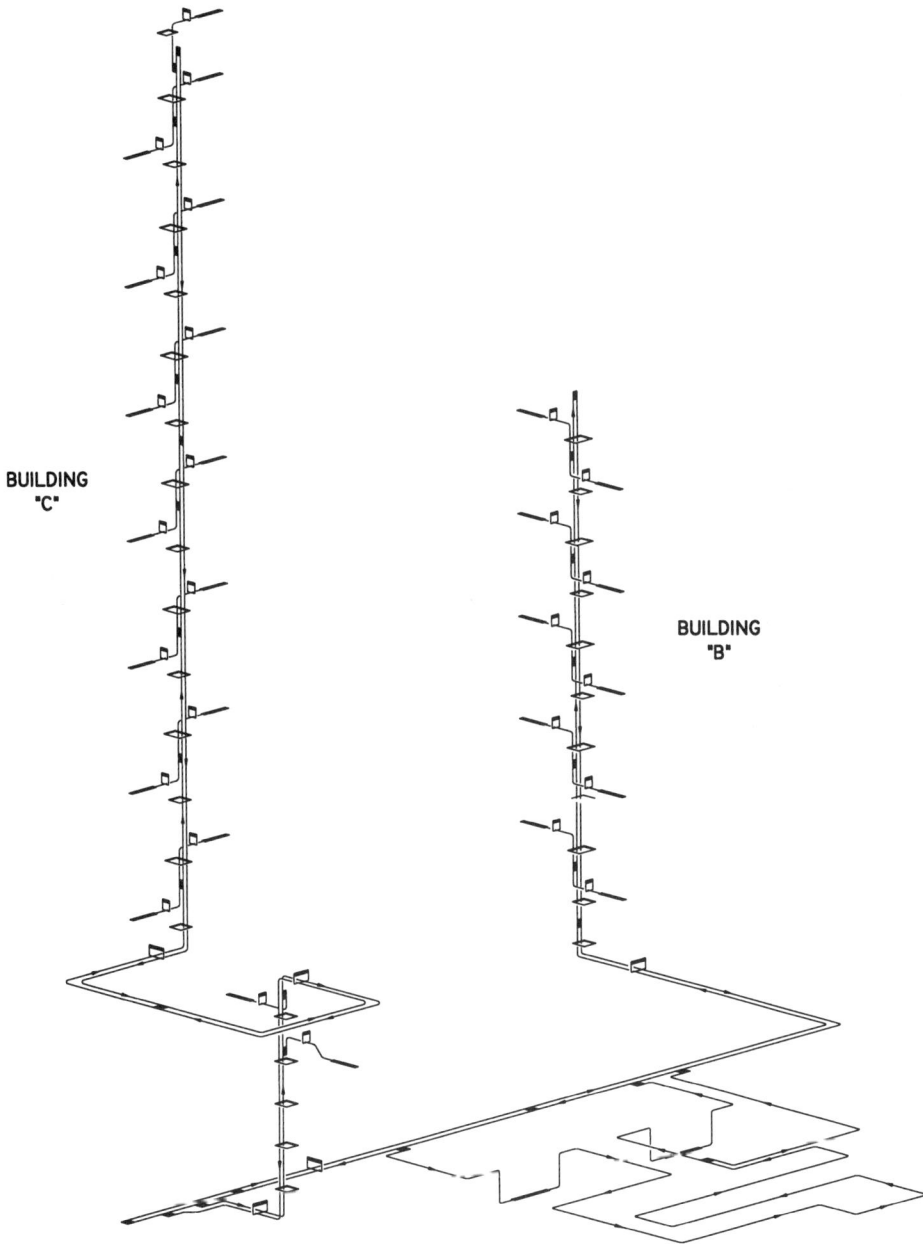

Figure 15.14. Home Depot store support center, Atlanta, Georgia: TELECAR system computerized supervision indicating status of all stations, switches, occupancy of tracks, as well as means of color displays (Courtesy Teledynamics, L.L.C.).

Vertical fire-rated shaftways must be arranged to provide ample space for the rise of the system and the bends to be installed prior to the UL-rated fire door mounted vertically in the front wall. Consideration must be made for the release of the fire doors either by means of a 165°C fusible link or by connecting them to the Class E life safety system with a fail-safe release device for release upon detection of smoke, minimizing the stack effect.

An essential system design criterion for both the selective vertical conveyor (SVC) and the electric track conveyor (ETC) is the layout of the main ancillary support function area, whether it is the mail room, central supply, reproduction center, pharmacy, or any combination. An in-depth study of functions and requirements must be made to determine the proper layout of the tracks to service the operator's points of service, storage facilities for empty carriers, and so on. Computer simulations are often performed to prove that the proposed design allows for the ultimate number of transactions within a given period. The results of this simulation along with the final design identifies possible bottlenecks in the design prior to installation.

A rule of thumb in regard to the number of carriers needed is that if a system has 10 receiving stations and it is assumed that 2 carriers are needed per station, 2 or 3 carriers should be provided in the central area to be loaded for each station; thus, this minimum system requires at least 40 carriers. With selective vertical conveyors, the investment in totes is small owing to the carriers' not being motorized. Another factor to consider is that because the tote boxes are not captive to the system, additional quantities are usually ordered inasmuch as many operators "hoard" the totes. The electric track conveyor system uses the theory of a captive carrier; however, this investment is considerably larger because of the price and complexity of the motorized carriage.

In both systems, space should be allocated for routine preventative and corrective maintenance, repairs, and storage of spare parts required for periodic overhauls.

A centralized facility such as the main mail room located in an office setting, or the security office in a health care facility, is an ideal location for the system's central computer. Via computer graphics problem areas are pinpointed, and via radio transmission maintenance personal can be alerted and dispatched at a moment's notice. A full menu of displays indicates various areas of the operation on display screens, such as carrier traffic, station operation schedules, alarms, and so forth.

The value of any material handling system is its ability to decrease the number of employee hours spent by unsupervised people hand-carrying items from place to place. Another advantage is the elimination of pushcarts in the elevator system, which can disrupt the traffic flow and cause undue interior damage to the elevator cabs. Material handling systems eliminate the need for batching information, which creates unnecessary peaks and unproductive valleys in a daily operation. Material handling systems provide the interdepartmental connectivity required for a smooth, reliable, and efficient work flow.

TOTE BOX LIFTS

As mentioned in the introduction to this chapter, the first application of an automatic transfer device on a lift was an automatic unloading dumbwaiter. The device, which was mounted in the dumbwaiter, automatically ejected the tote box onto a table after the power-operated doors opened and, once the transfer was completed, the doors closed and the dumbwaiter returned for the next load (Figure 15.15).

Later developments improved the operation so that the transfer device on the dumbwaiter not only automatically unloaded the tote box, but was arranged to automatically pull a waiting tote box in for the next delivery. This was done in two ways: the first, with a transfer device that reached out under the box, raised a pusher bar behind the box, and

Figure 15.15. Tote box showing transfer table and lift (Courtesy Courion Industries, Inc.).

pulled the box into the dumbwaiter car. The second means was to provide a belt on the car and a second belt on the loading table, which was driven by the motor driving the car belt. Unloading was done by turning the belt in one direction, and loading by turning it in the opposite direction. These refinements led to the fully automated tote box lift.

The tote box lift is valuable in offices, hospitals, laboratories, or any place where tote box loads have to be sent from floor to floor on an intermittent basis. It is ideal in a multifloor law office, where briefs and law books can be kept at a central location and sent to people who may need them. Similarly, in libraries, the tote box lift is valuable where less frequently used books can be stored on upper floors and recovered when required. The tote box feature and automatic unloading provide the means to keep the lift in operation as opposed to the necessity of manually loading or unloading a dumbwaiter and losing that productive time.

Many variations can be found. A double-deck tote box lift has been built, the bottom deck to accept loads and the top to discharge, using gravity, to a table. Lifts with front and rear openings have been built so that loads can be picked up and/or discharged at the same floor level. For example, a bank uses such a lift to provide secure delivery of negotiable paper form a vault area to the banking floor. A double-deck lift with a tote box on top and a cart lift on the bottom may also be considered.

A tote box lift and a dumbwaiter require the same hoistway, pit, and overhead space. The tote box lift usually has an entrance of lesser height than a dumbwaiter, because there is no need for a person to put his or her hands in the lift. Unlike a dumbwaiter, a tote box lift requires a table for loading and unloading in front of the lift. The table is equipped with an electrical presence detector so that a transfer will not be attempted if the table is full and a call buzzer that is intended to summon an attendant will sound.

Both automated cart lifts and tote boxes are in fully enclosed hoistways with fire-rated entrance assemblies that are kept closed and opened only when transfer operation occurs. All the rules of the A17.1 Elevator Code apply to such lifts, and Part XIV of that code provides special rules with the intent of maintaining safety requirements.

AUTOMATED CART LIFTS

Often a tote box is not large enough to transport bulky documents required in the everyday activity of a firm. For example, in an engineering firm large rolls of drawings are commonly sent from place to place; in a power company headquarters, boxes containing hand-held meter reading recorders are often transported to a central point for data processing; in other firms, bulk mail in boxes is a common requirement. All these items, being larger than the largest tote box and often too heavy to be handled in a single tote box by an individual, require cart transportation.

In hospitals, loads of linen, medical supplies, and surgical equipment often require a single cart each. It was in hospitals that the automatic unloading cart-carrying dumbwaiter was introduced and subsequently refined into an automated material handling system that automatically loads and unloads carts on a dumbwaiter or elevator and that can be programmed to perform prearranged deliveries or pickups. This class of equipment is called "lifts with automatic transfer devices."

In a delivery cycle, loaded carts are positioned in front of the entrance to the lift and a destination floor is registered on an operating panel. The lift doors are automatic power-opened, and a transfer device is driven out under the cart, couples, and injects the cart into the lift. The doors automatically close; the lift travels to the destination floor; doors open, and the cart is ejected out by the transfer device, which then uncouples and retracts; doors close, and the lift is free for the next assignment. This cycle is referred to as the "Dispatch" program.

In a pickup cycle, the empty carts are left in front of the entrance and a call is registered. When the lift is programmed to pick up, it will travel to each floor, pick up the waiting cart, and deliver it to the central area. This cycle is referred to as the "Return" program.

If the lift is equipped with both front and rear openings, dispatch and return programs can be performed on the same trip, the cart delivered to the front entrance, and any empty cart picked up at the rear entrance and returned to an entrance opposite the central loading entrance.

With a single opening, the cart lift is often arranged with a program selection switch to establish the delivery or pickup program. By proper arrangement at the floor landings, sensing devices can be used to indicate waiting carts or areas where delivered carts have to be removed from in front of the lift. This is especially important if there are space limitations in the unloading area.

Proper layout of a cart lift system proceeds from an initial determination of the size of the cart and its maximum load. If the gross area covered by the cart is 9 ft^2 (2700 mm) or less, the cart is no more than 4 ft (200 mm) high, and the cart weighs 500 lb (230 kg) or less when loaded, a floor-stopping dumbwaiter arranged as an automated cart lift can be used. Any dimension in excess of these limitations requires an elevator design including a car safety device and safe means for maintenance operation. Part XIV of the A17.1 Elevator Safety Code outlines the requirements and limitations.

Once the width and depth of the platform necessary to accommodate the cart is established, the rest of the space required for a hoistway is similar to that for a conventional elevator. Pit and overhead vertical space is established and will be a function of cart lift speed.

Floor space, both at the central area and on each floor, is an essential consideration. The horizontal distance in front of the cart lift entrance at each floor must be at least the length of a cart, preferably two, before any obstruction is encountered. If the cart lift opens into a public space such as a corridor, a minimum of 4 ft (1200 mm) plus the length of a cart must be provided. In many local areas, protective rails are necessary to prevent a person from walking in front of the lift entrance, which may open and discharge a cart at any time.

The central loading and cart storage area must be designed to accommodate all the carts expected to be staged for a dispatch or return cycle.

All carts must be of identical chassis or base dimensions and have the coupling device located in the same location. It is recommended that the cart chassis be ordered along with the cart lift itself, because the superstructure can be designed for the intended use and later placed on the standardized chassis. Caster wheels are necessarily of large diameter, 5 in. (125 mm) minimum, with one pair arranged to swivel.

As part of the cart lift installation contract, a floor plate with slightly depressed tracks should be provided and integrated with each cart lift entrance. In this way, distortions and out-of-level landing areas are avoided, the cart lift installer aligns the floor area with the transfer mechanism, and dependable cart lift transfer operation is ensured. Once the floor plate is aligned, it can be permanently set in concrete. The slightly depressed tracks in the plate ensure that the cart and casters are properly positioned, and even the most careless attendant can readily push the cart up to the lift entrance without serious misposition. Figure 15.16 shows a front view of a single cart lift elevator size and capacity, Figure 15.17 shows a transfer mechanism inside the lift and landing floor tracks, while Figure 15.18 shows a cart in position ready for dispatch. This also ensures in-line ejection of the cart at the receiving floors.

Nominal cart sizes in the United States and Canada are 24 in. (600 mm) wide by 5 ft (1500 mm) long, and the general height is 5 ft (1500 mm), so that an attendant can see over the top while pushing. In hospitals, carts are made of stainless-steel construction and designed so that they can be washed and sterilized. For office buildings, carts built up of racks are often used, and the racks allow reconfiguration for various tasks. Figure 15.19 shows the layout of cart lifts in a building.

TRACK SYSTEMS—OVERHEAD

Extensive development of "powered and free" track conveyor systems has been accomplished for industrial material handling applications, and some of these systems have been applied to hospital material handling. A powered and free track conveyor is an overhead track on which carriers are moved horizontally by a power-driven chain and may be released at any point to accumulate or to travel down an incline by gravity. The innovation that allowed track conveyors to be interfaced with vertical travel was the development of a transfer device that would accept a waiting carrier at the entrance to a lift and transfer it onto the lift. The lift would then travel to another level, and the transfer device would automatically discharge the carrier at the new level where it would engage the powered track to be horizontally transported to a final destination.

The power chain is such that the carriers can travel only in one direction, so both front and rear entrances are required on the lift. The lift is a modified elevator, reinforced to withstand the cantilevered load effects of the transfer device's reaching out to pick up the carrier and designed so that the mounting of the transfer device is at the top of the car. Car and hoistway doors are provided, the hoistway doors usually requiring a slot so that the track extension can be aligned with the transfer device with a minimum horizontal gap.

The carrier attached to the track is designed to grasp the top of the cart by means of hooks (Figure 15.20). The track at the pickup and discharge points is depressed so that the cart can be rolled off when the hooks are released or rolled into position for pickup. During horizontal travel as the carrier moves along the track, the cart wheels are off the floor a short distance.

Figure 15.16. Cart lift with cart being loaded (or discharged) at a typical floor station (Courtesy Jaros, Baum & Bolles).

The disadvantage of the system is the fixed installation of the track system and the space required, making it difficult to add to or rearrange the system. The obvious labor saving is effected by the rapid vertical and horizontal travel without an attendant. Carts must be manually attached or released from the carrier, which requires an attendant.

TRACK SYSTEMS—IN-FLOOR

Tow chain conveyors, which feature a continuously moving chain embedded in the floor whereby a cartlike carrier can be engaged by dropping a pin into a slot, have been ac-

Figure 15.17. Cart lift transfer device, "ferry slips" on the entrance and guidance rollers in the car. Extended rollers actuate switches to indicate cart position (Courtesy Jaros, Baum & Bolles).

Figure 15.18. Cart on vertical lift ready to be ejected. Swivel casters will be aligned by floor-mounted tracks as cart moves out (Photo courtesy Jaros, Baum & Bolles; transfer device, tracks, and doors by Courion Industries, Inc.).

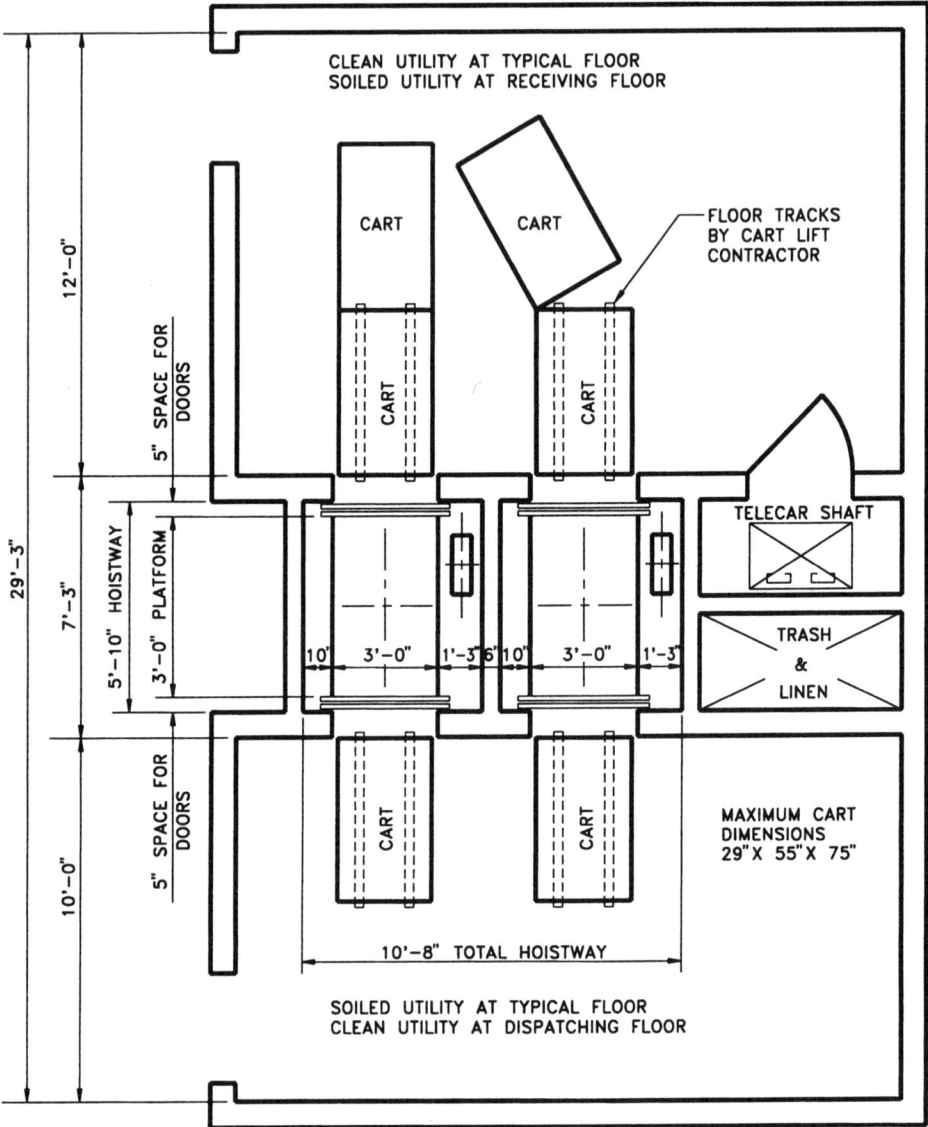

Figure 15.19. Hospital application of a cart lift system layout indicating front and rear entrances and separation of carts for clean and soiled articles (Courtesy Horst Hopf).

Figure 15.20. Overhead powered and free conveyor track system and lift interface (Courtesy Acco Industries, Inc.).

cepted horizontal conveying devices for many years. They are used extensively in factories and mail order warehouses, are heavy-duty and rugged, and if two or more floor travel is required, can be arranged to be moved up a ramp. Switching means have been developed, and magnetic readers on the carts can be used to route the vehicle to various loading and unloading stations.

Moving from floor to floor by ramp requires considerable horizontal distance and valuable warehouse or factor space. These requirements led to the development of an in-floor transfer device that is mounted in the platform of a lift and used to move carts onto or off a lift.

The traveling cart signals the lift by means of a magnetic or photoelectric encoder mounted on the cart and a reader in the floor. The cart is brought to the lift entrance and stopped (see Figure 15.21a). The lift is signaled, the doors open, and a device on the lift reaches out into the floor slot in front of the lift and pulls the cart by means of its engagement pin into the lift. Doors close, the lift travels, and the cart is discharged by a

Figure 15.21. Tow chain conveyor and elevator interface: (*a*) vehicles about to enter elevator; (*b*) vehicles exiting from elevator.

discharge transfer device onto the destination floor (Figure 15.21*b*), with the cart's pin in a back floor slot, where it is picked up by the powered chain and taken to its destination.

The system is adapted for heavy industrial use and may be found in a number of mail order warehouses in the United States. Its main functions are warehouse stocking and order picking, wherein loads of up to 3000 lb can be moved by one cart to and from the shipping area. The lift is a freight-type elevator with vertical biparting doors and can be designed for single or parallel track operations. Front and rear entrances are required since tow chains can operate in only one direction and the towed cart is engaged only by a pin at the front end.

AUTOMATED GUIDED VEHICLE (AGV) SYSTEMS

Automated guided vehicle systems utilize self-propelled, battery-powered transportation devices in the robotics family. The vehicles travel at floor level horizontally and interface with passenger or service elevators for vertical travel.

An AGV system may be composed of multiple vehicles (100 +/−) operating within a facility, traveling simultaneously in vertical and horizontal directions, with all movement controlled by a host computer system. A system can also be as modest as a single vehicle traveling a "bus route" with predetermined stops on one floor—or any arrangement between the two extremes.

AGV systems have value as a transportation solution in such venues as health care facilities, commercial buildings, light and heavy industry. A typical AGV payload can be as varied as small tote boxes of mail, parts, trays of supplies, loads of cartons, cases of food, and so forth (Figure 15.22). Such systems seem to find a place in the material handling scheme of things somewhere between the smallest, speediest Pneumatic Tube System (PTS) and Electric Truck Conveyor (ETV), and large bulk handling.

A typical payload varies from a 5 lb (2.27 kg) tray to a 2000 lb (900 kg) pallet, with the size related to the weight. It is usually composed of product that is too large or heavy for PTS or ETV at the low end and is too small for bulk, heavy handling at the high end.

AGV systems follow a guidance path that is formed by floor-imbedded wire, invisible noninvasive paint, laser, or some type of inertial method. Vehicles usually travel in one direction but have the capacity to travel in reverse direction by dead reckoning or standard two-direction travel. Most vehicles are of a tricycle design with transmission and steering in the front wheel while the rear wheels are inert.

Systems can carry independent carts that are picked up and delivered at stations, totes or pallets that are automatically handed off/on at fixed stations, or product that can be carried directly on the AGV for manual load and unload functions.

Vehicles/systems have all the necessary noncontact and redundant safety systems to allow unattended operation in such sensitive locations as hospital corridors, high-pedestrian-volume areas such as office buildings, and high-throughput production areas in industry.

Elevator interface is accomplished by interlocking the AGV system controls with the elevator controllers and capturing an elevator between passenger tasks. In addition, dedicated, nonpassenger elevators are sometimes used, depending on the application of the system.

PALLET LIFTS

Bulk material is commonly packaged on pallets for transportation. The entire material handling industry is pallet oriented: industrial forklift trucks are especially designed to handle pallets, and completely automated warehouses are in operation that will automatically store and retrieve pallets and provide storage facilities for thousands of pallets.

Simplified pallet handling systems utilizing an elevator equipped with automatic pallet loading and unloading means can provide effective low-cost material handling in a multi-

Figure 15.22. Robotic vehicle (Courtesy Bell & Howell Mailmobile Co.).

story production facility. This operation was briefly described in Chapter 13 and shown in Figure 13.10. Pallet-loads of raw material can be placed on a receiving pallet conveyor, programmed for a destination floor, and automatically unloaded on pallet conveyors at that floor. Conversely, finished products can be placed on pallets, loaded on the pallet conveyor at various floors, and automatically transferred to the loading dock floor for subsequent shipment. Forklift trucks are used at each floor, but waiting time for the freight elevator to return and the cost of an attendant to unload it are eliminated. Modular-type pallet conveyors have been developed, allowing a system to be created for a particular application. The lift is a conventional freight elevator arranged so that the pallet transfer device is placed on the platform. The pallet transfer device and the pallet conveyor consist of powered rollers with limit switches so that the pallets can be transported and positioned.

Biparting freight-elevator-type doors are used to close the hoistway to provide fire integrity together with a fully enclosed hoistway. a programmable controller interfaces with both the elevator and conveyor controllers to direct their movements. As pallets are received, the attendant registers their destination and the pallet joins the queue. As the lift delivers them, the memory in the controller directs each operation. A printout can be produced that lists the various movements and provides inventory control.

ECONOMICS

The Value of Automated Materials Handling

The addition of an automated mail- and materials-handling system in an existing single-tenant building can be an extremely cost-effective and elevator traffic reducing step. A thorough survey of the number of people involved in carrying mail and supplies throughout the building must be made and the benefit of an automated materials-handling system calculated. This analysis must include the capital cost, maintenance cost, and the number of employee equivalents required or the number that can be reduced by the use of the new system. Various consolidations of mail handling, supply room, reproduction center, print shop, or other frequently used facilities must also be considered. For example, the difference in cost of having infrequently used reproduction machines at each floor versus the cost of an intensively used machine at a central location related to the cost of transportation of documents to and from that center needs to be determined.

Full-time Employee Equivalents

Any job that requires an employee to be on duty full time requires more than one employee to provide the full-time coverage. This is true of any essential personnel, such as a security guard, elevator operator, or mail delivery personnel upon whom the day-to-day functioning of the business depends.

The determination of a full-time employee equivalent (FTE) is calculated in the following manner.

1. Actual working time per employee

Days per year		365
Less: Saturdays and Sundays	104	
Vacation	10	
Sick time	3	
Personal time	2	
Holidays	11	
	119	119
Actual days worked		235
At 40 hr per week		× 8
Hours worked		1880

2. Efficiency: coffee breaks, wasted time, waiting for assignments. Estimated at $70\% = 70\% \times 1880 = 1316$, actual hours worked per employee.

3. If a job requires full-time attention, hours required 8×250 days (5 days per week less holidays) 2000 hr.

4. Therefore to fulfill the time requirements, 1.5 employees $(2000 \div 1316)$ are required for each full-time job and the cost of a full-time employee equivalent is 1.5 times the cost of a single employee. Stated otherwise, it requires 1.5 employees to man a position that requires full-time manning.

For a mail-handling function, the number of hours required to perform the activity is determined by survey. The cost of the function can then be established by multiplying the

	1994 $(000)	1995 $(000)	1996 $(000)	1997 $(000)	1998 $(000)	1999 $(000)	2000 $(000)	2001 $(000)	2002 $(000)	2003 $(000)
Manual Delivery of Mail, Reproduction and Office Supplies by Cart to Entire Building										
Capital cost	15									
Direct labor with fringes (41%) (9%/annum escalation of 7 persons)	173	188	205	223	243	265	289	315	343	373
Total	(188)	(188)	(205)	(223)	(243)	(265)	(289)	(315)	(343)	(373)
Robot Vehicle With Elevator and Special Guide Path										
Capital cost	485									
Maintenance costs	60	65	71	77	84	92	100	109	119	130
Total	(54)	(65)	(71)	(77)	(84)	(92)	(100)	(109)	(119)	(130)
Annual cash flow savings	358	123	134	146	159	173	189	206	224	243
DCF @ 14%	1.00	0.88	0.77	0.67	0.59	0.52	0.46	0.40	0.35	0.31
Net DCF	(358)	108	103	98	94	90	87	82	78	75
Cumulative DCF	(358)	(250)	(147)	(49)	45	135	224	306	384	459

(a)

Figure 15.23. Return on investment analysis of various automated material-handling systems versus manual delivery of mail and office material and supplies. (a) Robot (AVG) vehicle versus manual.

	1994 $(000)	1995 $(000)	1996 $(000)	1997 $(000)	1998 $(000)	1999 $(000)	2000 $(000)	2001 $(000)	2002 $(000)	2003 $(000)
Manual Delivery of Mail, Reproduction and Office Supplies by Cart to Entire Building										
Capital cost	15									
Direct labor with fringes (41%) (9%/annum escalation of 7 persons)	173	188	205	223	243	265	289	315	343	373
Total	(188)	(188)	(205)	(223)	(243)	(265)	(289)	(315)	(343)	(373)
Selective Vertical Conveyor (SVC), Tote Boxes and Standard Guide Path										
Capital cost	382									
Maintenance costs	34	37	40	44	48	52	57	62	68	74
Total	(416)	(37)	(40)	(44)	(48)	(52)	(57)	(62)	(68)	(74)
Annual cash flow savings	(228)	151	165	179	195	213	232	253	275	299
DCF @ 14%	1.00	0.88	0.77	0.67	0.59	0.52	0.46	0.40	0.35	0.31
Net DCF	(228)	133	127	120	115	110	107	101	96	93
Cumulative DCF	(228)	(95)	(32)	152	310	420	527	628	724	817

(b)

Figure 15.23. (*Continued*) (*b*) Selective vertical tote box conveyor versus manual.

	1994 $(000)	1995 $(000)	1996 $(000)	1997 $(000)	1998 $(000)	1999 $(000)	2000 $(000)	2001 $(000)	2002 $(000)	2003 $(000)
	Manual Delivery of Mail, Reproduction and Office Supplies by Cart to Entire Building									
Capital cost	15									
Direct labor with fringes (41%) (9%/annum escalation of 7 persons)	173	188	205	223	243	265	289	315	343	373
Total	(188)	(188)	(205)	(223)	(243)	(265)	(289)	(315)	(343)	(373)
				Automated Cart Lift						
Capital cost	516									
Maintenance costs	47	51	56	61	66	72	79	86	94	102
Total	(563)	(51)	(56)	(61)	(66)	(72)	(79)	(86)	(94)	(102)
Annual cash flow savings	(375)	137	149	162	177	193	210	229	249	271
DCF @ 14%	1.00	0.88	0.77	0.67	0.59	0.52	0.46	0.40	0.35	0.31
Net DCF	(375)	121	115	109	104	100	97	92	87	84
Cumulative DCF	(375)	(254)	(139)	(30)	74	174	271	363	450	534

Figure 15.23. (*Continued*) (*c*) Automated cart lift versus manual.

(*c*)

	1994 $(000)	1995 $(000)	1996 $(000)	1997 $(000)	1998 $(000)	1999 $(000)	2000 $(000)	2001 $(000)	2002 $(000)	2003 $(000)
Manual Delivery of Mail, Reproduction and Office Supplies by Cart to Entire Building										
Capital cost	15									
Direct labor with fringes (41%) (9%/annum escalation of 7 persons)	173	188	205	223	243	265	289	315	343	373
Total	(188)	(188)	(205)	(223)	(243)	(265)	(289)	(315)	(343)	(373)
Telecar System										
Capital cost	1815									
Maintenance costs	92	100	109	119	130	142	154	168	183	200
Total	(1907)	(100)	(109)	(119)	(130)	(142)	(154)	(168)	(183)	(200)
Annual cash flow savings	(1719)	88	96	104	113	123	135	147	160	173
DCF @ 14%	1.00	0.88	0.77	0.67	0.59	0.52	0.46	0.40	0.35	0.31
Net DCF	(1719)	77	74	69	67	64	63	59	56	54
Cumulative DCF	(1719)	(1642)	(1568)	(1498)	(1431)	(1367)	(1304)	(1245)	(1189)	(1135)

(d)

Figure 15.23. (*Continued*) (*d*) Telecar system versus manual.

449

hours required by the FTE. The budget cost of the automated facility plus the number of personnel hours required to service the automated facility is then determined and a study of the financing and a discounted cash flow is developed.

Figures 15.23*a* to *d* show such a study made to determine the cost effectiveness of a material-handling system requiring various capital investments from about $500,000 to $1,800,000 but capable of replacing 7 full-time equivalent employees with an annual total payroll of approximately $25,000 per employee including all benefits and indirect costs.

SUMMARY

It is difficult to accept the idea that an office is becoming more and more like an industrial factory where material (paper) is received, processed, sorted, and shipped. Although the increased use of computers and information processing systems may indicate the paper-free office in the future, the present requires the systematic movement of hard copy.

This chapter has described state-of-the-art automated materials handling in public places, that is, where people are not specifically trained to interface with an automated system. When such conditions exist, all the facets of line safety are a necessary consideration when the material handling system is planned. In the industrial environment, training is expected and the individuals are more aware of hazards. The line between a conveyor as regulated by the ANSI B20 Conveyor Code and a material handling system as regulated by the ANSI/ASME A17.1 Elevator Code is rather fine and, hopefully, will be defined with greater precision in the coming years.

The description of each system given here has been brief and designed to provide an overview with a minimum of specifics.

Other important aspects in this regard are the relationships between the capital cost of the equipment, its annual maintenance, direct labor expense, and the reduction of man-hour equivalents the equipment affords. Too often an elaborate system may look attractive, but may not return the investment fast enough to meet financial criteria, as shown in Figure 15.23 a-d.

ABOUT THE AUTHOR

WAYNE A. GILCHRIST began his career in the material handling industry in 1968 with the Mosler Safe Company, a wholly owned subsidiary of American Standard, concentrating on the company's Telelift system design to automate delivery of mail and packages throughout an office or hospital complex. His experience includes project management, product management, and sales management.

Mr. Gilchrist has had extensive experience in design, installation, maintenance, and management, having engaged in precontractual design engineering, cost justification, systems consulting and overseeing well over 300 material handling systems nationwide for the past 28 years. His contributions have yielded more than 200 million dollars in sales volume, making him the recipient of numerous "Winners Circle" and "Top Gun" awards. He is currently with Teledynamics, L.L.C., in Newfoundland, New Jersey, where he is vice president of Sales and Marketing. He participates in various health care, commercial, and industrial conferences and organizations nationwide.

Mr. Gilchrist is a charter member of the Mail Systems Management Association (MSMA) and guest speaker for the National Association of Vertical Transportation Professionals (NAVTP) and annual MAILCOM conferences.

16 Codes and Standards

EDWARD A. DONOGHUE

INTRODUCTION

In the United States there are some 44,000 local jurisdictions having a population of 2,500 or more, which author their own codes or adopt a model code with or without local modifications. In 1975, the American Institute of Architects (AIA) reported that "the design professional is in much the same position as motorists were in Europe before Common Market nations began highway standardization . . . having to drive on different sides of the road after crossing political boundaries; trying to follow a bewildering variety of road signs with different symbols and instructions in many languages; and even facing two measuring systems for speed limits."[1] The AIA report went on to address the need for uniform construction codes, including the need for uniform enforcement of the codes.

Seven years later the President's Commission on Housing also addressed this issue and published a report[2] recommending that "the Board for the Coordination of Model Codes (BCMC) should be encouraged in its efforts to resolve differences between building codes and fire safety standards." The Commission's report was primarily directed at resolving conflicts between the model codes, but also discussed the need for cooperation between the various disciplines of code enforcement personnel.

In the ensuing period of time, there has been considerable progress toward development of a single national set of construction codes and standards. Local jurisdictions can no longer afford the costs associated with developing unique regulations. These costs are both direct (such as for the required research, engineering expertise, and so on, in developing home-grown regulations) and indirect (such as the cost of business going elsewhere owing to the high expense associated with unique construction requirements). Most jurisdictions are now adopting the model codes and standards. Unfortunately, many still include local modifications. However, progressive jurisdictions also have means to obtain variances for the use of new methods and/or materials.

The consensus codes and standards organizations, such as ASME, BOCA, SBCCI, ICBO, NFPA, ASTM, and so forth, have also recognized that the market is not big enough to support competing documents. Many code and standard writing groups are cooperating in the development of a single national code or standard. In some instances cooperation extends across national political boundaries. This chapter presents an overview

---

1. American Institute of Architects, *One Code: A Program for Building Regulatory Reform.* (Washington, DC: A1A, 1975).
2. *Report of the President's Commission on Housing* (GPO, Washington, DC, April 29, 1982,), 219.

The Vertical Transportation Handbook, Third Edition, Edited by George R. Strakosch
ISBN 0-471-16291-4 © 1998 John Wiley & Sons, Inc.

of the status of codes and standards, as well as guidelines on how to use such information. The section on enforcement may be of particular interest. The management of any project must be aware of the applicable codes and how they are enforced. An ongoing working relationship with the relevant enforcement authority is always advised.

HISTORY

Building codes have a long history. The first is credited to King Hammurabi when, at about 1700 B.C., a law was enacted whereby a builder could be executed if a house he built collapsed, resulting in the death of the owner. In one form or another, building codes have been around since Hammurabi's time. Most were created after the occurrence of a fire or other disaster that resulted in the death and/or injury of many people or caused heavy property damage or a general disruption of life in general. Public outcry after these disasters provoked governments to pass laws that they believed would protect the public from recurrences of such events.

OVERVIEW OF APPLICABLE CODES AND STANDARDS

A brief overview of the consensus regional, national, and international codes and standards pertinent to the manufacturing, installation, inspection, maintenance, and repair of elevators, escalators, moving walks, and dumbwaiters follows. The authority having jurisdiction should be consulted to determine which, if any, of the following codes and standards have been legally adopted, the applicable edition, and the existence of modifications or additions to the consensus code or standard.

Elevator Codes and Standards

In the United States the first edition of the *Safety Code for Elevators and Escalators* was published in January 1921. It was developed by the American Society of Mechanical Engineers (ASME) Committee on the Protection of Industrial Workers with the assistance of elevator manufacturers, insurance carriers, regulatory bodies, and technical societies. Since then 15 editions have been published, the latest in 1996, with 32 supplements. Currently, major revisions are made every third year and supplements publish annually.

The documents described in the following paragraphs are published by the American Society of Mechanical Engineers (ASME) and can be obtained from:

American Society of Mechanical Engineers (ASME)
Order Department
22 Law Drive/Box 2300
Fairfield, NJ 07007-2300
800-THE-ASME (in New Jersey (201) 882-1167)
http://www.asme.org

The *Safety Code for Elevators and Escalators,* ASME A17.1, covers the design, construction, operation, inspection, testing, maintenance, alteration, and repair of elevators, escalators, dumbwaiters, moving walks, material lifts and dumbwaiters with automatic transfer devices, vertical and inclined wheelchair lifts, and stairway chair lifts. At the time this chapter is being written, the responsibility for code requirements for vertical and inclined wheelchair lifts and stairway chair lifts was in the process of being transferred to another ASME code committee. Eventually, a new code, tentatively identified as the *Safety Code for Platform and Stairway Chair Lifts,* ASME A18.1, will be published. The target date for publication is 1998. At that time the requirement for this equipment will be removed from ASME A17.1.

The American Society of Mechanical Engineers also publishes the *Handbook for the Safety Code for Elevators and Escalators,* ASME A17.1. The ASME A17.1 *Handbook* contains the rationale for Code requirements; explanations, examples, and illustrations of the implementation of the requirements; and excerpts from other nationally recognized standards that are referenced by the Code. The information was compiled from A17 Committee minutes, correspondence, and interpretations, as well as conversations with past and present committee members.

Safety Code for Existing Elevators and Escalators, ASME A17.3, covers retroactive requirements for electric and hydraulic elevators, dumbwaiters, escalators, and moving walks.

Safety Code for Elevators and Escalators, ASME A17.1, requires electrical equipment to be certified to the Elevator and Escalator Electrical Equipment Standard CSA B44.1/ASME A17.5.

Inspectors' Manual for Electrical Elevators, ASME A17.2.1, *Inspectors' Manual for Hydraulic Elevators,* ASME A17.2.2, and *Inspectors Manual for Escalators and Moving Walks,* ASME A17.2.3, are guides for the inspection of elevators, escalators, and moving walks, based on the requirements of ASME A17.1 and ASME A17.3. They also include pertinent information on the inspection of equipment installed under earlier editions of ASME A17.1, useful to the inspector. One-page abbreviated versions of the checklist shown in the *Inspectors' Manuals* are also published in convenient-sized pads. ASME A17.1, A17.2.1, A17.2.2, A17.2.3, and A17.3 are also published in a CD-ROM format.

Interpretations of the various ASME A17 Codes and Standards are published periodically. The interpretations of ASME A17.1 approved by the ASME A17 Committee from June 14, 1972, through June 14, 1979, were published in a book in 1980. A second book of interpretations covering the period from June 1979 through May 1989 was published in 1989. Both interpretations books are available from ASME. Starting in 1981, interpretations have been published with each new edition and addenda of the applicable standard.

In the early 1980s a number of prominent elevator industry associations met to discuss the need to establish minimum standards for the qualifications of elevator inspectors. The group recognized a trend wherein other than employees of jurisdictional authorities were being required to inspect elevators and escalators. The concerns of the group were taken to the American Society of Mechanical Engineers, which established the Qualification of Elevator Inspectors Committee. The first edition of the *Standard for the Qualification of Elevator Inspectors,* ASME QEI-1, was published in 1984. In the early 1990s the *Safety Code for Elevators and Escalators,* ASME A17.1, established a requirement that elevator and escalator inspectors must be certified by an ASME QEI-accredited organization.

Electrical Code and Other NFPA Publications

The following documents are published by the National Fire Protection Associations (NFPA) and can be obtained from:

National Fire Protection Association (NFPA)
Batterymarch Park
Quincy, MA 02269
(617) 770-3000 (in Massachusetts)
(800) 344-3555
http://www.nfpa.org

In 1881, the National Association of Fire Engineers proposed developing the *National Electrical Code.*® The first edition was published by the National Board of Fire Underwriters (now the American Insurance Association) in 1895. The National Board of Fire Underwriters continued to publish this code until 1951. In 1911 the National Fire Protection Association (NFPA) assumed sponsorship and control of the *National Electrical Code* and has been publishing it since 1952. The *National Electrical Code* is the most widely adopted code in the world. Sales are approaching a million copies for each edition.

The *National Electrical Code*, ANSI/NFPA 70, covers the installation of electric conductors and equipment in buildings and other structures. Article 620 of this Code pertains specifically to elevators, escalators, and related equipment. The *Safety Code for Elevators and Escalators,* ASME A17.1, requires all electrical equipment and wiring to conform to the requirements of the *National Electrical Code.*

When the term "fire codes and standards" is used, one does not think of elevator, building, or electrical codes. One immediately thinks of NFPA and its codes and standards. That is correct most times, but fire prevention codes are also published by BOCA, ICBO, and SBCCI. To add further confusion, the building codes, electrical codes, and elevator codes all have requirements that address fire protection. However, the following discussion of fire codes and standards will be confined to those regulations normally enforced by fire fighters. The National Fire Protection Association, founded in 1896, publishes more than 260 nationally recognized codes and standards, including the *Life Safety Code*® (formally the *Building Exits Code*) the *Fire Prevention Code,* building fire equipment standards (i.e., sprinklers, fire alarms, etc.), fire service standards, fire service training programs, and public fire safety education materials.

The *Life Safety Code,* ANSI/NFPA 101, addresses new and existing construction, protection, and occupancy features necessary to minimize danger to life from fire, smoke, fumes, or panic. The code includes requirements for elevators and escalators relating to general egress and accessible means of egress. The 1997 edition also recognizes that elevators can be used as a means of egress in some occupancies, a trend that is likely to expand in the future.

In addition to the aforementioned codes and standards, requirements affecting elevators can be found in a number of other nationally recognized fire protection standards, including:

- *National Fire Alarm Code,* ANSI/NFPA 72
- *Standard for the Installation of Sprinkler Systems,* ANSI/NFPA 13

The NFPA also publishes a *Fire Prevention Code,* ANSI/NFPA 1. This typically is enforced by fire department officials as a minimum requirement for existing construction.

Handbooks are available for the NFPA's most popular codes and standards, including the following:

- *National Electrical Code Handbook*
- *Life Safety Code Handbook*
- *National Fire Alarm Code Handbook*
- *Sprinkler Standard Handbook*

Building Codes

Today most of the building codes in the United States are based on one of the three model building codes. The *National Building Code,* promulgated by the Building Officials and Code Administrators International (BOCA), is widely adopted in the Northeast and the Midwest. Before the 1984 edition, this was known as the *Basic Building Code,* first published in 1950. The 1984 edition was titled the *Basic/National Building Code.* The changes in title reflect an agreement between BOCA and the American Insurance Association, which from 1905 through 1976 published the *National Building Code.* The *National Building Code* was the first model building code in the United States. It was referenced by the first edition of the *Safety Code for Elevators,* A17.1-1921, for hoistway fire protection requirements. The *Standard Building Code,* promulgated by the Southern Building Code Congress International (SBCCI), is widely adopted in the South. This code was originally published as the *Southern Standard Building Code* in 1945. The *Uniform Building Code* is promulgated by the International Conference of Building Officials (ICBO) and is widely adopted west of the Mississippi. The *Uniform Building Code* is the oldest of the model building codes, still being published by its original sponsor. The first edition was published in 1927.[3]

Building codes include requirements relative to the construction of hoistways, movement of smoke through elevator hoistways, standby (emergency) power, and elevators used for normal and accessible means of egress.

Code commentaries (handbooks) are available for each of the model building codes.

The model building codes also publish companion fire prevention codes. Fire prevention codes prescribe minimum requirements that establish a reasonable level of fire safety and protection of property from fire. Typically, they include requirements to provide fire fighters' elevator operations and the fire resistance rating of hoistway enclosures for both new and existing structures. The model fire prevention codes are as follows:

- *National Fire Prevention Code*
- *Standard Fire Prevention Code*
- *Uniform Fire Prevention Code*

The model codes can be obtained from:

Building Officials and Code Administrators International (BOCA)
4051 West Flossmoor Road
Country Club Hills, IL 60477
(312) 799-2300
http://www.bocai.org

---

3. Edward A. Donoghue, *ASME A17.1 Handbook* (New York: American Society of Mechanical Engineers, 1996).

Southern Building Code Congress International (SBCCI)
5200 Montclair Road
Birmingham, AL 35213
(205) 591-1853
http://www.sbcci.org

International Conference of Building Officials (ICBO)
5360 South Workman Mill Road
Whittier, CA 90601
(213) 699-0541
http://www.icbo.org

Accessibility Standards

Two documents include specifications for making elevators accessible to, and usable by, the physically disabled persons.

The *American National Standards, Accessible and Usable Buildings and Facilities,* CABO/ANSI A117.1, has been adopted by the three model building codes. The standard can be obtained from any of the three model building code organizations listed earlier or from:

American National Standards Institute (ANSI)
11 West 42nd Street, 13th Floor
New York, NY 10036
(212) 642-4900
http://www.ansi.org/home.html

Compliance with the *Americans with Disabilities Act Accessibility Guidelines for Buildings and Facilities* (ADAAG) is mandated by the Americans with Disabilities Act (ADA) regulations. ADAAG can be obtained from:

United States Architectural and Transportation Barriers Compliance Board (ATBCB)
1331 F Street N. W., Suite 1000
Washington, DC 20004-1111
(202) 272-5434
http://www.access-board.gov

ADA and Vertical Transportation is a handbook giving extensive coverage to accessibility requirements for vertical transportation. The handbook compares the numerous accessibility regulations applicable to vertical transportation and provides the reader with detailed information on compliance with the regulations. An analysis of the regulations clearly identifies the most stringent, which if adhered to would ensure compliance with all. The regulations for elevators, wheelchair lifts, and escalators are also analyzed, with explanations, examples, drawings, checklists, and excerpts from the cited documents. This handbook is available from:

Elevator World, Inc.
P. O. Box 6507
Mobile, AL 36660
(334) 479-4514
http://www.elevator-world.com

Testing Standards

Compliance with numerous testing standards, such as fire resistance rating, is required by the codes and standards previously specified. These testing standards include, for example:

- *Standard Test Method for Surface Burning Characteristics of Building Materials,* ASTM E-84
- *Fire Test of Door Assemblies,* UL 10B
- *Standard Test Method for Evaluating Room Fire Growth Contribution of Textile Wall/Covering,* SBCCI SSTD 9-88

Testing standards are available from organizations such as the American National Standards Institute (ANSI), model building codes, and the following:

American Society for Testing and Materials (ASTM)
1916 Race Street
Philadelphia, PA 19103
(215) 299-5400
http://www.astm.org

Underwriters Laboratories Inc. (UL)
333 Pfingsten Road
Northbrook, IL 60062
(312) 272-8800
http://www.ul.com

In addition to the resources mentioned in this chapter for locating codes and standards, the reader should also review the reference document section in the base document. Typically, a directory of references is included in the document. For example, ASME A17.1, Section 4, contains a list of all referenced documents, in excess of 50, including procurement information.

Architectural, Engineering, and Performance Standards

Architectural, engineering, and performance criteria can be found in *Vertical Transportation Standards.* Included are architectural layouts, definitions of performance terms, power requirements, including power confirmation forms, and modernization guidelines. *Vertical Transportation Standards,* which in the next edition will be renamed *Building Transportation Standards and Guidelines,* is available from Elevator World and, as well as from:

National Elevator Industry, Inc.
185 Bridge Plaza North, Room 310
Fort Lee, New Jersey 07024
201-944-3211

Canadian Codes and Standards

The *Safety Code for Elevators, Escalators, Dumbwaiters, Moving Walks, and Freight Platform Lifts,* CAN/CSA B44, has been adopted by all the Canadian provinces. The requirements are very similar to those found in ASME A17.1. The first edition of the *Canadian Elevator Code* was published in 1938. Eight editions have been published to date. The *Canadian Elevator Code* requires electrical wiring to conform to the *Canadian Electrical Code*, CAN/CSA C22.1. All electrical equipment must also conform to the *Elevator and Escalator Electrical Equipment Standard*, CAN/CSA-B44.1/ASME A17.5. These documents may be obtained from:

> Canadian Standards Association
> 178 Rexdale Boulevard
> Rexdale (Toronto), ON, Canada M9W 1R3
> (416) 747-4000
> http://www.csa.ca

Canada has a single model building code, the *National Building Code of Canada* (NBCC), which is enforced in most of the Canadian provinces. Requirements for accessibility can be found in the *National Building Code of Canada*, which references Appendix E of the *Safety Code for Elevators, Escalators, Dumbwaiter, Moving Walks and Freight Platform Lifts*, CAN/CSA-B44. The Canadian fire prevention code is the *National Fire Code of Canada*.

Canadian building and fire prevention codes can be obtained from:

> National Research Council Canada
> Institute for Research in Construction
> Ottawa, ON, Canada K1A 0R6
> (613) 993-2463
> http://www.irc.nrc.ca

Compliance with numerous standards are specified in the *Safety Code for Elevators, Escalators, Dumbwaiters, Moving Walks and Freight Platform Lifts*, CAN/CSA-B44. A complete list of these standards can be found in Clause 1.3 of CAN/CSA-B44. Likewise, the *National Building Code of Canada* references numerous standards that address fire resistance, alarm systems, and similar topics. A complete list can be found in Table 2.7.3.2 of the NBCC.

The *National Elevator Industry Vertical Transportation Standards* is also utilized in Canada.

European and International Codes and Standards

In the common market countries of Europe, a code has been developed and designated CEN to replace the individual national standards. Many CEN (Comité Européen de Normalisation) standards adopt International Standards Organization (ISO) standards.* The CEN code for elevators, or lifts as they are known in Europe, is CEN 81. The CEN 81 standards are composed of the following parts:

> *Part 1.* Safety rules for the construction and installation of electric lifts
> *Part 2.* Safety rules for the construction and installation of hydraulic lifts

---

* A discussion of ISO standards can be found at the end of this section.

Part 3. Electric service lifts

Part 4. (Reserved for future use)

Part 5. Specification for dimensions of standard lift arrangements (implementing ISO 4190/1 and ISO 4190/3)

Part 6. Code of practice for selection and installation

Part 7. Specification for manual control devices, indicators, and additional fittings (implementing ISO 4190/5)

Part 8. Specification for eyebolts for lift suspension

Part 9. Specification for guide rails (implementing ISO 7465)

Part 10. Specification for testing and inspection of electric and hydraulic lifts

Part 11. Recommendations for the installation of new and the modernization of electric lifts in existing buildings

Part 12. Recommendations for the installation of new and the modernization of hydraulic lifts in existing buildings

Part 13. Code of practice for vandal-resistant lifts

English versions of the preceding standards are available from:

British Standards Institution
389 Chisnick High Road
London W1A 2BS England
(44) 181-9969000

Work in the international arena has concentrated on standardization of equipment. Because the initial ISO standards were developed with minimal input from the North American elevator industry, ISO elevator standards did not truly reflect the views of the worldwide elevator industry. The situation has since changed, and a number of new and revised ISO standards will reflect not only the European market influences but the North American market influences as well. This will also be evident in the next editions of *NEII Building Transportation Standards and Guidelines,* which will highlight hard metric standards (ISO) for the North American market. The ISO dimensional standards applicable to elevators (lifts) is composed of the following parts:

- ISO 4190-1 Lift Installation—Part 1: Lifts of Classes I, II, and III (passenger and hospital)
- ISO 4190-2 Passenger Lifts and Service Lifts—Part 2: Lifts of Class IV (freight)
- ISO 4190-3 Passenger Lift Installations—Part 3: Service Lifts Class V (dumbwaiters)
- ISO 4190-5 Lifts and Service Lifts (USA: elevators and dumbwaiters)—Part 5: Control Devices, Signals, and Additional Fittings
- ISO 4190-6 Lifts and Service Lifts (USA: elevators and dumbwaiters)—Part 6: Passenger Lifts to Be Installed in Residential Buildings—Planning and Selection

Other ISO standards of interest include:

- ISO 9589 Escalator Building Dimensions
- ISO 4344 Steel Wire Ropes for Lifts

- ISO 7465 Passenger Lifts and Service Lifts—Guide rails for lifts and counter-weights—T-type
- ISO 8383 Lifts on Ships—Specific requirements
- ISO/TR 11071-1 Comparison of Worldwide Lift Safety Standard—Part 1: Electric Lifts (elevators)
- ISO/TR 11071-2 Comparison of Worldwide Lift Safety Standards—Part 2: Hydraulic Lifts (elevators)

As the world becomes increasingly smaller, international standards will play an important role. ISO 9000, for example, is a standard series on quality system registration and related issues that has become very popular recently. The American National Standards Institute (ANSI) is the United States sales agent for all standards issued by the International Organization for Standardization (ISO) and the International Electrotechnical Commission (IEC). A catalog of ISO and IEC standards is available for purchase from ANSI. In addition, ANSI is the best source in the United States for standards published by foreign governments and private entities. Readers in other countries should be able to purchase ISO and IEC standards from their respective national standards organizations. If not, they should contact ISO and IEC in Geneva, Switzerland. ISO can be contacted at:

International Organization for Standards
Case Postale 56
CH-1211 Geneva 20
Switzerland
http://www.iso.ch/welcome.html

CODE ADOPTION

In the United States and Canada all of the codes and standards previously mentioned, with the exception of the Americans with Disabilities Act Accessibility Guidelines (ADAAG), are advisory. The documents (with the exception of Model Building Codes) are developed by consensus committees, in which all stakeholders are offered an opportunity to participate and no one interest group can dominate the committee membership. All interested parties are given the opportunity to comment, and due process is guaranteed.

Typically, these codes and standards form the basis for local regulation. They may be adopted in whole or part and may be modified to meet specific needs of a local community. In years past most regulations were drafted by the authority having jurisdiction. Recently this practice has changed, as the benefits accrued from the adoption of a consensus standard are significant. For example, the benefits gained in adoption of ASME A17.1 include the following:

- National input improves the quality of the regulations. The ASME A17.1 Code is developed and maintained annually by a national organization whose collective expertise represents a broad scope of interests, including governmental bodies, manufacturers, consultants and specialists, contractors, maintainers, suppliers, owners, and labor.
- The administrative costs of code development and revision can be drastically reduced, providing considerable savings to the taxpayers. The ASME A17.1 Code is

developed, maintained, interpreted, and revised by the ASME A17 Code Committee at no cost to the authority having jurisdiction.

- The ASME A17 Code Committee maintains close liaison with other national code and standard-developing organizations, such as NFPA, the Secretariat of the National Electrical Code, and developers of fire codes and the three model building codes.
- The integration of expertise from several highly specialized fields ensures an up-to-date safety code embodying the latest thinking in safety applicable to vertical and horizontal transportation products, whether using traditional or new technology
- Substantial savings for building owners may be realized, as elevator manufacturers can provide uniform and less costly equipment because of the resulting standardization of products. Elevator companies, as a rule, design, test, and manufacture products in conformance with the latest state-of-the-art national consensus standards, ASME A17.1.
- The purchasers and riding public can be confident that products will be designed, manufactured, installed, and inspected in accordance with code requirements embodying the latest in highly developed standards of safety.

This trend has led to the adoption of ASME A17.1 by all but two jurisdictions with elevator codes in the United States. One of these other jurisdictions is actively pursuing adoption in lieu of its home-brewed elevator code. The other typically grants variances for installations that comply with ASME A17.1. All Canadian provinces have adopted the Canadian Elevator Code.

Enforcement varies throughout the country and the world. In some areas and countries enforcement is a governmental function with inspectors who are civil servants. Other jurisdictions have delegated this activity in part (routine and/or periodic) or in whole to private organizations. Private organizations contract to perform the work with either the government or the elevator owner.

How inspection and enforcement are performed also varies widely. It behooves anyone intending to install or maintain equipment in a particular area to become acquainted with and interact with the agency or individual who will enforce the local code. The authority having jurisdiction usually presumes that the provisions of the local code are understood.

ENFORCEMENT CONFLICTS

In reviewing the scopes for the preceding lists of codes and standards, it will be obvious that there are numerous conflicts and duplicate claims for jurisdiction over elevator installations. The ASME A17.1 scope claims that it covers the "design, construction, operation, inspection, testing, maintenance, alteration and repair of elevators." Elevators use electricity to operate, and thus they fall under the National Electrical Code, whose scope states, "This code covers installation of electrical conductors and equipment." Another example can be found in the area of fire protection. The ASME A17.1 scope states, "This code of safety standards covers . . . hoistways. . . ." Upon reading the scope of the building codes it is clear that they intend to cover hoistway construction. The Life Safety Code claims jurisdiction thus: "This code addresses those construction procedures, protection, and occupancy features necessary to minimize danger to life from fire, smoke, fumes, or panic." This surely includes elevator hoistways and elevator operation in regard to fire fighters service. The NFPA Fire Prevention Code states, "The provisions of this Code are applicable to (a) the inspection of buildings, process, equipment," and through its reference to other NFPA codes and standards, such as the Life Safety Code and National Electrical Code, it indirectly regulates elevators. The model building code and Life Safety Code all claim jurisdiction over elevators as they consider an elevator just another portion of a building that has to be regulated to ensure public safety.

It should be obvious that enforcement conflicts pose a potentially serious problem, especially if the different codes address the same feature with conflicting requirements. An excellent example is a major city in the Northeast. The state has an elevator code that prohibits sprinklers in elevator machine rooms and hoistways. The city fire code, on the other hand, mandates sprinklers in these two areas. In reviewing this conflict with elevator contractors, a question often asked is, How do you meet these two diametrically different requirements? The typical answer is that sprinklers are installed and the fire department is invited to make its inspection. Once fire department approval is obtained, the sprinkler contractor returns and the sprinkler heads are removed and the pipe capped. The site is now ready for the elevator inspector. Is this the proper response for such a situation? The building owner has paid for a sprinkler system that in all likelihood has not been reactivated after final approval is obtained from the elevator inspector. The building owner and all contractors involved have serious product liability exposure. If the sprinkler system is not operational and a fire occurs, all involved will be cited for noncompliance with the fire code. On the other hand, if the sprinkler pipe leaks on the elevator equipment, causing it to respond in an unsafe manner, all involved will be cited for installing a sprinkler system in violation of the elevator code. This is a no-win situation and is not the proper way to respond to conflicting code requirements.

CONFLICT RESOLUTION

The designer, architect, elevator consultant, and elevator contractor should take steps to avoid jurisdictional conflicts. In the overall architectural and structural planning of a building, the factors that affect the elevator installation must be recognized. These include, but are not limited to, the following:

- The fire rating of the hoistway and the interface between the elevator landing entrances and the wall construction must be in compliance. This is usually a building code requirement and an architectural detail.
- Glass used to enclose observation elevator hoistways must conform to the structural requirements in the building code as well as the elevator code requirements that the glass be laminated glass. The elevator code also has requirements that address ledges inside the hoistway owing to the placement of glass in a frame.
- The venting of the hoistway or pressurization of a hoistway is a building code requirement, which most often is designed by the HVAC (heating, ventilating, and air-conditioning) engineer. Coordination between their design and the elevator code, as well as the electrical code, in providing suitable clearance between duct work and equipment is necessary.
- Sprinklers are the province of the fire protection engineers, and their installation is regulated by the building code. Conflict over locating sprinklers in machine rooms

and hoistways is always present and must be resolved between building officials, fire officials, and elevator consultants, contractors, and inspectors. Their relationship to elevator operation and fire fighters' emergency operation must be resolved in the early stages of building design.

- Many projects employ a specialist known as a fire protection or life safety engineer. The responsibility is to coordinate various aspects of life safety, such as sprinklers, alarm systems, smoke detectors, fire command stations, standby power requirements, elevator recall, etc. Elevator design coordination with that specialist is essential. Similarly, engaging a security consultant, if one is available, should be considered, inasmuch as many of the aspects of security can affect both elevator and life safety. The security consultant must also be aware of the elevator code requirements relating to security when Phase II Emergency Elevator Operation is in effect.

- The electrical engineer plays an essential part in coordinating any of the code-related aspects of elevators. Proper standby power and dedicated elevator requirements appear in building codes. ASME A17.1 requires power to be removed from elevators prior to machine room sprinkler action. This and many other aspects, such as specialized power supplies, have to be coordinated between the elevator and electrical contractors. Power confirmation data forms should be employed by the elevator contractor to relay requirements to the electrical engineer. Industry standard power confirmation data forms can be found in the *NEII Vertical Transportation Standards* (soon to be known as the *Building Transportation Standards and Guidelines*). These data forms are developed for the various types of elevator motor control systems. In the *Vertical Transportation Standards* an explanation and cover document is provided and a sample of the actual form for a DC Static drive system (SCR Control) can be seen in Figure 16.1.

- As time goes on, an increasing number of relationships between building design and elevatoring will develop. Discussions concerning areas of refuge and building evacuation in emergencies are ongoing; however, some eager politicians and local jurisdictions have been precipitous in passing laws that require such provisions. Too often, concern for disabled persons overshadows sensible engineering judgment, and early resolution of conflict is essential.

The NEII Vertical Transportation Standards (soon to be known as the Building Transportation Standards and Guidelines) will assist in identifying areas in which coordination between the elevator designer/contractor and other trades is essential in planning a trouble-free installation. Many of those areas are also discussed in detail in other chapters in this book. The problems that usually occur with enforcement of the various codes that apply to an elevator installation include determination of the following questions:

- Which code has jurisdiction?
- Who should enforce the requirements of the various codes?
- When there are conflicts, what code requirement governs and who makes the decision to that end?

The motto of the National Association of Elevator Safety Authorities International (NAESA) is, "In the Public's Interest." Through the ASME Qualification of Elevator Inspectors (ASME/QEI) accreditation program, organizations such as NAESA and Lift

POWER SUPPLY
CONFIRMATION DATA FORM

Type of Motor Control: STATIC DC

Prepared By:_____Date:_____

Location, City:_____State:_____

Building Name:_____

Elevator Numbers:_____ Job Number:_____

NOTE: This form to be used in conjunction with NEII Power Standards.

MAIN ELEVATOR FEEDERS:

_____VOLTS_____PHASE_____HERTZ

Voltage Tolerance +5%, −10%. Frequency Tolerance ±2% with slew rate of 1 Hz/sec maximum (continuous) 3.5 Hz/sec maximum (10 cycles maximum transient).

Voltage Balance: Phase to Phase 5%. Phase to Neutral 5%.

Note: Standby (emergency) power to conform to these tolerances.

CONDUCTORS:

Electrical contractor to supply copper conductors and circuit protective devices from the building service to elevator control equipment, in compliance with NFPA 70 and/or local code requirements, based on the electrical elevator load data below.

ELECTRICAL LOAD DATA:

Car Numbers				
Elevator Controller Nameplate (kw)/(acA)				
Elevator Controller Nameplate (hp)				
Running Up Full Load (line acA)				
Accelerating Up Full Load (line acA)				
Transformer Magnetizing In Rush (A-Peak)				

Figure 16.1. Data form for a DC Static drive system (SCR Control) (Reprinted with permission from *NEII Vertical Transportation Standards* 7th Edition, including 1994 Supplement. Copyrighted © 1992 and 1994, National Elevator Industry, Inc., Fort Lee, New Jersey. This reprinted material is not the complete and official position of National Elevator Industry, Inc. on the referenced subject which is represented only by the standard in its entirety).

BRANCH CIRCUITS: [GFCI protection (Items 2 and 3) per NFPA 70, Article 620]

1. Lighting supply (125 Vac-1 Phase) with the following current ratings:

Car Numbers				
Current (A)				

2. To each group supervisory panel, one single phase branch circuit, at feeder voltage, separately fused rated at 15 A.
APPLICABLE ELEVATORS:_____

3. To each console panel 125 Vac, 15A-1 phase. APPLICABLE ELEVATORS:_____

4. Other: _____

HEAT EMISSION

1. The machine room temperature must be controlled in order to maintain ambient room temperature of 12°C to 32°C (55°F to 90°F) with a maximum relative humidity of 80% non-condensing.

2. Approximate heat emission per elevator is as follows:

Car Numbers				
Heat Emission (Btu/h) At 35% Duty Factor				

STANDBY (EMERGENCY) POWER REQUIREMENTS AND NOTES:

1. The maximum electrical demand on the standby (emergency) power system is dependent upon the number of elevators to be simultaneously started. The acceleration power required per car is given in the chart below.

2. One power transfer signaling contact rated at 125 Vac, 3 A minimum must be provided to each machine room. This contact to be closed on normal power, open on standby (emergency) power.

Car Numbers			
Required Acceleration Power (kw)			
Required Acceleration Power Time (s)			
Maximum Regenerated Power (kw)			

DISCONNECTING MEANS:

1. Numbered, sized and located in accordance with NFPA 70.

2. When sprinklers are provided in elevator machine room or hoistway, additional means (i.e. shunt trip) is required.

ADDITIONAL RECOMMENDATIONS: (To be noted by elevator contractor)

CONFIRMATION/APPROVAL:

Electrical contractor/project engineer please confirm, approve and return this data sheet.

_____
Elevator Contractor (Title)   Date

_____
Electrical Contractor (Title)   Date

Figure 16.1. *(Continued)*

Technologies Inc. have been instrumental in certifying inspectors who are knowledgeable of the various code documents and their overlapping requirements. These do not provide the only answer, as knowledge does not ensure the elimination of conflict. There are times when an elevator inspector will be faced with other inspection authorities claiming jurisdiction over portions of the elevator system such as electrical wiring, fire, safety, and so forth. It is in the public's interest to ensure that when jurisdictional disputes arise, the building owner and contractors are not put in a position similar to that previously described. No one group or inspector has the learning or expertise to know it all. The code writers, architects, building owners, consultants, elevator contractors, and inspection departments and inspectors must all work together to protect the public interest.

The elevator inspector should take the lead in elevator inspections, but he or she should realize that coordination with other inspection authorities is needed: Work with them, don't fight them. If they are wrong, iron out the differences and then present a unified set of requirements to the building owner and elevator contractor.

This is not to suggest that the elevator industry accept at face value the positions of every enforcing authority. The elevator industry should keep an open mind and when it disagrees with an enforcing authority, should present a knowledgeable rebuttal. One has only to look at how all the codes are developed to see that cooperation is possible. Compromise and willingness to collaborate among various groups, with differing goals, are responsible for the development of codes and standards for safe buildings, including elevator equipment, not only in the United States but throughout the world. Enforcing authorities should have a program of coordination with other local authorities that have overlapping responsibility for an elevator system. A sound approach is to review the areas of responsibility and iron out conflicts before they become a problem.

CODE COORDINATION

In the mid-1970s the elevator industry began to see a trend: the different code organizations all claiming control of elevator installations, many of them adopting requirements that were in conflict with those in the Safety Code for Elevators and Escalators, ASME A17.1. A major goal of the industry was to develop a working relationship with the various organizations and to establish guidelines to eliminate overlapping and conflicting requirements. As a result, the ASME A17 Main Committee established the A17 Code Coordinating Committee. This Committee has representation not only from the ASME A17 Committee but also from the three model building code organizations, BOCA, ICBO, SBCCI, as well as NFPA.

The Committee established an informal agreement delineating responsibility for the various components of an elevator installation and developed a matrix of overlapping requirements in the ASME A17.1 Code, Model Building Codes, and NFPA 101 Code. In September 1986 the Committee agreed on the following delineation of responsibilities:

Building Codes: Cover building construction and fire protection

Elevator Codes: Cover elevator construction and operation

Dual Responsibility: Covers building operation

A matrix of overlapping requirements was prepared and is maintained by the Committee. This matrix can be found in the *ASME A17.1 Handbook*.[4] The Committee also con-

4. Edward A. Donoghue, *ASME A17.1 Handbook*, 1996.

cluded that a number of provisions in the model building codes and Life Safety Code should be revised to eliminate potential conflicts and to establish uniformity and clarity in all the documents. Proposals were prepared, submitted to the appropriate code-writing organization, and subsequently adopted. All the organizations involved agree that this is an ongoing project and are of the opinion that we cannot afford to not succeed.

FUTURE TRENDS

The world is becoming smaller every day. One needs only look at Europe. Borders are disappearing as the European Common Market takes hold. In Europe, national standards are being replaced by European Common Market Standards (CEN). On the North American continent, Canada, Mexico, and the United States are signatories to the North American Free Trade Agreement (NAFTA). The elevator industry, especially in Canada and the United States, is consolidating. Products manufactured in one country are often installed in the other. The elevator industry in Canada and the United States were quick to realize that national codes and standards were barriers to free trade.

In the mid-1980s the Canadian and United States elevator code committees initiated a project to develop a binational elevator electrical equipment testing standard that would be jointly managed and approved. The project was successful, and a binational standard was initially published in 1991. This led to discussion between all interested parties on the feasibility of publishing a single North American code for elevators and escalators. In January 1995, the American Society of Mechanical Engineers, the ASME A17 Main Committee, the Canadian Standards Association (CSA), and the CSA B44 Technical Committee agreed to harmonize the ASME A17.1 and CSA B44 codes. The binational elevator and escalator code is targeted for publication in 1999. The technical content of the binational code is essentially complete. At this time, issues of administration, joint committee activity, adoption, enforcement—plus all the other aspects of a binational code—have not been finalized.

In the United States the three model building codes have historically had a major influence on specific regions of the country: the National Building Code in the Northeast and Midwest, the Standard Building Code in the South and Southwest, and the Uniform Building Code west of the Mississippi. Each operated independently; thus, the requirements were sometimes quite different from one code to another. In the early 1990s the architectural community approached the model building code organizations and convinced them that this arrangement was confusing, expensive, and not justified. The three organizations agreed to a common code format. For example, elevators would be addressed in Chapter 30 in all three model building codes. In 1993, the National Building Code was the first to publish an edition using the common code format. In 1994 the other two model building code organizations published new editions using the common code format. Once the model building codes were all using the common code format, it was apparent that the next step should be a single model building code. In 1995 the three model building code organizations chartered the International Code Council and agreed to harmonize their respective codes into a single set of codes, to be known as the International Building, Fire Protection (etc.) Code. It was recognized that the most difficult code to harmonize would be the building code. Thus, it was mutually agreed to harmonize that document last. At the time this chapter is being written, work on both the International Plumbing and Mechanical Codes has been completed and those documents published.

Work has commenced on the International Building Code, with publication scheduled for the year 2000.

LOCATING CODES AND STANDARDS

Locating codes and standards can be frustrating and time-consuming. However, in recent years it has become increasingly easier to locate and purchase documents using the World Wide Web. Web addresses are given for most of the sources of information in this chapter. Updates with current Web addresses can be found at http://www.elevator.world.com. Mandatory codes and standards can usually be obtained from the authorities promulgating them, or they can provide you with procurement information.

CONCLUSION

It is commonly stated that the applicable code in an elevator installation is:

- A local home-brewed code,
- A model consensus code, adopted by the authority having jurisdiction,
- A model consensus code, with local modifications, adopted by the authority having jurisdiction, or
- The code according to the inspector at the moment.

The enforcement of codes, standards, and regulations is considered by many in the elevator industry to be nothing but a hindrance. Many do not even take the time to review the appropriate code requirements applicable to a specific application. They rely on the inspector's finding any violations. When violations are found, changes are made to satisfy the inspector. With a little research up front, however, violations can be avoided before costly corrections are required.

Conversely, when a violation notice is incorrect, a knowledgeable contractor can politely point out the correct requirement to the authority having jurisdiction, thus avoiding costly changes, that expose the contractor to potential liability for noncompliance.

ABOUT THE AUTHOR

EDWARD A. DONOGHUE, CPCA, is president of Edward A. Donoghue Associates Inc., Code and Safety Consultants, Salem, New York. From 1976 to 1989 he was manager of Codes and Safety for the National Elevator Industry, Inc. (NEII). He is chairman of the ASME Qualification of Elevator Inspectors Committee and in 1986–87 chaired the Safety Division of ASME. He was the 1992 recipient of the ASME Safety Codes and Standards Medal.

Mr. Donoghue is a member of the ASME A17 Safety Code for Elevators and Escalators Main, Hoistway, Emergency Operations, Construction Elevator, Existing Installation, Maintenance, Dumbwaiter, Inspectors' Manual, International Standards, Sidewalk Elevator, Limited Use/Limited Application Elevator, and Evacuation Guide committees and chairs the Editorial and Code Coordination committees. He has been a licensed master electrician and building contractor and is the author of the *Handbook A17.1 Safety Code for Elevators and Escalators* and *ADA and Vertical Transportation*.

He has been certified as a Professional Code Administrator by the National Academy of Code Administration and as an Elevator Inspector by the National Association of Elevator Safety Authorities (NAESA), Building Officials and Code Administrators International (BOCA), and Southern Building Code Congress International (SBCCI). Mr. Donoghue is a member of American Society of Mechanical Engineers (ASME), International Association of Elevator Engineers (IAEE), American Society for Testing and Materials (ASTM), BOCA, SBCCI, International Conference of Building Officials (ICBO), National Association of Elevator Contractors (NAEC), NAESA, National Fire Protection Association (NFPA), and the National Conference of States on Building Codes and Standards (NCSBCS).

17 Elevator Specifying and Contracting

JOSEPH MONTESANO

PREPARING TO BID

Once preliminary designs of a building have been approved, the task of preparing contract documents consisting of drawings and specifications for the various building systems is initiated.

As a current practice, the structural steel and elevator contracts are the first to be bid and awarded owing to the long lead time of both. This requires that the elevator layouts be firmly established and that the architect has worked with the elevator consultant so that adequate space is available.

Space allocation is an extremely important aspect of elevator contracting, especially if some innovative design is expected to be used. Outside observation-type elevators, underslung elevators, and even building elevators of a size different from manufacturers' standards have to be planned and reviewed with knowledgeable individuals. One project had to be completely redesigned because an assumption was made that an elevator would fit; foundations were started, steel ordered, and when the elevators were bid it was discovered that the space and location of the counterweights for the elevators could not be built as shown.

Once all the variants of space allocation in both plan and elevation are finalized, elevator specifications can proceed. Elevator bidding may take two forms, scope or final design. Scope is used for budgetary reasons and generally proceeds from a simple outline of the project. The scope specifications are intended to be refined after working drawings are completed.

SPECIFICATIONS

Once the method of bidding is determined as based on either scope or specifications, specifications must be complete if final bidding is expected or may be a simple outline if the bidding is preliminary. The areas that must be covered in the specifications are, first, special conditions of the particular project, such as restrictions on the contractor, methods of payment, insurance requirements, necessary reports, and safety precautions. These are expected in construction projects, and many standard forms of terms and conditions have been issued.

The Vertical Transportation Handbook, Third Edition, Edited by George R. Strakosch
ISBN 0-471-16291-4   © 1998 John Wiley & Sons, Inc.

Second, the scope of the work that includes any services the elevator contractor is expected to offer is covered, such as the type and extent of postinstallation maintenance, availability of elevators for temporary use by the building before actual completion, particular hazards, plus anything that may be out of the ordinary in the particular installation. Included in the scope are the owner's responsibilities. This includes construction of the hoistway, any necessary patching and cutting, setting of inserts for rail brackets provided by the elevator contractor, and providing power for tools and for the preliminary operation of the elevator, as well as any other item that is normally the owner's responsibility or may be so on the particular job. Statements as to the procedure of approval of shop drawings and samples of finishes should be included in this section. The scope also includes an outline of the elevator equipment required, such as capacity, speed, rise, number of entrances, type of machinery expected, signals, type of door and door operation, cab description or dollar allowance for a cab design,[1] and any items considered necessary for the particular job. The foregoing is the first main section of an elevator specification.

The second main section of an elevator specification includes a description of various equipment components: machine installation, rails, pit structure, fixture description, operating panels, intercom system, cab specification or elaboration as to what the allowance does or does not include, details of entrance construction, plus any special entrance fire test or installation requirements. Any item that includes the installation of a fixture or equipment should be included in this section. If a specific operation is desired, any hardware needed should be specified in section 2 and the expected operation detailed in section 3.

The third section of an elevator specification should be devoted to operations both basic and special. The individual elevator or group elevator operations should be generically described and further elaborated to specify the exact features desired. For example, the A17.1 Elevator Code defines Group Automatic Operation but does not describe program and operating features such as up-peak, load dispatch, load bypass, two-way traffic, down-peak, parking of elevators, and other desirable operations. Unless the specifications call for such features, the individual elevator contractors have a perfect right to provide what they consider their standard.

Special operations must be spelled out in detail. Fire recall as required by local codes must be furnished by the elevator contractor. What may be lacking from the local code is the extent to which each elevator must be equipped or any additional feature that may be desired. These aspects must be distinctly specified. In computer jargon, section 2 of an elevator specification can be called "hardware" and section 3 "software."

Section 4 of an elevator specification is entitled "Execution" and should describe details of painting, electrical installation, tests, and final acceptance procedures. Any special wiring, such as shielded intercommunication and telephone systems wiring or coaxial cable for TV cameras, should be included in this section. Painting specifications should include any extraordinary painting that is to be done by the elevator contractor, such as painting the divider beams in the hoistways. Tests may include recorded accelerometer tests to be performed to qualify the ride characteristics of the elevators.

---

1. An allowance for a car enclosure is a means of deferring the decision on the architectural treatment or design, just as a decision on the colors to paint the doors or other features could be deferred. Because elevator equipment takes considerable time to manufacture before actual installation begins in a building, details of visual treatment may not be firm when specifications are first being prepared. By using an allowance, the probable cost of the installation is established and the architect is allowed latitude in final design. If what is designed costs more or less than the allowance, the difference is negotiated at the later date. Allowances can also be used for lobby floor entrance assemblies or for special signal fixtures.

Section 5 describes any alternatives required to the base elevator bid or added to the specifications if, for example, the base bid price is low enough to warrant extra expenditure on higher-speed equipment. An alternative may be used to call for additional equipment that may be desirable but not necessary. Alternatives may be requested should the owner decide to build part of a facility now and the rest at a future date. In any event, the alternative should clearly state the intent and, if for future installation, the expected delay.

Often it is desirable to provide an illustration to supplement a word description of an operating panel, fixture, or method of installation. This can be done in section 6, entitled "Illustrations." The descriptive paragraphs in other sections should indicate a reference to the illustration, such as "See illustration, section 6." It is best not to reference illustrations by figure numbers because any change, such as adding or subtracting paragraphs, may destroy the sequence and result in a multitude of changes.

Escalator and material-handling system specifications follow the same basic outline as described for elevator specifications.

CONSTRUCTION SPECIFICATION INSTITUTE

The Construction Specification Institute, a national organization in the United States, has been active in organizing specification wiring and attempting to develop standards for all specification writers. Its basic specification is developed in three parts: Part 1. General, Part 2. Products, and Part 3. Execution. This three-part arrangement is very satisfactory for most building trades, but in elevatoring, operation is the main goal of the installation and should be treated separately. Such operations are often affected by other building systems, and close coordination or interface must be considered. Elevators have unique operations related to the people in the building, the electrical system, and the life safety system. Elevator specifications require an additional section describing the operation expected.

The Construction Specification Institute has set aside Division 14 as the division used to describe conveying systems. The general headings assigned to elevators, escalators, and material-handling systems are as follows:

14000—Conveying system–general

14100—Dumbwaiters

14200—Electric elevators

14700—Escalators

Section 14200 must be further broken down if parts of the elevator are separated. For example, in some localities it is an accepted practice to let separate contracts for the elevator entrances and car enclosures. For this purpose, Sections 14220 for the doors and 14250 for the cabs are created.

The use of the Construction Specification Institute format will depend on the preference of the building architect. In many areas, the designations 14.1, 14.2, and so on, are used, and in others, 14A, 14B, 14C. At least Division 14 seems to prevail.

CODES AND STANDARDS

The universally accepted safety code for vertical transportation in the United States is contained in the ASME-A17.1 Code, and its official title in 1997 is the Safety Code for

Elevators, Escalators, Dumbwaiters, and Moving Walks. Copies may be purchased from the ASME, 345 East 47th Street, New York, NY 10017-2392 (800-843-2763) or, from *Elevator World Magazine* (see Appendix). Many localities, basic building codes, the federal government, and others use this code as the minimum basis of elevator and escalator requirements. It is continuously updated, and either supplementary or new editions are issued annually. The code is administered by a main committee that meets quarterly. All changes are subject to public review, and any question or additional interpretation can be answered by writing to the ASME at the aforementioned address.

The A17.1 Code is an advisory code and mut be adopted by a locality before it can be enforced as a building code. City and state elevator inspectors are charged with code enforcement and usually use a companion publication, the *A17.2 Inspector Manual,* as a guide in their elevator inspections.

Some localities have written their own codes, often based on the A17.1 rules. Unfortunately, these local codes are often not updated to reflect current practices and particular attention must be paid to the local code since, even though something may be new and better, its use may not be allowed because of local requirements. Of course, the reverse may also be true. Local code authorities can react to situations faster than national bodies with their extensive review process. This was especially true when fire safety rules were developed and caused the comment, "A fire is different in every political subdivision," which reflects the many different approaches local code authorities used to establish rules for elevator recall.

In Canada, the B44 Code is the counterpart to the A17.1 Elevator Code. Liaison members are on both the A17.1 and B44 committees, so there is a close correlation. In the common market countries of Europe, the CEN81 Code is being developed with the finalized versions of the elevator and escalator sections currently (1998) available. The European Common Market Standards (CEN) Code is not as extensive as the A17.1 Code since it is relatively new and European countries do not have the extent of high-rise construction found in the United States and Canada.

Scheduled for 1999 is a harmonized version of the American A17.1 Code and the Canadian B44 Standard. The purpose is to create a code that will be applicable in both countries and eventually replace the existing documents. It is expected that a considerable time lapse will occur before it is universally adopted, inasmuch as many jurisdictions will have to rewrite their existing regulations. The chapter on codes and standards details this effort, and it is probable that most of the content will not be radically different from that currently in use.

PERFORMANCE CRITERIA

If a building has been elevatored on the basis of performance criteria as described in Chapters 3, 4, 5, and 6, these criteria should be included as part of the elevator specifications. Lesser performance usually requires less advanced equipment and may cause the building's elevator service to suffer. Such criteria should include door time, floor-to-floor time, dwell-time, plus any other measurable quantities desired.

Performance criteria, that is, the way an elevator is expected to perform within limits of acceleration, deceleration, vibration, noise, and so on, is the subject of extensive effort by the National Elevator Industries through its Performance Standards Committee. The Committee has developed guidelines that can be used in elevator and escalator specification for many of the characteristics that a buyer of an installation can expect. This work is published in the book *NEII Vertical Transportation Standards,* details of which can be found in the appendix to this book.

Bidders should be qualified before they are invited to bid; they should be able to demonstrate their competence to install equipment of the nature required. The architect or owner's representative should investigate all the installations cited as qualifying examples. If elevator service in a particular building is expected to be critical, as in a hospital or hotel, the bidder should also have adequate maintenance facilities in the immediate area. Failure to qualify on this point may result in each shutdown's requiring excessive travel time for which the owner will have to pay and be without elevator service.

PROPOSAL BIDDING

The alternative to specification bidding is to outline briefly the elevator requirements and request various elevator concerns to submit a detailed proposal describing what they intend to furnish. In this case, the owner must evaluate all the criteria for a suitable installation.

Unless the invitation to bid specifically states that the lowest bid will be accepted, the owner is under no obligation to accept a low bid. This fact has been established by litigation, and the term "lowest responsible bidder" is often used. It should be cautioned that, with proposal bidding, failure to include necessary items in the proposal may result in unexpected expense.

NEGOTIATION

Often elevators are not bid separately but as part of the general contract for construction of a building. The general contractor then awards the elevator subcontract, seeking those who promise to meet the plans and specifications at the lowest price. If the owner does not approve of the elevator contractor who submits the lowest bid, the owner may have to add to the general contract the difference between the lowest bid and that of the preferred elevator subcontractor.

No matter what plans and specifications are used to describe the elevator work, their scope and completeness provide only a partial guide to a good elevator installation, which depends, essentially, on the reputation and the ability of the installer. For this reason it behooves the architect and owner to exercise care in choosing invited bidders.

ESTABLISHING A CONTRACT

When the bidding negotiation is complete, a contract is established. This may be a letter of intent until final details of terms and conditions are agreed on, a signed proposal submitted by the elevator company, or the owner's signed contract. The work of manufacturing and installing the elevator equipment can then begin.

One of the first steps is taken by the elevator contractor in preparing and submitting a layout drawing for approval. This layout should show details of the elevator installation in the building in the space allocated, coordinated with the building's plans. The architect, elevator consultant, and structural engineer must check the layout with the contract documents and approve it. The general contractor must coordinate this drawing with other trades and allocate the preparatory work that must be accomplished. Once the layout is approved, the elevator contractor can get the necessary building permit from municipal authorities to proceed with the installation. Simultaneously, the owner is expected to confirm the power to be supplied by the local utility in the completed building.

Once an approved layout and power confirmation are obtained, the elevator equipment is manufactured. The elevator company's construction superintendent will arrange for the rail brackets or inserts to be placed while the building is rising. At the proper time rails will be delivered and installed, and the elevator installation will proceed.

Meanwhile, other drawings are submitted to the architect or owner, including approvals for operating fixtures and finishes, cars and entrances, and any other items having optional treatment. By a certain time the elevator contractor will need power to start up the elevator and complete the installation. This will be well before the elevator is complete, because the moving platform is often used as a scaffold to install the doors and to finish interior hoistway equipment. The electrical contractor must supply power in the machine room, plus wiring to the elevator controller. Additional circuits for car lighting, intercommunication, fire safety, and other functions must also be provided in the machine room.

TEMPORARY OPERATION

The general contractor or the owner may wish to use an elevator for temporary service long before the building is complete. This understanding must be established at the time of bidding and the time of temporary service determined, usually a stated number of days or weeks after the hoistway and power become available. The extent of temporary operation may be such that the elevator contractor has to provide a temporary hoisting machine or the general contractor must provide a temporary location for the housing machine at some point below the permanent machine room level.

Labor contracts have recognized that construction workers should not be required to walk up more than a certain number of floors. This requirement is met by a temporary elevator, usually the regular elevator platform with a temporary plywood cab and wooden doors on the hoistway. An operating safety test should be performed and the temporary operation licensed. To allow work to proceed on adjoining elevators, the hoistway is enclosed in wire screening or otherwise protected. The temporary elevator is operated by an attendant and often stops only at every third floor or so.

The number of floors to be served by such temporary elevators must be established early because concentrated effort is usually required to get the equipment on the job and running. The number of temporary elevators is based on the number of people expected to be on a job at a particular time. Calculations can be made to determine how quickly people can be moved, and elevator requirements established to meet the temporary needs. If an outside hoist is used, a temporary hoist machine will not be required since the outside hoist will be in use until permanent elevator machines are in place.

ACCEPTANCE

Once a permanent elevator is complete, it may be turned over to and accepted by the owner, at which time an elevator being used for temporary operation can be completed by the elevator contractor. Elevator constructors are often the earliest on the job and the last to leave, starting with the foundations and finishing only when tenants are moving in. To complete their job, elevator installers must have all the elevators in a group running so that group operations may be adjusted. With the general contractor wanting to use a car to complete construction work, tenants wanting to use one to move in, and passengers requiring the others, the final stage of completion is often drawn out.

The contract is essentially complete when the architect or owner accepts each individual elevator. By this time a regulatory authority has made a safety test and the elevator has been certified. The contractor has made a check and any items of deviation noted have been cleared up. The elevator is turned over when the owner signs a final acceptance, at which point the new installation service begins.

NEW INSTALLATION SERVICE

New installation service is a maintenance service included in the original contract. The elevator installer agrees to maintain the equipment for a minimum period of three months, making necessary adjustments and seeing that the elevator is operating properly. During this period the management of the building is expected to make arrangements for continued maintenance of its equipment, either by its own staff or through a maintenance contract with the elevator company. In either event, if a major part failure occurs, its replacement is warranted by the elevator installer for a usual period of one year, based on normal use of the elevator and reasonable maintenance. For this reason it is wise to include a year's service with the initial contract.

Because elevators are completed, turned over, and accepted over a considerable period of time, an actual date for the start of new installation service for an entire installation can be controversial, especially when the initial service period is concluded and a formal maintenance contract commences. This problem is overcome through establishing an equalized final acceptance date by gives and takes. For example, if one elevator is completed 30 days before a certain date, one completed on that date, and the other 30 days after that date, an equalized final acceptance date could be easily established on the middle date. If one year's initial service was included in the original contract, an equalized final acceptance date will be essential to start the continuing elevator maintenance contract. In a major complex, the equalized price is based on the daily maintenance cost per unit as compared with the total monetary amount in the total project included for new installation service. This amount would have been stated in the original contract.

ACTUAL ELEVATOR CONSTRUCTION

The foregoing sections simply highlight the process of installing an elevator in a building. The actual work depends on close scheduling with the other parts of the building so that an elevator installation proceeds smoothly and without interference with other trades. Step-by-step installation of an elevator has been thoroughly outlined in The *Elevator Erection Manual*.[2] Designed primarily as a handbook for elevator constructors, this book covers many of the considerations necessary for installing elevators. These consist of

---

2. Available from *Elevator World Magazine,* PO Box 6506, Mobile, AL 36606.

reading elevator layouts, rigging and hoisting of equipment, fastening to steel and concrete, setting brackets and rails, erecting the car frame and platform, setting the machine and overhead sheaves, roping and electrical wiring, plus all the other facets of an elevator installation.

Coordination and timing are the key to the success of any construction project. To this end, various means of scheduling are used, generally Gantt charts and the Critical Path Method (CPM), converted to computer software.

Gantt charts are simple bar graphs charted against a time base and showing the time required to deliver and complete necessary steps in an elevator installation. CPM is a more sophisticated approach and requires constant updating, utilizing a computer. Scheduling specialists are used by contractors to maintain construction schedules, and the information they provide has to be updated by inputs on actual field and material delivery conditions. Specialized texts on such procedures are available.

Each project is totally different, with its complexity, size, and height determining the time required. As can be imagined, installation scheduling for a multigroup elevator project will be exceedingly complex as compared with a smaller system. Similarly, an apartment house elevator installation is a relatively simple situation, and formal schedules will probably not be prepared as they can be discussed verbally and depend on a hoistway ready date. In a commercial or industrial building, however, where considerable tie-in with other trades may be necessary, formal scheduling may be essential and should be considered before bidding, possibly with dates and delivery time included in the contract terms.

APPROVALS

One of the essential aspects of maintaining an elevator installation schedule is prompt approvals. These include layout drawings, design of fixtures, entrances, cabs, color selections, and the myriad choices that must be made to make the elevator appear and operate as desired. Although it may be difficult for an owner to decide a year or so in advance what type of metal fixture is wanted in the building, these fixtures are essential parts of the elevator and so must be ordered early. For example, the landing button must be mounted in an electrical box imbedded in the corridor wall. Because the ultimate style of the landing button fixture and its mounting box must be coordinated, the design must be firm long before the walls are built.

Similarly, hoistway doors are generally made with a factory-applied baked enamel finish. Because the doors must be installed before the locking mechanism, door color is one of the earliest approvals. This can be deferred by installing elevator entrances with a prime finish only and applying the final finish and color when the building is complete.

With any special elevator equipment, such as an outside glass cab, the manufacturing process may be six months or longer in addition to on-site installation. No one will start manufacture before an approved design is received, and a glass cab usually must be completely custom built and, often, must be final fitted by hand. In addition, the owner may wish to see a mock-up of the final product in the cab manufacturer's shop. After dissembly, packing, and shipping, six or more months may elapse after the drawings are approved. Once at the site, the elevator contractor may require a month for erection and attachment of all the operating accessories before the final adjustment of the elevator is accomplished. The time lapse may be a year or more to get a cab delivered on an elevator which, if it did not have a glass cab, may have taken only six months to complete.

DELIVERY SCHEDULING

As plans for a building become firm and actual construction begins, the need to have certain items installed at certain times become apparent. Some equipment, including elevators, may be critical to the overall progress of the job and require considerably study and consultation with manufacturers' representatives to determine when items are necessary.

For example, an observation tower is to be built for an exposition expected to open at a specific date. At least two years before that date the design of the tower must be made firm. Its essential elements are the structure and the elevators, because the tower must be built before the elevators are installed and the elevators must be installed before the furnishing of, for example, a restaurant on the top can proceed. The architect, in planning the tower, must consult with structural steel and concrete contractors to determine a practical delivery and erection schedule for the tower itself. The architect must likewise consult with the elevator manufacturer to determine delivery and installation times for the somewhat special equipment the tower requires. Let us assume that the tower structural work will require a year before elevator machinery can be set on the top.

The elevator manufacturer may require a year before the design and delivery of the necessary machinery can be accomplished and, say, six months to complete its installation. No matter how many people are placed on the job, certain time requirements are irreducible because people can work only on one part at a time. The machines, for example, cannot be finally set until the rails are installed.

If the elevator contract is delayed for six months, or until 18 months before completion of the project, no time is left for contingency and the schedule is critical. If the delay is longer, the elevator project must be considered for overtime and costs will skyrocket. If further delay is encountered, it may be impossible to complete in the given time; no inducement will make it possible.

The first essential for any construction or elevator installation is to establish a realistic original schedule. To avoid delay during the progress of a job, two courses are open. One is to provide the schedule and the necessary personnel to coordinate all trades to maintain such a schedule, which is accomplished by progress reports, follow-up, expediting, and prompt approvals by all concerned. This procedure is usually the most satisfactory and is used by the knowledgeable contractors in the field.

The second means is to establish an expected delivery date a given number of days after receipt of the contract and impose a charge on the contractor if the date is not met. If the time is reasonable and the charge is not excessive, the elevator contractor can accept, provided liability is limited to only those delays subject to control and responsibility. If the time is too short, there are three choices: to refuse to participate in the job; to include, if possible, the necessary overtime to complete within the time allowed; or to estimate how many more days the job will take than the allotted time and include the cost of these liquidated damages in bidding.

The architect can rest assured that if such a job is accepted, every delay, every inaccuracy, each interference between trades, delays owing to weather and labor difficulties, and so on, will be recorded, and any delay, including time to receive approvals, will be imposed to extend the introduction of a penalty. The cost in paperwork and argument may far exceed the gain expected by the penalty. As may be surmised, it is far better to schedule a job properly and make necessary allowances for contingencies than to try to substitute penalties for poor scheduling.

RELATION OF ELEVATOR WORK TO OTHER TRADES

Because an elevator, escalator, materials-handling system, or moving walkway becomes an integral part of a building when it is completed, considerable coordination is required between equipment installation and construction of the other parts of the building. Recognizing this, the general contractor should keep track of all the trades so that interference is minimized and the work flows smoothly. It also behooves the elevator installer to get equipment in place at the proper time so that the other trades will not interfere.

For example, in a steel building, the rail brackets should be set and fastened before the beams are fireproofed. If not, the elevator contractor may have to cut away fireproof material before the brackets can be fastened. The machine beam supports must be in place before the elevator machine beams are set. If the other trades fail to complete this task, the installation is delayed. Unless the elevator contractor gets door sills in place before the finished floors are poured, considerable cutting and patching will be necessary. Similarly, boxes for landing button and lantern fixtures must be set before walls are built. Proper power is necessary early enough so that the running elevator can be used.

When the job is ending, a considerable amount of cleaning up must be done. Gypsum dust and loose masonry abounds and must be cleared away. The hoistways can act like flues every time and elevator hoistway door opens and draw in dirt and dust from every floor. As the air-conditioning in the building is balanced, door locks must be adjusted, for the conditioned floors will be at a different air pressure from the hoistway and stack effect will tend to hold doors open.

In a large building the elevator contractor may leave the system in a state of adjustment designed to meet the initial traffic conditions. As the building becomes occupied and tenants' habits become established, additional tuning to adjust performance to traffic is necessary. This may include adjusting the load weighing that causes elevator dispatch and bypass operation, the time when peak traffic operations are established, and the traffic responsiveness of the group of elevators.

If the building is slow in being occupied, these adjustments must necessarily be delayed—a condition that emphasizes one of the advantages of having the elevator manufacturer maintain the equipment on a contract basis. Adjustments that must be made will then be done at the necessary time. The alternative, with owner maintenance, is either to have the owner's mechanic perform such adjustments or to issue a separate order to the elevator manufacturer.

Representatives of the elevator manufacturer should call on the building management as the building is being occupied. They should give the owner instruction for maintenance, parts leaflets, aid in overcoming operational problems, and all the services necessary make the installation a credit to the manufacturer as well as the owner. A reputable elevator company should strive to satisfy a customer so that he or she will recommend the installation to others in the market for elevators.

CHANGES AND ALTERATIONS

Occasionally, an unpredictable development may require interrupting progress of an installation to make a change. The variety of such changes is immeasurable. A prospective tenant of the building may want an additional entrance on the elevator, floor designations may be changed to suit somebody's fancy, and additional control features may be warranted, as well as a host of other alterations.

If the need for a change become apparent early enough, it can often be made without delaying the job. If the change is late, scrapping of ready-worked material or extensive changes to installed equipment may be necessary. Needless to say, changes should be carefully investigated to limit them to only those features that involve minimum scrapping or rework.

Once equipment is installed it may be desirable to add further operating features, possibly to handle a traffic situation unanticipated at the time of planning or to meet the changed requirements of a particular tenant.

With microprocessor operation of the elevators such changes can be accomplished with minimum interference to interim building operation. The equipment is designed to be reprogrammed, and elevator manufacturers with good service organizations should be able to make such program changes in a short time. Hardware changes are another matter. New fixtures or changed signs may have to be ordered and, possibly, custom built.

CONTRACT COMPLETION AND ACCEPTANCE

If the owner or the architect has engaged a consulting elevator engineer, that engineer will play a considerable role in the entire elevator design, specification, contracting, review, and acceptance procedure of the elevator installation.

The consultant has written the specifications; hence, the review and approval of shop drawings are tasks that should be performed by the consultant.

When the job is considered complete and ready to be accepted, the consultant should make thorough observations of the equipment and installation practice as an initial step. Any discrepancies in the installation of the equipment or in the interior condition of the hoistway should be noted and recorded in a job report, which then becomes a punch list for the elevator and general contractors. When the elevators are operating, the consultant should be a witness to tests by local authorities for elevator licensing and may require additional testing to assure the owner that the installation meets the building requirements.

Operational tests have to be made and the performance times observed and recorded. Any interfaces with a life safety system must be tested. For example, the smoke detectors in the elevator lobbies should be tested to ensure they provide a signal so that elevators are recalled.

The riding characteristics of the elevators must be observed and tested. This consists of taking a number of trips and noting any unusual noises or bumps. If an elevator appears reasonably smooth, further testing is accomplished by use of a recording accelerometer to provide a visual record of the ride. A second such test should be made about a year later after the building has gone through a heating and cooling cycle, has become reasonably occupied, and, it is hoped, fully compressed. The first and second readings must be compared and corrective action taken, if necessary, to equalize them.

When solid-state control has been provided on an elevator, it may be necessary to address any complaints of electrical line pollution in the building that can affect tenants' computer equipment. Necessary corrective measures such as the installation of filters may be required. Similarly, if structure- or air-borne noise from the elevator equipment is apparent, corrective measures are required. Power factor correction capacitors may substantially reduce adverse electrical line pollution.

If the installation has a standby generator designed to provide emergency operation of the elevators, a thorough test of that system must be accomplished when all the systems are ready. Generator tests often have to be scheduled for middle-of-the-night hours because during the test practically all the vital building systems are shut down for a period of time.

In addition to overseeing the necessary physical tests, the elevator consultant usually works with the owner to aid in establishing a maintenance contract and to help solve operational or elevator programming problems that may have been unforeseen. These may include problems caused by building sway and wind, as described in an earlier chapter, or material and mail handling problems if such systems were part of the elevator contract.

ELEVATOR MODERNIZATION

The foregoing sections describe most of the steps required for the installation of elevators in a new building. In new buildings, a general contractor is usually involved as the coordinating agency, and the architect as the representative of the developer. With an existing building many of the aforementioned steps are necessary, but relationships change dramatically.

Elevator modernization is an ongoing affair. The first of the major efforts in modernization occurred during the 1950s, when updating elevators to eliminate elevator operators was emphasized. The older buildings, those built during the 1920–1930 era and before, usually required a complete update, and that became a major activity of elevator contractors, many of whom were not the original manufacturers. Elevator installations built during the postwar years needed additions to the existing control systems and relatively minor modifications to the operating systems—notably, door protective devices—which were usually performed by the original manufacturers. After most of the major buildings become "operatorless," such peak modernization activity settled down and became a substantial business for the nonmanufacturing elevator contractors. These agencies could buy the necessary equipment from the growing elevator supply industry consisting of a variety of independent specialists.

Encouraging the growth of the nonmanufacturing elevator contractor involved in modernization was the unprecedented new building boom during the 1960s and 1970s that provided the major manufacturers with sufficient work that they tended to ignore the modernization market. The situation changed in the 1980s, however, as new building projects almost disappeared and modernization became a prime activity.

Fueling the modernization market in the 1980s were a number of factors. The control systems of the elevators updated in the 1950–1960 period were becoming obsolete, because microprocessors could improve elevator service dramatically. Life safety requirements were increasing, and local laws mandated updating building protective systems such as fire safety, which also affected elevators. Moreover, the employees with skills to maintain the vintage equipment were being retired. Many building operators requested proposals from various companies to upgrade their elevators, and often the diversity of responses tended to confuse. The result was the need for an independent elevator consultant who understood the changing requirements and could advise a building operator as to what was needed to update the elevator system.

Because only the elevators were involved in such situations, the need for a general contractor and, possibly, an architect was eliminated or minimized, and the consultant became the lead person in the update. Surveying the building to determine what was needed, appraising proposals from various elevator companies, meeting with financial people to discuss budgets and costs, and preparing the specifications for the update—all became part of the consultant's activity.

Contracting to modernize elevators is a much more complicated task since the need for continuing building service is essential. If the current maintenance contractor is not the one who will do the modernization, arbitration between the two companies may be required. This factor, plus the expected upset of having a contractor on the site while normal building operations take place, adds to the considerations an elevator consultant must address.

The typical steps involved in preparing a bid for modernization include the following:

1. *Surveying the present installation to determine what must be replaced and what can be retained.* Because substantial investment is required for modernization, an estimation of the degree of improvement in service that can be expected from the updated elevators is required. It is possible that an elevator may be removed from service based on the improvement a modernization may afford. This latter activity may involve a traffic study, which can be combined with the next step.

2. *Traffic study to determine the effect of having elevators out of service for an extended period.* The consultant must apprise the building operators of this effect and advise them as to the arrangements they may have to make with the tenants regarding the expected inconvenience.

3. *Physical changes needed to ensure proper operation of the modernized elevators.* These may include additional machine room space, upgraded electrical power, air-conditioning of machine rooms, and considerations of any legislation or changes to elevator codes that may affect the elevators. This last item will include possible updating of life safety systems, security systems, accessibility to meet the handicapped requirements, and any other items needed for the upgraded elevators to comply with the latest code enforced in that jurisdiction.

4. *Preparation of specifications which, in general, will follow the same format as those for a new installation,* with added provisions to adhere to building conditions. These may include contractor employee facilities, security, enhanced protection of work areas since the building is in daily use, storage space, removal of old material, plus the multitude of possibilities to cover instances when "strangers" are in the building.

5. *Working with the building owner to appraise the bids received and to reconcile differences between various contractors.*

6. *Award of contract and the ongoing activity of scheduling,* approvals, unforeseen conditions, progress reports and approvals for progress payments, etc.

7. *Acceptance observations and testing of the completed units,* plus arranging for continued maintenance of both the new and existing elevators.

8. *Once work is completed, a traffic survey* to appraise the expected improvement in service that may havxe been predicted before the modernization was started.

The foregoing is only a broad outline of some of the steps in the overall modernization process. The elevator consultant and the building operator become partners in this activity, and, as may be seen, a great deal of project management is involved. It would be unusual to have the job go smoothly without unforeseen problems; however attention plus prompt action can often minimize such difficulties. A reward for a successful job is seldom given, but is often recognized by "no complaint."

SUMMARY

As may be surmised, elevator specifying and contracting, plus the acceptance of a completed elevator installation, can be an involved process. Similar requirements exist for escalators, materials handling, and moving walk installations.

Experience is of unestimable value. If the elevator contractor with a large contract appoints a knowledgeable project manager as a representative, a first essential step is taken. The project manager should be conversant in all aspects of the installation, be the source of information, attend the job meetings, and bear the brunt of the expected trouble. An astute project manager becomes the liaison between the general contractor and the elevator contractor on a new building, and between the elevator contractor and either the building owner or the elevator consultant on a modernization project.

The elevator consultant often assumes the role of project manager and becomes both the owner's representative and the advocate for the elevator contractor. Yet no matter what the role involves, elevator specifying and contracting requires the complete understanding of the process as outlined in this chapter.

ABOUT THE AUTHOR

JOSEPH MONTESANO is president and founder of DTM Inc., an elevator consulting firm located in College Point, NY, and operating, primarily, in the New York metropolitan area. They specialize in the modernization and evaluation of existing buildings and have been the consultants on a number of major new projects in the area. Mr. Montesano was formerly service manager and a prime project manager for the Westinghouse Elevator Company for 33 years before founding his own company in 1982.

His experience includes the management of the modernization of Rockefeller Center, and serving as the engineering and project manager of the Pan Am building in New York City, and the Sears Tower in Chicago as well as numerous 40- to 60-story buildings constructed during his tenure with Westinghouse. Other notable projects include various major hospitals in the New York area.

His firm has grown from a one man operation to the present staff of fourteen people, including elevator engineers and mechanics. Current clients include major building owners and managers in the area as well as architects and general contractors. The firm has undertaken major projects, including the conversion of the Gimbels Store on 34th Street into a major shopping center that includes numerous observation type elevators and escalators, plus major additions to the World Trade Center complex.

18 Economics, Maintenance, and Modernization

LEN LEVEE

COST

There are three cost factors that must be considered in designing vertical transportation systems for any type of building. Of initial importance is the cost of the installation. Of parallel importance is the cost of operating the vertical transportation system. The third, but perhaps less definitive, factor is the return that may be anticipated on the initial investment.

Installation Costs

The installation cost of any vertical transportation system obviously cannot be estimated until the system duty requirements have been established and the system has been quantified. In today's complex world, establishing the parameters of a vertical transportation system is no longer simply a function of how many floors and how many rentable square feet per floor will make up a building. There are many other factors, both external and internal to the building, that can have a significant impact on the selection of the vertical transportation system.

Topography of the building site can be a major consideration. If the structure is to be built on sloping ground, consideration must be given to how many levels will serve as entrances to the building. There may be one, two, three, or more levels of external access. One or more of these levels may be accessed by employee parking lots or by visitor parking lots. If the new facility is in an urban location, there may be a public transportation stop immediately adjacent to one or more of the entrance levels. The site may be suited to a long, relatively low-rise facility that would be better served by several separated small groups of elevators, as opposed to a mid- to high-rise facility with a smaller footprint that is best served by a central core of elevators.

Local zoning regulations may impose height restrictions on the building, which could force the building designers into working with short floor heights. This, coupled with the anticipated tenant profile, can affect the potential choices of elevator speed. Parking regulations may force the building designers into consideration of underground parking facilities within the confines of the structure. Building security may then become a factor, which may require consideration of a parking shuttle elevator system.

The Vertical Transportation Handbook, Third Edition, Edited by George R. Strakosch
ISBN 0-471-16291-4 © 1998 John Wiley & Sons, Inc.

The anticipated tenant profile can also have an effect on the vertical transportation system selection process. A single-tenant building will develop entirely different traffic patterns than a multiple-tenant building will. Consideration of 2- to 3-stop dedicated internal elevators may become necessary in single-tenant buildings or in multiple-tenant buildings in which a single firm occupies several adjacent floors. Similar considerations must be given to the unique security needs that develop in facilities such as law enforcement agencies, university libraries, and administration buildings.

Traffic peaks and patterns in an office building will differ extensively from those experienced in hotels or residential buildings. Office buildings develop fairly well-regulated up and down traffic peak times of relatively short periods. Moderate two-way traffic is consistently encountered between the peaks. Hotels and residential buildings develop down-peaks of traffic over a longer period of time in the mornings as guests and residents leave for work. During the rest of the day traffic in these facilities is relatively light.

Traffic in a hospital environment is an entirely different matter. Although pedestrian traffic does develop peaks at shift change and meal times, there is always a heavy volume of opposing traffic. Subsequently, heavy two-way interfloor traffic is the result for most of the normal hospital workday. In addition to pedestrian traffic, patient and service traffic must be considered as separate entities. Special transport needs, such as dedicated food delivery and dedicated surgical supply delivery systems, must also be discussed with the materials management consultants.

Once these various external and internal factors have been analyzed and an appropriate weight has been applied to their impact, a functional analysis of the facility must be performed in close association with the building designers and other system consultants. The result of these discussions will be a vertical transportation system that is a compromise between the ultimate elevator system, the construction budget, and available space allocations.

Operational Costs

Among the factors considered by building management in considering modernization of a vertical transportation system are the need to reduce downtime of the system, rising maintenance costs, and diminishing system efficiency. Often overlooked in these considerations is the reduction of energy consumption.

Although it may still be necessary in some facilities to maintain motor generator control, the use of solid-state power conversion units should always be explored in modernizing older vertical transportation systems. Over the course of a year of operation, power consumption by a vertical transportation system will show a marked reduction through the use of solid-state converters. Some savings in maintenance costs can also be achieved through the elimination of the motor generator rotational equipment.

The introduction of the microprocessor has improved system operational efficiency levels to such an extent that the actual number of elevator movements during the course of a normal workday is reduced. Although this reduction is not easily quantified, it does exist and will contribute to a further reduction in power consumption.

RETURN ON INVESTMENT

The factors noted in the preceding section can be reasonably evaluated in appraising a potential return on investment from the modernization of any vertical transportation system.

Perhaps less tangible, but of equal importance to the building management, is the improvement in building image and tenant satisfaction. Dissatisfaction with a vertical transportation system is always at the top of the list of tenant complaints, along with heating and air-conditioning. The improvement in equipment reliability after modernization will result in a higher level of tenant satisfaction achieved by a reduction in downtime and far fewer entrapments, and a much lower potential for liability lawsuits.

It should be noted that visual evidence of modernization in the form of cosmetic improvements to car enclosures and entrances is essential to eliminate tenants' negative conception: "All that inconvenience was for nothing. These are the same old elevators."

Although it is difficult to place a dollar value on these improvements, they clearly are a major source of better tenant relations, which subsequently result in higher occupancy levels.

VERTICAL TRANSPORTATION ARRANGEMENTS

Variations in Configuration

As discussed in earlier chapters, there are no set configurations for vertical transportation systems. Each system must be tailored to the projected needs of the building. Choices must be made between single full rise systems, high rise/low rise systems, and shuttle elevator systems serving parking garages or exclusive penthouse levels. Extremely tall buildings that are intensely populated may require consideration of double-deck elevator systems. Escalators must also be factored in for visible very high trafficked levels or in conjunction with double-deck elevator systems.

The need for free interfloor access by tenants may require designers to consider sky lobbies connecting low rise systems to mid or high rise systems.

Full Rise vs. Low Rise, High Rise Systems

In a full rise system all elevators in a group serve all levels of the building. There may be special exceptions to full service, such as intermediate mechanical floors that may be served by one or two cars in the group through secure access by means of key-operated switches or magnetic swipe card readers. Other exceptions may be encountered in specialized facilities such as educational or other institutional buildings where security needs may override free access to levels containing libraries, records, or administrative functions.

There are no hard-and-fast rules establishing the limits of travel for a single full rise elevator system. The traffic study will establish those limitations for each individual building. Generally, the upper limits for a single full rise system fall between 15 and 20 floors. Average intervals and handling capacities of single full rise systems beyond these limitations usually do not meet the agreed-upon performance criteria for the facility under study.

Deficiencies in the average interval and handling capacity projected by the traffic study can occasionally be overcome by increasing the number of elevators and/or the rated speed. Obviously, these solutions have their own practical limitations. If the traffic study for a 22-stop office building indicates that an eight-car group with a rated speed of 700 fpm (3.5 mps) will fulfill the criteria requirements, consideration should also be given to a low rise, high rise option.

Under the single full rise option, the eight-car system will require 176 entrances. A low rise system serving floors 1 through 12 coupled with a high rise system serving the first floor, skipping floors 2 through 12, and then serving floors 13 through 22, will require only 92 entrances. The cost of the installation will be reduced by 84 entrances. Future maintenance costs will also be reduced because of the reduction in the number of entrances.

The 4 low rise cars will operate efficiently on a lower speed, perhaps 500 fpm (2.5 mps). Installation costs will again be reduced because of the lower speed of the four low rise elevators. The cost of the initial installation of the power supply will be reduced by the reduction in horsepower requirements for the lower-speed hoisting machines. Future operational costs will also be less because of the reduced power consumption by the lower-speed hoist motors.

Assuming that both the low rise and high rise groups have a rated capacity of 3500 pounds (1600 kg), the eight-car full rise group will require 7400^2 ft (725 m^2) of core space more than required for the low rise/high rise system of four elevators each. If the low rise/high rise solution is accepted, it will enable the building designer to increase the available rentable space on each upper floor or to obtain a better ratio of usable to nonusable space (see Figure 18.1).

The single full rise eight-car group will present one advantage and one disadvantage to the passengers using the system. With eight cars available, the waiting interval for passengers will be relatively short. Conversely, once in an elevator, passengers may experience a long destination time as their trip will be interrupted more often, because the probable stop curve is greatly increased over the full 22-floor rise.

Building management can realize several advantages from a single full rise group. During periods of nonpeak traffic, one car may be removed from group operation for maintenance or for building service traffic without detriment to the efficient operation of the remaining seven cars. The elevator core space remains in the same location throughout the building. This may provide the building designer with an easier solution to the location of other building utilities. The full rise group requires only one machine room located at the top of the building, as opposed to the introduction of an intermediate mechanical floor or the loss of rentable space on one or two floors for the machine room requirements of a low-rise group. With the microprocessor group operational systems now available, the complexity of an eight-car operational system as opposed to a four-car system no longer poses a disadvantage to the building management.

The low rise/high rise configuration of four cars each also presents advantages and disadvantages for the passenger. Although the number of available elevators in each group is reduced by 50%, the increase in waiting interval is generally is more than offset by the reduction in travel time, owing to a much lower probable stop curve. An added advantage is that the cars will be less crowded because of the separation of low rise and high rise traffic.

Although traffic movement is enhanced and installation and operational costs are reduced by the low rise, high rise configuration, there are building functional disadvantages. The operational efficiency of either four-car group is more likely to be adversely affected by the removal of one car from group operation for maintenance or building service traffic. Flexibility in tenant location and tenant space expansion can become restricted to the travel zone in which the tenant is initially located, whereas with the full rise group the entire building becomes accessible. The low rise/high rise configuration opens additional rental areas on the upper floors above the low-rise hoistways. This can be a major advantage to building management, as these upper floors often provide a higher rental per square foot.

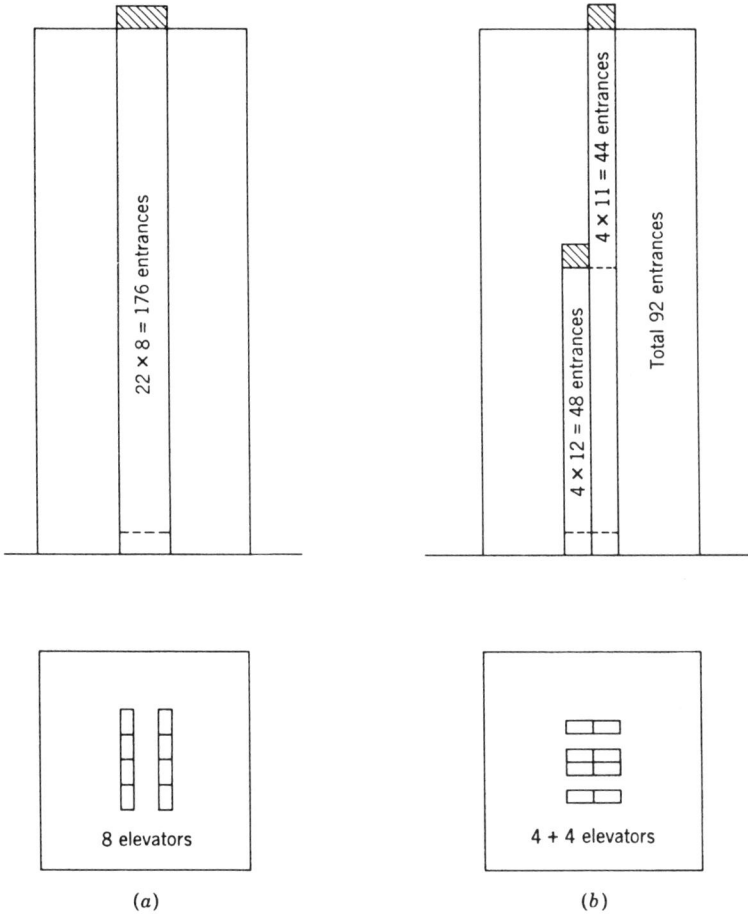

Figure 18.1. Eight elevators serving all floors versus four low rise and four high rise elevators. Approximate space required by the elevators and elevator lobbies: (*a*) 30,000 ft^2 (2800 m^2); (*b*) 22,600 ft^2 (2075 m^2).

A similar comparative analysis can be applied to taller buildings. Figure 18.2 shows two potential arrangements for a 30- to 40-floor building.

Based on similarly configured elevators, the arrangement of three groups of 6 elevators each, for a total of 18 elevators, may prove more economical in both initial installation and future operational costs than two groups of 8 elevators each, a total of 16 elevators. As previously discussed, the initial cost savings are generated by the reduced number of openings, the reduced speed of the lower-rise elevators, and the recovery of rentable space on the upper floors. Reduced structural, electrical, and mechanical loads imposed on the building by the elevators may also result in a reduction in the cost of these necessary building support functions. Operational cost reductions are achieved through lower maintenance costs and reduced power consumption. Depending on projected building population profiles and the potential need for interfloor travel between the low, mid, and high rise zones, it may be necessary to overlap elevator travels to provide sky lobbies for transfer between the various vertical zones.

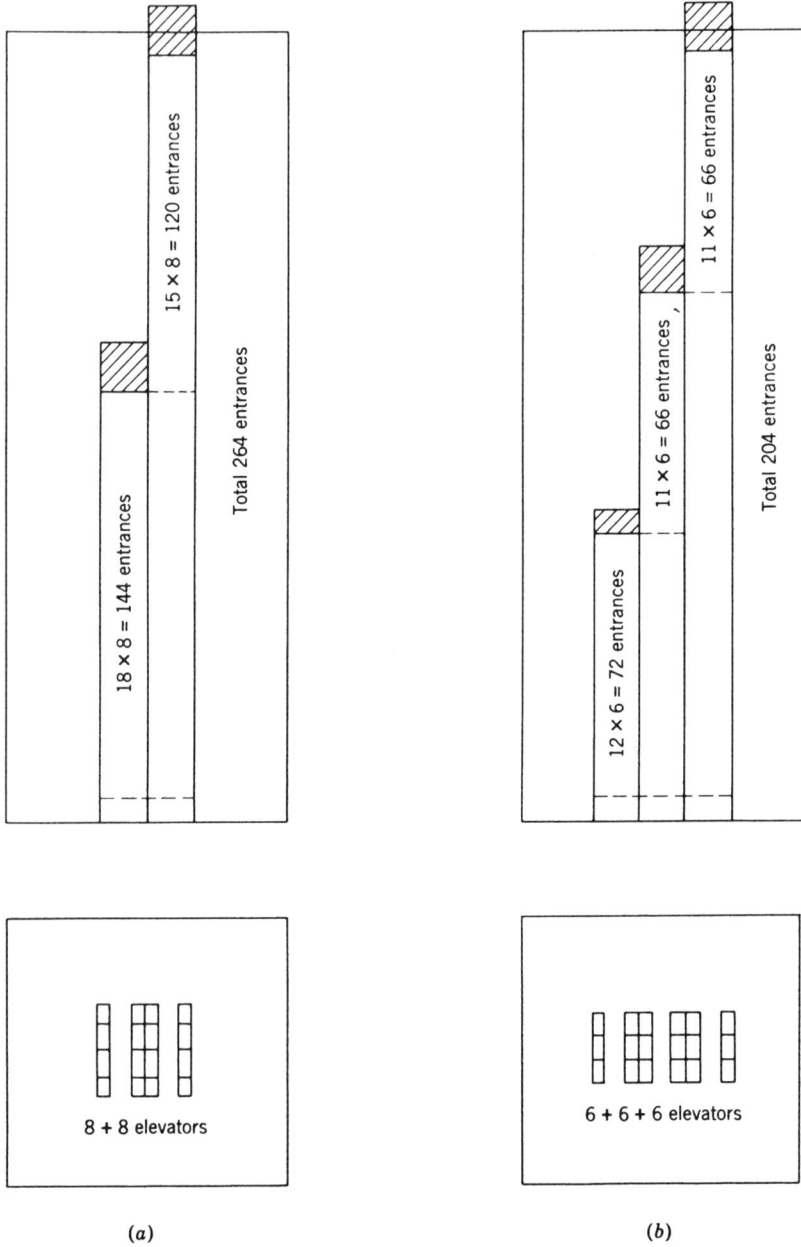

Figure 18.2. Two groups of eight elevators versus three groups of six elevators for the same building. Approximate space required by elevators and elevator lobbies; (a) 60,000 ft² (5580 m²); (b) 58,000 ft² (5400 m²).

These are not the only potential scenarios that bear consideration. Sky lobbies may be brought into play in a somewhat different manner (see Figure 18.3). Instead of low, mid, and high rise groups, three groups of local elevators would serve the three vertical zones of the building. The first would travel from the ground level to the first sky lobby at the

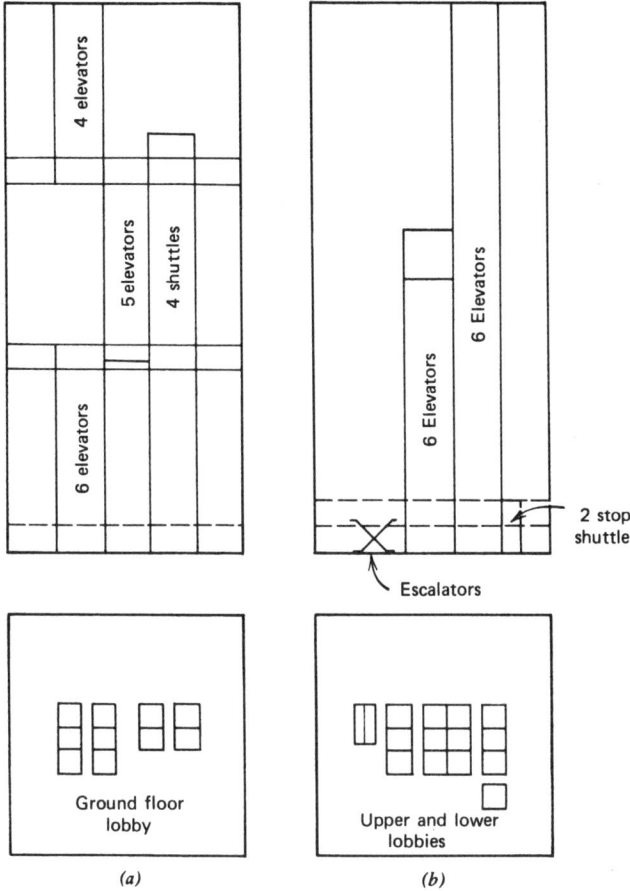

Figure 18.3. (*a*) Sky lobby arrangement with shuttle elevators to upper stops and local elevators from the two sky lobbies serving upper floors. (*b*) Double-deck elevators.

12th floor. The second would travel from the first sky lobby to the second sky lobby at the 23rd floor. The third would complete the travel from the second sky lobby to the top landing of the building. To avoid the inconvenience and delays that would be encountered by the mid- and upper-zone tenants in transferring between the three elevator groups for access to and egress from the building, shuttle elevators would directly connect the 12th and 23rd floors with the street level. Assuming that there is the same population mix as proposed in the original schemes, the low-rise group would still require six elevators. By eliminating the express runs and their time penalties for the mid and high rise groups, the mid-rise group may be reduced to five elevators and the high rise group may be reduced to four elevators. The shuttle group could potentially include 4 elevators. Although the number of elevators increases to 19 elevators, as opposed to the 18 and 16 employed in the previous schemes, the actual cost of the installation could be reduced considerably because of the lower-speed equipment required in the mid and high rise groups. Although rentable space is recovered in the low rise section of the building, an offsetting loss of rentable space must be considered owing to the introduction of the sky lobbies. Studies of this nature must thoroughly compare all potential scenarios not only for system performance, but for installation and future operating costs as well, and must also include a study of the potential rentable space pluses and minuses.

Another alternative scenario is the introduction of double-deck elevators into the design mix. The use of double-deck units may present as many design problems as it solves. Although a potential reduction in elevators to two groups of 6 elevators each is possible for our theoretical building, the cost of double-deck elevators is considerably higher than that of conventional single-deck units. The recovery of rentable space is obvious, as the 12 double-deck units require approximately one-third of the floor area required by more conventional units. However, the rentable space recovery must be weighed against the need for a double-level main lobby that will require escalators to make both levels accessible for access to and egress from the building. Furthermore, current regulations of the Americans with Disabilities Act (ADA) will require the introduction of shuttle elevators serving the double lobby. The building designers must also be aware that all floor heights throughout the building must be identical in order to make use of double-deck elevators. This may present intolerable design constraints.

Shuttle Elevators

The need for shuttle elevators can be generated by a number of circumstances. As noted in the preceding discussion on double-deck elevators, shuttle elevators would be required to provide handicapped access between the two levels of the main lobby. Parking garage facilities on the underground levels of the building may also generate the need for shuttle elevators should security concerns require that all elevator traffic from the parking levels funnel through the main lobby levels rather than having direct access to the upper floors. Certain tenants occupying two or more consecutive floors may require an internal shuttle elevator system dedicated solely to their use. High-rent penthouse units may also be best served by a private shuttle system from the top landing of the general-use elevators.

Where shuttle elevators are required, hydraulic elevators are generally the most practical units. Their location within the building will dictate whether these units are direct piston, holeless, or roped hydraulic elevators.

Escalators

The primary function of escalators is to move large volumes of pedestrian traffic between floors in the shortest time possible. Escalators offer the using public several advantages. They are always available. If traffic does queue up for boarding, delays are counted in seconds and may seem insignificant, since the queuing passenger can always see movement. In department stores and shopping malls, they allow passengers to survey the areas around them and get their bearings. In like manner, escalators permit management to display featured items or special programs where more people will see them.

Because escalator use is limited to pedestrian traffic, people in wheelchairs, parents with strollers, or workers with handcarts must find alternative means of transportation between floors. Escalators occupy large areas of space that might otherwise provide income for building management. Since it is estimated that about 10% of the potential passengers are either unable or unwilling to use escalators, prominent elevator service should be provided as an adjunct to an escalator location.

In general, traffic volumes in excess of 2000 passengers per hour of operation must be anticipated in order to justify the use of escalators. However, there are exceptions to any rule-of-thumb statement. In the case of the double-deck elevator lobbies previously discussed, escalators provide an essential link between the lobby levels even if traffic volumes are not that heavy. Escalators also serve well when they are used to funnel segments of traffic to specific destinations. Hospitals provide prime examples of this type of use, in which escalators move incoming outpatients and clinical visitors directly to their destinations without involving them in the main vertical transportation system of the hospital. Hotel casinos also use escalators to direct incoming casino patrons to the proper floors, reducing the possibility of non-hotel guests finding their way to residential floors. Escalators also serve well in hotels with convention centers by guiding outside visitors directly to the convention halls without involving them in the normal hotel guest traffic.

EQUIPMENT SELECTION

The basic selection of either geared, gearless, or hydraulic units is predicated on the specific needs developed by the traffic studies. Beyond these basic choices, the designer is faced with a choice between "pre-engineered" units and custom units.

Pre-engineered Elevators

Pre-engineered elevators are available from most manufacturers. Pre-engineering offers quality elevators at reduced cost as compared with custom-designed elevators. These units are designed to be manufactured for stock. They consist of components sized to match standardized hoistway dimensions, door types and sizes, cab heights and design, operation, and signal fixtures. The designer will find that a few options do exist for car enclosure finishes.

By standardizing layout and ordering procedures, manufacturing and installation times are reduced. The ultimate customer benefits from these costs savings as well as the economics obtained from quick availability and jobs shipped in total without delaying shortages. Models offered by the various manufacturers vary greatly. Furthermore, offerings vary from year to year. Intensive comparison of the manufacturers' most current brochures is an essential step in equipment selection.

The purchaser must match the needs of the building as closely as possible to the relatively limited choices available in selecting pre-engineered units. Because these units are often purchased through a catalog, the architect, contractor, or owner may select a unit that is visually appealing or has the lowest cost without regard to the needs of the building. Securing the services of an experienced elevator sales representative or an experienced consultant during the selection process will ensure the purchase of equipment that most closely meets the needs of the building.

Custom-design Elevators

Facilities such as hospitals, high-rise office buildings, special-purpose buildings, industrial complexes, and the like, have such unique needs that custom-design elevators are required. Special platform sizes, oversized entrances, front and rear openings, corner post requirements, special operational features, two-compartment car enclosures (prisoner elevators), and so forth, are not uncommon in these buildings. Any one or any combination of these special features will require the use of custom-design elevators.

It is essential that the vertical transportation consultant prepare clear, concise, and well-detailed specifications to provide the client with a vertical transportation system that exactly meets the unique needs of the individual facility. To accomplish this, the consultant must be confer at length with the architect and representatives of the owner to establish all of the special vertical transportation requirements. Where special entrance and cab details and/or signal fixtures are required, the consultant and the architect should work together to ensure that the ultimate design does not violate the code, and that it can be manufactured at the most reasonable cost.

There are no clear-cut guidelines to assist with the choice between pre-engineered and custom-designed vertical transportation systems. Each project has its own personality and must be judged on that basis.

ELEVATOR MAINTENANCE AND REPAIR

Although the vertical transportation equipment available today is well designed and manufactured in a quality manner, once in service, like any other electromechanical device, it requires periodic maintenance. A systematic maintenance program is essential if the equipment is to continue to perform at peak efficiency. As the equipment ages, even with quality maintenance, wear and tear will take its toll, requiring parts replacement. The maintenance program must be designed to meet the specific needs of the facility. Where heavy continuous traffic is encountered, the program should call for weekly maintenance. Where use is moderate, semimonthly examinations are usually sufficient. In lightly loaded buildings, a monthly program may suffice.

Replacement of small parts such as contacts and brushes should be accomplished as needed during systematic examinations. Major replacements, such as wire rope renewal, must be scheduled for downtime and can occur every eight or ten years, depending on the design, use, atmospheric conditions, and similar factors. Major component failures, such as motor or generator burnouts, bearing failures, or other large-item breakdowns, must be corrected when they occur.

The industry offers a variety of maintenance programs to provide for the care and replacement of equipment. These services range from full maintenance, which is the most desirable and most costly, through several lesser forms of service.

Full Maintenance

Except for the most sophisticated installation where the operational and motion control systems are proprietary, full maintenance is available, not only from the manufacturers, but from a wide range of independent contractors. Full-maintenance agreements should effectively place the responsibility for continued, efficient operation of the elevator system in the hands of the maintenance contractor.

Under the typical industry full-maintenance contract agreement, the contractor periodically (weekly, semimonthly, etc.) provides an examiner who will check all safety and operational features. The examiner will adjust door operation, leveling, acceleration and deceleration, and so on, as necessary. In addition, during the course of these routine examinations damaged or worn small parts such as contacts, brushes, signal lamps, and so forth, will be replaced. Lubrication will be provided as required. Major components will be examined, and a condition report, if warranted, will be forwarded to the contractor's office for further action. Generally, a full-maintenance contract includes the care and replacement of all components of the elevator system with the exception of the car enclosure, car floor covering, and car lighting.

Full-maintenance contracts are written for an extended period of time, generally five years. The agreed-upon monthly charge is adjusted annually in accordance with the metals and labor indices. The monthly charge may also include overtime callback options if the building operation warrants the additional expense.

Lesser Forms of Service

A wide variety of lesser forms of service are available. These range from contracts that cover specific parts to those that include only an examination and lubrication. These lessor forms of service normally are applicable only to older, less sophisticated elevators. The owners must judge for themselves whether the savings in cost compensate for the additional risk assumed with a lesser form of service.

Owner's Responsibilities

The act of signing a full-maintenance contract does not relieve the building owner or manager of all responsibility.

Before accepting the proffered contract, the owner's representative must be aware not only of what is included but of what is not mentioned in the contract language. Any component that is not specifically included in the list of what is covered is automatically excluded from coverage. For instance, if motor armatures are not included in the list, a motor failure would require the owner to assume the substantial repair or replacement cost.

Unfortunately, not all companies view a full-maintenance contract as a partnership with the building operator to provide the best possible elevator operation. There are a few that regard elevator maintenance as means to get as high a return as possible until major repairs become imminent. As the time of these repairs approaches, the contractor either cancels the contract or reduces service levels to the extent that the building operator cancels the contract. Another tactic used by some contractors is to declare the equipment obsolete and unserviceable and then approach the building operator with a costly proposal for replacement with modern equipment.

The building operator has a responsibility to monitor the ongoing maintenance program. Logs of callbacks and breakdowns must be kept and periodically reviewed with the contractor. Means for the examiner or callback mechanic to check in and out must be established. At least one person from the building operator's staff should be assigned to check the operation of the elevators on a regular basis. Should repeated callbacks occur, the building operator's representative should meet on-site with the contractor's superintendent to discuss and resolve the problems.

Despite the building operator's best efforts, it sometimes becomes necessary to change maintenance contractors. If the equipment was allowed to deteriorate extensively, this can, at worst, involve a costly premaintenance repair proposal from the new company or, at the least, a list of prorated components such as hoist ropes, motors, motor generators, and the like. The monthly cost of a prorated contract may compare favorably with the contract being canceled. However, as the prorated components reach the point of replacement, the building operator will be required to share in the possibly inflated cost of replacement.

Guidelines for Good Maintenance Contracting

There are no easy answers to good elevator maintenance, but there are a few guidelines to sensible contracting. The building operator should deal with a company of good reputation, well known in the area, and which has a substantial financial rating.

For smaller, less complex installations the industry standard maintenance contract will generally suffice. Specific terms such as overtime and/or weekend callback service and required periodic safety testing must be discussed and included or deleted, depending on the specific needs of the building.

For more complex installations, the industry standard contract may not be as inclusive as may be required. It may be desirable to include as part of the contract terms items such as the number of hours that a mechanic is to spend on-site, specific periods when a helper is required, how and where replacement parts are to be stocked, specific hours as to when certain maintenance tasks can be performed, billing rates for work not covered, a performance review process, a process to follow in the event of disputes, replacement of an incompetent mechanic, credit for excessive equipment downtime for repair or replacement, credit for elevators taken out of service, credit for light elevator usage should the occupancy of the building be significantly reduced, the owner's right to obtain a second opinion or proposal should work beyond the contract be requested by the contractor, or a myriad of other requirements, which may include insurance, indemnification, on-site parking, security clearances, and so on.

All full-maintenance contracts should include the costs involved in performing safety and operational tests as required by the local jurisdiction having authority over elevator usage or by the owner's insurance company. They should not include the cost of making changes to or replacement of equipment either because of obsolescence or changes in laws or codes governing elevators. The elevator contractor does, however, have a responsibility to notify the building owners of new devices that would improve the operational safety and apprise them of the cost of such items.

PROPOSALS FOR ELEVATOR MAINTENANCE

Typical Industry-Generated Contract Coverage

The following two sections show examples of maintenance coverage. The italics highlight differences.

Extent of Coverage.   Under the terms and conditions of this agreement subsequently set forth we will maintain the elevator equipment described in this proposal, using skilled elevator personnel whom we directly employ and supervise. This proposal includes 24-hour emergency adjustment callback service only at no additional charge.

We will systematically and regular examine, adjust, lubricate as required, and, when we consider it necessary, repair or replace components including the following:

MACHINE	*Worm shaft bearings—thrust bearing—gear shaft bearings—* slack cable device—machine limit device—*regroove* drive sheave—gear oil—packing
MOTOR	Armature—commutator—field coils—interpole coils—brushes—brush holders—rotor—stator windings—slip rings—motor bearings—motor oil

MOTOR-GENERATOR	Armature—commutator—field coils—interpole coils—brushes—brush holders—rotor—stator windings—slip rings—motor bearings—motor oil
BRAKE	Linings—magnet coils—linkage-bushings—contact springs—plunger—dashpot—adjusting screws
CONTROLLER	Magnet coils—copper contacts—carbon contacts—shunts—springs—fuses—insulators—resistors—rectifiers—timing devices—solid-state devices—overload heaters
SELECTOR	Sheaves—drive chains—cables and tapes—contacts—floor bars—springs—shunts—reduction gears—selector carriage—advance motors—coils
WIRING	Traveling control cables—hoistway control wiring
HOISTWAY	Hoistway limit switches—car limit switches—speed governor—bearing boxes—governor tension sheave—interlocks—door rollers—buffers—*counterweight guide shoes*—bottom door guides
CAR	Guide shoes—guide shoe linings—guide shoe rollers—leveling units—car door operator motor—belts—gears—cables—sheaves—gate switch—safety edge—photoelectric devices
ROPES	*On a pro-rata basis:* hoist—drum counterweight—car counterweight—operating—governor—car safety rope
SIGNALS	Pushbuttons—contacts—indicator lights—hall lanterns

Extent of Coverage. Under the terms and conditions of this agreement subsequently set forth we will maintain the elevator equipment described in this proposal, using skilled elevator personnel whom we directly employ and supervise. This proposal includes 24-hour emergency adjustment callback service only at no additional cost.

We will systematically and regularly examine, adjust, lubricate as required, and, when we consider it necessary, repair or replace components including the following:

MACHINE	*Worm gear—worm shaft*—worn shaft bearings—thrust bearing—gear shaft bearings—slack cable device—machine limit device—drive sheave—gear oil—packing
MOTOR	Armature—commutator—field coils—interpole coils—brushes—brush holders—rotor—stator windings—slip rings—motor bearings—motor oil
MOTOR-GENERATOR	Armature—commutator—field coils—interpole coils—brushes—brush holders—rotor—stator windings—slip rings—motor bearings—motor oil
BRAKE	Linings—magnet coils—linkage—bushings —contacts—springs—plunger—dashpot—adjusting screws
CONTROLLER	Magnet coils—copper contacts—carbon contacts—shunts—springs—fuses—insulators—resistors—rectifiers—timing devices—solid-state devices—overload heaters
SELECTOR	Sheaves—drive chains—cables and tapes—contacts—floor bars—springs—shunts—reduction gears—selector carriage—advance motors—coils

WIRING	Traveling control cables—hoistway control wiring
HOISTWAY	Hoistway limit switches—car limit switches—speed governor—bearing boxes—governor tension sheave—interlocks—*door hangers*—door rollers—*buffers—counterweight—guide shoes*—bottom door guides.
CAR	Guide shoes—guide shoe linings—guide shoe rollers—leveling units—car door operator motor—belts—gears—cables—sheaves—gate switch—safety edge—photoelectric devices
ROPES	Hoist—drum counterweight—car counterweight—operating—governor—car safety ropes
SIGNALS	Pushbuttons—contacts—indicator lights—hall lanterns

Owner-Generated Contract Coverage

It is recommended that the following items be considered for inclusion in any owner-generated maintenance contract coverage. Only the topics are listed; specific details must be developed to suit the needs of the owner and/or the facility.

1. Number of hours of work to be performed per week or month by a journeyman mechanic for preventive maintenance (Trouble calls and overtime callbacks should not be considered as part of this time allowance.)
2. Hours of work
3. Job material inventory on-site
4. Spare parts inventory off-site
5. Parts storage on-site
6. Space provided by owner for storage of supplies
7. Contractor's responsibilities
8. Code requirements
9. Job cooperation with owner's personnel and other on-site contractors
10. Monitoring of work by owner
11. Extent of coverage
12. Performance evaluation
13. Nonperformance evaluation and cure
14. Safety tests
15. Periodic traffic analysis
16. Adjustments
17. Credits
18. Exclusions
19. Conditions of service
20. Wiring diagrams
21. Obsolete equipment
22. Assignment of contract
23. Cancellation
24. Additional work

25. Reports
26. Responsibility and liability
27. Invoicing
28. Prorations
29. Premaintenance repairs
30. Insurance coverage
31. Automobile liability
32. Performance bond
33. Contractor's qualifications
34. Price adjustments

ELEVATOR MODERNIZATION

Although progress in design has continuously been evident from the turn of the century through the end of the 1970s, very little from that era can be compared to the virtual explosion in design innovations that have occurred in the 1980s and 1990s. The switch from relay logic to microprocessor operational and motion control alone has revolutionized the entire industry. Subsequent to these developments, the industry has produced new concepts in drive systems, new door protective devices, and new signal fixture systems. An awareness of the needs of the physically challenged members of society has produced a host of improvements to all passenger-use devices.

This rapid progress in elevator component design, as well as changing tastes in architectural treatment, has led to the complete modernization of elevators in the 15- to 20-year age range. Systems 30 years of age or older are often undergoing their second major modernization.

Why Modernize?

There are several obvious reasons to consider modernization. A primary consideration would be a continuous record of callbacks that are the direct result of aging equipment. An operational system that can no longer cope with the traffic demands of the building would also bring building management to the conclusion that modernization is required. A more subtle reason may be an effort on the part of management to reduce energy costs through a more efficient drive system. Operational "gimmicks" that claim amazing improvements in system efficiency clearly require thorough investigation, including visits to several existing installations and interviews with the building operators.

When a building undergoes a total renovation, modernization of the vertical transportation system generally becomes an integral part of the renovation project. Under these conditions, it is not unusual for new car enclosures, partial or complete new entrances, and, possibly, new hoisting machines to be included in the modernization of the elevators (see Figure 18.4a–d).

Benefits

In the 1950s and early 1960s elevator modernization primarily meant the replacement of operator-controlled car switch elevators with relay-logic group automatic operational systems.

Figure 18.4. Before and after photographs of modernized elevator equipment. (*a*) Gearless hoisting machine with auxiliary leveling machine and brake. (*b*) Auxiliary leveling machine replaced by a new main drive machine brake. (*c*) Old and new elevator controllers. (*d*) Old and new elevator safety governors. (Courtesy *Elevator World Magazine*.)

This change produced an immediate cost benefit to building management through a reduction in the number of employees, as elevator operators and lobby starters were no longer required.

Today the primary benefit that building management expects from an elevator modernization is a major reduction in elevator downtime. This is accomplished through greater reliability of the new components.

Figure 18.4. *(Continued)*

Replacement of a relay-logic operational system with a microprocessor operational system will also provide an increase in system efficiency and handling capacity in the range of 20 to 25%.

In the case of a building whose population has outgrown the capacity of the original elevator operating system, this increase in handling capacity will reduce employee travel time over the course of a normal workday, thus reducing the operating costs of the building occupants. Building management also benefits from these cost savings through better-satisfied tenants and a subsequent higher occupancy rate.

In those buildings where population has not outgrown the handling capacity of the original operating system, it may be possible to eliminate one or more elevators. Not only does this reduce operating costs for building management, but it will free up space for other critical building needs such as air-conditioning ducts, new electrical feeder systems, computer network communication lines, and the like. However, this is a venture that should not be entered into lightly. Extensive studies of elevator traffic, anticipated elevator system performance, and projections of building use and population must be performed and satisfactorily support the elimination of elevators before such decisions can be made. This approach can be very effective where the nature of the building changes from a high-density single-tenant occupancy to a widely diversified, low-density multiple-tenant occupancy.

In those cases in which the old motor generators are replaced with solid-state power converters, a slight reduction in maintenance costs may result because of the elimination of the motor generator rotating equipment. This replacement should not be done without a complete understanding of the potential complications that may develop from the use of solid-state power converters. In large medical centers and highly developed computer centers, past experience has shown that solid-state power converters may interfere with the proper operation of sensitive patient monitoring and medical diagnostic equipment or with mainframe computers. These problems can be caused by transients generated by the solid-state power converters that subsequently emanate from the existing elevator feeders, polluting the building power system. The feeder system should be examined by an electrical engineer to determine whether such line contamination is possible. If the feeders are suspect, replacement of the elevator feeders with new feeders properly isolated from the building feeder system should be included with the modernization program.

EQUIPMENT SURVEY

Once building management has determined that a modernization program of the vertical transportation system may be advantageous to the continued efficient operation of the building, an intensive survey of the physical condition and traffic-handling ability of the existing elevator system is absolutely necessary.

Equipment Evaluation

A complete survey of the condition of the vertical transportation system by qualified elevator personnel is the first step in developing a modernization program for building management. The survey must include an evaluation of all major components, such as hoisting machines, motors, power converters, controllers, floor selectors, overspeed governors, machine room wiring, counterweights, car frames and safety devices, car platforms, car enclosures, entrances, door-operating equipment, buffers, compensation, hoist ropes, all hoistway electrical protective devices, hoistway and car wiring, and signal fixtures. Nonelevator components that should be examined are main line disconnects and feeders, machine room ventilation, hoistway ventilation, machine room and hoistway construction, sprinkler system, and emergency power provisions.

A primary objective of any modernization program is to reuse as much of the existing elevator equipment as possible without sacrificing the efficiency and reliability of the modernized system. Therefore, the surveyor and/or the analyst who prepares the recommendations and report must be capable of establishing reasonable life expectancy limits for any planned retention of equipment. Subsequently, the specification writer must be able to develop specifications that will ensure compatibility of the new equipment with the retained existing equipment.

Maintenance Evaluation

An evaluation of the maintenance performance level should be performed simultaneously with the equipment survey. This survey will, in all likelihood, uncover several areas, such as car speed, uniform door function times, leveling accuracy and so forth, that no longer meet the design criteria of the original installation.

Many of the relay-logic group operating systems have set multiple programs that the systems automatically change to adapt to the varying load demands or because of specific times of the day. Past experience has shown that these multiple program systems often become "locked" into one program or another, thus defeating whatever flexibility was originally designed into the system. If this condition is found, the maintenance contractor should be required to find and correct the cause of such lockouts and return the system to full operational capabilities.

Quite often, readjustment of these items will produce a temporary upgrade in system performance prior to the start of the modernization program. Coincidentally, these improvements will assist with the problem of reduced handling capacity of the system as one car at a time is taken out of service for modernization.

Other Survey Essentials

An examination of the trouble logs maintained by building management, as well as available copies of callback time tickets issued prior to the surveys, can assist the surveyor in determining where to concentrate the inspection. Interviews with any building personnel connected with elevator operation are also helpful.

An on-site traffic study will establish existing traffic patterns. This will assist the analyst and spec writer in establishing the proper criteria to ensure that the modernized system will meet the present and projected needs of the building.

Evaluation Report

The evaluation report should be clear and concise, not "sold by the pound." It should clearly establish the recommendations in easily understood layman's language. If options are available, these should be listed in their proper priority. Budgets should be established. Items that are not generally included in the elevator contractor's work should be clearly delineated.

SPECIFICATIONS

A small or partial modernization project may require only an outline of the work to be done. If this is the case, proposals should be solicited only from a carefully selected list of reputable contractors. Once received, these proposals must be evaluated as to content as well as price. When the contractor is selected, not necessarily the lowest bid, a contract agreeable to both parties can be entered into.

Larger, more complex projects generally require a full complement of bidding documents including terms and conditions, work schedules, specific materials, and performance requirements. The work schedule should establish the number of elevators that can be removed from service for the specified work and the order in which they are to be removed from service. Work to be performed by other contractors, such as machine room or hoistway ventilation, new power feeders, security and emergency systems, and so on, should be outlined. When non-elevator work is minimal, it may be included as part of the elevator contractor's work.

Site conditions must be detailed. These should include material storage, removal of old materials and trash, when and where new materials can be delivered, safety barricades in corridors and hoistways between cars, conduct of the elevator contractor's employees, what facilities are available or "off limits," compliance with building security regulations, and any other condition that may affect the relationship between the elevator contractor and building management.

Specifications for elevator modernization require a more intense level of detail than comparable new elevator specifications. Equipment to be retained and overhauled or modified must completely spelled out. Building conditions must be thoroughly described. All items to be removed must be clearly delineated.

SAFETY AND LIABILITY

Passenger safety and building management liability are inseparable. If passenger safety problems are ignored, then building management can be faced with liability for injuries, lost time from work, and punitive damages. If an incident involving a personal injury reaches the court system, the attorney for the plaintiff will charge negligence on the part of both building management and the elevator maintenance contractor for any indication of repeated malfunctions that are not corrected after the first or second attempt.

When a pattern of repeated callbacks for leveling, door operation, cars in the pit or overhead begins to develop, building management must, in writing, require the maintenance contractor to thoroughly investigate the causes and take action for correction. If these problems cannot be satisfactorily corrected through normal maintenance procedures, a replacement upgrade program must be considered. The extent of such an upgrade may range from simply replacing a failing leveling system to a complete modernization.

Full responsibility for maintaining safe elevators does not lie completely with the maintenance contractor. Management has its responsibilities. Platform floor coverings that are worn or loose are generally the responsibility of building management, as is lighting in the cars and lobbies. Lobby carpets that become loose or torn, especially near elevator entrances, also are the obligation of building management to repair or replace.

In any event of this nature, a clear and concise paper trail of action requested and action taken may save management from an excessive settlement in a future liability case. Building management should maintain a log of trouble calls and require that callback as well as regular maintenance procedures be clearly spelled out on the time tickets prepared by the mechanic. If tickets are illegible or lack sufficient information, management should reject them and require them to be rewritten. A file of all correspondence between management and the maintenance contractor should be accurately maintained.

When new products are developed that can enhance the safe operation of an elevator system, the maintenance contractor has an implicit obligation to make building management aware of their availability and cost. Failure to do so could place the contractor in a poor light in any subsequent litigation involving an incident that may have been prevented by such equipment upgrades.

ROLE OF THE CONSULTANT

Through the 1950s and 1960s major elevator manufacturers provided design consulting services, generally at no cost to the client. As competition became more intense and costs rose, this free consulting service all but disappeared. As a result, a new segment of the industry began to develop. A few pioneering firms were already breaking ground in the vertical transportation consulting field before the end of the 1950s.

These few early firms set the standards in methods and ethics for the growing group of present-day consultants who are having an increasingly significant impact on the industry. The consultant today provides an essential link between architects, structural and electrical engineers, building management, code-enforcing authorities, and the elevator industry. To be successful, a consultant must be conversant in the "language" of each these disciplines. He or she can be called on to perform any or all of the following tasks:

- Conduct traffic studies of proposed buildings
- Participate in the entire design process, assisting architects, structural and electrical engineers
- Prepare specifications unique to each facility under design
- Obtain variances from local code authorities when necessary
- Participate in the bidding process
- Review shop drawing submittals
- Negotiate disputes between the elevator contractor and the owner
- Perform contractual compliance inspections
- Where code authorities do not exist, perform code testing and acceptance procedures

Consultants also provide valuable services directly to building management. Building management generally engages a consultant to evaluate a maintenance program only when it is perceived that the program is failing. In this role, the consultant is in a position to mediate the deteriorating management/contractor relationship through an objective evaluation of the condition of the elevators as well as the personal relationships that have developed between the two parties. The consultant can also assist building management in the development of its long-term capital expenditure costs through the evaluation of existing systems and the establishment of recommendations and budgets for modernization.

A growing role pursued by many consultants lies in the area of legal consulting. Such activities range from review of contractual conditions to expert witness testimony. These services can be provided to building management, to member firms of the industry, and to plaintiffs in injury cases.

SUMMARY

Once vertical transportation systems are installed, a complete maintenance program is essential to keep them operating as close to the level of the initial installation as possible.

Efficient passenger service and high equipment reliability can then be ensured. As a result, both building management and passengers are benefited through trouble-free, safe vertical transportation.

Like any electromechanical device, vertical transportation systems cannot avoid obsolescence. New technology, improvements to existing components, changing public requirements, new architectural treatments, and, always, age play their parts on the inevitable path to obsolescence. Most buildings have an economic life of at least 60 years. Although the primary components, machines, rails and structural components of a vertical transportation system can easily last as long as the building, operational and motion control systems, signal systems, door-operating equipment, entrance panels, and car enclosures will require replacement two to three times during the life of the building. With each proper modernization program, an essentially brand new vertical transportation system will result.

Although elevators are an extremely safe form of transportation, accidents can occur. Proper attention to maintenance, operation, and timely repair is crucial to the efforts to minimize the possibility of accidents, maintain an excellent safety record, and reduce the cost of liability insurance.

ABOUT THE AUTHOR

LEONARD G. LEVEE, JR. has been associated with the elevator industry in various capacities for the past 50 years. The first 23 years were spent with the General Elevator Co. of Baltimore in the manufacturing, engineering, sales, and branch management phases of the industry. The next 11 years were spent in the Office of Construction of the Veterans Administration as chief of the Automatic Transport Systems Division. During these years, Mr. LeVee was responsible for everything that moved vertically and horizontally in the 172-hospital system.

In 1982, Mr. LeVee began his career as a consultant with Lerch Bates & Associates. In 1987, he became a partner in the firm of International Elevator Consultants, Ltd. of Detroit, Michigan, before founding the firm of LeVee & Associates, Inc., Vertical Transportation Consultants in 1990.

Mr. LeVee is active in the National Association of Elevator Safety Authorities (he holds certified elevator inspector certificate #425); the National Association of Vertical Transportation Professionals, where he is executive director; and the National Association of Elevator Contractors, where he acts as the liaison between the professional group NAVTP and the Board of Directors.

Mr. LeVee is currently a member of the National Interest Review Committee of the ASME A17.1 Safety Code for Elevators and Escalators. Past code activities have included membership on four A17.1 subcommittees, the ASME B20.1 Safety Standards for Conveyors, and the Code Making Panel 12 of the National Electrical Code.

Mr. LeVee has had articles and essays published in *Elevator World Magazine, Mainline,* the newsletter of the National Association of Elevator Contractors, *China Elevator* trade magazine, and is a regular contributor to the NAVTP newsletter.

19 Traffic Studies and Performance Evaluation

LEN LEVEE

ROLE OF THE CONSULTANT

As the cost of labor continues to be an ever increasing concern to those responsible for the "bottom line" of any commercial, health care, or governmental facility, time spent by employees in travel within the building becomes a significant, although ill-defined, cost factor. Excessive elevator waiting times, malfunctioning equipment, and/or extensive downtime of individual elevators eat away at the productive levels of employee time. Although difficult to actually quantify, time lost by those employees who must use the vertical transportation system in the performance of their duties can become a major source of tenant dissatisfaction.

Building management seldom has the resources in-house to satisfactorily determine the causes of the deteriorating efficiency of a vertical transportation system. Outside help becomes an essential management tool to determine what these problems are and to provide the most effective solution, along with establishing the necessary budgetary information to assist building management with its long-term capital expenditure program.

An experienced vertical transportation consultant should be engaged to provide an independent, unbiased evaluation of the traffic generated by the building occupants, the performance of the equipment, and the level of maintenance provided by the elevator contractor. All three of these factors must be in place, and properly related to one another, in order to obtain a complete vertical transportation system evaluation. Once the program for the upgrading of either the level of maintenance or the equipment itself is established, the consultant can continue to assist building management by providing bidding specifications and construction oversight services until the established program is complete.

ELEVATOR PERFORMANCE VS. BUILDING PERSONALITY

Although any two given buildings may appear to be similar in area and number of floors, each will develop its own definitive personality. The tenant profile is the major factor in the development of the personality of any building. The building on the left may have either a single tenant or just a few major tenants, each occupying several floors. The building on the right may be occupied by many different small firms on each floor. The vertical

The Vertical Transportation Handbook, Third Edition, Edited by George R. Strakosch
ISBN 0-471-16291-4   © 1998 John Wiley & Sons, Inc.

transportation traffic developed in each of the buildings will be significantly different. The building on the left will produce a demand for heavy two-day interfloor traffic, whereas the building comprised of many small tenants will see little or no interfloor traffic.

If we assume that each of these buildings is served by a four-car relay-logic group automatic operational system with the equipment equally adjusted and maintained, the system with the heavy two-way interfloor traffic demand will provide considerably lower response time than the system with little or no interfloor traffic. The probable stop curve increases significantly where heavy two-way interfloor traffic is the prime mover, thus extending the average round-trip time for each elevator in the system.

Therefore, it is essential to understand that no vertical transportation system can be judged solely on the square foot area of the building and the number of floors served. The total building and its occupant mix must be thoroughly studied if an accurate evaluation of the functioning of the vertical transportation system is to be obtained.

PUBLIC CONCEPTS OF ELEVATOR USAGE

Although criteria have been developed by which individual elevator and system performance can be judged, there are no hard-and-fast rules for judging performance that can be uniformly applied to all vertical transportation systems. Each vertical transportation system should be evaluated in conjunction with an analysis of the specific traffic demands generated by the population of the building. As noted earlier, each building develops a distinct personality.

Management's judgment of a system's performance is often by default. If there are few or no complaints by the tenants, the service is considered good. Should tenant complaints begin to multiply, service is perceived to be poor. As the list of complaints grows, management is quick to call the maintenance contractor into account.

Service complaints can be initiated by a number of events. Excessive buildup of waiting passengers in the main entrance lobby at the start of the workday generates one of the more common complaints from tenants. Employees are prevented from arriving at their workstations on time and blame their tardiness on the elevators. Excessive waiting times on the upper floors during the course of the workday causes lost time for those employees who must use the elevator system as a normal function of their employment.

There is a segment of society that has a definite fear of elevators. In many cases this anxiety is caused by claustrophobia. Others may simply be uncomfortable riding in any device over which they feel they have little or no control. These anxieties can quickly turn to complete panic should these persons be caught in a stalled elevator.

Elevator manufacturers and servicing companies can do a great deal to alleviate these fears. Quietness of operation and smoothness of ride are very important factors. Nothing is more disconcerting for nervous passengers than riding in an elevator that rattles, bangs, and squeaks as it travels through the hoistway. Even a composed rider can become frightened should the elevator strike a door lock, which can result in a panic stop. Car position indicator lights that move in direct relation to the car movement can also be reassuring to the timid rider. Car position indicators that advance immediately to the floor where the car is programmed to stop can give the impression that the car is over speeding, especially when this occurs in the down direction. In an office building where passengers are often hurried or in a hospital where passengers' thoughts are elsewhere, excessive pre-opening of the car and landing doors can not only be disconcerting, but can present a serious tripping hazard.

Architectural features, simple, clear signage, and digitized voice announcements that provide floor information and travel direction also contribute to passenger comfort and safety. Well-lighted car interiors that are designed in keeping with the architectural style of the building do much to overcome the strangeness encountered by many passengers when they enter "that box." Visual, audible, and tactile signals must now comply with the requirements of the Americans with Disabilities Act (ADA) so that all passengers, regardless of their physical capabilities, can use the vertical transportation system with ease and comfort.

Double-deck elevators present their own set of unique concerns. Clear signage in the main entrance lobby that directs incoming passengers to the proper level for their desired floors is mandatory. Ease of access, generally by escalator, to the mezzanine floor that serves the upper decks of these cars is also essential. The car and landing doors should open on both decks, each time the car stops, to alleviate the discomfort of passengers on the deck other than that where the car is programmed to stop.

EVALUATING ELEVATOR PERFORMANCE COMPLAINTS

Complaints about poor elevator service generally fall into two distinct categories, complaints of record and perceived complaints. Complaints of record are those reported to building management personnel, who then physically observe the problem, record the problem in a log, and notify the elevator maintenance contractor. Perceived complaints are those heard while waiting an elevator lobby or riding in an elevator car. The most often expressed of these gripes are "I waited 10 minutes for this elevator!" and "These are the world's slowest (worst) elevators!"

The trouble log must be kept accurately and current. Vague notes with no specific description of an incident are of no help to either the servicing mechanic or the personnel who must evaluate the performance of the vertical transportation system.

Perceived complaints, however, do provide the evaluator with a profile of how the users perceive the day-to-day operations of the system. It is this profile, which may or may not present an accurate picture of the functioning of the elevator system, that can assist the evaluator in understanding the thrust of the tenant reaction to the vertical transportation system. This understanding can be a valuable additional tool that the evaluator uses to guide his or her studies.

EVALUATION OF EXISTING TRAFFIC CONDITIONS

There are several different methods of analyzing vertical transportation traffic. Lobby head counts at the major peak periods—at the start of the workday, at the noon hour, and at the close of the workday—are an essential part of any traffic study. The lobby counts should be supplemented by observation of the distribution of traffic, that is, where passengers board, where they exit the elevators, and what they are pushing or carrying. This is especially important in older medical facilities where a lack of dedicated systems inhibits traffic separation. Current technology provides several different electronic traffic analyzer devices that provide a wealth of information on the performance of any vertical transportation system. Depending on the intensity of the analysis required, any or all of these methods can be combined to complete the study.

Regardless of which methods of traffic analysis are implemented, it is essential that the study begin with interviews of the management team that is most concerned with traffic movement throughout the facility. A complete study must include not only the vertical movement, but an analysis of the horizontal traffic patterns in the facility.

How the incoming traffic is generated is the most logical starting point.

1. Is there an integral parking facility in the building? Does the parking facility transport system serve the entire building, or is there a shuttle elevator system that brings passengers from the garage directly to some security checkpoint in the main lobby?
2. If the main entrance is at grade level, is there a mass transit stop in the immediate vicinity that will cause large groups of incoming personnel to enter the lobby each time a bus empties?
3. Is there a major entry point at a level other than the grade entrance, such as a bridge or tunnel from another building or an elevated mass transit system station?

Other information essential to the study can be obtained from these interviews.

1. Do all personnel start and quit at the same time, or do they have staggered work hours?
2. Is there any type of security system affecting the elevators? If so, what are the hours the security system is in operation?
3. What is the occupancy makeup? Is this a single-tenant building? Are there just a few major tenants, each occupying several floors? Or do the tenants consist of many small unrelated organizations?

Once sufficient information has been obtained from building management, the physical count programs can be scheduled so that the maximum traffic counts can be gathered. The lobby count should be performed by at least one observer for every three to four elevators. These observers record the number of people exiting and entering each elevator as it arrives at the main lobby. The tallies should be segregated for each five-minute period of the scheduled survey time. While gathering these tallies, the observer should also make note of other events, such as waiting passengers backing up in the lobby; bunching of cars, that is, two or more elevators arriving or departing together or within seconds of each other; interference with normal elevator operation by passengers; any of the several overnight delivery services boarding the elevators with full handcarts; internal freight movement; and any other unusual traffic or operational occurrence.

Using information as shown in Figure 19.1, the results of the lobby count during the various periods of the study can be analyzed as follows.

Assuming that the period from 9:00 A.M. to 9.15 A.M. generated the heaviest traffic observed, the peak traffic load was approximately 75 people in 5 min. and the elevators had an average loading time of 49 sec. The loading interval must be averaged over a period of time, inasmuch as it is difficult for any observer to stop counting at the exact end of the given 5-min period and begin a simultaneous start of the next line. Average interval is calculated by dividing the total time period in seconds by the total number of trips. The peak traffic load of 75 persons must always be considered an approximate count, as a change of one carload more or less could have a significant impact. The brackets show

UP ELEVATOR TRAFFIC

NO. OF ELEVS.	HOUR	MINUTES	NUMBER OF PASSENGERS ENTERING AT 1ST FLOOR (BY CAR LOADS)										NO. PASS. ENTERING
4	9	00 – 05	6	10	11	14	10	6	16				75
		05 – 10	4	8	10	3	2						27
		10 – 15	2	10	10	7	14	16					59
		15 – 20											
4	12	20 – 25	8	4	5	7	8	9	6	2	10	12	72
		25 – 30	2	8	8	3	6	7	10	12			56
		30 – 35	10	12	5	8	7	2	4	2	0		50
		35 – 40											
4	1	40 – 45	3↑	1↑	4↑	5↑	2↑	3↑	1↑				15↑ 4↓
		45 – 50	8↑	9↑	2↑	0↑	0↑	1↑	3↑				21↑ 2↓
		50 – 55	6↓	10↑	2↑	10↑	2↑	0↑	1↑				23↑ 8↓
		55 – 60											
		TOTALS											

DOWN ELEVATOR TRAFFIC

NO. OF ELEVS.	HOUR	MINUTES	NUMBER OF PASSENGERS LEAVING ELEVATOR CAR										NO. PASS. LEAVING	TOTAL NO. PASS.
4	9	00 – 05	0	1	3	2	0	1	2				10	85
		05 – 10	2	3	0	1	0	0					6	33
		10 – 15	4	2	1	2	4	2					15	74
		15 – 20												
4	12	20 – 25	3	2	0	0	1	2	4	8	2		25	99
		25 – 30	10	11	12	10	0	0	2	9	3	8	62	118
		30 – 35	6	5	7	2	3	0	3	0	8		24	74
		35 – 40												
4	1	40 – 45	0↑	1↑	2↑	0↑	1↓	6↓	6↓	8↓	8↓		8↑ 8↓	23↑ 12↓
		45 – 50	0↑	0↑	0↑	1↓	2↓	6↓	8↓	3↓			8↑ 3↓	29↑ 5↓
		50 – 55	3↓	0↑	0↑	2↓	3↓	0↓	1↓	3↑	6↓		3↑ 6↓	26↑ 14↓
		55 – 60												
		TOTALS												

Figure 19.1. Tabulation of pedestrians on and off elevators at a lobby floor.

511

that three cars left within seconds of each other and that two cars arrived very closely together. Both of these observations clearly indicate that the elevator cars are "bunching." Frequent bunching is a clear indication that the dispatching system adjustment requires further investigation.

Assuming that the cars are rated at 3500 lb (1600 kg) capacity, the car that was dispatched with 16 passengers must be considered a full car. If that elevator did not leave the lobby promptly upon reaching full load, the adjustment of the load dispatch device requires readjustment.

Using the lobby head counts of 75 people traveling up and 10 people traveling down, a calculation can be made to determine the potential average interval, which can then be compared with the observed interval to assist in evaluating the efficiency of the system. If the actual occupancy of the building is known for the day of observation, the lobby head count of 75 people can be used to establish the percentage of handling capacity at which the system is performing. If the typical average daily occupancy is relatively low as compared with the actual building census, the handling capacity percentage can be adjusted to predict the performance of the elevator system on those days, such as payday, when all but a few of those included in the actual census total will be in the building.

Although peak traffic studies are essential to any traffic analysis, the picture is not complete without the inclusion of two-way traffic. Basing the two-way traffic analysis on the same population figures as those of the up-peak, the indications are clear that a substantially higher percentage of the building population uses the elevators during the noontime rush. This may indicate that morning starting times may be staggered. As previously noted, the information gathered in the opening interviews can be used to confirm these observations. The difference in interval times between up-peak and two-way traffic, when considered in conjunction with the greater numbers of passengers in two-way traffic, indicates that there may be other deficiencies during the up-peak operation or that a considerable amount of counterflow traffic may be occurring, creating a significant interfloor traffic condition during up-peak. This could create the possibility that some cars do not always return to the lobby. Additional investigation and traffic studies may be required to resolve these questions.

The tabulations bracketed as Section C in Figure 19.1 show the results of observations taken at an upper floor, such as a floor used for the transfer between groups of elevators, at a lobby floor where there is a basement(s) served by the group, or at an upper-level cafeteria floor. These results must be compared with counts obtained at other floors in order to develop a complete profile of the traffic. The final results may indicate the need to provide additional operational features that will ensure additional elevators at that floor to accommodate the traffic generated there. If the floor where these observations were taken is below a main-level lobby, it may be necessary to consider the installation of separate elevator service to these lower levels if the main system is at full capacity.

In single-, two-, or three-tenanted buildings, two-way interfloor traffic can have a significant impact on the efficiency of the elevator group operational system. Under these conditions, it is essential that the observers ride the elevators and survey the traffic. Collecting data on the distribution of traffic should be done in 15-min increments. Observers ride each car in the group during the same time periods. A tally sheet, such as that shown in Figure 19.2, is used to register the data. Each entering and exiting passenger is registered at the floor of entry to and/or exit from the car. The data thus collected allows the evaluator to visualize the traffic patterns. The data can also be used to provide an actual probable stop curve should it be necessary to generate a computer model of the traffic.

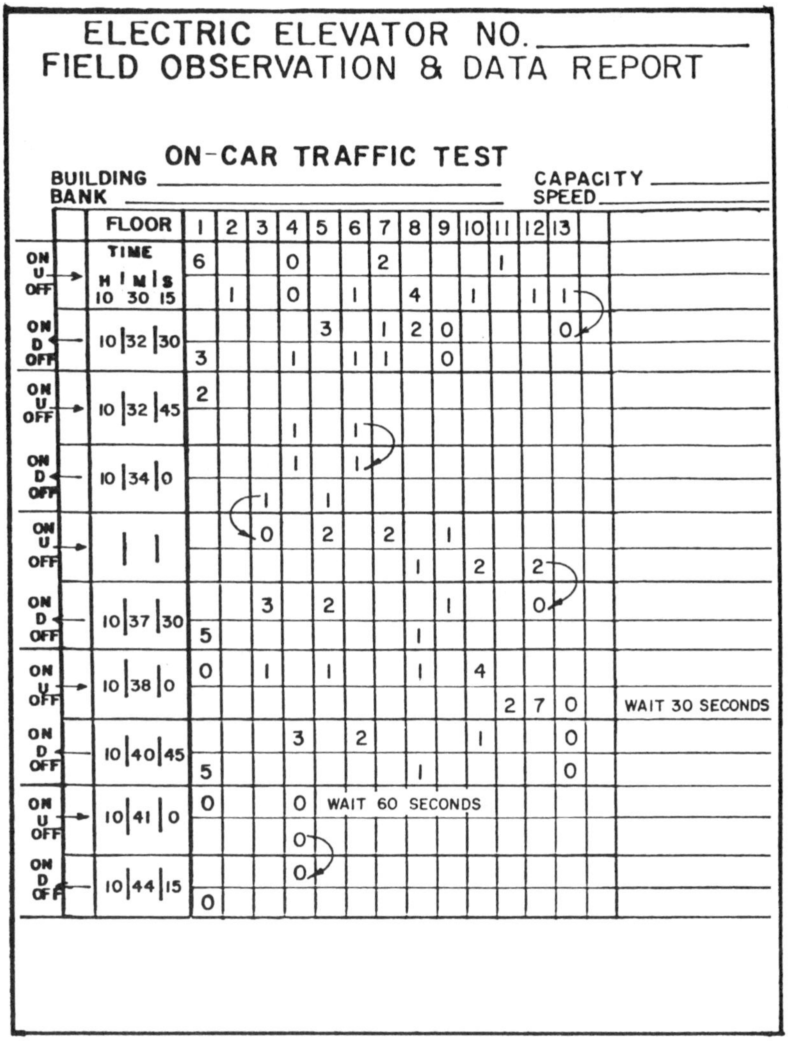

ELECTRIC ELEVATOR NO._____
FIELD OBSERVATION & DATA REPORT

ON-CAR TRAFFIC TEST

BUILDING _____ CAPACITY_____
BANK _____ SPEED_____

	FLOOR	1	2	3	4	5	6	7	8	9	10	11	12	13	
ON U OFF	TIME H M S 10 30 15	6	1	0	0	2	1	4	1		1	1	1		
ON D OFF	10 32 30	3			3 1	1	2	1	0 0			0			
ON U OFF	10 32 45	2				1	1								
ON D OFF	10 34 0				1	1	1								
ON U OFF				1	0	2	2	1		1	2	2			
ON D OFF	10 37 30	5			3	2			1 1			0			
ON U OFF	10 38 0	0	1		1			1	4		2	7	0		WAIT 30 SECONDS
ON D OFF	10 40 45	5			3	2		1	1			0 0			
ON U OFF	10 41 0	0		0 0	WAIT 60 SECONDS										
ON D OFF	10 44 15	0		0											

Figure 19.2. Tabulation of pedestrian movement on and off an elevator observed from within the elevator.

Supplemental periods of observation may be required in facilities with heavy interfloor traffic.

The form shown in Figure 19.2 has been marked to show some typical traffic solutions, along with some notations used to show what is actually happening at the given moment. As with the lobby counts, any unusual occurrence should be noted for possible further investigation.

Each observer should have a watch that indicates hours, minutes, and seconds. Sufficient copies of the form should be given to each observer so that he or she can complete the full 15-min period of observation. It may be necessary to schedule several 15-min periods of traffic observation in order to obtain a complete typical daily traffic profile.

Passengers are counted moving on and off the elevator in each direction at their respective floors. The time that an elevator leaves a terminal landing is noted. It is also important to note the elapsed time(s) that an elevator may park during the period of observation. All stops must be recorded. Zeros indicate that no one entered or exited the elevator. If the elevator reverses before it reaches either terminal landing, an arrow is used to indicate the change in direction.

Comparison of the surveys of an entire group of elevators over an extended period of time will assist in the determination of the extent of interfloor traffic. An evaluation of how well the system is performing in accordance with the original design specifications can also be made from the analysis of these surveys. With operating systems that were in vogue before the development of microprocessor-controlled operational systems, it is entirely possible that the cars may not always travel to a terminal landing unless there is a demand for them to do so. This operation favors interfloor traffic over the one-way peaks. These older systems may require the addition of time clocks to accommodate the heavier one-way peak traffic. The surveys may also contribute knowledge of other ineffective or nonfunctional features of the operating system.

The counting of actual on/off traffic within the cars is especially important in hospitals and hotels, where many functions essential to the operation of the facility may be located above or below the main floor. Current hospital design strategy aims, where possible, for a complete separation of pedestrian, patient, and service traffic. In older existing facilities, this separation is not always available. With heavy two-way interfloor pedestrian traffic and frequent movement of patients, carts, and equipment, in-car traffic surveys are essential to assist the hospital management in the proper scheduling of these movements to reduce the competition for elevator capacity to a minimum. Similarly, the service elevators in a hotel may require in-car traffic counts to resolve conflicts between the movements generated by the main kitchen, housekeeping, maintenance, and guest luggage transport.

Of equal importance in the analysis of the traffic-handling ability of any system is the length of time prospective passengers must wait for response to their landing calls. In an earlier chapter, the relationship between interval and the distribution of waiting times was shown. This distribution can be confirmed by measuring the actual waiting times during a busy traffic period and comparing them with a calculation of what the distribution of waiting time should be.

Several years ago the most common means of obtaining the actual waiting times was through the use of an event recorder connected to the hall call circuitry. The event recorder produced a paper tape that noted the duration of registration of each hall call at the floors being monitored. The value of the results of the event recorders could be limited if the number of floors in the building exceeded the number of channels the recorder could monitor. The resulting tape produced a graphic representation of each hall call registration, rather than the actual time in seconds. Therefore, the final results of such a study depended entirely on the accuracy of the tape reviewer. Although these results were not always "to the second" in accuracy, they did present a reasonable picture of the traffic conditions and system responses for a limited number of hours.

Present technology has made several types of portable computerized traffic analyzers available. These are especially useful in gathering traffic data over continuous periods of time up to seven consecutive days. These devices will record every hall call registered in both directions for a preset period of observation. The length of time each call is in registration is computed in seconds to two decimal places and is stored for later printout. The systems can also record such occurrences as the number of cars in service, the number of times a stop button is used, and so on. Some of the more so-phisticated devices also have trouble-shooting programs to assist in tracing those "you have to be there and see them" malfunctions. These devices provide a variety of for-mats in the final printout, as shown in the various parts of Figure 19.3. They are most useful in obtaining a concise summary of the hall call waiting times, broken down by floor and direction. The printouts also provide the distribution of calls in selected time segments, as well as the total number of calls by floor, by hour, and by day. Many of the current microprocessor operational systems have versions of these traffic analyzers built in. With these built-in systems, it is possible to observe what is actually happen-

DELTA 5049 - ELEVATOR TRAFFIC STUDY

BUILDING Name and Address
  Hurly Medical Center
  N. Washington Street
  Boston, Ma. 02111

SYSTEM Description
  Haughton 4-Car Group
STUDY Identification
  Passenger cars 1-4
STUDY 4 of 7 Daily
REPORT Status - PRINTED
DATA FILE status - OK
CIS Polarity - SAME

STUDY Commissioned By
  Associated Management Inc.
No. of Cars - 04
No. of Stops - 08
SCAN Date - 05/08/92
SCAN Time - 06:00:00   No. Int - 24
REPORT Date - 05/16/92
SCAN Interval - 60 Minutes
SYSTEM Return - NEGATIVE
INPUT Threshold - 40 Volts
TRANSFER Status - VERIFIED
PEN No. & Function - 10 LB UP

			- NUMBER OF CALLS ANSWERED IN SECONDS -							
TIME	SLOT	# CALLS	0-10	11-20	21-30	31-45	46-60	61-90	91+	L/CALL	AVG W/T
06:00	07:00	28	15	5	3	1	0	3	1	95	20.00
07:00	08:00	31	13	3	5	3	4	3	0	80	25.84
08:00	09:00	37	8	8	6	4	4	6	1	117	31.65
09:00	10:00	44	15	4	4	8	3	7	3	167	35.20
10:00	11:00	38	8	5	4	7	4	6	4	148	40.16
11:00	12:00	36	6	6	4	5	6	6	3	106	42.11
12:00	13:00	36	8	2	7	7	3	6	3	108	38.69
13:00	14:00	35	8	2	7	4	2	8	4	144	43.60
14:00	15:00	36	9	4	2	6	4	7	4	133	44.50
15:00	16:00	29	6	5	3	1	3	7	4	158	49.59
16:00	17:00	27	11	4	1	5	2	3	1	104	26.33
17:00	18:00	29	16	6	3	3	0	0	1	135	15.52
18:00	19:00	26	9	1	4	3	2	5	2	120	35.81
19:00	20:00	23	10	5	2	3	1	1	1	110	21.35
20:00	21:00	12	9	1	1	1	0	0	0	31	7.92
21:00	22:00	5	4	0	1	0	0	0	0	24	7.60
22:00	23:00	14	8	2	1	1	1	1	0	61	16.07
23:00	00:00	11	9	0	0	1	0	1	0	71	11.36
00:00	01:00	1	1	0	0	0	0	0	0	1	1.00
01:00	02:00	1	1	0	0	0	0	0	0	1	1.00
02:00	03:00	1	1	0	0	0	0	0	0	4	4.00
03:00	04:00	1	0	0	1	0	0	0	0	23	23.00
04:00	05:00	1	1	0	0	0	0	0	0	1	1.00
05:00	06:00	3	3	0	0	0	0	0	0	1	1.00
TOTALS		505	179	63	59	63	39	70	32	167	32.04

(a)

Figure 19.3. (a) A 24-hr recording of all up calls at the lobby floor (Courtesy Delta Elevator Ser-vice Corporation).

BUILDING Name and Address
 Hurly Medical Center
 N. Washington Street
 Boston, Ma. 02111

SYSTEM Description
 Haughton 4-Car Group
STUDY Identification
 Passenger cars 1-4
STUDY 4 of 7 Daily
REPORT Status - PRINTED
DATA FILE status - OK
CIS Polarity - SAME

STUDY Commissioned By
 Associated Management Inc.
No. of Cars - 04
No. of Stops - 08
SCAN Date - 05/08/92
SCAN Time - 06:00:00   No. Int - 24
REPORT Date - 05/16/92
SCAN Interval - 60 Minutes
SYSTEM Return - NEGATIVE
INPUT Threshold - 40 Volts
TRANSFER Status - VERIFIED
PEN No. & Function - Composite ALL

- NUMBER OF CALLS ANSWERED IN SECONDS -

TIME	SLOT	# CALLS	0-10	11-20	21-30	31-45	46-60	61-90	91+	L/CALL	AVG W/T
06:00	07:00	115	63	20	13	5	6	6	2	134	18.15
07:00	08:00	173	67	25	24	17	20	17	3	103	26.21
08:00	09:00	184	64	33	20	25	16	19	7	133	28.77
09:00	10:00	234	59	28	29	36	33	35	14	217	37.21
10:00	11:00	215	35	24	27	39	26	43	21	162	44.38
11:00	12:00	229	53	23	23	33	23	40	34	209	47.79
12:00	13:00	234	45	31	25	37	28	44	24	147	43.15
13:00	14:00	234	46	23	22	43	28	47	25	151	44.29
14:00	15:00	253	54	32	25	30	35	43	34	194	45.83
15:00	16:00	224	32	27	31	28	22	43	41	201	53.48
16:00	17:00	189	66	31	23	30	19	15	5	130	27.49
17:00	18:00	171	86	28	19	21	8	7	2	135	18.25
18:00	19:00	193	70	27	27	16	20	22	11	139	30.03
19:00	20:00	160	57	36	22	20	12	10	3	110	23.74
20:00	21:00	117	64	20	11	13	2	5	2	92	16.90
21:00	22:00	76	58	13	2	2	0	1	0	68	8.46
22:00	23:00	92	52	17	8	8	3	4	0	83	15.33
23:00	00:00	66	49	10	2	3	0	2	0	73	10.32
00:00	01:00	27	21	2	3	0	1	0	0	48	9.26
01:00	02:00	16	9	6	1	0	0	0	0	23	9.44
02:00	03:00	37	23	3	2	3	3	2	1	118	19.32
03:00	04:00	18	14	1	2	0	1	0	0	48	10.72
04:00	05:00	22	19	1	2	0	0	0	0	30	7.18
05:00	06:00	15	15	0	0	0	0	0	0	9	3.93
TOTALS		3294	1121	461	363	409	306	405	229	217	33.18

(b)

BUILDING Name and Address
 Hurly Medical Center
 N. Washington Street
 Boston, Ma. 02111

SYSTEM Description
 Haughton 4-Car Group
STUDY Identification
 Passenger cars 1-4
STUDY 4 of 7 Daily
REPORT Status - PRINTED
DATA FILE status - OK
CIS Polarity - SAME

STUDY Commissioned By
 Associated Management Inc.
No. of Cars - 04
No. of Stops - 08
SCAN Date - 05/08/92
SCAN Time - 06:00:00   No. Int - 24
REPORT Date - 05/16/92
SCAN Interval - 60 Minutes
SYSTEM Return - NEGATIVE
INPUT Threshold - 40 Volts
TRANSFER Status - VERIFIED
PEN No. & Function - Sys. DEMAND

SYSTEM DEMAND by LOCATION

FLOOR	UP Calls	Avg UP W/T	DN Calls	Avg DN W/T	TOT Calls	Avg W/T	% Calls
BT	420	35.72	0	0.00	420	35.72	12.7 %
LB	505	32.04	67	39.14	572	32.87	17.3 %
02	306	33.57	197	32.71	503	33.23	15.2 %
03	201	30.17	289	35.17	490	33.12	14.8 %
04	119	26.25	193	38.40	312	33.77	9.4 %
05	65	34.52	303	32.99	368	33.26	11.1 %
06	33	31.24	341	31.00	374	31.02	11.3 %
07	0	0.00	255	31.96	255	31.96	7.7 %

(c)

Figure 19.3. *(Continued)* (*b*) A 24-hr composite record of all hall calls at all floors; (*c*) a 24-hr distribution of all hall calls by floor (Courtesy Delta Elevator Service Corporation).

DELTA 5049 - ELEVATOR TRAFFIC STUDY

BUILDING Name and Address
    Hurly Medical Center
    N. Washington Street
    Boston, Ma. 02111

SYSTEM Description
    Haughton 4-Car Group
STUDY Identification
    Passenger cars 1-4
STUDY 4 of 7 Daily
REPORT Status - PRINTED
DATA FILE status - OK
CIS Polarity - SAME

STUDY Commissioned By
    Associated Management Inc.
No. of Cars - 04
No. of Stops - 08
SCAN Date - 05/08/92
SCAN Time - 06:00:00 No. Int - 24
REPORT Date - 05/16/92
SCAN Interval - 60 Minutes
SYSTEM Return - NEGATIVE
INPUT Threshold - 40 Volts
TRANSFER Status - VERIFIED
PEN No. & Function - Graph 3 of 9

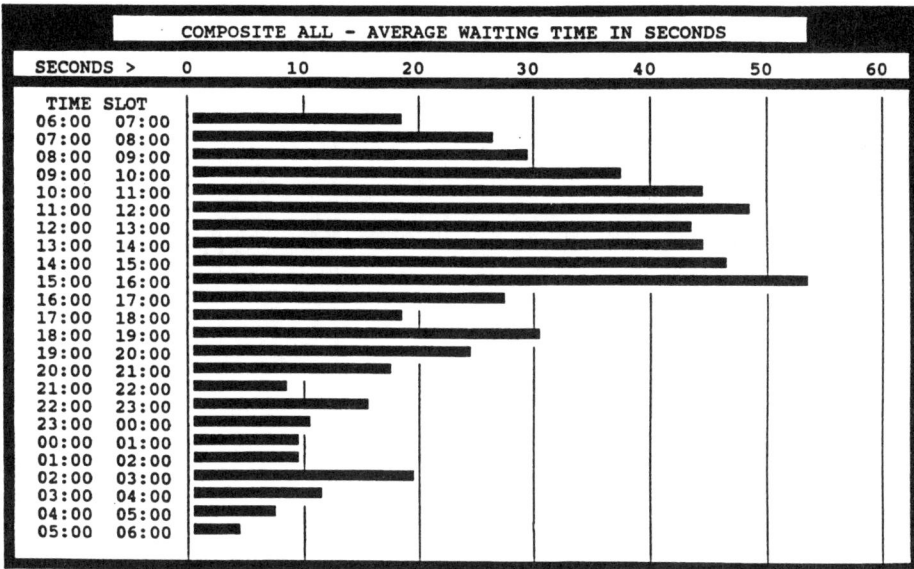

(d)

Figure 19.3. (*Continued*) (*d*) A bar chart composite of average hourly wait times (Courtesy Delta Elevator Service Corporation).

ing in real time through the TV-type display. Like the portable units, the built-in ana-lyzers store the data for later printout.

Regardless of which method or combination of methods is used in analyzing traffic in an existing facility, the observations should be conducted over a period of several days to level out any anomalies that may occur.

If the distribution of waiting times shows a high percentage of long call waits, it is an indication that the cars are bunching and are not equitably responding to demands for hall call service. These printouts, with a written analysis, should be presented to the mainte-nance contractor for use in determining the extent of readjustment required to return the operating system to an efficient level of functioning.

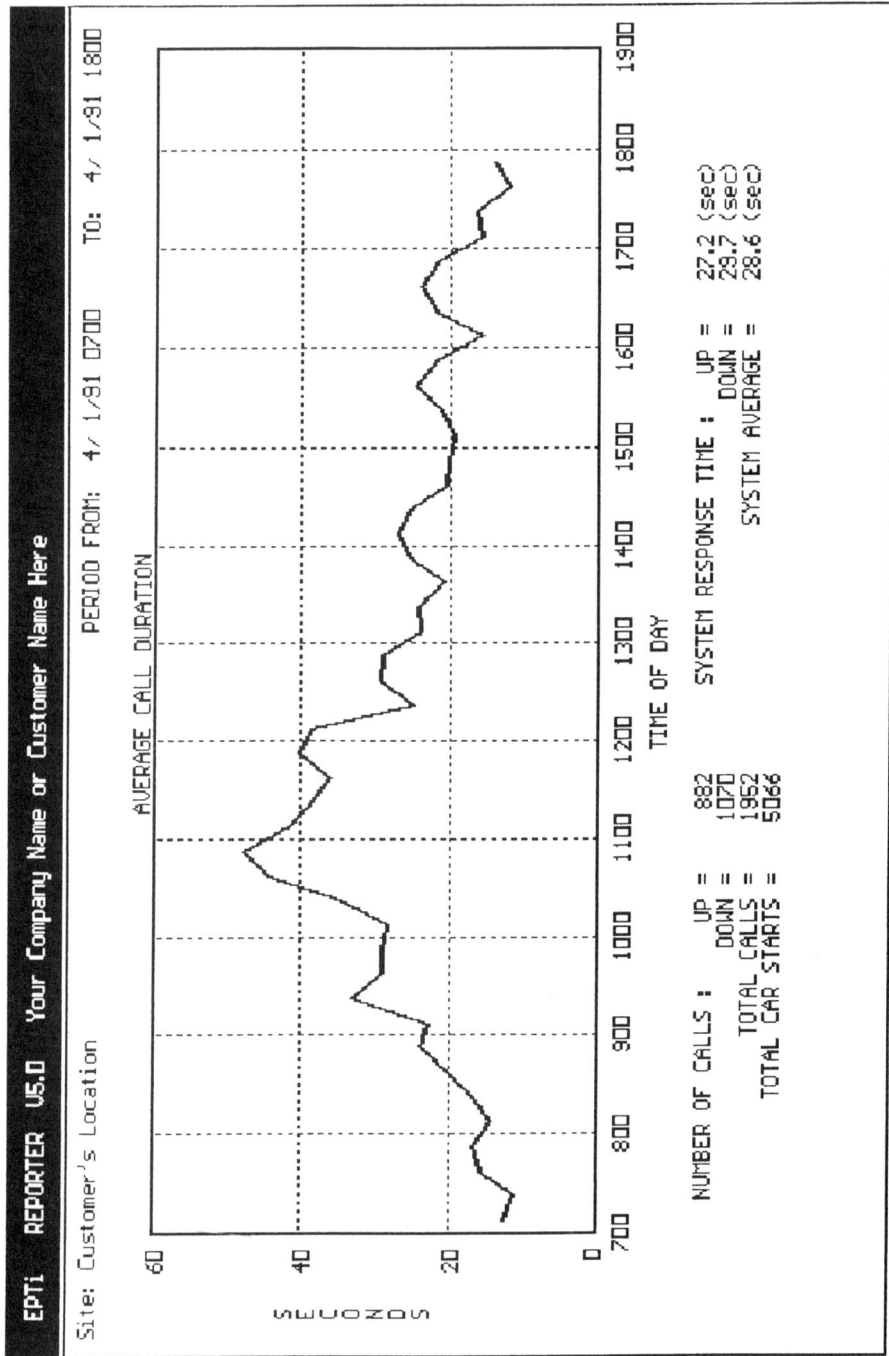

Figure 19.3. (*Continued*) (*e*) A line graph composite of average hourly wait times (Courtesy DigiMetrix, Inc.).

```
+-------------------------------------------------------------------------------+
|** ANALYZER DATA DUMP     *********************** RECORD NUMBER     :   11 **|
|** DATE PRINTED   6/21/96 *********************** DATE  4/01/91 TIME : 0945 **|
+-------------------------------------------------------------------------------+
```

CALL STATISTICS:
================

```
             TOT MAX          CALL WAIT COUNTS IN 15 MIN. INTERVALS
   FLOOR CNT SEC SEC    15  30  45  60  75  90 105 120 135 150 165 180 180+
   ----- --- ---- ---   ---  ---  ---  ---  ---  ---  ---  ---  ---  ---  ---  ---  ----
     CUP   3   49  22     1   2   0   0   0   0   0   0   0   0   0   0   0
     GUP  12  497  86     1   3   3   3   0   2   0   0   0   0   0   0   0
     GDN   1   43  43     0   0   1   0   0   0   0   0   0   0   0   0   0
     1UP   7  197  66     2   3   1   0   1   0   0   0   0   0   0   0   0
     1DN   8  172  50     4   1   2   1   0   0   0   0   0   0   0   0   0
     2UP  11  258  46     4   4   2   1   0   0   0   0   0   0   0   0   0
     2DN   5  143  45     1   1   2   1   0   0   0   0   0   0   0   0   0
     3UP   5  134  32     0   4   1   0   0   0   0   0   0   0   0   0   0
     3DN   6  130  37     2   2   2   0   0   0   0   0   0   0   0   0   0
     5UP   3   91  37     0   1   2   0   0   0   0   0   0   0   0   0   0
     5DN  10  235  50     2   6   1   1   0   0   0   0   0   0   0   0   0
     6UP   0    0   0     0   0   0   0   0   0   0   0   0   0   0   0   0
     6DN   4   93  55     2   1   0   1   0   0   0   0   0   0   0   0   0
     7UP   1   30  30     0   0   1   0   0   0   0   0   0   0   0   0   0
     7DN   5  215 102     1   1   1   1   0   0   1   0   0   0   0   0   0
     8DN   2  103  56     0   0   0   2   0   0   0   0   0   0   0   0   0
         --- ----         --- --- --- --- --- --- --- --- --- --- --- --- ----
  TOTALS  83 2390         20  29  19  11   1   2   1   0   0   0   0   0   0
```

SUMMARY: UP   DOWN   OVERALL
=======   ----------   ---------- ----------
   TOTAL NUMBER OF CALLS: 42   41   83
   AVERAGE CALL (sec)   :   29.9   27.7   28.8 (for ALL calls)
   LONGEST CALL (sec)   :   86   102   102

GROUP PROGRAM : PROGRAM TIME (seconds):   BALANCED =   0   UP   =   0
==============   DOWN   =   0   ZONE =   0

CAR STATUS :   CAR   O=OUT   I=INDEPENDENT   (MIN =   0)
============   CAR   STATUS   (SEC)   (SEC)
      \   ---   ------   -----   -----
                    1   0   0
                    2   0   0
                    3   0   0
                    4   O   4   0

CAR STARTS : CAR 1 =   49 CAR 2 =   37 CAR 3 =   44 CAR 4 =   44
============

(f)

Figure 19.3. *(Continued)* *(f)* A 15-min composite of call waits by floor (Courtesy DigiMetrix, Inc.).

PERFORMANCE EVALUATION

When it becomes necessary to evaluate elevator performance, the most logical starting point is the trouble log. To gain an accurate understanding of the system's performance, the log for the previous 12 months should be studied. Once this has been completed and the appropriate notes taken, those building personnel involved with the day-to-day operations of the elevator system should be interviewed, not only as to the problems of record but as to their opinions and observations of the system operations.

Only when the evaluator is satisfied that the gathering of background material is complete and accurate should evaluation of the physical performance of the vertical transportation system begin.

```
------------------ ~SPL047F.TMP follows --------------------
YALE NEW HAVEN HOSPITAL
CT-78311
NEW HAVEN  CT                      Comp. All
4  Cars
Scan Start Date  02-06-96
12 Landings


              - Number Of Calls Answered In Seconds -

   Time Slot    #Calls 0-10  11-20 21-30 31-45 46-60 61-90  91+   C
 IS  Avg W/T
 ------------------------------------------------------------------
 ------------
 07:00 -> 07:30  0098  0071  0017  0004  0005  0001  0000  0000  4.
 000   08.4
 07:30 -> 08:00  0106  0040  0014  0020  0017  0006  0008  0001  4.
 000   23.5
 08:00 -> 08:30  0083  0050  0013  0009  0007  0003  0001  0000  4.
 000   13.7
 08:30 -> 09:00  0097  0056  0019  0009  0005  0007  0001  0000  4.
 000   13.8
 09:00 -> 09:30  0100  0038  0031  0011  0010  0007  0003  0000  4.
 000   18.6
 09:30 -> 10:00  0117  0050  0015  0013  0020  0007  0008  0004  3.
 878   24.7
 10:00 -> 10:30  0108  0063  0026  0006  0009  0003  0001  0000  3.
 801   12.7
 10:30 -> 11:00  0120  0079  0020  0007  0009  0002  0003  0000  4.
 000   12.1
 11:00 -> 11:30  0129  0077  0033  0010  0008  0001  0000  0000  4.
 000   10.7

 11:30 -> 12:00  0130  0078  0029  0011  0008  0002  0002  0000  4.
 000   12.4
 12:00 -> 12:30  0124  0077  0030  0006  0007  0003  0001  0000  4.
 000   11.4
 12:30 -> 13:00  0137  0066  0029  0019  0018  0003  0001  0001  4.
 000   16.1
 13:00 -> 13:30  0153  0048  0036  0031  0018  0011  0007  0002  4.
 000   22.8
 13:30 -> 14:00  0155  0076  0023  0019  0016  0009  0011  0001  4.
 000   19.7
 14:00 -> 14:30  0131  0075  0038  0011  0006  0001  0000  0000  4.
 000   11.3
 14:30 -> 15:00  0152  0078  0029  0021  0015  0007  0002  0000  4.
 000   15.5
 15:00 -> 15:30  0140  0071  0031  0015  0006  0012  0005  0000  4.
 000   17.0
 15:30 -> 16:00  0126  0071  0029  0015  0009  0001  0001  0000  4.
 000   12.8
 16:00 -> 16:30  0073  0061  0006  0004  0002  0000  0000  0000  3.
 410   06.6
 16:30 -> 17:00  0084  0048  0020  0009  0001  0004  0002  0000  3.
 000   12.7
 17:00 -> 17:30  0079  0052  0019  0006  0001  0001  0000  0000  3.
 000   09.0
 17:30 -> 18:00  0094  0055  0021  0006  0006  0002  0004  0000  3.
 000   14.3
 18:00 -> 18:30  0085  0053  0016  0011  0004  0001  0000  0000  3.
 000   10.4
 18:30 -> 19:00  0080  0046  0017  0008  0006  0000  0003  0000  2.
 609   13.4
                 ---------------------------------------------------

 ------------
   Totals       2701  1479  0561  0281  0213  0094  0064  0009  3.
 737   14.8
```
(g)

Figure 19.3. *(Continued) (g)* A 12-hr composite of all hall call waits from a built-in traffic analyzer (Courtesy Montgomery KONE).

Car performance tests are perhaps the easiest to conduct. They can be done at any time and require only a stopwatch. An observer rides the car and records the times required for the closing of the doors, for the car to start, the floor-to-floor travel, the full opening of the doors, and the amount of time spent at the floor (noted as dwell time or transfer time). These timing records should be taken in both directions. In older installations where speed regulation may not be as consistent as that found in modern solid-state speed regulation systems, these time functions should be measured with the car empty (observer only), with a balanced load, and with a nearly full load. Care should be taken to record these function times at floors that are approximately equal in height so as to achieve typical floor timing.

The results of these performance tests are then compared with the design criteria used in calculating performance, as described in earlier chapters of this book. If any of the car operational function times exceed the design criteria for these functions, readjustment will be required to return these features to an acceptable level. Dwell or transfer times should be compared with the traffic patterns that were observed during the on-site traffic studies and adjusted accordingly to ensure a continuously smooth flow of traffic in and out of the elevators. Most group systems installed within the past 30 years provide for different dwell-times in response to car or hall calls. If there is no difference or if the dwell-time for car calls is longer than that recorded for hall calls, readjustment is necessary.

Figure 19.4 shows a sample form that can be used for on-car performance evaluations. The various function times are developed by direct readings or by difference between events. For example, to establish the time it takes the car to start from the time the doors are fully closed requires the measurement of the elapsed time of the full door closing function and the elapsed time from the start of the door closing function until the car starts. The difference between these two time functions provides the time it takes the car to start after the doors are fully closed.

If the doors are arranged to preopen, that is, start to open before the car is level at the floor, the preopening time can be recorded by using chalk to mark the car door sill at the position of the doors when the car is level. At the next stop, the time from when the doors pass the chalk mark until they are fully open can be measured. Subtracting this time from the time the doors start to open until they are fully open will establish the preopening time. If the resulting preopening time is excessive, the car should be observed from the landing side to determine the distance from the floor at which the doors start to open. If this is in excess of 3 in., an adjustment is necessary. A visual check can be made to judge that the car is stopped before the doors are open about 2 ft. Caution should be exercised in the use of preopening in hospitals and residential facilities that cater to older passengers. In a hospital environment, passengers are often preoccupied with personal concerns; they may not be aware that the car has not completely leveled and may trip as they attempt to exit before the car levels with the floor. Similarly, in residential retirement facilities, older passengers with impaired vision may not be able to properly judge when a car is level.

There is an argument that claims that as the doors are "checked" or slowed prior to their fully open or fully closed position, door-open and door-close times should be measured from a three-quarter open or closed position. Checking, a normal component of the door operational function, is included to provide a smooth and quiet operation, free of slamming at the end of the open or close movement. As checking is an integral part of the total door operation function, the full travel must be timed. The tables shown in earlier chapters include the total travel and should be used for comparison.

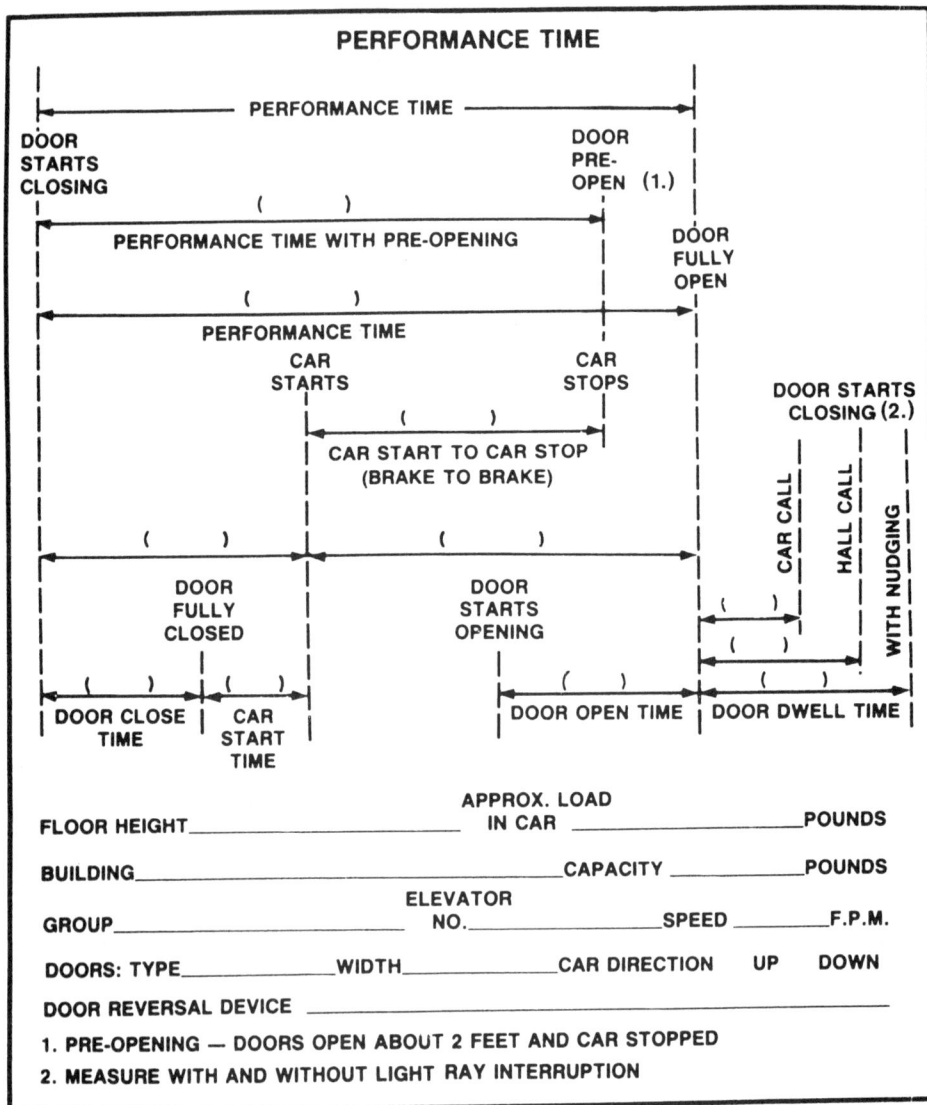

Figure 19.4. Performance time chart for an elevator floor-to-floor travel showing the various elements of a one-floor elevator trip (Courtesy *Elevator World Magaine*).

The total performance of an elevator system can be recorded and presented in graphic form. By connecting recording meters which produce a tape on a time base to each elevator, a profile of elevator action showing travel up, stopping at a floor, continued travel and stops, reversal, and downward travel and stops can be recorded. This tape can be matched with similar tapes of other elevators all on the same time base to develop a composite of all the various trips a group of elevators make over a given period of time.

A sample of such a recording is shown in Figure 19.5. This enables the consultant to analyze the interaction of the various elevators and suggest corrective actions that may be taken if deficiencies are apparent.

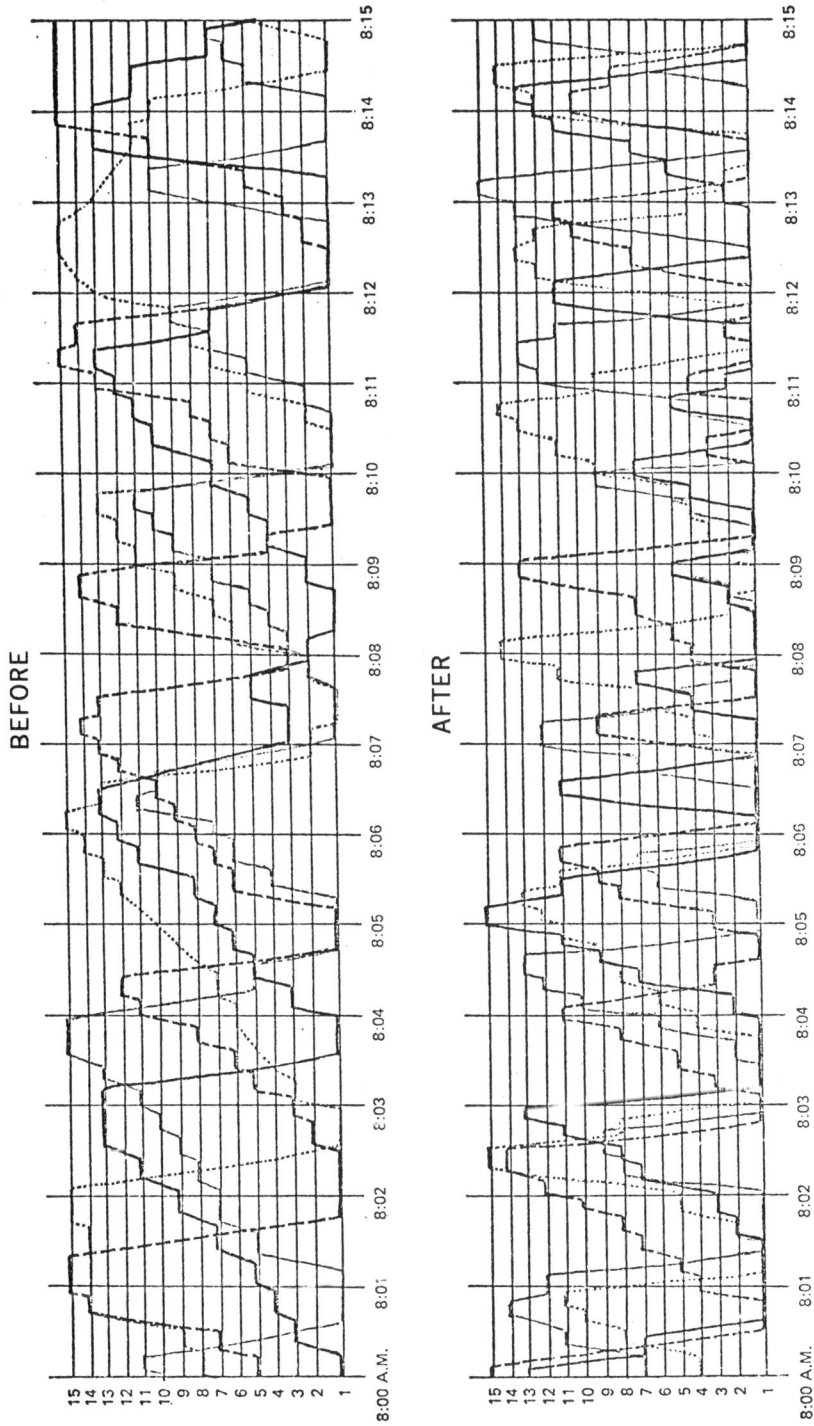

Figure 19.5. Taped recording of elevator travel up and down showing time spent on floor-to-floor travel and time spent stopped at the various floors. "Before" chart shows serious elevator "bunching." "After" chart shows performance after corrective measures have been taken (Courtesy Performance Profiles, Inc.).

All of the data collected through the traffic analyzers, the lobby counts, the on-car traffic counts, and the car and door functions, when collated present a complete picture of the performance of the elevator system. These can then be compared with established criteria to obtain a judgment of the efficiency of the elevator system. Graphic presentations of the results of the various studies and data gatherings should be included in the report to the building management. Once a program of corrective actions has been established and completed, the data gathering process can be repeated to provide a record of system efficiency improvement.

No less important then the gathering of performance data is an evaluation of the ride quality of the elevator system. Acceleration and deceleration should be truly stepless. Final stops should be soft and almost unnoticeable. The sway of the cars on high-speed runs should be minimal. Noises encountered during the movement through the hoistway require investigation and possible realignment of the offending components.

EQUIPMENT EVALUATION

An evaluation survey of the components of an elevator system can often provide additional insight into causes of inefficiencies in system operation or ineffective car, door, or leveling functions that may be observed during traffic studies. Such a component evaluation should include surveys of the machine room, hoistway, pit, car tops, and all equipment contained therein.

In surveying the machine room, car speeds should be measured with an electronic tachometer in both directions and under varying loads. These recorded speeds can then be compared with the rated speed of the equipment. Any difference noted between the actual speed and the rated speed can then be used to assist in the judgment of the overall system operation. The condition of the machine bearings, worm and gear (in the case of geared machines), and brakes should be noted. The commutators and brushes of all rotating equipment should be examined for wear, threading, brush grooving, arcing under load, and so forth. Governors should be checked for cleanliness, test date tags, and seals. Ropes should be examined for wear or excessive lubrication. Controllers and selectors should be carefully checked for excessively worn contacts, inoperative relays, temporary jumpers, relays that have been blocked in or out, and loose or damaged wiring. The condition of the machine room itself should be examined for excessive accumulation of dirt, trash, rags, broken parts, and the like. The lighting, ventilation, and physical condition of the machine room should be noted. The condition of the feeders and disconnects should also be noted.

The hoistway should be examined throughout its length from the car top. Means of, or lack of, hoistway ventilation should be noted. The general condition of the hoistway walls should be examined for broken blocks or tiles or damaged drywall. Any buildup of lint and grease in the hoistway should be noted as a potential fire hazard. This is of special concern in hospitals because of the volume of lint in the air. Rail brackets and their fastenings should be checked. If the building is located in a seismic 2, 3, or 4 zone, all seismic protection devices should be examined. If they do not exist, the building management should be advised as to what the current code requires for seismic protection in that particular zone. The condition of limit switches, door locks, door hangers, closers, and the door operator should be noted. The security of the car emergency exit panels should be checked.

The pit must be checked for an accumulation of trash. Special care must taken in hospitals, as discarded needles, broken glass, and other potentially infectious materials may have found their way into the pit. The condition of the spring or oil buffers, limit switches, pit stop and light switches, pit ladder, governor tension sheaves, compensating ropes or chains, and the car traveling cables should be noted. The presence of water or dampness should also be noted. The condition of the underside of the car and the safety device should be examined.

In examining the car, special attention should be given to the operation of all signal devices, car leveling, and door operation. The condition of the car enclosure and the car flooring should be noted. The presence of worn or damaged floor tile can pose a potential tripping hazard. Broken or torn car enclosure trim or wainscoting also can be a potential hazard to passengers.

STUDY REPORTS

Traffic study reports must include a completely detailed written analysis of the observations and the subsequent conclusions, as well as recommended corrective measures. These reports should be logically arranged so the reader is easily led from one phase of the study to the next. A table of contents should be included to enable the reader to quickly find the areas of prime interest. The graphs and informational charts mentioned earlier in this chapter should be included in a logical manner, not in an appendix at the end of the report.

The equipment evaluation should be covered in a separate report or, at the very least, in a separate section of the traffic study report. Again, a logical arrangement beginning with the machine room, followed by the information on the hoistway, the pit, and, finally, the car, will enable the reader to better grasp those areas that require further attention.

Where recommendations are made for adjustments, alterations, or system upgrades, they should be clear and concise, written in language easily understood by the layman. Each recommendation should be accompanied by a budget estimate for the cost of accomplishing the recommended work. These recommendations should be prioritized as to their relative importance: safety hazards, code violations, ADA concerns, and system improvements.

SUMMARY

This chapter has outlined some of the many varied traffic and equipment studies that can be performed on vertical transportation systems in all types of buildings. It also suggests strategies to evaluate the results of such studies.

The primary reason for a vertical transportation system study is to determine how far the system has strayed from its original design criteria and what remedies may be available to return the system to peak efficiency. This chapter, as well as other chapters in this book, has presented a wide range of solutions that may be appropriate to any given situation. It must be emphasized that the available solutions may not completely resolve traffic problems in a building that was originally under elevatored. It is essential for the study reports to indicate clearly when such a situation exists. Building management must be made fully aware that although the suggested solutions will offer improvement in efficiency and equipment reliability, it is not possible to make an existing four-car system do the work of the six-car or eight-car system that the traffic studies show should have been included in the original design.

ABOUT THE AUTHOR

LEONARD G. LEVEE, JR. has been associated with the elevator industry in various capacities for the past 50 years. The first 23 years were spent with the General Elevator Co. of Baltimore in the manufacturing, engineering, sales, and branch management phases of the industry. The next 11 years were spent in the Office of Construction of the Veterans Administration as chief of the Automatic Transport Systems Division. During these years, Mr. LeVee was responsible for everything that moved vertically and horizontally in the 172-hospital system.

In 1982, Mr. LeVee began his career as a consultant with Lerch Bates & Associates. In 1987, he became a partner in the firm of International Elevator Consultants, Ltd. of Detroit, Michigan, before founding the firm of LeVee & Associates, Inc., Vertical Transportation Consultants in 1990.

Mr. LeVee is active in the National Association of Elevator Safety Authorities (he holds certified elevator inspector certificate #425); the National Association of Vertical Transportation Professionals, where he is executive director; and the National Association of Elevator Contractors, where he acts as the liaison between the professional group NAVTP and the Board of Directors.

Mr. LeVee is currently a member of the National Interest Review Committee of the ASME A17.1 Safety Code for Elevators and Escalators. Past code activities have included membership on four A17.1 subcommittees, the ASME B20.1 Safety Standards for Conveyors, and the Code Making Panel 12 of the National Electrical Code.

Mr. LeVee has had articles and essays published in *Elevator World Magazine, Mainline,* the newsletter of the National Association of Elevator Contractors, *China Elevator* trade magazine, and is a regular contributor to the NAVTP newsletter.

20 The Changing Modes of Horizontal and Vertical Transportation

WILLIAM STURGEON

INTRODUCTION

In this chapter, as in the previous edition of *Vertical Transportation,* we speculate on the future and changing trends in the application of elevators and escalators. In addition, observations regarding the growing field of automated people movers and their relation to elevator technology are made. The current emphasis is on building the world's tallest building, even while exploring the more mundane challenge of effectively and economically overcoming the short rise of a few feet, thus providing access from street level to the raised entrance of a building.

The race to build the tallest building erupts every decade or so. This was apparent during the early part of the twentieth century when downtown New York City saw the growth of the Woolworth Building, the Metropolitan Life Tower, and the Singer Building. The race was repeated in the 1930s as the Empire State and the Chrysler Tower competed for the title of world's tallest building. It was repeated again in the 1960s, between the World Trade Center in New York and the Sears Tower in Chicago. The title has stayed with Sears for the past two decades, the challenge now being accepted in the Far East.

Hardly a few months go by without announcement of a taller building within a developing country. These announcements lead to speculation of how to both construct and serve the "Mile High Building." Ideas, and some developments of self-propelled elevators emerge along with announcements of elevators that move both horizontally and vertically, the intention of being multiple units in one vertical shaft. This is a prime necessity because a conventional elevator approach would mean lower floors containing only hoistways!

In previous chapters the more conventional approaches have been discussed—sky lobbies, shuttle elevators, and double-deck elevators. In this chapter we consider how we can use innovation. The real challenge, which remains beyond the mechanical and electrical aspects, lies in overcoming the major antagonist, Mother Nature. She rebels against sudden horizontal acceleration of humans, uses wind and seismic disturbances to fight height, as well as sunshine, the warmth of which can cause a building to lean as it is heated on one side.

The short level change is another challenge, more universal than achieving the "tallest" building. This has created an entire industry in the past decade or so. Developed to accommodate wheelchairs, it has grown to serve all sorts of needs. An aging popula-

The Vertical Transportation Handbook, Third Edition, Edited by George R. Strakosch
ISBN 0-471-16291-4 © 1998 John Wiley & Sons, Inc.

tion will have difficulty with stairs. Safe, reliable, and economical accessibility equipment is being offered to meet the need, some of which was alluded to in the chapter on residential use.

The following is but an overview. Certainly, the next decade will bring other "changing modes." These will add to the vastness of the elevator industry. Where else can an individual ride a vintage conveyance 75 to 100 years old in one building and experience the latest in the building next door?

TRADITION

The vertical elevator, propelled by a traction machine and equipped with a counterweight, developed over the past century, is established as a safe, reliable, and relatively standard product. Many firms engage in its production, and safety codes have been written to ensure its proper installation. The public has accepted that elevators based on the traction principle provide the standard of good elevator service.

The hydraulic-type elevator enjoys a similar reputation. To the public, its appearance is the same as that of a traction elevator because the doors, cab, controls, and signals are common features.

Both traction and hydraulic elevators are developed as very rigid, permanent installations in a building, seldom moved from one building to another. In addition, it is difficult to install most traction elevators and direct-plunger-type hydraulic elevators in an existing building because stringent pit, overhead, hoistway, and structure requirements must be met. It is also difficult to adapt either type to certain extreme vertical lifting requirements because certain physical limitations are present. These require expensive additions to meet the requirements. For example, using a traction elevator to serve a 1000-ft (300-m)-high television antenna tower requires roping that will not be affected by extreme wind.

Alternate approaches to elevator drives have been developed to overcome some of the necessary restrictions of conventional traction and hydraulic elevators. Moreover, conventional hydraulic and traction elevator machinery has been adapted to other than pure conventional enclosed elevator applications to meet special lifting requirements. The hydraulic elevator has been adapted to lifting stages in theaters, and the traction elevator machine to moving vehicles horizontally. A class of new propelling devices consisting of racks and pinions, screw lifts, and scissor jacks has been adapted to elevator lifting means. In addition, some elevators, both hydraulic and traction, are capable of unusual approaches to transporting personnel.

An entirely new field is opening up, borrowing a great deal of elevator people-moving discipline and activity. The vast experience of the elevator industry in opening, closing, and protecting people from moving doors during transfer is being adapted to the horizontal people-moving field. Even escalator and moving walk technology is being adapted to provide a new vehicle—the accelerating moving walk—which can transport people from a walk, accelerate them to 10 mph (4.5 mps), and then gradually slow them down to a walk.

These aspects are briefly discussed in this concluding chapter, and some of the current applications introduced in Chapter 14 are expanded to introduce innovations. During the next decade it is expected that some ideas will be more extensively developed and will broaden the entire realm of personnel transportation.

SCREW LIFTS

Screw lifts are not new—they were known by the ancient Greeks. What may be new is their refinement into higher-speed devices and passenger elevator application. Industrial screw lifts have been used to lift extremely heavy loads at very low speeds, usually measured in inches per minute (millimeters per second). Recent developments have made screw jacks available that can move moderate loads of up to 2000 lb (1000 kg) at 100 fpm (0.5 mps). Such equipment can be used to move a small elevator operating with the screw in tension and being driven by a motor on the top. As the screw is turned, the elevator travels up or down in its guides (Figure 20.1). By using the screw in tension, the bending that could occur if the screw were used to push from below is avoided.

Large-size screw jacks are used to lift platforms and can be exceedingly valuable in raising a section of floor to form a platform, because a screw jack has a natural tendency to lock the load when the turning power is off. Unlike a hydraulic lift, which may creep as the oil cools, or if a slight leak develops, a screw jack raised platform can be left in place indefinitely. If more than one screw jack is required to lift the load, the screw jacks

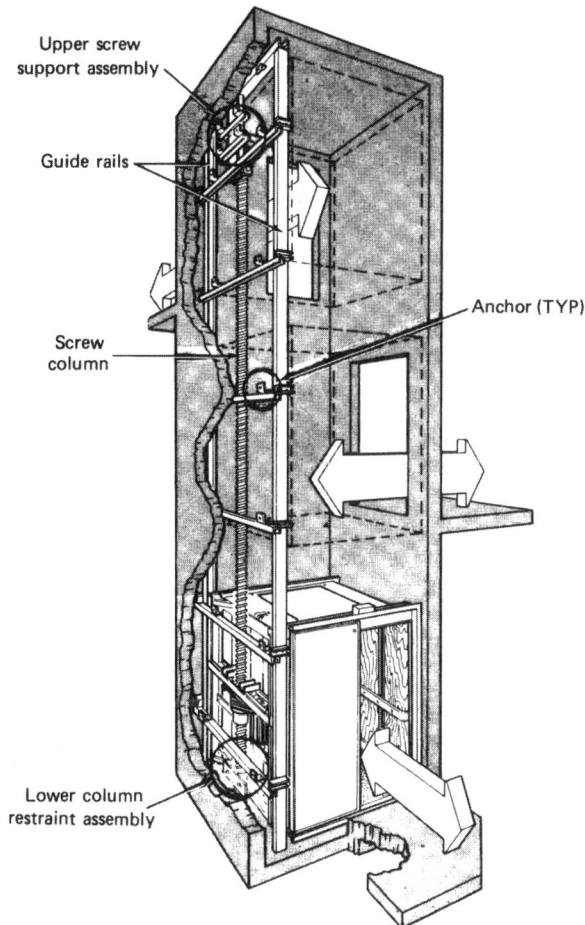

Figure 20.1. A screw lift elevator.

have a further advantage of being self-equalizing if driven through gearing by a single-drive motor.

Such screw jack arrangements are used to lift the bottoms of swimming pools to change depth from diving to wading, lift building walls to close off entrances during nighttime hours, and to move heavy vault doors down and out of the way when a vault is opened.

SCISSOR JACKS

If one visualizes two X's placed one on top of the other so that the touching points are connected and moved apart or closer together by a hydraulic piston or screw mechanism, the principle of the scissor lift is apparent (Figure 20.2). Imagine the X's 10 or 20 ft (3 or 6 m) high and made of heavy steel with a platform on top. Such devices are used to form elevators to move trailer trucks or other large loads. One such application is on shipboard where advantage is taken of the extremely low profile created when the scissor lift is collapsed. Other applications consist of multiple scissor lifts mounted on a wagon to raise small platforms 50 ft (15 m) or more to service ceiling fixtures in a high hall.

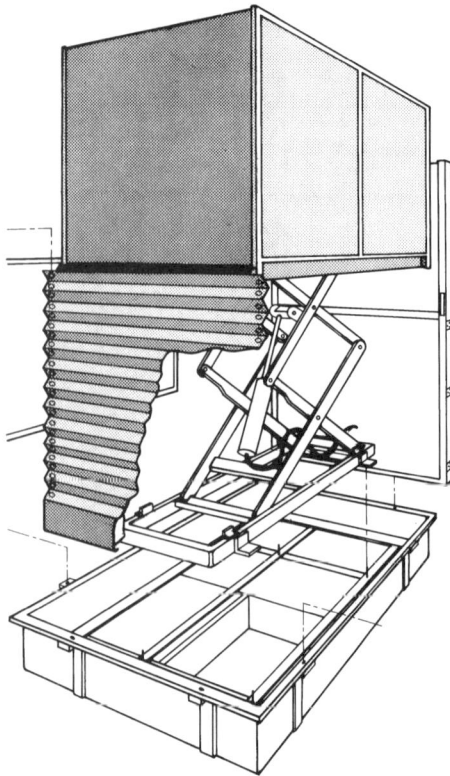

Figure 20.2. A scissor lift.

ACCELERATING MOVING WALKS

In 1980 three prototype systems provided horizontal moving walkways that allowed individuals to enter at a walk, accelerate to a high speed, and decelerate to walking speed. Costs were excessive, and the economics could not support continuing interest; hence, further development of the walks was suspended. The mechanical and human aspects were almost insurmountable and, as with any device intended to carry people, concern about safety and liability discouraged continuing effort. This situation becomes apparent when the approach shown in Figures 20.3*a* and *b* are critically analyzed.

The mechanical problem lies in developing steps that start at a reasonable entry speed, accelerate, and then slow down at a terminal to allow exit at normal walking speed. The objective is to have people walk on and off at their normal speed without hesitation, as is done with conventional, single-speed moving walks. Another serious mechanical problem is synchronizing the handrail with the speed of the steps, implying a "stretchable" handrail, which does not exist.

A "people problem" remains. If individuals were not social and would space themselves along the fast portion of the moving walk, they would have sufficient room to exit when the walkway is slowed. However, individuals tend to get close to the one in front, and, as the walkway slows, the space between them decreases and a bumping develops. This problem may prove insurmountable.

Figure 20.3. An accelerating moving walkway. (*a*) Prototype model showing step tread and handrail arrangement (Courtesy *Elevator World Magazine*).

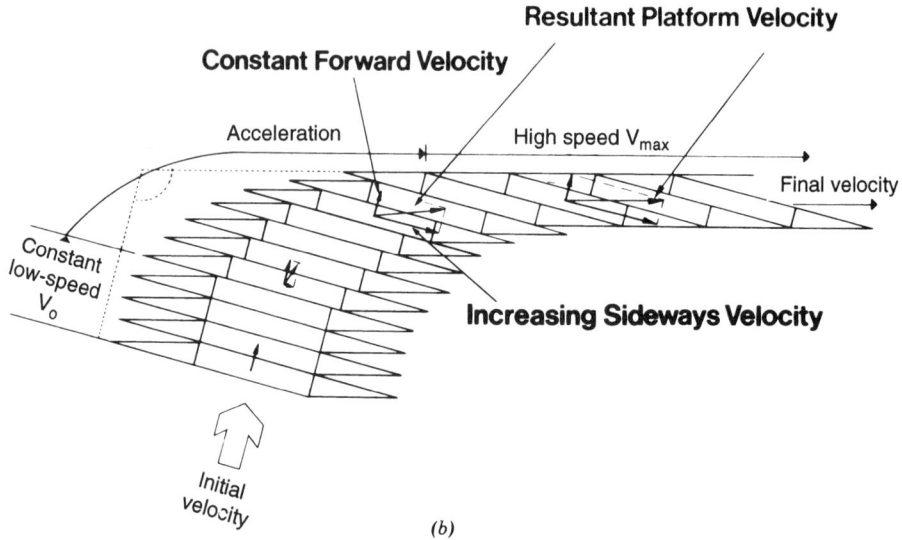

Figure 20.3. *(Continued)* *(b)* Diagrammatic view of changing step configuration as people are accelerated.

An advisory guidance code, established as part of ASME A17.1 in Appendix G, sets up some preliminary safety considerations for the future design of accelerating moving walks. If an active program gets underway to develop a working model for public demonstration and use, and becomes successful, it may introduce a new era in short-range personnel transportation.

Why is there a need for such a development? The need comes from the increasing population and congestion in the world, especially in transportation terminals. As more people use such facilities, the larger and more spread out the facilities become. Walking is accomplished at about 120 to 240 fpm (0.6 to 1.2 mps). A moving walkway moves at 160 fpm (5.0 mps), and a person walking on a moving walkway can travel at 400 fpm (2 mps). In some terminals the center is 2500 ft (800 m) from the loading areas, which can be a tiring 10-min walk. In other facilities, parking is 5000 ft (1600 m) away, which requires bus transportation and vehicle and driver expense. An accelerating moving walk at 1000 fpm (5.0 mps) can reduce a 500-ft (1600-m) travel distance to 5 min. Short, single-speed moving walkways from that point can make parking or transportation loading quickly accessible.

ESCALATORS: TRANSIT STATIONS

Escalators remain the major means of moving large groups of people vertically in an increasing variety of applications. Transit lines are being built and expanded both in the United States and throughout the world. As they become deeper, escalator rises exceed 100 ft (30 m).

On exceedingly long escalators the lower landing is often invisible from the top. A solution to eliminating this visual impact is to have a flat section in the middle. This is illustrated in Figure 20.4.

Figure 20.4. Escalator with an intermediate flat section (Courtesy Schindler).

In some cities inclined elevators have been installed adjacent to escalators with success. A number of patrons, estimated at 10%, are either unable or unwilling to use an escalator. The inclined elevator provides the same access top and bottom. A vertical elevator would have to be installed remotely, thus compromising the entry and exiting areas. The vertical elevator is a necessity for the disabled, hence an architectural dilemma exists.

An escalator has been developed to handle a wheelchair. The escalator is stopped, and a number of steps are reconfigured to form a platform. The utility is questionable, inasmuch as the time required to reconfigure and restore detracts substantially from the people-moving aspects which prompted the escalator installation initially.

The paradox is that the escalator for a deep transit station is desired for the movement of people. Yet the time people must spend waiting and riding can become excessive. There is a limit to how fast an escalator can travel, which currently is 120 fpm (0.6 mps) in the United States and Canada, and upward to 200 fpm (1.0 mps) in other parts of the world. Individuals in wheelchairs and parents with strollers cannot be accommodated, and the alternate vertical elevator is inconvenient. Even so, few consider vertical or inclined elevators as the viable alternative.

The following section proposes a vertical elevator arranged in a unique way as an alternative. In 1994 a major step in the application of the inclined elevator was made with the introduction of high-speed inclined elevators operating at an angle of 39° at the Luxor Hotel in Las Vegas. Chapter 14, on nonconventional elevators, describes this application in detail. With the experience of this installation, an inclined elevator operating at 30° to match an escalator becomes a viable alternative.

ELEVATOR-TRANSIT STATIONS

Most developers of transit systems envision hordes of people entering and exiting each station as packed trains arrive and depart. It often follows that the only way to serve those people is by escalators, but the design of the station may be compromised to accommodate the escalators even though well-designed elevator systems can accomplish the same people-moving objective.

As a 10-car train arrives, people are discharged over a horizontal distance of about 600 ft (180 m). To effectively use an escalator system, all these people must walk to a central location. The natural tendency is for people to want to board the train at a point that will be near the escalator at their destination so that they can enter it when they arrive.

The escalator operates at a 30° incline, so the boarding point is 1.7 times the rise of the escalator from either end. In a deep station, the boarding or disembarking point may be a block or two away from entry or exit point.

Consider elevators spaced along the platform. People are near an elevator no matter where they get off the train, and the vertical elevator exits right above the station.

The argument continues that people must wait for an elevator and an elevator ride is too long. A trainload of people must also wait for an escalator, and a station that is 200 ft (60 m) deep will require an almost 3-min ride by escalator.

An elevator system that can serve such a deep station and carry almost as many people as an escalator, with a much faster travel time, is practical, as shown diagrammatically in Figures 20.5a and b. This system was submitted by the editor for use in a proposal for a subway system that could be economically bored out of solid rock below a city, as opposed to conventional subway systems that are built by cut and cover. The elevator system increases the utilization of one hoistway approximately three times over that of a conventional elevator and provides proportionally greater elevator-handling capacity for a given area of hoistway. This was an important consideration, because the hoistway must also be bored vertically out of rock. Figure 20.5c shows the time-cycle usage of the vertical shaftway. An elevator cab is traveling up or down while the cabs at the floors are loading or unloading.

The concept of the entire elevator cab transferring on and off the elevator can also lead to consideration of a completely integrated horizontal and vertical transportation system. Visualize tenants on a floor of a future super apartment house entering the elevator of tomorrow, digitally coding their destination, traveling up or down, being transferred horizontally across town, traveling up or down, and arriving at the desired office building floor a few minutes later.

This idea, proposed in the second edition of *Vertical Transportation* in 1983, was thought to be farfetched at the time, but in 1996 both the Otis Elevator Company with its "Odyssey" system, and a consortium of Japanese companies with a linear drive system, proposed similar approaches. The Japanese developed an experimental prototype of their

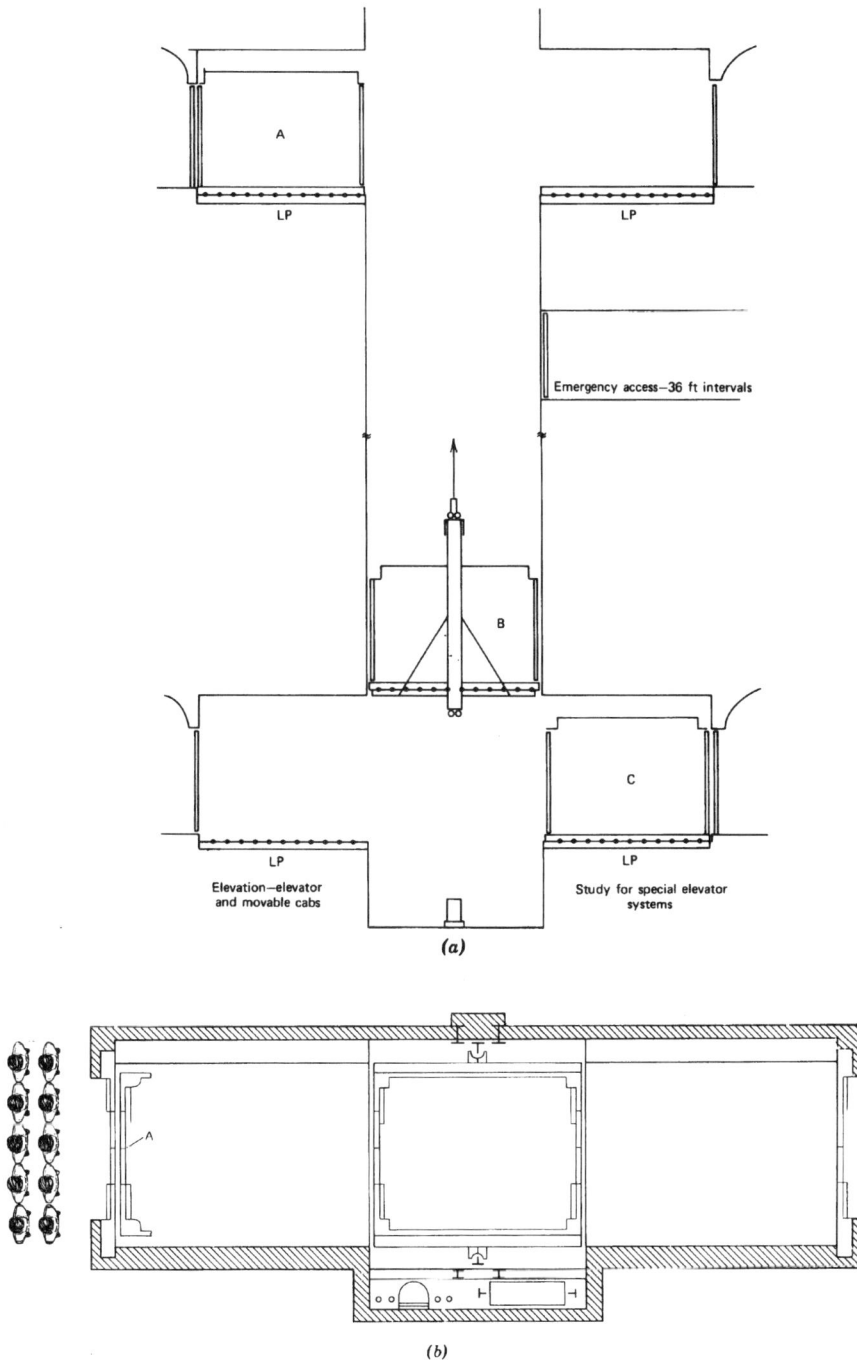

Figure 20.5. Deep transit station elevator system. (*a*) Elevation sketch of a high-capacity elevator system with three elevator cabs and a single hoistway. (*b*) Plan sketch of the top and bottom landing floor arrangement and passenger loading area.

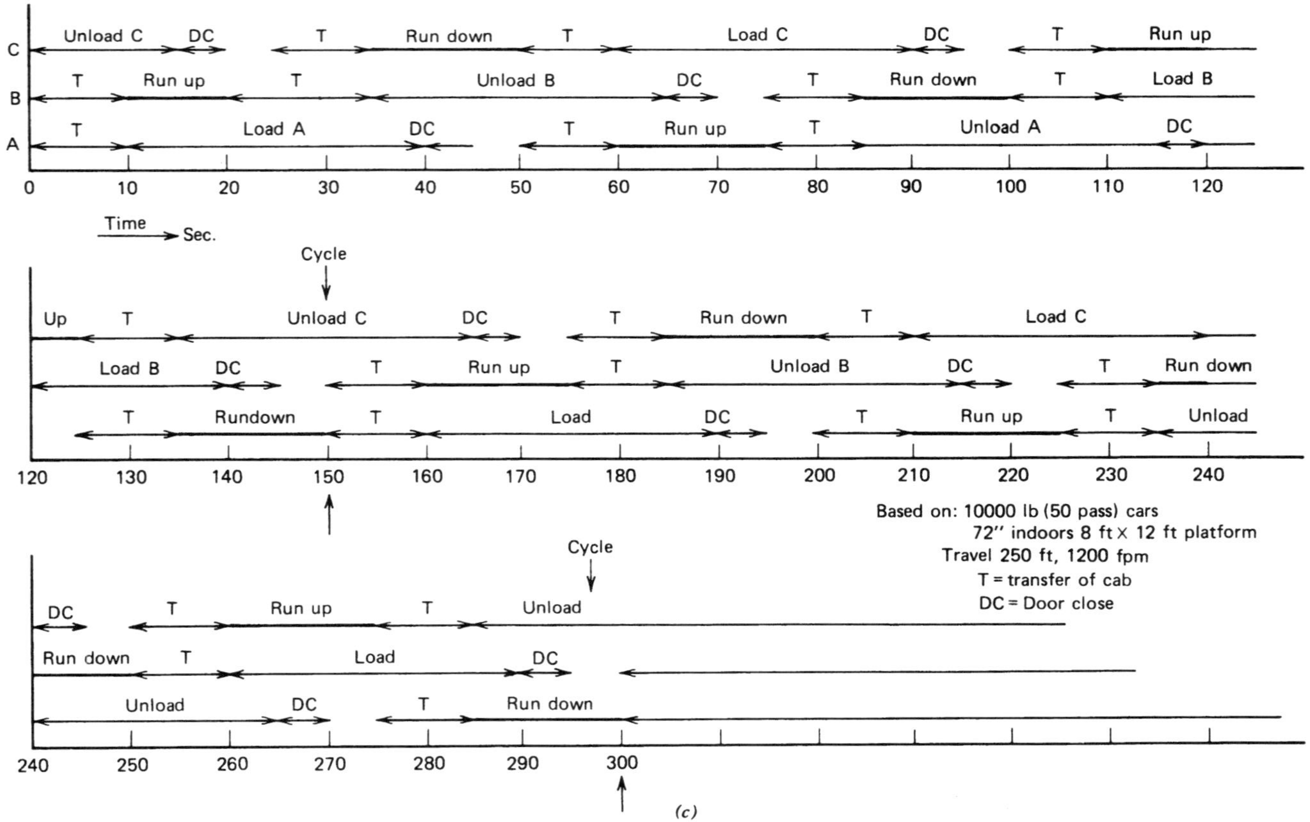

Based on: 10000 lb (50 pass) cars
72″ indoors 8 ft × 12 ft platform
Travel 250 ft, 1200 fpm
T = transfer of cab
DC = Door close

(c)

Figure 20.5. (*Continued*) Deep transit station elevator system. (*c*) Diagram of elevator travel and loading-unloading time cycle.

system, and Otis expanded a development it had used for a tourist attraction in a Disney-land facility.

The principle underlying both approaches is that passengers can board a vehicle and be transported both horizontally and vertically to their destination. Use of vertical hoistways by a number of such vehicles is feasible and overcomes the greatest inhibition to construction of supertall buildings. With more cars traveling vertically in the same hoistway, fewer hoistways are required. Our deep transit station concept used a utilization factor of 3. These advanced systems can possibly double or triple this factor. Instead of being delayed in an expressway traffic jam, our commuter can now be delayed in a lateral-elevator traffic jam!

PEOPLE MOVERS

Solving the problem of moving people from place to place safely and reliably is one of the goals of developing an accelerated moving walkway. This goal has been accomplished in other ways and is in active public use in many airports, in amusement parks, and between buildings in a medical center. This approach is the people mover whereby passengers are encapsulated in an automated moving vehicle traveling on its own dedicated guideway.

Two major systems are in use. One is an automated "trolley car," a self-propelled vehicle operating on rubber tires and traveling on a dedicated guideway (Figure 20.6a). Motive power is picked up from a trolley wire raceway in the guideway, and stopping and starting are controlled by controller logic located on the vehicle. At the terminals, automatic doors with safety protective edges open and close and are timed to permit passenger transfer. Operation can be programmed to be on schedule—vehicles traveling at time periods designed to maintain a headway timing between vehicles—or "on call"—operating when someone indicates a service requirement by operating a call button at a station. The vehicles are large and designed to accelerate and decelerate at about 0.1 g (3 fps² or

(a)

Figure 20.6. (*a*) High-capacity, horizontal automated personnel rapid transit vehicle (Courtesy Westinghouse Electric).

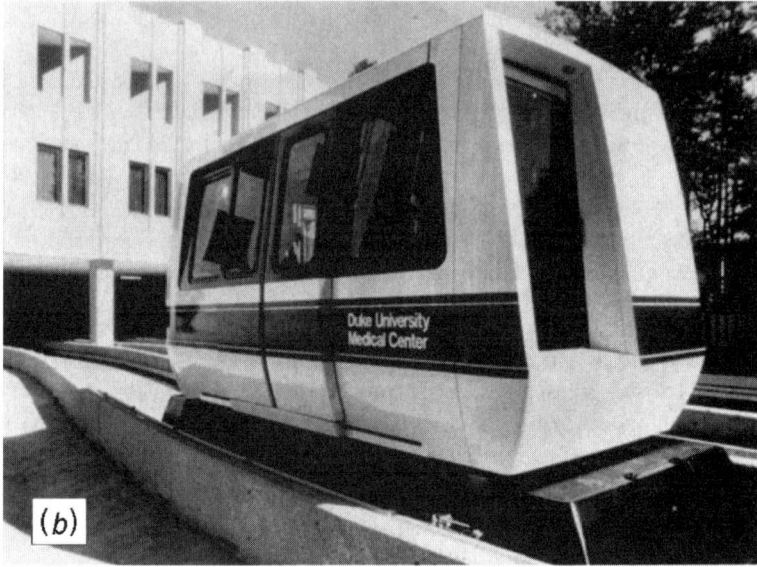

Figure 20.6. *(Continued)* (*b*) Air-cushioned, linear induction propelled, automated personnel rapid transit vehicle (Courtesy Otis Elevator).

0.9 mps^2). Emergency stopping is limited to about 0.3 g (10 fps^2 or 3 mps^2), although that rate is rather severe. Passenger capacity is based on vehicle design, and the usual vehicle is designed for a comfortable load of about 50 people in 150 ft^2 (4 m^2), with a maximum load based on 2.3 ft^2 (0.2 m^2) per person.

A smaller people mover, designed close to elevator capacities of 15 or 20 people, employs an air-cushion suspension and is propelled by linear induction motors (Figure 20.6*b*). This approach allows the lateral movement of the vehicles so that loading and unloading can be accomplished off the main guideway and other vehicles can pass. An elevator-type door is also employed, and the system can be programmed for on-call or scheduled operation. The advantages of the lateral docking include the ability to use a multiplicity of vehicles and provide increased service when traffic demands increase. Moreover, the air cushion and linear induction motor drive provide an easy way to switch from one track to another and to negotiate curves without the wheel or tire squeal that occurs with wheeled vehicles. The disadvantage is the need for a relatively smooth and flat guideway.

SKYWALKS

Second-level access is being extensively applied in colder cities in North America, even though the automated vehicles are not yet provided. Downtown areas in at least three major cities have a great number of buildings connected by "skywalks," which protect patrons from extreme weather and serve to make the second floors of stores major retail areas. The street levels are reserved for bus and auto transport and have narrow sidewalks. From an elevatoring standpoint, the skywalks also create a second elevator lobby, which can add to elevator round-trip time. Escalators between the street and the skywalk level are therefore recommended. In addition to necessary elevators to serve people unable or unwilling to use escalators as well as wheelchairs and strollers.

CABLE-DRIVEN PEOPLE MOVERS

In North America the quest is for an automated people mover to avoid the cost of operating attendants, whereas in other parts of the world cable cars, tramways, and funiculars are common and are all operated by an attendant posted in a stationary control cabin. There are some notable installations in the United States, most of which are used primarily as tourist attractions, and at least one as a commuter transportation means in New York City (Figure 20.7a).

(a)

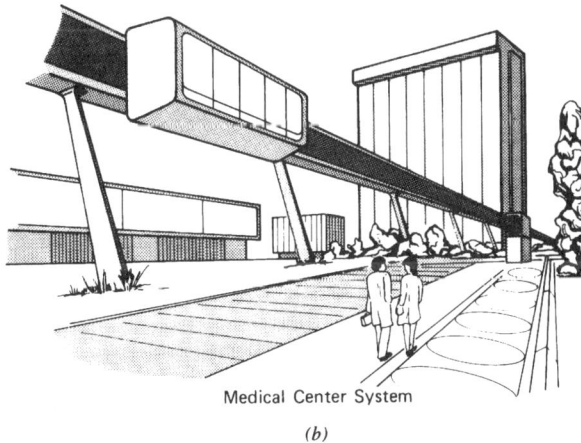

Medical Center System

(b)

Figure 20.7. (a) Suspended cable car (tramway), Roosevelt Island, New York (Courtesy VSL Corp.). (b) Cable-driven people mover—horizontal elevator (editor's design).

Ropeways, as cable cars are often called, are the product of a highly refined technology, a specialized field of their own. The vehicle is suspended on a wire rope designed to support the traveling load and propelled by driving ropes connected to a traction drive. The cars are interconnected so that the down-traveling cars or those traveling in one direction counterbalance the up-traveling cars or those traveling in the opposite direction. At the terminals the cars may be freed from the load-bearing rope system for loading or unloading or may be captive so that when one stops, all stop.

A system wherein the car is suspended from rails, similar to a monorail with a lower guide, and propelled by cables driven by elevator-type traction machines has been proposed by the editor (Figure 20.7*b*). The system can be described as a "horizontal elevator" and has all the attributes of an elevator system. It would be ideal for short-range transportation in airport terminals to serve satellite terminals, and in shopping centers between parking areas and the mall. It could also be used to bridge busy thoroughfares and in industrial applications to bring workers, for example, from a mine field entry point to the head of the mine.

A variation of the cable-driven approach may be found in Tampa, Florida, where a cable-driven vehicle using an elevator traction machine moves along a guideway between two terminals. The vehicle is air-cushioned similarly to those installed with linear induction motors. The monorail approach has also been implemented and, as time goes by, we are seeing more and more installations expanding the development of the various approaches. This is an expanding technology, and the potential is limitless. Application has increased dramatically since the subject was first introduced in the 1983 edition of *Vertical Transportation* and has evolved into a distinct field of its own, aptly called APMs, or "automated people movers."

AUTOMATED PEOPLE MOVERS

Since the early 1980s numerous installations of horizontal vehicles, distinct from the common attended trolley cars or transit systems, have been developed. Some notable installations are in airports such as Atlanta's Hartsfield. McCoy in Orlando, and Seattle's Sea-Tac. Such vehicles have been installed in dozens of other locations over the past decade or so. At these installations automated vehicles stop at various stations, doors open, recorded announcements advise destinations and safety instructions, and the vehicles proceed, unattended, to the next stop.

Variations in design include linear drives, rubber tires, steel wheels, air-cushion support, cable drives, and other exotic features. A full treatment of this rapidly expanding field is beyond the scope of this book. The interested reader is directed to the *APM Industry Guide,* available from Trans21, P.O. Box 249, Fields Corner Station, Boston, MA 02122; phone 617-825-2318.

WATER TOWERS

Many towns have a water tower to supply local needs, but few, if any, have taken advantage of the possibilities such towers present. Perhaps water does not enjoy the same importance in the Northern Hemisphere as it does in the more arid parts of the world, inasmuch as water towers are usually no more than tanks on stilts. In the Mideast, where water is valuable and well respected, the water tower is a prominent and important feature, so much so that water towers have been developed into centers and main attractions.

Figure 20.8. Water tower with public space developed on the top of the water storage are (Courtesy Schindler).

The top of the water tower is leveled off, and the area used for recreation, restaurants, libraries, civic offices, or a gathering place for the citizens (Figure 20.8). All of this activity is accomplished by a simple elevator traveling through the tower to serve the upper level. Perhaps as water shortages persist, water towers in North America may become symbols of civic pride and be more intensively developed.

THE FUTURE

Improvements in the state-of-the-art and the introduction of new technology into the vertical transportation industry are constant. Major changes occurred from the later 1970s to the present, with microprocessors essentially replacing all relay-type controllers applied to new elevators, as well as the many thousands modernized during that period. The motor generator of previous eras has virtually disappeared, and various forms of solid-state motor drives have replaced that approach on both new and modernized equipment. Mechanical floor-selecting devices have disappeared, and later developments foretell more radical change.

We are looking forward to a time when the rope brake will replace the on-car safety device that initially made possible the elevator industry. The performance history of wire rope and its reliability have obviated the need for an on-car catching device, because rope failure is practically unknown. Overspeed remains a concern, but it is not limited to downward travel, for which the traditional car safety provides protection, but includes both up and down travel. The rope brake acting on the hoisting ropes affords that extra measure of protection. The elimination of car safeties leads to more extensive engineering change. Rails can be lighter, because they are no longer structural but merely guides. Car structure can be lighter, encouraging self-propelled approaches. We are certain the opportunities for change will increase with the changing technology, but we cannot predict the directions such change may take.

The development of vertical transportation systems combining both horizontal and vertical travel has been initiated. We look forward to the expansion of this concept and its application to the taller buildings that will be constructed in the future. Newer cities will proliferate, as is currently demonstrated in Asia. Older cities will change as major existing buildings are recycled, the marginal ones replaced, and obsolete urban areas redeveloped.

Our advice given in the 1983 edition of *Vertical Transportation* is still valid. The student of vertical transportation technology must keep abreast of these developments and evaluate them in view of past practices and the overall philosophy of what is to be accomplished. The challenge is to change the seven famous last words of any organization from "We've always done it that way before" to "What new and better way can it be done?" The approaches and methods outlined in the previous chapters are part of an ongoing development and represent a refinement and stricter application of principles given in both the first and second editions of this book. In the future it is sure that the methods and approaches will be somewhat different and greatly expanded. We look forward to the changes.

ABOUT THE AUTHOR

WILLIAM (BILL) STURGEON and I collaborated on this chapter. Bill, the founder and editor emeritus of *Elevator World Magazine* has been a student of the changing nature of the vertical transportation industry, and I value his views and insights. He and I have frequently discussed the changing nature of both the product and the personnel involved and their impact on the past and future aspects of elevators and escalators. The foregoing and the input of the many contributors to the various chapters of this book provide a reasonable appraisal of both the state of the art and what can be expected.

Appendix: Literature on Elevators and Escalators

Although the available literature on elevators and escalators was very limited in the past few years, a number of interesting publications became available for those who wish to pursue or increase their knowledge of the many aspects of vertical transportation. A listing of those familiar to the author will be given below.

An additional source is the promotional publications and magazine article reprints available from the various manufacturers, and which can usually be obtained by a written request. The Otis Elevator Company publishes two excellent booklets designed for lay people. They are entitled *Tell Me About Elevators* and *Tell Me About Escalators*. Single copies can be obtained free of charge by writing to The Otis Elevator Company, Public Relations Department, Farmington, CT. 06032.

The Elevator and Escalator Safety Foundation was established in 1991 to educate the public on the safe and proper use of elevators and escalators through educational programs. To that end they have prepared two programs consisting of informational videos and booklets: one for school children, "Safe-T-Rider," and one for seniors, "A Safe Ride." These are available to those who are interested by contacting the foundation at P.O. Box 6273, Mobile AL 36660–0273 (phone 334-479-2199).

Elevator World Magazine, P.O. Box 6507, Mobile, Alabama 36606-0507, (phone 800-730-5093 or 334-479-4514), William Sturgeon, founder, and Ricia Hendrick, publisher. This is a monthly publication published continuously since 1953. It contains articles and picture stories of unique, timely, or technical aspects of elevators and escalators as well as articles about the people and companies in the vertical transportation business. Once a year, "The Elevator World Source"© is prepared and describes, in detail, the participants in the elevator industry and other related activities.

William "Bill" Sturgeon is the founder and former editor and publisher of *Elevator World Magazine,* and its publication has been his pride since he gave up his own repair and service company. He is still active and I value him as a good friend and I am proud to serve as a member on *Elevator World Magazine*'s board of directors.

Most of the following publications can be obtained from *Elevator World Magazine,* through their educational division. Over the years, they have expanded their service by retailing books related to vertical transportation and republishing significant books that were out of print. A brief description of some of these follows. The complete *Elevator World Magazine* catalog of publications can be obtained by contacting them at the address given above.

The Vertical Transportation Handbook, Third Edition, Edited by George R. Strakosch
ISBN 0-471-16291-4   © 1998 John Wiley & Sons, Inc.

Electric Elevators, by Fred A. Annett, originally published by McGraw-Hill, New York, 1960, and republished by *Elevator World Magazine.* This book was one of the first efforts to compile a volume on the many aspects of elevators, especially from a technical standpoint. Mr. Annett was an editor of *Power Magazine* during the 1950s and 1960s, and the first edition of *Electric Elevators* was a compilation of various articles published in that magazine. Later editions of the book were updated with data and information gained as a result of Mr. Annett's visits and discussions with the engineers in various companies. It represents an excellent commentary on the state of the art at the time of the edition's publication. Since elevators seldom wear out, much of the equipment described in *Electric Elevators* is still in every day use.

Pedestrian—Planning and Design, by Dr. John J. Fruin, Metropolitan Association of Urban Designers and Environmental Planners, 1971. Republished by *Elevator World Magazine,* and currently available. This volume is considered a vital reference for anyone engaged in the design and layout of any facility where people must be served in any quantity—such as terminals, lobbies of buildings, exhibits, stadiums, or museums. This work is the only source known to the author wherein the area where people can walk or assemble is clearly defined in terms of how much space must be allocated related to pedestrian traffic volume. It is a source of information on how many people (and at what rate) can use stairs, corridors, revolving doors, sidewalks, subway turnstiles, gates, doors, and so on. It is an ideal companion work to *Vertical Transportation* if all the aspects of pedestrian accommodation in any facility are considered.

The National Elevator Industries Incorporated Installation Manual (NEII Installation Manual), originally published by NEII and now republished by *Elevator World Magazine.* This book could almost be called a "do it yourself" guide to installing elevators. It gives descriptive detail of the various steps required to install an elevator in a building, starting from the setting of rail brackets to the finished, operating elevator. Construction managers and architects would find the book instructive, as would anyone engaged in the vertical transportation field who may not have had the opportunity to observe installation progress at a construction site.

The book is fully illustrated with photographs and diagrams showing each of the major steps. It is also a guide to the unique elevator language where words such as rails, shaft, DBG, and others have highly specialized meanings.

Historically notable are *Elevators* by John J. Jallings and *Electric Elevators* by Fred Hymans. *Elevators,* originally published in 1919, describes in detail the state of the art of that era. The basis of many of the engineering principles of elevators can be found in the book *Electric Elevators.* Originally presented as a correspondence course by the International Textbook Company, this book has been republished and can be considered a handbook of the mechanical design of elevators.

Contemporary books from *Elevator World Magazine* are *Defensive Elevatoring* by Dee Swerrie, former chief elevator inspector for the state of California, *The Inspection Handbook* by Zack McCain, *ADA and Vertical Transportation* by Ed Donoghue, and numerous others. A full catalog is available by contacting the magazine.

A notable *Elevator World Magazine* publication is the *Elevator World Guide to Elevatoring* a self-study course on the many aspects of elevators and escalators that any participant in the industry should know. This course, developed by me is based on the second edition of *Vertical Transportation,* which is expanded in this third edition. The guide includes audio cassettes introducing each section plus numerous checkcharts and is supplemented by a series of questions and answers that makes it ideal for a company training program. The guide could be considered the study program and this book as the reference source.

An American National Standard, The ASME (American Society of Mechanical Engineers) Safety Code for Elevators and Escalators, A17.1 and *The Inspector's Manual, A17.2.* Published by the American Society of Mechanical Engineers (ASME), 345 East 47th St., New York, N.Y. 10017, and available both from them and from *Elevator World Magazine.* The A17.1 elevator code has been in continuous publication since 1922 and has served in an advisory capacity as a model for elevator safety standards throughout the world. It is updated every three years and supplements are issued yearly. Each edition includes revisions to reflect the state of the art or clarification of previous rules. The code is supplemented by a tabulation of official interpretations that are consensus opinions based on inquiries to the code committee on how the various rules should be applied in specific situations. New interpretations can be obtained by writing ASME if a question arises about a rule and its application. A standard inquiry format appears in the front section of the code.

Supplementing the A17 code is the *A17.1 Handbook,* a commentary and added explanation of the various rules found in the code plus practical examples of their application. This volume is prepared by Edward Donoghue, our contributor to Chapter 16, "Codes and Standards," of this edition. The book is available from both the ASME and from *Elevator World Magazine.* It is updated with each new three-year edition of the code.

Among the many organizations concerned with vertical transportation is The International Association of Elevator Engineers (IAEE), which is based in London and holds an annual meeting wherein numerous papers on many of the technical aspects of elevators and escalators are presented. These are published in their proceedings and copies are available from *Elevator World Magazine.* The subjects are too numerous to mention, but one paper, "Statistical Analysis of Modern Elevator Hall Call Response Time," by Jon Halpern, our contributor to Chapter 7, "Operation and Control," is notable and is cited in that chapter as a full explanation of the concept of elevator waiting time.

The preceding is but a sample of the literature available. In addition, numerous organizations, both in the United States and abroad, have literature and publications—a listing would overwhelm this section. Anyone interested is invited to obtain a copy of the "Elevator World Source," which has a complete listing of the organizations and associations throughout the world, as well as of the hundreds of companies engaged in some aspect of the industry plus those who offer themselves as consultants. Starting in 1987, when I joined *Elevator World Magazine* as their director of educational services, the effort was made to make *Elevator World Magazine* the major source of elevator information. With all modesty, between that source and both this edition and the previous editions of *Vertical Transportation,* we've done a good job.

Finally, in this day of on-line and World Wide Web services additional information can be had at http://www.elevator-world.com/

Index of Tables and Charts

Index of Examples

Subject Index